The Nature
and Properties of
SOILS

NINTH EDITION

The Nature and Properties of

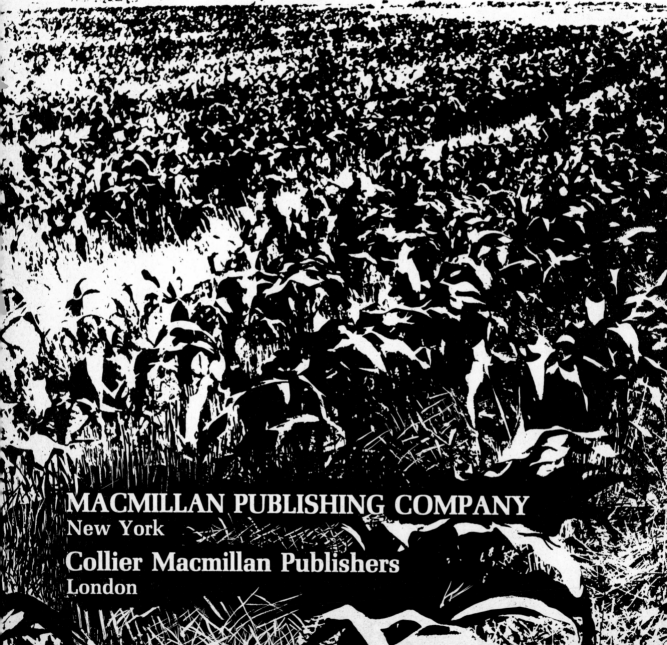

MACMILLAN PUBLISHING COMPANY
New York

Collier Macmillan Publishers
London

SOILS

NYLE C. BRADY
Cornell University
and
United States Agency for International Development

Macmillan Publishing Company
866 Third Avenue, New York, New York 10022

Collier Macmillan Canada, Inc.

Library of Congress Cataloging in Publication Data

Brady, Nyle C.
    The nature and properties of soils.

    Includes bibliographies and index.
    1. Soil science. I. Title.
S591.B79 1984   631.4   83–19545
ISBN 0-02-313340-6 (Hardcover Edition)
ISBN 0-02-946030-1 (International Edition)

Printing: 4 5 6 7 8   Year: 7 8 9 0 1 2

ISBN   0-02-313340-6

PREFACE

Unprecedented population increases have occurred since World War II, mostly in those countries with low agricultural productivity and low per capita incomes. Population growth coupled with rising energy costs and worldwide inflation has taxed the ability of farmers to meet human food requirements. In fact, in some areas, such as sub-Saharan Africa, the per capita food production has actually declined in the past decade. In other heavily populated regions, including South Asia, per capita food production has remained reasonably steady, but at levels so low as to dictate widespread nutritional deficiencies and constant hunger for tens of millions of people. Meeting human food requirements continues to be a major worldwide challenge, not only for the developing nations but for all of us.

Soils play a critical role in meeting human food needs. Their conservation and wise use are essential today and will be even more critical in the future as population pressures increase. These pressures are already forcing the use of lands for food production that good judgment says should be kept in forests and rangelands. This change in land use encourages accelerated erosion and runoff losses that reduce soil productivity and simultaneously force sediment-burdened floodwaters on downstream populations. Political leaders and development analysts alike agree that at no time in the history of humankind has the welfare of people been so dependent on the wise use and management of soils.

Effective conservation and management of soils require an understanding of these natural bodies and of the processes going on within them. These processes, which are vital to the production of plants, also influence the many other uses made of soils. The primary purpose of this text is to help the reader to gain a better understanding of the nature and properties of soils and to learn ways of making soils more useful to humankind. I hope that it will be as useful to the student preparing to work in tropical areas as in the United States or Europe.

This ninth edition recognizes the role soils must play in enabling the world to feed itself. But most of the changes from the eighth edition are dictated by the numerous recent advances in soil science. For example, specific empha-

V

sis is placed on new and innovative "conservation tillage" systems, which have expanded so rapidly in the United States in recent years. These systems maintain soil cover and thereby reduce soil erosion and water runoff. Attention is given to the effects of these systems on soil properties as well as on soil productivity in a variety of agro-climatic zones, including those in the tropics.

New knowledge is presented of nitrogen transformations both in well-drained upland soils and in soils of wetland paddy fields where much of the food for developing countries is produced. Our improved understanding of the nature of soil colloids, including the soil clays in tropical areas, is reflected in several chapters. Likewise, recently acquired knowledge of the charge properties of different soils and of the nature of soil acidity is included.

The text has been rather extensively reorganized. The introductory chapter has been expanded to include general information on the soil as a source of nutrients for plants. Chapters 2–8 cover the basic physical, chemical, and biological properties of soils. There follow three chapters (9–11) dealing with the essential plant nutrients and three (12–14) concerned with soil formation, classification, and use. Chapter 13 on soil classification has been revised to conform with recent changes in *Soil Taxonomy*.

Chapters 14–19 deal with the practical management of soils, first through improved water conservation and management and then through use of lime, fertilizer, and organic supplements. The treatment of pollution in Chapter 20 has been updated to include information on the dangers of domestic and industrial wastes containing excessive quantities of heavy metals. Finally, Chapter 21 focuses on the role of soils in meeting human food needs. All chapters have been revised to incorporate new knowledge of soil processes.

The format of the text has been improved to make it more attractive and readable. A photograph at the beginning of each chapter gives the reader a general idea of the subject to be covered. The charts and graphs have been redrawn, and many new illustrations, both drawings and photographs, have been added to better explain the text material. Ray Weil and Wybe Kroontje have written a Study Guide that will greatly aid the student in mastering the concepts involved in soil processes.

I am indebted to many soil and crop scientists for advice and counsel in preparing this text and for permission to reproduce their graphs and charts. I also appreciate the helpful suggestions of Macmillan's reviewers: Wybe Kroontje, Virginia Polytechnic Institute and State University; Murray H. Milford, Texas A&M University; George Van Scoyoc and William McFee, Purdue University; R. L. Thomas, University of Guelph; and Ray Weil, University of Maryland. I am especially grateful to my wife, Martha, for her encouragement and for her help in all stages in the preparation and checking of the figures and manuscript. Dr. Joyce Torio was also most helpful in finding backup material, including photographs, and in typing, checking, and proofreading the figures and parts of the manuscript.

N. C. B.

CONTENTS

4 Soil Air and Soil Temperature

5 Soil Colloids: Their Nature and Practical Significance

6 Soil Reaction: Acidity and Alkalinity

7 Organisms of the Soil

8 Organic Matter of Mineral Soils 253

9 Nitrogen and Sulfur Economy of Soils 283

10 Supply and Availability of Phosphorus and Potassium

327

16 Soil Erosion and Its Control 533

17 Lime and Its Soil–Plant Relationships 571

18 Fertilizers and Fertilizer Management

19 Recycling Nutrients Through Animal Manures and Other Organic Wastes

20   Soils and Chemical Pollution   653

21   Soils and the World's Food Supply   689

The Soil in Perspective

People are dependent on soils—and to a certain extent good soils are dependent upon people and the use they make of them. Soils are the natural bodies on which plants grow. Society enjoys and uses these plants because of their beauty and because of their ability to supply fiber and food for humans and for animals. Standards of living are often determined by the quality of soils and the kinds and quality of plants and animals grown on them.

But soils have more meaning for humankind than as a habitat for growing crops. They underlie the foundations of houses and factories and determine whether these foundations are adequate. They are used as beds for roads and highways and have a definite influence on the length of life of these structures. In rural areas soils are often used to absorb domestic wastes through septic sewage systems. They are being used more and more as recipients of other wastes from municipal, industrial, and animal sources. The deposition of undesirable silt in municipal reservoirs makes the protection of soils in upstream watersheds as important to city dwellers as to their counterparts on the farm or in the forest. Obviously, soils and their management are of broad societal concern.

Great civilizations have almost invariably had good soils as one of their chief natural resources. The ancient dynasties of the Nile were made possible by the food-producing capacity of the fertile soils of the valley and its associated irrigation systems. Likewise, the valley soils of the Tigris and Euphrates rivers in Mesopotamia and of the Indus, Yangtze, and Hwang Ho rivers in India and China were habitats for flourishing civilizations. Subject to frequent replenishing of their fertility by natural flooding, these soils provided continued abundant food supplies. They made possible stable and organized communities and even cities, in contrast to the nomadic, shifting societies associated with upland soils and with their concomitant animal grazing. It was not until the value of manures and crop residues was discovered that people were able to make extensive use of upland soils for sustained crop culture.

Soil destruction or mismanagement was associated with the downfall of some of the same civilizations that good soils had helped to build. The cutting of timber in the watersheds of the rivers encouraged erosion and topsoil loss. In the Euphrates and Tigris valleys, the elaborate irrigation and drainage systems were not maintained. This resulted in the accumulation of harmful salts, and the once productive soils became barren and useless. The proud cities that had occupied selected sites in the valleys fell into ruin, and the people migrated elsewhere.

History provides lessons that modern nations have not always heeded. The wasteful use of soil resources in the United States during the first century of intensive agricultural production by the early settlers and the following generations provides such an example. Even today there are many who do not fully recognize the long-term significance of soils. This may be due in part to widespread ignorance of what soils are, what they have meant to past generations, and what they mean to us and will mean to future generations.

1.1 What Is Soil?

Concepts of Soil. Part of the lack of concern for soils may be attributed to different concepts and viewpoints concerning this important product of nature. For example, to a mining engineer the soil is the debris covering the rocks or minerals that he must quarry. It is a nuisance and must be removed. To the highway engineer the soil may be the material on which a roadbed is to be placed. If its properties are unsuitable, that particular soil will need to be removed and replaced with rock and gravel.

The average homeowner also has a concept of soil. It is good if the ground is mellow or loamy (Figure 1.1). The opposite viewpoint is associated with "hard clay," which resists being spaded into a good seedbed for a flower garden. The homemaker can differentiate among variations in the soil, especially those relating to its stickiness or tendency to cling to the shoe soles and eventually to carpets.

The farmer, along with the homeowner, looks upon the soil as a habitat for plants. The farmer makes a living from the soil and is thereby forced to pay more attention to its characteristics. To the farmer the soil is more than useful—it is indispensable.

A prime requisite for learning more about the soil is to have a general concept of what it is. This concept must encompass the viewpoints of the engineer, the homeowner, and the farmer. It has developed through the practical and scientific discoveries of the past.

FIGURE 1.1 A hard soil clod (left) that would be difficult to work into a good seedbed. A friable soil intermixed with organic matter (right) is much more easily handled. [*Courtesy USDA National Tillage Machinery Laboratory.*]

1.2 Evolution of Modern Concepts of Soil

There are two basic sources of our current knowledge of soils. First, there is the practical knowledge gained by farmers through centuries of trial and error. This was the only information available before the advent of modern science, which now provides the second source of facts about soils and their management.

Experience of the Cultivator. The earliest recorded history contains evidence that, through trial and error, humans learned to distinguish differences in soils. They also learned the value of treating soils with plant and animal wastes. More than 42 centuries ago the Chinese used a schematic soil map as a basis for taxation. Homer, in his *Odyssey*, said to have been written about 1000 B.C., makes reference to the use of manure on the land. Biblical references are made to the "dung hill" and to the beneficial practice of "dunging" around plants. Greek and Roman writers described a reasonably elaborate system of farming that involved leguminous plants and the use of ashes and sulfur as soil amendments. This evidence suggests that by the time of the early Roman civilization many of the practical principles governing modern agriculture, including soil management, had been discovered and put to use by farmers and livestock owners.

Further development and application of these principles were halted by the barbarian invasions of Rome. Even so, Roman agriculture was the foundation for most European agriculture during the feudal Dark Ages. The cultural practices were passed from generation to generation even though the farmers were ignorant about why the practices were necessary. When in the seventeenth and eighteenth centuries there was a blossoming of scientific inquiry, the stage was set for the application of science to the improvement of agricultural systems, including those involving soils.

Early Scientific Investigations and Soil Productivity. From the seventeenth century until the middle of the twentieth century the primary occupation of soil scientists was to increase the production of crop plants. Jan Baptista van Helmont, a Flemish chemist, in his famous five-year willow tree experiment concluded that 164 pounds of dry matter came primarily from the water supplied since the soil lost no weight while producing the tree. This concept was altered by John Woodward, an English researcher, who found that muddy water produced more plant growth than rainwater or river water, leading him to conclude that the fine earth was the "principle" of growth. Others concluded that the principle was humus taken in by the plants from soil. Still others assumed that the principle was somehow passed on from dead plants or animals to the new plant. Jethro Tull early in the eighteenth century demonstrated the benefits of cultivation, thinking erroneously that stirring the soil would make it easier for plants to absorb small quantities of fine earth.

It remained for the French agriculturist J. B. Boussingault, through a series of field experiments starting in 1834, to provide evidence that the air and rain were the primary sources of carbon, hydrogen, and oxygen in plant tissues. But his investigations were largely disregarded until 1840, when the eminent German chemist Julius von Liebig reported findings that crop yields were directly related to the content of "minerals" or inorganic elements in the manures applied to the soil. Liebig's reputation as an outstanding physical chemist was instrumental in convincing the scientific community that the old theories were wrong. He proposed that mineral elements in the soil and in added manures and fertilizers are essential for plant growth.

Testing Liebig's Concepts. Liebig's work revolutionized agricultural theory and opened the way for numerous other investigations. Those of J. B. Lawes and J. H. Gilbert at the Rothamsted Experiment Station in England are most noteworth since they put Liebig's theories to the test in the field. While they confirmed many of his findings, they identified two errors. Liebig had theorized that nitrogen came primarily from the atmosphere rather than the soil. He further assumed that salts could be fused before being added as fertilizers; apparently he gave no thought to what this drastic action would do to the solubility and "availability" to plants of the nutrient elements. The Rothamsted field research proved that nitrogen applications to the soil markedly benefited plant growth. It showed further that mineral elements must be in an "available" form for optimum uptake by plants. The investigations led to the development of acid-treated phosphate rock or "superphosphate," which is still an important commercial fertilizer source of phosphorus.

While the work of Gilbert, Lawes, and others pinpointed weaknesses in components of Liebig's theory, the primary concept is still considered basically sound. For example, Liebig stated what has since been called the law of the minimum: "By the deficiency or absence of one necessary constituent, all the others being present, the soil is rendered barren for all those crops to the life of which that one constituent is indispensable." The significance of this finding will be more apparent later as soil fertility and plant nutrition interactions are considered (see Section 1.16).

After Liebig, unraveling the complexities of nitrogen transformations in soil awaited the emergence of soil bacteriology. J. T. Way, who demonstrated the cation adsorption properties of soils, discovered in 1856 that nitrates are formed in soils from ammonia-containing fertilizers. Twenty years later R. W. Warington demonstrated that this process was biological, and in 1890 S. Winogradski isolated the two groups of bacteria responsible for the transformation of ammonia to nitrate. Coupled with the discovery in the 1880s that nitrogen-assimilating bacteria grow in nodules of the roots of legumes, these findings provided background information for sound soil and crop management practices.

Early Research in the United States. The European investigations on soil fertilizers were found to be quite applicable to the United States when they were tested in the late nineteenth century. Upon being tilled, the soils along our eastern and southern seaboards were easily depleted of nutrients by the percolation of rainfall and by crop removal. Edmond Ruffin, a Virginia farmer, grasped the concepts of nutrient depletion and in his writings was especially critical of those who did not properly care for and replenish their soil. Unfortunately, the abundance of open land to the west encouraged the abandonment of the "worn-out" soils in the east rather than the adoption of more realistic management systems.

The establishment of the U.S. Department of Agriculture in 1862 and the state Agricultural Experiment Stations in 1886 gave a decided boost to both field and laboratory investigations on soils. Numerous field trials were initiated to test the applicability of findings of the European investigators. Unfortunately, in most of the trials soil was not considered to be a dynamic medium for crop growth. Instead the soil was considered merely as a "storage bin," in keeping with Liebig's concepts. Exceptions existed, such as the work of men like F. H. King of Wisconsin, who studied the movement and storage of water in soils in relation to root penetration and crop growth. Also C. G. Hopkins of Illinois developed effective soil-management systems based on limestone, rock phosphate, and legumes. Milton Whitney of the U.S. Department of Agriculture urged greater consideration of properties of soils in the field and initiated the first national soil survey system.

Field Soil Investigations. Liebig's concepts thoroughly dominated the thinking of soil scientists in the late nineteenth and early twentieth centuries. Furthermore, except for the field testing of crop response to fertilizer, much of the research was done in the laboratory and greenhouse. Inadequate attention was given to the variable characteristics of the soils as found in the field. Nor was much significance given to the differences in soils as related to the climate in which they were found. Soils were considered as geological residues on the one hand and as reservoirs of nutrients for plant growth on the other.

As early as 1860, E. W. Hilgard, then in Mississippi, published his findings, which called attention to the relationships among climate, vegetation, rock materials, and the kinds of soils that develop. He conceived of soils not merely as media for plant growth but as dynamic entities subject to study and classification in their field setting. Unfortunately, Hilgard was ahead of his time. It was necessary for many of his concepts to be rediscovered before they were accepted.

Parallel to Hilgard's investigations were those made by the brilliant team of soil scientists in Russia led by V. V. Dokuchaev. These scientists found unique horizontal layerings in soils—layerings associated with the climate, vegetation, and underlying soil material. The same sequence of layering was found in widely separated geographical areas provided the areas had similar climate and vegetation. The concept of soils as natural bodies was well developed in the Russian studies, as were concepts of soil classification based on field soil characteristics.

The Russian studies were under way as early as 1870. Unfortunately, language barriers prevented effective communication of the Russian concepts to scientists in Western Europe, Asia, and the Americas until 1914, when they were published in German by K. D. Glinka, a member of the Russian team of soil scientists. These concepts were quickly grasped by C. F. Marbut of the U.S. Department of Agriculture, who had been placed in charge of the U.S. National Soil Survey by Dr. Whitney. Marbut and his associates developed a nationwide soil classification system based to a great extent on the Russian concepts. Consideration of soils as natural bodies has led to further modifications in soil classification systems, which will receive attention in later chapters.

1.3 The Approach—Edaphological Versus Pedological

The previous section suggests that two basic concepts of soil have evolved through two centuries of scientific study. The first considers soil as a natural entity, a biochemically weathered and synthesized product of nature. The second conceives of the soil as a natural habitat for plants and justifies soil studies primarily on that basis. These conceptions illustrate the two approaches that can be used in studying soils—that of the *pedologist* and that of the *edaphologist.*

The origin of the soil, its classification, and its description are involved in *pedology* (from the Greek word *pedon,* which means soil or earth). Pedology considers the soil as a natural body and does not focus primarily on the soil's immediate practical utilization. A pedologist studies, examines, and classifies soils as they occur in their natural environment. These findings may be as useful to highway and construction engineers as to the farmer.

Edaphology (from the Greek word *edaphos,* which also means soil or ground) is the study of the soil from the standpoint of higher plants. It considers the various properties of soils as they relate to plant production. The edaphologist is practical, having the production of food and fiber as an ultimate goal. Simultaneously, the edaphologist must be a scientist to determine the reasons for variation in the productivity of soils and to find means of conserving and improving this productivity.

In this textbook the dominant viewpoint will be that of the edaphologist. Pedology will be used, however, to the extent that it gives a general understanding of soils as they occur in nature and are classified. Furthermore, since studies of the basic physical, chemical, and biological characteristics of soils contribute equally to edaphology and pedology, it is not possible to separate these approaches fully. This is illustrated in the following section, which deals with soils as they are found in the field.

1.4   A Field View of Soil

Someone has said that the soil is to the earth as the peel is to the orange. This analogy is acceptable but should be modified to stress the great variability in soil from site to site on the earth's surface. Even a casual examination of road cuts from one geographic area to another suggests differences in soil depth, color, and mineral makeup. The trained eye of the soil scientist, however, can identify common properties of soils from areas as distant as Hawaii, India, and the continental United States. The common properties will receive immediate attention; the variations will be treated in succeeding chapters.

Soil Versus Regolith.   Views such as that shown in Figure 1.2, in which unconsolidated materials are found on underlying rocks, are quite familiar. Above bedrock some unconsolidated debris is present almost universally. This material, known as *regolith*, may be negligibly shallow or hundreds of feet thick. It may be material that has weathered from the underlying rock or it may have been transported by the action of wind, water, or ice and deposited upon the bedrock or upon other material covering the bedrock. As might be expected, the regolith tends to vary in composition from place to place.

An examination of the upper 1–2 meters of the regolith shows that it differs from the material below. It is higher in organic matter since plant residues

FIGURE 1.2   Relative positions of the regolith, its soil, and the underlying country rock. Sometimes the regolith is so thin that it has been changed entirely to soil; in such a case, soil rests directly on bedrock. [*Photo courtesy Tennessee Valley Authority.*]

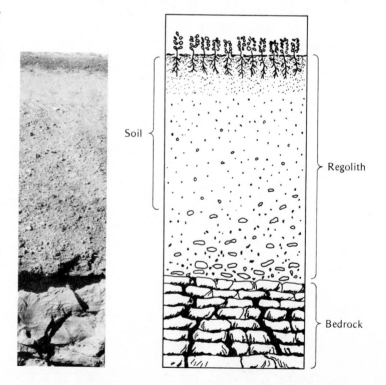

deposited originally on the surface have been incorporated by earthworms and other animals and have been decomposed by microorganisms. Being near the surface, it has been more subject to weathering. Products of this weathering, especially if they have moved vertically, give rise to characteristic layering.

This upper and biochemically weathered portion of the regolith[1] is the soil. It is the product of both destructive and synthetic forces. Weathering and microbial decay of organic residues are examples of destructive processes, whereas the formation of new minerals, such as certain clays, and the development of characteristic layer patterns are synthetic in nature. These forces have given rise to a distinctive entity in nature called the soil, which, in turn, is comprised of a large number of individual soils.

The Soil Versus *A* Soil. Characteristics of the soil vary widely from place to place. For example, the soil on steep slopes is generally not as deep and productive as soil on gentle slopes. Soil that has developed from sandstone is more sandy and less fertile than soil formed from rocks such as limestone. The properties of a soil that has developed in tropical climates are quite different from those of a soil found in temperate or arctic areas.

Scientists have recognized these soil variations from place to place and have set up classification systems in which the soil is considered as composed of a large number of individual soils, each having its distinguishing characteristics. Therefore, *a* soil, as distinguished from *the* soil, is merely a well-defined subdivision having recognized characteristics and properties. Thus, a Cecil clay loam, a Marshall silt loam, and a Norfolk sand are examples of *specific soils,* which collectively make up the overall *soil* covering the world's land areas. The term "soil" is a collective term for all soils, just as "vegetation" is used to designate all plants.

1.5 The Soil Profile

Examination of a vertical section of a soil in the field reveals the presence of more or less distinct horizontal layers (Figure 1.3). Such a section is called a *profile,* and the individual layers are regarded as *horizons.* These horizons above the parent material are collectively referred to as the *solum* (from the Latin legal term *solum* meaning soil, land, or parcel of land). Every well-developed, undisturbed soil has its own distinctive profile characteristics, which are utilized in soil classification and survey and are of great practical importance. In judging a soil one must consider its whole profile.

Soil Horizons. The upper layers or horizons of a soil profile generally contain considerable amounts of organic matter and are usually darkened appreciably

---

[1] Where the original regolith was relatively uniform, the material below the soil is considered to have a composition similar to the parent material from which the soil was formed.

FIGURE 1.3 Field view of a road cut that reveals the underlying layers of a soil. The closeup emphasizes soil layering and the distinctive character of the *soil profile*. The surface layer is darker in color because of its higher organic matter content. One of the subsurface horizons (point of pick) is characterized by a distinctive structure. The existence of layers such as those shown is used to help differentiate one soil from another.

because of such an accumulation. When a soil is plowed and cultivated, these layers are included in what is termed the surface soil or the *topsoil*. This is sometimes referred to as the *furrow slice* because it is the portion of the soil turned or "sliced" by the plow.

The underlying layers (referred to as the *subsoil*) contain comparatively less organic matter than the topsoil. The various subsoil layers, especially in mature, humid region soils, present two very general belts: (a) an upper transition zone characterized by the loss of minerals and by some organic matter accumulation and (b) a lower zone of accumulation of compounds such as iron and aluminum oxides, clays, gypsum, and calcium carbonate.

The solum thus described extends a moderate depth below the surface. A depth of 1–2 meters is representative for temperate region soils. Here, the noticeably modified lower subsoil gradually merges with the less weathered portion of the regolith whose upper portion is geologically on the verge of becoming a part of the lower subsoil and hence of the solum.

The various layers comprising a soil profile are not always distinct and well defined. The transition from one to the other is often so gradual that the establishment of boundaries is rather difficult. Nevertheless, for any particular soil the various horizons are characteristic and their properties greatly influence the growth of higher plants.

1.6 Topsoil and Subsoil

Topsoil, being near the surface, is the major zone of root development. It carries much of the nutrients available to plants, and it supplies a large share of the water used by crops. Also, as the layer that is plowed and cultivated, it is

FIGURE 1.4 Root system of a corn plant growing in a deep open soil. Roots of crops such as alfalfa or of trees probably penetrate even further. [*Courtesy USDA National Tillage Machinery Laboratory.*]

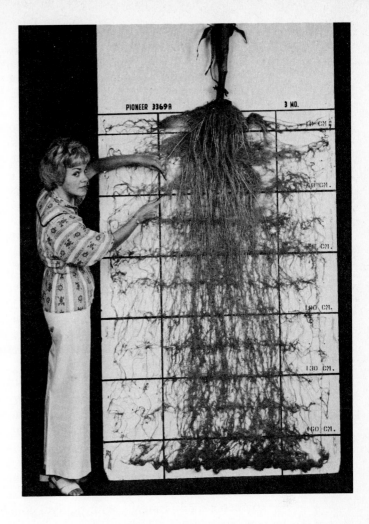

subject to manipulation and management. By proper cultivation and the incorporation of organic residues, its physical condition may be modified. It can be treated easily with chemical fertilizers and limestone, and it can be drained. In short, its fertility and to a lesser degree its productivity[2] may be raised, lowered, or satisfactorily stabilized at levels consistent with economic crop production.

This explains why much of the soil investigation and research has been devoted to the surface layer. Plowing, cultivation, liming, and fertilization are

---

[2] The term *fertility* refers to the inherent capacity of a soil to supply nutrients to plants in adequate amounts and in suitable proportions. *Productivity* is related to the ability of a soil to yield crops and is the broader term since fertility is only one of a number of factors that determine the magnitude of crop yields.

applied essentially to the furrow slice. In practice, then, the term "soil" usually denotes the surface layer, the "topsoil," or the furrow slice.

Even though the subsoil cannot be seen from the surface, there are few land uses that are not influenced by subsoil characteristics. Certainly crop production is affected by root penetration into the subsoil (Figure 1.4) and by the reservoir of moisture and nutrients held therein. Likewise, the selection of building sites and the location of roadways are influenced by subsoil characteristics. These observations are of practical significance since, unlike the topsoil, the subsoil is subject to little human alteration except by drainage. As a consequence, land-use decisions are often more dependent on the nature of the subsoil than on topsoil characteristics.

1.7 Mineral (Inorganic) and Organic Soils

The profile generalizations just described relate to soils that are predominantly *mineral* or *inorganic* in composition. Even in their surface layers mineral soils are comparatively low in organic matter, generally ranging from 1 to 10%. In contrast, soils in swamps, bogs, and marshes commonly contain 80–95% organic matter. Organic soils include all soils with more than 20% (by weight) organic matter along with soils that are continually saturated with water and contain 12–18% organic matter (depending on clay content). When drained and cleared, organic soils are most productive, especially for high-value crops such as fresh market vegetables. Organic deposits may also be excavated, bagged, and sold as organic supplements for home gardens and potted plants. The economic significance of organic soils is considerable in localized regions.

Because they occupy such a high proportion of the total land area, mineral soils are of greater importance than organic soils and will deservedly receive major attention in this text. The origin, character, and agricultural use of organic soils are considered as a unit in Chapter 14. Until then the discussion will be concerned primarily with mineral soils.

1.8 General Definition of Mineral Soils

Mineral soils have already been denoted as the "upper biologically, chemically, and physically weathered portion of the regolith." When expanded by profile and horizon study, this statement presents a pedological concept of soil origin and characterization. Keeping in mind the role of soils in plant production, we may define the *soil* as "a collection of natural bodies developed in the unconsolidated mineral and organic material on the immediate surface of the earth that (a) serves as a natural medium for the growth of land plants and (b) has properties due to the effects of climate and living matter acting upon parent material, as conditioned by topography, over a period of time."

1.9 Four Major Components of Soils

Volume Composition of Mineral Soils. Mineral soils consist of four major components: mineral materials, organic matter, water, and air. Figure 1.5 shows the approximate proportions of these components in a representative silt loam surface soil in optimum condition for plant growth. Note that this soil contains about half solids and half pore space (water and air). Of the total soil volume, about half is solid space, 45% mineral matter and 5% organic matter. At optimum moisture for plant growth, the pore space is divided roughly in half; 25% of the volume is water space and 25% is air. The proportions of air and water are subject to rapid and great fluctuations under natural conditions, depending on the weather and other factors.

It should be emphasized that the four major components of a typical soil exist mainly in an intimately mixed condition. This encourages interactions within and between the groups and permits marked variation in the environment for the growth of plants.

The volume composition of subsoils is somewhat different from that just described. Compared to topsoils they are lower in organic matter content, are somewhat lower in total pore space, and contain a higher percentage of small pores. This means they have a higher percentage of minerals and water and a considerably lower content of organic matter and air.

FIGURE 1.5 Volume composition of a silt loam surface soil when in good condition for plant growth. The air and water in a soil are extremely variable, and their proportions determine in large degree the soil's suitability for plant growth.

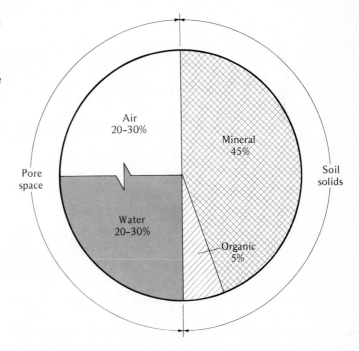

1.10   Mineral (Inorganic) Constituents in Soils

A casual examination of a sample of soil illustrates that the inorganic portion is variable in size and composition. It is normally composed of small rock fragments and minerals[3] of various kinds. The rock fragments are remnants of massive rocks from which the regolith and, in turn, the soil have been formed by weathering. They are usually quite coarse (Table 1.1). The minerals, on the other hand, are extremely variable in size. Some are as large as the smaller rock fragments; others, such as colloidal clay particles, are so small that they cannot be seen without the aid of an electron microscope.

TABLE 1.1   Four Major Size Classes of Inorganic Particles and Their General Properties

Size fraction	Common name	Means of observation	Dominant composition
Very coarse	Stone, gravel	Naked eye	Rock fragments
Coarse	Sands	Naked eye	Primary minerals
Fine	Silt	Microscope	Primary and secondary minerals
Very fine	Clay	Electron microscope	Mostly secondary minerals

Quartz and some other primary minerals such as micas and feldspars have persisted with little change in composition since they were extruded in molten lava. Other minerals, such as the silicate clays and iron oxides, have been formed by the weathering of less resistant minerals as soil formation progressed. These minerals are called secondary minerals. In general, the primary minerals dominate the coarser fractions of soil, whereas secondary minerals are most prominent in the fine materials, especially in clays. Clearly, mineral particle size has much to do with the properties of soils in the field.

1.11   Soil Organic Matter

Soil organic matter represents an accumulation of partially decayed and partially synthesized plant and animal residues. Such material is continually being broken down as a result of the work of soil microorganisms. Consequently,

[3] The word "mineral" is used in this book in two ways: (1) as a general term to describe soils dominated by inorganic constituents and (2) as a more specific term to describe distinct minerals found in nature, such as quartz and feldspars. More detailed discussions of the common soil-forming minerals and the rocks in which they are found are given in Section 1.19 and in Chapter 12.

it is a rather transitory soil constituent and must be renewed constantly by the addition of plant and/or animal residues.

The organic matter content of a soil is small, varying from 2 to 6% by weight in typical well-drained mineral soils. Its influence on soil properties and consequently on plant growth, however, is far greater than the low percentage would indicate. Organic matter functions as a "granulator" of mineral particles, being largely responsible for the loose, easily managed condition of productive soils. Also, it is a major soil source of two important mineral elements, phosphorus and sulfur, and essentially the sole soil source of nitrogen. Through its effect on the physical condition of soils, organic matter also increases the amount of water a soil can hold and the proportion of this water available for plant growth (Figure 1.6). Finally, organic matter is the main source of energy for soil organisms, both plant and animal. Without it, biochemical activity would come nearly to a standstill.

FIGURE 1.6   Soils high in organic matter are darker in color and have greater water-holding capacities than do soils low in organic matter. The same amount of water was applied to each container. As the lower photo shows, the depth of water penetration was less in the soil at the left because of its greater water-holding capacity.

Soil organic matter consists of two general groups: (a) original tissue and its partially decomposed equivalents and (b) humus. The original tissue includes the undecomposed roots and the tops of higher plants. These materials are subject to vigorous attack by soil organisms, which use them as sources of energy and tissue-building material.

The more resistant products of this decomposition, both those synthesized by the microorganisms and those modified from the original plant tissue, are collectively known as *humus*. This material, usually black or brown in color, is colloidal in nature. Its capacity to hold water and nutrient ions greatly exceeds that of clay, its inorganic counterpart. Small amounts of humus thus augment remarkably the soil's capacity to promote plant production.

1.12   Soil Water—A Dynamic Solution

Two major concepts concerning soil water emphasize the significance of this component of the soil in relation to plant growth.

1. Water is held within the soil pores with varying degrees of tenacity depending on the amount of water present and the size of the pores.
2. Together with its dissolved salts, soil water makes up the *soil solution*, which is so important as a medium for supplying nutrients to growing plants.

The tenacity with which water is held by soil solids determines to a marked degree the movement of water in soils and its use by plants. For example, when the moisture content of a soil is optimum for plant growth (Figure 1.5), plants can readily absorb the soil water, much of which is present in pores of intermediate size. As some of the moisture is removed by the growing plants, that which remains is present in only the tiny pores and as thin films around the soil particles. The attraction of the soil solids for this water is great, and they can compete successfully with higher plants for it. Consequently, not all the soil water is available to plants. Much of it remains in the soil after plants have wilted or died as a consequence of water shortage.

The soil solution contains small but significant quantities of dissolved salts, many of which are essential for plant growth. Nutrients are exchanged between the soil solids and the soil solution and then between the soil solution and plants. These exchanges are influenced to a degree by the concentration of salts in the solution, which, in turn, is determined by the total salts in the soil, by the makeup of the soil solids, and by the content of soil water. Such are the dynamic nature and importance to plant life of this solute-bearing water.

1.13   Soil Air—Also a Changeable Constituent

Soil air differs from the atmosphere in several respects. First, the soil air is located in the maze of soil pores separated by soil solids. This fact accounts for its variation in composition from place to place in the soil. In local pockets, reactions involving the gases can greatly modify the composition of the soil air. Second, soil air generally has a higher moisture content than the atmosphere; the relative humidity of soil approaches 100% when the soil moisture is optimum. Third, the content of carbon dioxide is usually much higher and that of oxygen lower than those found in the atmosphere. Carbon dioxide is often several hundred times more concentrated than the 0.03% commonly found in the atmosphere. Oxygen decreases accordingly, and in extreme cases may be no more than 10–12% as compared to about 20% for normal atmosphere.

The content and composition of soil air are determined to a large degree by the water content of the soil. The air is found in those soil pores not occupied by water. After a rain, large pores are the first to be vacated by the soil water, followed by medium-sized pores as water is removed by evaporation and plant utilization. Thus the soil air first occupies the large pores and then, as the soil dries out, those intermediate in size.

This explains the tendency for soils with a high proportion of tiny pores to be poorly aerated. In such soils, water dominates, and the soil-air content is low, as is the rate of diffusion of the air into and out of the soil from the atmosphere. The result is high levels of CO_2 and low levels of O_2, conditions unsatisfactory for best plant growth. The dynamic nature of soil air is apparent. The tendency for rapid changes in air content and composition has marked effects not only on the growth of economic plants but also upon the soil organisms, both plant and animal.

1.14   The Soil—A Tremendous Biological Laboratory

Mineral soils harbor a varied population of living organisms. Both animals and plants are abundant in soils. The whole range in size from the larger rodents through worms and insects to the tiniest bacteria commonly occurs in normal soils. Moreover, most organisms vary so much both in number and in amount as to make precise statements impossible. For example, the number of bacteria alone in 1 gram of soil may range from 100,000 to several billion, depending on conditions. In any case, the quantity of living organic matter including plant roots is sufficient to influence profoundly the physical and chemical trend of soil changes (Figure 1.7). Virtually all natural soil reactions are directly or indirectly biochemical in nature.

Activities of soil organisms range from the largely physical disintegration of plant residues by insects and earthworms to the eventual complete decomposition of these residues by smaller organisms such as bacteria, fungi, and actinomycetes. Accompanying these decaying processes is the release of several

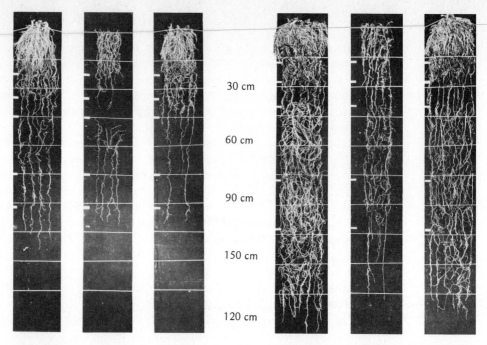

30 cm

60 cm

90 cm

150 cm

120 cm

FIGURE 1.7 Plant roots tell us something about soil characteristics and the treatment the soil has received. The corn crop was grown on an Illinois (Cisne) soil that received no fertilizers or crop residues (left) and that received both fertilizers and crop residues (right). [*Courtesy J. B. Fehrenbacher, University of Illinois.*]

nutrient elements, including nitrogen, phosphorus, and sulfur, from organic combination. By contrast, conditions in nature are such that organisms need these elements for their growth and a reversal occurs; that is, the elements are converted again into organic combinations not available to higher plants. The total process is an excellent example of *biocycling,* through which residues and wastes are incorporated into soils, disintegrated and decomposed, and pertinent products are then taken up by plants to stimulate further biomass production.

Humus synthesis, exclusively a biochemical phenomenon, also results from soil organism activity. This material is certainly one of the most useful products of microbial action.

1.15 Clay and Humus—The Seat of Soil Activity

The dynamic nature of the finer portions of the soil—clay and humus—has been indicated. Both these constituents are characterized by extremely small individual particles, the smallest of which are *colloidal* in size. They have large surface area per unit weight, and the presence of surface charges to which ions and water are attracted.

The chemical and physical properties of soils are controlled largely by clay and humus. They are centers of activity around which chemical reactions and nutrient exchanges occur. Furthermore, by attracting ions to their surfaces, they temporarily protect essential nutrients from leaching and then release them slowly for plant use. Because of their surface charges, they also act as "contact bridges" between larger particles, thus helping to maintain the stable granular structure that is so desirable in a porous, easily worked soil.

On a weight basis, the humus particles have greater nutrient- and water-holding capacities than does clay. However, clay is generally present in larger amounts, and its total contribution to the chemical and physical properties will usually equal that of humus. The best agricultural soils contain a balance of the properties of these two important soil constituents.

The clay and humus along with other soil solids, soil water, and soil air determine the suitability of soils for all kinds of use, most important of which is to sustain plant growth and development. We will turn briefly to a consideration of soils as a habitat for plants and as a source of essential mineral elements for these plants.

1.16   Soils as a Habitat for Plants

Plants are dependent on a favorable combination of some six environmental factors: (a) light, (b) mechanical support, (c) heat, (d) air, (e) water, and (f) nutrients. With the exception of light, soils can supply each of these factors.

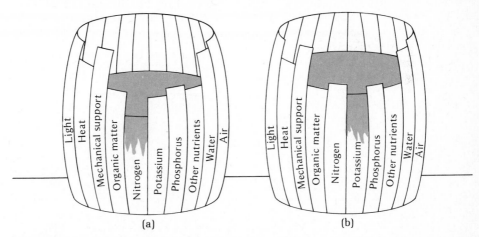

FIGURE 1.8   An illustration of the law of the minimum. The level of water in each barrel above represents the level of crop production. (a) Nitrogen is represented as being the factor that is most limiting. Even though the other elements are present in more adequate amounts, crop production can be no higher than that allowed by the nitrogen. (b) When nitrogen is added, the level of crop production is raised until it is controlled by the next most limiting factor, in this case potassium.

But only when they are supplied in the right combination is best plant growth obtained. The factor that is least optimum will limit this growth (Figure 1.8). The principle, called the *law of the minimum* (Section 1.2), may be stated in a practical way as follows: *The level of plant production can be no greater than that allowed by the most limiting of the essential plant growth factors.* One of the principal reasons for studying soils is to ascertain which factor is least optimum and how its limitation to plant growth can be removed.

The physical properties of soils, especially as they affect water, air, and heat, will receive attention in the next three chapters. We shall consider briefly in this chapter the supply and availability of plant nutrients and the general mechanisms by which soils can make these nutrients available.

1.17   The Essential Elements

Starting with the work of Liebig and other nineteenth-century scientists discussed in Section 1.2, research has shown that certain elements are essential for plant growth and that each element must be present in a specific concentration range for optimum plant growth (Figure 1.9). If the concentration of a given element in the plant root zone is too low, a deficiency of that element occurs and plant growth is restricted. Likewise, if the root zone concentration of that element is too high, toxicity occurs and plant growth is similarly limited. Only in a specific middle range of concentration is optimum plant growth attained. One of the principal objectives of soil scientists is to maintain concentrations of each essential element in this middle range.

Some seventeen elements have been found to be universally essential for plant growth. Three of them come from air and water and fourteen from

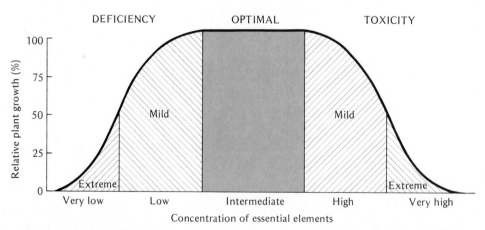

FIGURE 1.9   Relationship between plant growth and concentration in the soil of elements that are essential to the plant.

TABLE 1.2 Essential Nutrient Elements and Their Sources[a]

Used in relatively large amounts		Used in relatively small amounts
Mostly from air and water	From soil solids	From soil solids
Carbon	Nitrogen	Iron
Hydrogen	Phosphorus	Manganese
Oxygen	Potassium	Boron
	Calcium	Molybdenum
	Magnesium	Copper
	Sulfur	Zinc
		Chlorine
		Cobalt

macronutrients

[a] Other minor elements, such as sodium, fluorine, iodine, silicon, strontium, and barium, do not seem to be universally essential, as are the seventeen listed here, although the soluble compounds of some will increase the growth of specific plants.

soil solids (Table 1.2). Six of the fourteen are used in relatively large amounts and are called *macronutrients,* while the other eight, needed in only very small amounts, are called *micronutrients.* Even though the micronutrients are just as essential for plant growth as the macronutrients, they are required in such small quantities that most soils are able to provide them in sufficient quantities for normal plant growth. The conditions under which limited quantities of these nutrients needed by plants are made available will be given further attention in Chapter 11.

The macronutrients *nitrogen, phosphorus, potassium, calcium, magnesium,* and *sulfur* are absorbed by plants in sufficiently large quantities as to tax the ability of most soils to supply the plant's needs. Supplementary additions through fertilizers provide nitrogen, phosphorus, and potassium, and to a lesser degree, sulfur. Limestone is usually applied to help supplement the system's needs for calcium and magnesium.

1.18 Essential Macronutrient Content of Soils

The macronutrient chemical compositions for representative soils of three different regions are given in Table 1.3. It should be mentioned that these data do not fit particular soils, but represent a rough median of data available from the regions. Note that the soils from arid regions are highest in all constituents listed except nitrogen and organic matter. In turn, the representative culti-

TABLE 1.3 Amounts of Organic Matter and Primary Nutrients Present in Representative Cultivated Mineral Surface Soils from Humid and Arid Temperate Regions and a Humid Tropical Region

Constituent	Range ordinarily expected (%)	Humid temperate region soil (%)	Humid temperate region soil (kg/hectare–furrow slice)	Arid temperate region soil (%)	Arid temperate region soil (kg/hectare–furrow slice)	Humid tropical region soil (%)	Humid tropical region soil (kg/hectare–furrow slice)
Organic matter	0.40–10.0	4.00	90,000	3.00	67,500	2.50	56,250
Nitrogen (N)	0.02– 0.5	0.15	3,400	0.12	2,700	0.10	2,250
Phosphorus (P)	0.01– 0.2	0.04	900	0.07	1,570	0.03	675
Potassium (K)	0.17– 3.3	1.70	38,000	2.00	45,000	0.90	20,250
Calcium (Ca)	0.07– 3.6	0.40	9,000	1.00	22,500	0.15	3,375
Magnesium (Mg)	0.12– 1.5	0.30	6,700	0.60	14,400	0.10	2,400
Sulfur (S)	0.01– 0.2	0.04	900	0.08	1,800	0.03	675

As a supplement to the general figures above, the analyses of 12 soils from the southeastern part of the United States are presented as published by Lechler et al. (1981).

Soil	SiO_2	Al_2O_3	Fe_2O_3[a]	CaO	MgO	Na_2O	K_2O	MnO	TiO_2	P_2O_5	LOI[b]	Total
Beaumont	67.3	12.8	4.28	.79	.63	.18	.63	.048	.65	.11	13.15	100.57
Hiwassee	43.4	27.5	12.8	.028	.20	.011	.40	.025	1.38	.19	15.00	100.93
Houston	46.8	18.3	6.36	7.30	.77	.14	.71	.037	.98	.12	18.27	99.79
Houston Black	42.2	8.52	3.03	18.6	.91	.078	.56	.060	.36	.13	25.81	100.26
Iredell	50.6	15.6	12.0	1.15	4.07	.23	.26	.25	.98	.13	15.28	100.55
Katy	78.5	9.65	3.46	.17	.27	.097	.35	.006	.42	.086	6.99	100.00
Nacogdoches	63.1	11.6	15.1	.16	.28	.054	.60	.029	.45	.17	7.45	98.99
Oktibbeha	48.7	22.8	9.52	.50	.61	.064	.38	.008	.79	.15	15.97	99.49
Sango	79.4	7.16	7.21	.046	.24	.081	.37	.013	.60	.099	5.04	100.26
Sequoia	62.3	18.1	7.24	.081	.59	.21	3.03	.013	.95	.15	7.31	99.97
Varina	94.4	1.50	.58	.030	.030	.01	.07	.004	.57	.097	2.37	99.66
Whitestore	55.1	22.2	6.76	.067	.84	.30	2.30	.027	.60	.13	12.28	100.60

[a] Total Fe calculated as Fe_2O_3.
[b] LOI = loss on ignition: weight loss at 1050°C for 2 hr.

vated soil from the humid tropics is lowest in all constituents including nitrogen and organic matter. This is to be expected in view of the high temperature and rainfall conditions under which these well-weathered tropical soils have formed. In each of the regions the levels of nitrogen, phosphorus, and sulfur are much lower than for the other elements. And potassium contents are generally the highest.

Total Versus Available Nutrients. While the data in Table 1.3 provide some information on the *potential* supply of mineral elements from soils, they give no indication as to the proportion of any of the nutrients that is *available* to plants during a growing season. In fact, this proportion is generally quite low, no more than about 1–2% of the total quantity of most nutrients in a soil being readily available for crop uptake. This is a fact of great importance in the study of soils and plant nutrition.

1.19 Forms of Macronutrients in Soils

The original sources of macronutrients are the rocks and minerals from which soils form. Table 1.4 shows examples of these nutrient-containing minerals. Primary minerals such as feldspars, micas, hornblende, and augite, formed as molten magma cooled, contain small quantities of these macronutrients. As these primary minerals weather, secondary minerals such as silicate clay minerals, calcite, apatite, and gypsum are formed. They too contain and react with macronutrients.

The original and secondary minerals are examples of complex insoluble compounds in which macronutrients are found. These sources serve as reserves for these nutrients. They are complemented by simple, more soluble forms readily available to higher plants. As a result of the chemical and biological processes at work, elements in the soil commonly move slowly from the more complex to the simpler forms. The reverse also occurs, however, and available elements can revert to insoluble and unavailable forms.

The reactions each element undergoes in soils as it moves to or from a more available state are relatively specific for that element. Nevertheless, there are some general principles governing the reactions in soils of the cations, such as Na^+, K^+, Ca^{2+}, Mg^{2+}, and NH_4^+, on the one hand, and of the anions such as NO_3^-, HSO_4^-, and $H_2PO_4^-$ on the other. In this chapter we will discuss briefly some of these principles as they relate to the supply of nutrients for plants, leaving more specific reactions for subsequent chapters. We start with the cations, K^+, Ca^{2+}, Mg^{2+}, and NH_4^+.

TABLE 1.4   Examples of Original and Secondary Minerals in Soils

Note the macronutrients these minerals contain, as shown in their formulas.

Original minerals		Secondary minerals	
Mineral	**Formula**	**Mineral**	**Formula**
		Calcite	$CaCO_3$
Feldspar		Gypsum	$CaSO_4 \cdot 2H_2O$
Microcline	$KAlSi_3O_8$	Goethite	$FeOOH$
Plagioclase	$CaAl_2Si_2O_8$	Apatite	$Ca_5(PO_4)_3 \cdot (Cl,F)$
Micas		Clay minerals	Variable
Muscovite	$KAl_3Si_3O_{10}(OH)_2$		
Biotite	$KAl(Mg,Fe)_3Si_3O_{10}(OH)_2$		
Hornblende	$Ca_2Al_2Mg_2Fe_3Si_6O_{22}(OH)_2$		

1.20   Macronutrient Cations

To understand how macronutrient cations are made available to growing plants, attention must be given to the general inorganic makeup of mineral soils which contain these elements. While this makeup will vary considerably from soil to soil, most of the chemical and physical properties are related to three basic types of components. First, there is a *solid framework* that includes, in addition to small quantities of the essential macroelements, much larger quantities of silicon, oxygen, aluminum, and iron, in the case of inorganic materials, and of carbon, hydrogen, and oxygen in the case of organic materials (Figure 1.10). This framework dominates the soil in shear mass and holds most of the essential elements in relatively unavailable forms. It includes many of the coarse-sized primary minerals such as the feldspars, micas, and quartz, along with coarse organic particles. It also includes the finer clay and humus particles.

Second, there is an *associated assemblage of cations* held on the surface of clay and humus particles by negative charges that characterize these colloids (Figure 1.10). Cations thus held are not so significant in mass as the solid framework cations, but they dominate the chemical properties of soils in supplying essential elements to plants. Third are the cations in the *soil solution*, which are readily available for plant absorption or for leaching from the soil. These cations are present in smaller quantities than either those in the solid framework or those held on the colloidal surfaces.

Plants absorb soil solution cations, which, to a degree at least, are then replenished by cations held on the colloidal surfaces. In turn, cations held within the solid framework are released over a period of time by weathering and become adsorbed[4] or held on colloidal surfaces. The solid framework is thus seen to be a reservoir of nutrients for long-term future use. The release of these cations can be illustrated in a general way as follows.

$$\begin{pmatrix} \text{Primary minerals} \\ \text{framework cations} \end{pmatrix} \rightarrow \begin{pmatrix} \text{colloidal fraction} \\ \text{framework cations} \end{pmatrix} \rightleftharpoons \begin{pmatrix} \text{surface-adsorbed} \\ \text{exchangeable cations} \end{pmatrix} \rightleftharpoons \begin{pmatrix} \text{soil solution} \\ \text{cations} \end{pmatrix}$$

Note that, to a degree, the reactions are reversible, making it possible to add soluble nutrients and have them held in less available forms in the soil. This may be a desirable feature since the soil solution cations are subject not only to plant uptake but to leaching from the soil. The reaction also suggests that cations in the soil solution can exchange for cations adsorbed on colloidal surfaces. This process, called *cation exchange,* can be shown simply by illustrating how hydrogen ions in carbonic acid (H_2CO_3), formed by CO_2 and water in the soil, can exchange with Ca^{2+} adsorbed on the colloidal surface.

---

[4] Note the difference between *adsorption,* a phenomenon by which ions are held on the surface of colloidal particles, and *absorption,* the process by which these ions are taken into plant roots.

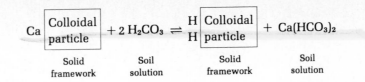

$$\text{Ca}\begin{array}{|c|}\hline \text{Colloidal} \\ \text{particle} \\\hline\end{array} + 2\,H_2CO_3 \rightleftharpoons {}^H_H\begin{array}{|c|}\hline \text{Colloidal} \\ \text{particle} \\\hline\end{array} + Ca(HCO_3)_2$$

Solid framework	Soil solution	Solid framework	Soil solution

This exchange of H^+ ions for Ca^{2+} ions and vice versa is of great practical significance in helping to maintain the proper acid–base balance in soils (see Section 1.22).

As previously emphasized, the bulk of the macronutrient elements is in unavailable forms as part of the solid framework of soil minerals. Figure 1.11 shows the proportion of potassium, calcium, and magnesium in exchangeable form compared to that in less available solid framework forms. Note that, especially for potassium, a very high proportion of the element is found in the relatively unavailable forms in the solid framework minerals. This emphasizes once again that the *total quantity* of a nutrient element in a soil is much

FIGURE 1.10 Diagram illustrating the three major components of the most chemically active parts of the soil. (a) The solid framework is made up of coarse particles and finer or colloidal particles, which comprise the bulk of the soil solids. (b) The adsorbed and exchangeable cations held on the surface of the colloidal particles are subject to exchange with (c) cations in the soil solution from which plants are able to absorb the essential elements. The coarse fraction is relatively inactive chemically, but over long periods of time can break down to yield colloid-sized particles and exchangeable cations. The colloidal fraction includes particles with negatively charged surfaces that are able to attract and hold (adsorb) the cations on their surfaces.

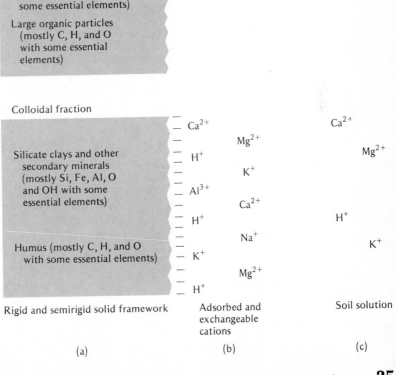

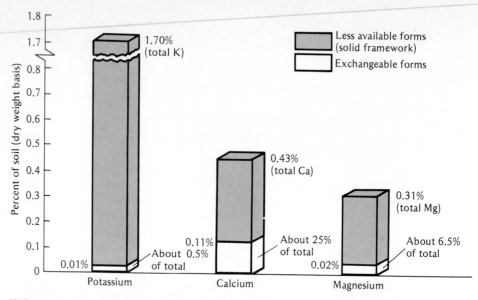

FIGURE 1.11 Percentages of total and exchangeable (adsorbed) K, Ca, and Mg present in a representative humid temperate region mineral soil. Note that the amount of exchangeable Ca is approximately ten times greater than the exchangeable K and five times that of the exchangeable Mg.

less important than is the amount that is *available* or that can be made readily available for plant uptake.

It is obvious from the general reaction described in this section that the supply of macronutrient cations is largely controlled by chemical reactions. It is also obvious that the finer fractions of the soil, clay and humus, are essential as focal points for these reactions, and that knowledge of these colloidal materials is essential to fully utilize the soil's ability to provide plant nutrients.

1.21 Macronutrient Anions

Nitrogen, sulfur, and phosphorus are the soil macronutrients that are absorbed most readily by plants in the forms of anions, NO_3^-, HSO_4^- and $H_2PO_4^-$ being examples of specific ions. Though each element undergoes specific reactions in the soil, there are some common types of reaction that illustrate how complex forms are simplified.

A major source of these anions is the soil organic matter. These elements are critical components of organic compounds such as proteins, amino acids, and nucleic acids found in the solid framework of plant materials, and in turn, of colloidal humus. In such organic compounds these macronutrients are not available for plant uptake. As with the cation macronutrients, these elements must be transformed to simple inorganic ions, before they can be taken up by plants. In this case, however, the reactions are *biochemical*, that is, they are accomplished by enzymes associated with microorganisms in the soil. The elements held as part of the organic solid framework are released in a manner which can be illustrated simply as follows.

$$R-NH_2 \; \underset{\longleftarrow}{\overset{\text{Microbial action}}{\longrightarrow}} \; NH_4^+ \; \underset{\longleftarrow}{\overset{\text{Microbial action}}{\longrightarrow}} \; NO_2^- \; \underset{\longleftarrow}{\overset{\text{Microbial action}}{\longrightarrow}} \; NO_3^-$$

$$R-S \; \rightleftarrows \; HS^- \; \rightleftarrows \; HSO_3^- \; \rightleftarrows \; HSO_4^-$$

$$R-SO_4 \; \rightleftarrows \; HSO_4^-$$

$$R-PO_4 \; \rightleftarrows \; H_2PO_4^-$$

Organic forms · · · · · · · · · · · · Inorganic forms (soil solution)

Since mineral or inorganic anions are the end products of these reactions, the process of their simplification is termed *mineralization*. The reverse of this process, whereby the inorganic forms are converted to organic compounds, is called *immobilization*, a process also accomplished by microorganisms. Sometimes immobilization is desirable since it may prevent leaching of soluble elements from the soil. At other times it is harmful because nutrients needed by plants are tied up in unavailable forms.

As the equations indicate, mineralization and immobilization of nitrogen, sulfur, and phosphorus are controlled by the action of microorganisms. Thus, higher plants are largely dependent on activities of these tiny microbes. In turn, microorganisms utilize plant residues as food and as sources of energy. The interdependence of soil microorganisms and higher plants is extremely important to crop production.

Mineral Forms of Phosphorus and Sulfur. A considerable portion of phosphorus and somewhat less of sulfur are held in mineral forms. For example, phosphorus ions (such as $H_2PO_4^-$) react with iron and aluminum compounds (in acid soils) or with calcium compounds (in neutral and alkaline soils) to form quite insoluble phosphates that are largely unavailable to plants. Likewise, much of the soil sulfur may be held as iron sulfides (pyrite) or calcium sulfate (gypsum) and must be solubilized before it can be used by higher plants.

Phosphorus may be changed from insoluble to soluble forms in the soil, although the speed of the process is often much too slow to provide adequate phosphorus for optimum plant growth. This release can be shown as follows, using $Ca_3(PO_4)_2$ as an example of the insoluble phosphate and carbonic acid (from $CO_2 + H_2O$) as the solubilizing agent.

$$Ca_3(PO_4)_2 + 4\,H_2CO_3 \; \rightleftarrows \; 3\,Ca^{2+} + 2\,H_2PO_4^- + 4\,HCO_3^-$$

Insoluble phosphate (solid) · · Carbonic acid · · · · · · · · Water-soluble phosphate (soil solution)

TABLE 1.5 Forms in Which Macronutrients Occur in Mineral Soils

Solid framework	Exchangeable cations	Soil solution ions
Potassium		
Original minerals such as feldspars and micas	K^+	K^+
Silicate clays and humus		
Calcium		
Original minerals such as feldspars, micas, hornblende	Ca^{2+}	Ca^{2+}
Limestone (calcite & dolomite)		
Magnesium		
Original minerals such as micas, hornblende, and serpentine		
Limestone (dolomite)	Mg^{2+}	Mg^{2+}
Silicate clays and humus		
Nitrogen[a]		
Organic: such as proteins and amino acids	NH_4^+	NH_4^+
		NO_2^-
		NO_3^-
Phosphorus[a]		
Organic: phytin, nucleic acid, etc.	—	$H_2PO_4^-$
Inorganic: Ca, Fe, and Al phosphates		HPO_4^{2-}
Sulfur[a]		
Organic: proteins, amino acids	—	HSO_3^-
Inorganic: gypsum, pyrite	—	HSO^-
		SO_3^{2-}
		SO_4^{3-}

[a] Anions of phosphorus and sulfur, and to a lesser extent, nitrogen, may be adsorbed on certain positively charged particles in the soil, but at much lower rates than those that characterize the adsorption of cations.

By this means, growing plants are supplied with available phosphorus, although at a very slow rate. Even when the phosphorus so solubilized is combined with that from the mineralization of organic phosphates, the amount available to plants is often too little for normal plant growth. An inadequate supply of both phosphorus and nitrogen is often the primary factor limiting optimum crop production.

Table 1.5 lists the forms of macronutrients in soils. The relatively unavailable forms in the solid framework, the more readily available exchangeable forms, and the ions in soil solution, which are immediately available to plants, can be seen from this table.

1.22   Soil Solution

The soil solution is simply water in which are found varying amounts of the cations and anions just discussed. Because of the different sizes and location of soil pores in which it is found, the soil solution is discontinuous, and the amount and kind of ions it contains differ greatly from place to place.

Concentration of Ions. At any one time, the soil solution contains only a tiny fraction of the total nutrients in a soil (Table 1.6). In humid regions the nutrient content in the soil solution in the plow layer of cultivated areas will likely amount to no more than about 500–1000 kg/ha. This is compared with the nearly 50,000 kg of *total* essential elements found in a hectare–furrow slice of the same soil.

The quantity of nutrient elements in the soil solution at any one time is far less than is needed to produce a crop. As plants absorb ions and draw down the soil solution nutrient level, additional ions are released either from the exchangeable form or by the biochemical breakdown of organic matter in the soil. Thus, the soil is seen to be a remarkable nutrient-supplying system. Only in modern times has it become necessary to provide substantial nutrients from alternate sources such as fertilizers to complement those provided by the soil.

Reaction of the Soil Solution. Another important property of the soil solution is its soil reaction; that is, whether it is *acid, neutral,* or *alkaline.* Acidity denotes an excess of H^+ ions over OH^- ions and alkalinity denotes the opposite (Figure 1.12). At neutral reaction the H^+ and OH^- ion concentrations are equal. Nutrient availability as well as plant growth is drastically affected by the relative concentrations of these two ions.

Soil reaction is customarily expressed as pH, which is the negative logarithm of the hydrogen ion concentration. It may be shown as follows, assuming $[H^+]$ is the concentration of the hydrogen ion.

$$pH = \log \frac{1}{[H^+]}$$

TABLE 1.6 Quantities of Several Essential Elements Commonly Found in Different Forms in a Representative Humid Region Soil

Note that most of the elements at any one time are in the solid framework, a small portion in exchangeable form and an even smaller portion in the soil solution. The soil solution ions are readily available to plants, the exchangeable cations moderately available and the solid framework ions very slowly available.

Element	Kilograms per hectare 15 cm deep		
	Solid framework	Exchangeable[a]	Soil solution
Ca	6,000	2000	50–100
Mg	6,600	400	10–30
K	33,800	200	10–30
P	800	—	0.5–1.0
S	800	—	3–15
N	3,000	—	5–25

[a] Small but significant quantities of N, P, and S are also exchangeable by a process called anion exchange. This will be considered in Chapter 5.

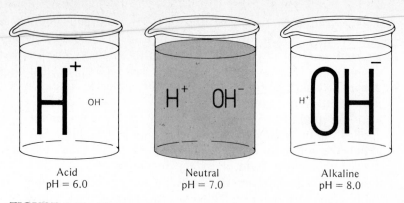

Acid	Neutral	Alkaline
pH = 6.0	pH = 7.0	pH = 8.0

FIGURE 1.12 Diagrammatic representation of acidity, neutrality, and alkalinity. At neutrality the H^+ and OH^- ions of a solution are balanced, their respective numbers being the same (pH 7). At pH 6, the H^+ ions are dominant, being ten times greater, while the OH^- ions have decreased proportionately, being only one-tenth as numerous. The solution therefore is acid, there being 100 times more H^+ ions than OH^- ions present. At pH 8, the exact reverse is true, the OH^- ions being 100 times more numerous than the H^+ ions. Hence, the solution is alkaline. This mutually inverse relationship must always be kept in mind when using pH data.

A measure of the pH of a solution gives an account of both H^+ ion and OH^- ion concentrations because of the mathematical relationship between these two ions. In any solution in which water is the solvent, the product of the concentrations of these two ions is 10^{-14} at 25°C. Thus

$$[H^+] \times [OH^-] = 10^{-14}$$

At a pH of 6.0 the H^+ ion concentration is 10^{-6} mole/liter and the hydroxyl concentration is $10^{-14}/10^{-6} = 10^{-8}$ mole/liter. Likewise, at pH 9.0 the concentration of OH^- is 10^{-5} mole/liter.

Range in pH. The pH of mineral soils varies from values of 3 or less in very acid soils of some coastal areas to more than 10 in alkali soils of some arid and semiarid areas (Figure 1.13). The range for productive cultivated soils is much narrower, however, generally 5–7 for humid region soils and 7–9 for soils of arid regions.

Nutritional Importance of Soil pH. At extreme pH values the hydrogen or hydroxyl ions can have some direct detrimental effects on plant growth. But perhaps the major effect of soil pH is on the concentration of inorganic ions in the soil solution, bringing about either deficiencies or toxicities of these ions. As the pH of some acid soils is raised from 5 to 7 or 8, nutrients such as iron, manganese, and zinc become unavailable whereas molybdenum availability increases. Phosphorus is never readily available, but its availability is generally highest in a range centering around pH 6.5.

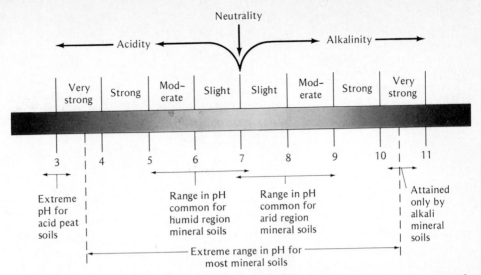

FIGURE 1.13 Extreme range in pH for most mineral soils and the ranges commonly found in humid region and arid region soils. Also indicated are the maximum alkalinity for alkali soils and the minimum pH likely to be encountered in very acid peat soils.

At pH values below 5, aluminum, iron, and manganese are often soluble in sufficient quantities to be toxic to plant growth. At very high pH values the bicarbonate ion can produce similar negative effects on certain plants.

Soil pH also determines the particular anionic species present for certain elements such as phosphorus and sulfur. For example, under very acid conditions the $H_2PO_4^-$ ion dominates. As the pH is raised, HPO_4^{2-} and finally PO_4^{3-} are found. The influence of soil pH on plant nutrition is obvious.

1.23 Soil and Plant Relations

The uptake of essential elements is determined not only by the availability of the soil-held nutrients but by their being in proximity to the root surface (Figure 1.14). Nutrients are supplied to the root surfaces in three ways: First, as roots penetrate the soil they come in contact with soil colloids on which nutrients are held. This is termed *root interception*. Second, some nutrients move to the roots with water as it is absorbed by plants. Such movement is called *mass flow*. Third, as nutrients are absorbed by plant roots, a concentration gradient is set up between the zone immediately around the root surfaces and the soil zones farther away. In response to this gradient, *diffusion* of ions toward the root surfaces takes place. For cations such as K^+ and Ca^{2+} diffusion is by far the most important means of supplying nutrients to plant roots. Diffusion is also important for anions such as NO_3^-, although mass flow also can be quite significant for these negatively charged ions.

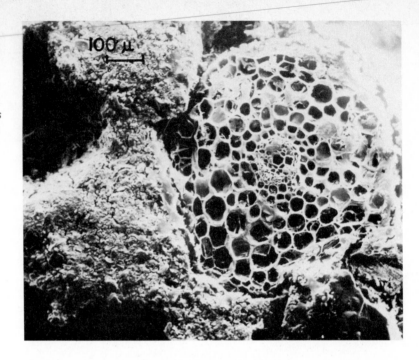

FIGURE 1.14 Scanning electron micrograph of a cross section of a peanut root surrounded by soil. Note the intimacy of contact. [*From Tan and Nopamornbodi (1981).* © 1981 The Williams & Wilkins Co., Baltimore, MD; used with permission.]

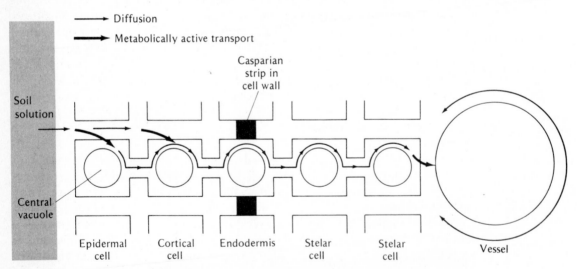

FIGURE 1.15 Cross-sectional diagram illustrating how inorganic elements move from the soil solution across the root to a vessel through which they can be transported upward to the stems and leaves. Note that through metabolic activity carriers (probably proteins) transport the ions across membranes into and out of the cytoplasm of the cells. Movement from cell to cell is by diffusion [*Redrawn from Epstein (1977). Copyright* © 1977 by the American Institute of Biological Sciences; used with permission.]

Nutrient Absorption. Root uptake of nutrients by plants requires intimate root–soil association. But research has clearly demonstrated that the plant does not simply absorb in a passive way essential nutrients presented to it. In the first place nutrient solubility is markedly affected by root exudates and by microbial activity in the vicinity of the roots (rhizosphere). Furthermore, entrance of soluble nutrients into root cells is stimulated by plant root metabolism (Figure 1.15). Chemical "carriers" (probably proteins) transport ions across membranes into cells. A combination of active transport and ion diffusion makes possible the movement of ions from the soil solution to the vessels that can carry the nutrients upward in the plant. Respiration by the root cells supplies energy for this nutrient absorption. Thus plant and microbial processes, coupled with the soil processes, implement the effective utilization of essential elements for crop production.

1.24   Summary

Soils are naturally occurring bodies whose origin, classification, and characterization are worthy of study. The pedologist is concerned with these aspects as well as with the variety of purposes, farm and nonfarm, for which soils may be used. The edaphologist, on the other hand, is concerned principally with one major use of soils—as a medium for the production of plants and particularly crop plants. This text will give deserved consideration to the pedological viewpoint, but most attention will be given to the edaphological aspects.

Of the six major factors affecting the growth of plants, only light is not supplied by soils. The soil supplies water, air, and mechanical support for plant roots as well as heat to enhance chemical reactions. It also supplies seventeen plant nutrients that are essential for plant growth. These nutrients are slowly released from unavailable forms in the solid framework of minerals and organic matter to exchangeable cations associated with soil colloids and finally to readily available ions in the soil solution. The ability of soils to provide these ions in a proper balance determines their primary value to humankind.

REFERENCES

Epstein, E. 1977. "The role of roots in the chemical economy of life on earth," *BioScience,* 27:783–787.

Lechler, P. J., W. R. Ray, and R. K. Leininger. 1981. "Major and trace element analysis of 12 reference soils by inductively coupled plasma–atomic emission spectrometry," *Soil Sci.,* 130:238–241.

Tan, K. H., and O. Nopamornbodi. 1981. "Electron microbeam analysis and scanning electron microscopy of soil–root interfaces," *Soil Sci.,* 131:100–106.

Some Important Physical Properties of Mineral Soils

2

Physically, a mineral soil is a porous mixture of inorganic particles, decaying organic matter, and air and water. The larger mineral fragments usually are embedded in and coated over with colloidal and other fine materials. Where the larger mineral particles predominate, the soil is gravelly or sandy; where the mineral colloids are dominant, the soil has clayey characteristics; and all gradations between these extremes are found in nature. Organic matter acts as a binding agent between individual particles, thereby encouraging the formation of clumps or aggregates.

Two very important physical properties of soils will be considered in this chapter: *soil texture* and *soil structure*. Soil texture is concerned with the size of mineral particles. Specifically, it refers to the relative proportions of particles of various sizes in a given soil. No less important is soil structure, which is the arrangement of soil particles into groups or aggregates. Together, these properties help determine not only the nutrient-supplying ability of soil solids but also the supply of water and air so important to plant life.

2.1 Soil Texture (Size Distribution of Soil Particles)

The size of particles in mineral soil is not subject to ready change. Thus, a sandy soil remains sandy, and a clay soil remains clay. The proportion of each size group in a given soil (the texture) cannot be altered and thus is considered a basic property of a soil.

To study successfully the mineral particles of a soil, scientists usually separate them into convenient groups according to size. The various groups are spoken of as *separates*. The analytical procedure by which the particles are separated is called *particle-size analysis*. It is a determination of the particle-size distribution.

As might be expected, a number of different classifications have been devised. The size ranges for four of these systems are shown in Figure 2.1. The classification established by the U.S. Department of Agriculture is used in this text.

Particle-Size Analyses. To make a particle-size analysis, a sample of soil is broken up and the very fine sand and larger fractions are separated into the arbitrary groups by the use of sieves and the weight of each group is determined. The silt and clay percentages are then determined by methods that depend on the rate of settling of these two separates from suspension. The principle involved in the method is simple. When soil particles are suspended in water, they tend to sink. Since there is little variation in the density of most soil particles, their rate of settling is roughly proportional to their size. The suspension of a sample of soil is therefore the first step. The particles are then allowed

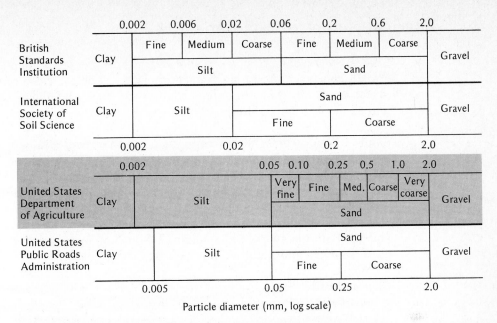

FIGURE 2.1 Classification of soil particles according to size by four systems. The U.S. Department of Agriculture system is used in this text.

to settle and the rate of settling determined either by monitoring changes in specific gravity of the suspension or by measuring directly the solids in suspension at given time intervals as settling occurs. Knowledge of the rate of settling of particles of known size makes it possible to calculate the percentage of each size of particle in a given sample, which in turn identifies the soil textural class (sand, silt, loam, etc.).

Although stone and gravel figure in the practical examination and evaluation of a field soil, they do not enter into the analysis of the finer particles. Their amounts are usually rated separately. The organic matter, ordinarily comparatively small in quantity, is usually removed by oxidation before the mechanical separation.

Sand, when dominant, yields a coarse-textured soil that has properties known to everyone as *sandy*. Such soils are sometimes referred to as *light* since they are easily tilled and cultivated. On the other hand, a fine-textured soil is made up largely of silt and clay, and its plasticity and stickiness indicate that it is likely to be difficult to till or cultivate; it is therefore termed *heavy*. The terms *light* and *heavy* refer to ease of tillage or manipulation and not to soil weight. As we shall see later, the dry weight of a cubic meter of sand is actually greater than that of a cubic meter of clay.

A particle-size analysis yields a general picture of the physical properties of a soil. It also is the basis for assigning the textural class name—that is, whether a soil is a sand, sandy loam, loam, and so on. This phase is considered in Section 3.5.

2.2   Physical Nature of Soil Separates

Coarse Fragments.   Fragments that along their greatest diameter range from 2 to 75 mm (up to 3 in.) are termed gravel or pebbles; those ranging from 75 to 250 mm (3–10 in.) are called cobbles or flags; and those more than 250 mm across are called stones.

Sands.   Sand grains may be rounded or quite irregular depending on the amount of abrasion they have received (Figure 2.2). When not coated with clay and silt, such particles are not sticky even when wet. They do not possess the capacity to be molded (plasticity) as does clay. Their water-holding capacity is low, and because of the large size of the spaces between the separate particles, the passage of water and air is rapid. Hence, they facilitate drainage and good air movement. Soils dominated by sand or gravel, therefore, possess good drainage and aeration, but may be drought prone.

Clay and Silt.   Surface area is the characteristic most affected by the small size and fine subdivision of silt and especially clay. Fine colloidal clay has about 10,000 times as much surface area as the same weight of medium-sized sand. The specific surface (area per unit weight) of colloidal clay ranges from about 10 to 1000 m²/g. The surface areas of the smallest silt particles and for fine sand are 1 and 0.1 m²/g, respectively. Since the adsorption of water,

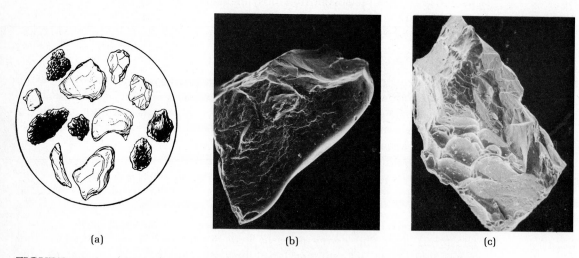

(a) (b) (c)

FIGURE 2.2   (a) Sand grains from soil. Note that the particles are irregular in size and shape. Quartz usually predominates, but other minerals may occur. Silt particles have about the same shape and composition, differing only in size. Scanning electron micrographs of sand grains show quartz sand (b) and a feldspar grain (c). The grains have been magnified about 40 times. [*Photos courtesy J. Reed Glasmann, Union Oil Company, Brea, CA.*]

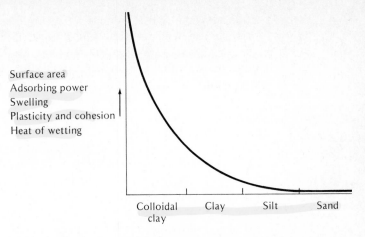

FIGURE 2.3 The finer the texture of a soil, the greater is the effective surface exposed by its particles. Note that adsorption, swelling, and the other physical properties cited follow the same general trend and that their intensities go up rapidly as the colloidal size is approached.

Surface area
Adsorbing power
Swelling
Plasticity and cohesion
Heat of wetting

Colloidal clay   Clay   Silt   Sand

nutrients, and gas and the attraction of particles for each other are all surface phenomena, the significance of the very high specific surface for clay is obvious. This relationship is shown graphically in Figure 2.3.

Clay particles commonly are platy in shape and highly plastic when moist. When clay is wetted, it tends to be sticky and is easily molded. Water tightly held on the surfaces of soil particles is in a lower energy state than free water. Consequently, energy in the form of heat is released from water as it is adsorbed by very dry soils. This *heat of wetting* is also related to particle size (Figure 2.3).

In contrast with the plate-like clay, silt particles are irregularly fragmental, diverse in shape, and seldom smooth or flat (Figure 2.2). They really are micro-sand particles; usually quartz is the dominant mineral. The silt separate, because it usually has an adhering film of clay, possesses some plasticity, cohesion (stickiness), and adsorptive capacity, but much less than the clay separate itself. Silt may lead to compaction and crusting of the soil surface unless it is supplemented by adequate amounts of sand, clay, and organic matter.

The presence of clay in a soil imparts to it a *fine texture* and slow water and air movement. Such a soil is highly plastic, becoming sticky when too wet, and hard and cloddy when dry unless properly handled. The expansion and contraction on wetting and drying usually are great, and the water-holding capacity of clayey and silty soils generally is high. As already mentioned, such soils are spoken of as *heavy* because of their difficult working qualities, in marked contrast to *light,* easily tilled sandy and gravelly surface soils.

2.3 Mineralogical and Chemical Compositions of Soil Separates

Although at this point our interest in soil particles is largely a physical one, a glance at their mineralogical makeup and chemical composition may be in order.

Mineralogical Characteristics. The coarsest sand particles often are fragments of rocks as well as minerals. Quartz (SiO_2) commonly dominates the finer grades of sand as well as the silt separate (see Figure 2.4). In addition, variable quantities of other primary minerals, such as the various feldspars (aluminosilicates) and micas (iron and aluminum silicates), usually occur. Gibbsite (hydrous oxide of aluminum), hematite, and goethite (hydrous iron oxides) also are found, usually as coatings on the sand grains. Hematite and goethite, because of their iron content, impart various shades of red and yellow if present in sufficient quantities. The soils of the southeastern part of the United States and well-oxidized tropical soils are good examples.

Some clay-size particles, especially those in the coarser clay fractions, are composed of minerals such as quartz and the hydrous oxides of iron and aluminum. But by far the most important are the complex silicate clays, which occur as plate-like particles comprised of thin sheets much like the pages of a book. In some sheets silicon and related minerals dominate, whereas in others aluminum and magnesium are prominent. These clays are classified in relation to the number of Si and Al/Mg sheets they contain in the crystal structure as follows: *1:1 (kaolinite), 2:1 (smectite, hydrous micas,* and *vermiculite),* and *2:1:1 (chlorite).* These groups vary markedly in plasticity, cohesion,

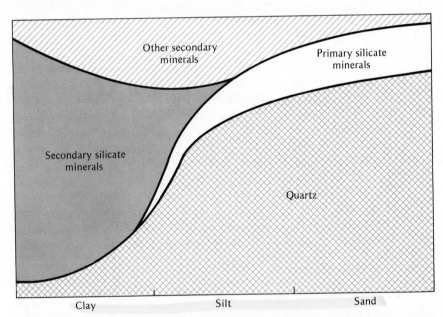

FIGURE 2.4 General relationship between particle size and kinds of minerals present. Quartz dominates the sand and coarse silt fractions. Primary silicates such as the feldspars, hornblende, and micas are present in the sands and, in decreasing amount, in the silt fraction. Secondary silicates dominate the fine clay. Other secondary minerals, such as the oxides of iron and aluminum, are prominent in the fine silt and coarse clay fractions.

and adsorptive capacity, the 1:1 group being lowest in each case and the 2:1 group highest. It is therefore important to know which clay type dominates or codominates any particular soil. These clays will be considered in greater detail in Chapter 5.

Chemical Makeup. Since sand and silt are dominantly quartz (SiO_2), a mineral known for its resistance to weathering, these two fractions are generally quite inactive chemically. Even the primary minerals that may contain nutrient elements in their chemical makeup are generally so insoluble as to make their nutrient-supplying ability essentially nil. An exception to this general rule is the silt fraction of certain potassium-bearing minerals, such as the micas, which have been known to release this element at a sufficiently rapid rate to meet some plant requirements.

Chemically, the silicate clays vary widely. Kaolinite is a relatively simple aluminosilicate. But the smectites, vermiculites, and chlorites contain in their crystal structures varying quantities of iron, magnesium, and other elements. The hydrous micas are basically potassium aluminosilicates. At the surfaces of all the silicate clays are held small but significant quantities of cations such as Ca^{2+}, Mg^{2+}, K^+, H^+, Na^+, and NH_4^+. The cations are exchangeable and can be released for absorption by plants.

In highly weathered soils, such as those in the hot, humid tropics, oxides of iron and aluminum are prominent if not dominant, even in the clay-size fraction. Thus, climate can have a profound effect on the chemical and mineralogical composition of soil separates.

TABLE 2.1 Average Phosphorus, Potassium, and Calcium Contents of Separates from Various United States Surface Soils[a]

	P (%)	K (%)	Ca (%)
Sand	.07	1.36	2.60
Silt	.10	2.01	3.21
Clay	.28	2.53	3.16

[a] From Failyer et al. (1908). Data in respect to the distribution of other elements through the various separates as reported by Joffe and Kunin (1942) are as follows.

Chemical Composition of Montalto Silt Loam Topsoil

Separate	SiO_2 (%)	Fe_2O_3 (%)	Al_2O_3 (%)	TiO_2 (%)	CaO (%)	MgO (%)
Sand	86.3	5.19	6.77	1.05	0.37	1.02
Coarse silt	81.3	3.11	7.21	1.05	0.41	0.82
Fine silt	64.0	9.42	12.00	1.05	0.32	2.22
Coarser clay	45.1	13.50	21.10	0.96	0.38	2.09
Finer clay	30.2	17.10	22.80	0.88	0.08	1.77

Since soil separates, from very coarse sand to ultrafine clay, differ so markedly in crystal form and chemical composition, it is not surprising that they also vary in their content of mineral nutrients. Logically, the sands, being mostly quartz, would be expected to be lowest and the clay separate to be highest. This inference is substantiated by the data in Table 2.1. The general relationships shown by these data hold true for most soils, although some exceptions may occur.

2.4 Soil Textural Classes

As soils are composed of particles varying greatly in size and shape, specific terms are needed to convey some idea of their textural makeup and to give some indication of their physical properties. For this, *soil textural class* names are used, such as sand, sandy loam, and silt loam. These names originated through years of soil study and classification and gradually have become more or less standardized. Three broad and fundamental groups of soil textural classes are recognized—*sands, loams,* and *clays.* Within each group specific textural class names have been devised.

Sands. The sand group includes all soils of which the sand separates make up at least 70% and the clay separate 15% or less of the material by weight. The properties of such soils are therefore characteristically sandy in contrast with the stickier nature of clays. Two specific textural classes are recognized— *sand* and *loamy sand*—although in practice two subclasses are used: *loamy fine sand* and *loamy very fine sand.*

Clays. To be designated a clay, a soil must contain at least 35% of the clay separate and in most cases not less than 40%. In such soils the characteristics of the clay separate are distinctly dominant, and the class names are *clay, sandy clay,* and *silty clay.* Note that sandy clays may contain more sand than clay. Likewise, the silt content of silty clays usually exceeds that of the clay fraction itself.

Loams. The loam group, which contains many subdivisions, is more difficult to explain. An ideal loam may be defined as a mixture of sand, silt, and clay particles that exhibits light and heavy properties in about equal proportions. Roughly, it is a half-and-half mixture on the basis of properties.

Most soils of agricultural importance are some type of loam. They may possess the ideal makeup described above and be classed simply as *loam.* Usually, however, the quantities of sand, silt, or clay present require a modified textural class name. Thus, a loam in which sand is dominant is classified as a *sandy loam* of some kind; in the same way there may occur *silt loams, silty clay loams, sandy clay loams,* and *clay loams.*

TABLE 2.2 General Terms Used to Describe Soil Texture in Relation to the Basic Soil Textural Class Names

U.S. Department of Agriculture classification system

General terms		Basic soil textural class names
Common names	**Texture**	
Sandy soils	Coarse	{ Sands { Loamy sands
	Moderately coarse	{ Sandy loam { Fine sandy loam[a]
Loamy soils	Medium	{ Very fine sandy loam[a] { Loam { Silt loam { Silt
	Moderately fine	{ Sandy clay loam { Silty clay loam { Clay loam
Clayey soils	Fine	{ Sandy clay { Silty clay { Clay

[a] While not included as class names in Figure 2.5, these soils are commonly treated separately because of their fine sand content.

Variations in the Field. The textural class names—*sand, loamy sand, sandy loam, loam, silt loam, silt, sandy clay loam, silty clay loam, clay loam, sandy clay, silty clay,* and *clay*—form a more or less graduated sequence from soils that are coarse in texture and easy to handle to the clays that are very fine and difficult to manage (Table 2.2). It is also obvious that these textural class names are a reflection not only of particle-size distribution but also of tillage characteristics and other physical properties. For some soils, qualifying factors such as stone, gravel, and the various grades of sand become part of the textural class name. Even *silt* and *clay* become qualifying terms in practice.

2.5   Determination of Soil Class

"Feel" Method. The common field method of classifying a soil is by its *feel*. Probably as much can be judged about the texture and hence the class of a soil merely by rubbing it between the thumb and fingers as by any other superficial means. Usually it is helpful to wet the sample in order to estimate plasticity more accurately. The way a wet soil "slicks out"—that is, develops a continuous ribbon when pressed between the thumb and fingers—gives a good idea of the amount of clay present. The slicker the wet soil, the higher the clay content. The sand particles are gritty; the silt has a floury or talcum-powder feel when dry and is only slightly plastic and sticky when wet. Persistent cloddiness generally is imparted by silt and clay.

The "feel" method is used in field operations such as soil survey and land classification. Accuracy in such a determination is of great practical value and depends largely on experience. Facility in class determination is one of the first things a field researcher should develop.

Laboratory Method. The U.S. Department of Agriculture has developed a method for naming soils on the basis of particle-size analysis. The relationship between such analyses and class names is shown diagrammatically in Figure 2.5. The diagram reemphasizes that a soil is a mixture of particles of different

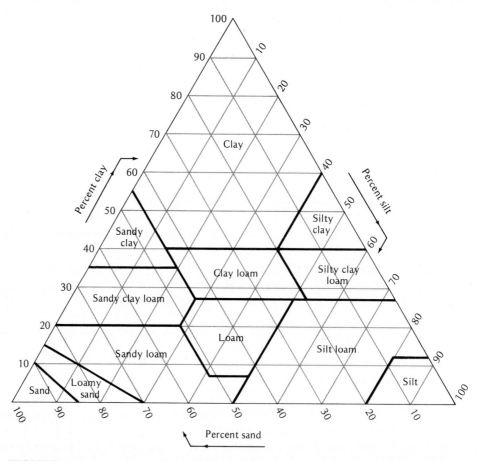

FIGURE 2.5 Percentage of sand, silt, and clay in the major soil textural classes. To use the diagram, locate the percentage of clay first and project inward as shown by the arrow. Do likewise for the percent silt (or sand). The point at which the two projections cross will identify the class name.

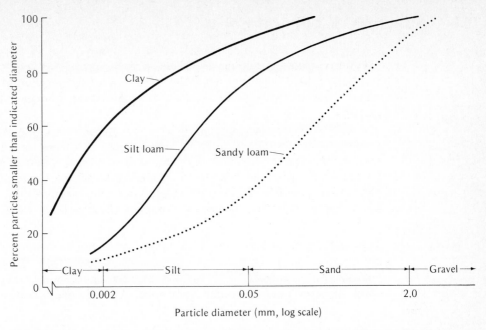

FIGURE 2.6 Particle-size distribution in three soils varying widely in their textures. Note that there is a gradual transition in the particle-size distribution in each of these soils.

sizes. It illustrates how particle-size analyses of field soils can be used to check the accuracy of the soil surveyor's class designations as determined by feel. A working knowledge of this method of naming soils is essential. The legend of Figure 2.5 explains the use of this soil texture triangle.

The summation curves in Figure 2.6 illustrate the particle-size distribution in soils representative of three textural classes. Note the gradual change in percentage composition in relation to particle size. This figure emphasizes that there is no sharp line of demarcation in the distribution of sand, silt, and clay fractions and suggests a gradual change of properties with change in particle size.

Changes in Soil Texture. The textural classes of soil horizons are not subject to easy modification in the field. For some garden and horticultural crops with high economic value, large quantities of sand may be added to a fine-textured soil to improve its tillage properties. In greenhouses, mixtures of different soils and organic materials are commonly used and the textural class of the mixture may be different from any of the components mixed. For most field crop and forest production areas, however, soil texture is not changed by cultural management. We now turn to a physical property of the soil that is subject to some change—the *soil structure*.

2.6 Structure of Mineral Soils

The term *structure* relates to the grouping or arrangement of soil particles. It is strictly a field term that describes the gross, overall combination or arrangement of the primary soil separates into secondary groupings called *aggregates* or *peds*.

A profile may be dominated by a single type of aggregate. More often, several types are encountered in the different horizons. Soil conditions and characteristics such as water movement, heat transfer, aeration, bulk density, and porosity are much influenced by structure. In fact, the important physical changes imposed by the farmer in plowing, cultivating, draining, liming, and manuring his land are structural rather than textural.

Types of Soil Structure. The dominant shape of peds or aggregates in a horizon determines their structural type. The four principal types of soil structure are *spheroidal, platy, prism-like* and *block-like*. The spheroidal, prism-like, and block-like types have two subtypes each. A brief description of each of these

FIGURE 2.7 Various structural types found in mineral soils. Their location in the profile is suggested. In arable topsoils, a stable granular structure is prized.

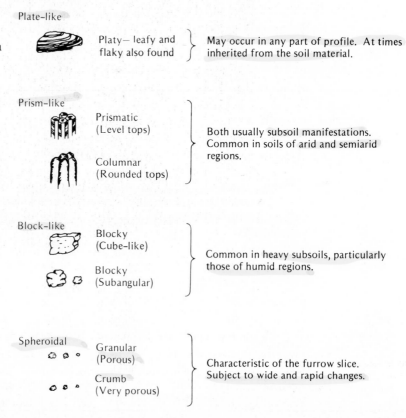

Plate-like

Platy— leafy and flaky also found } May occur in any part of profile. At times inherited from the soil material.

Prism-like

Prismatic (Level tops)

Columnar (Rounded tops)
} Both usually subsoil manifestations. Common in soils of arid and semiarid regions.

Block-like

Blocky (Cube-like)

Blocky (Subangular)
} Common in heavy subsoils, particularly those of humid regions.

Spheroidal

Granular (Porous)

Crumb (Very porous)
} Characteristic of the furrow slice. Subject to wide and rapid changes.

structural types with schematic drawings is given in Figure 2.7. A more detailed description of each follows.

1. Spheroidal (*granular* and *crumb* subtypes). All rounded peds or aggregates may be placed in this category, although the term more properly refers to those not over about 1 cm in diameter. These rounded complexes usually lie loosely and are readily shaken apart. Oridinarily the aggregates are called granules and the pattern *granular*. However, when the granules are especially porous, the term *crumb* is applied.

   Granular and crumb structures are characteristic of many surface soils, particularly those high in organic matter, and are especially prominent in grassland soils. They are the only types of aggregation that are commonly influenced by practical methods of soil management.

2. Plate-like (*platy*). In this structural type the aggregates (peds) are arranged in relatively thin horizontal plates, leaflets, or lenses. Platy structure is most noticeable in the surface layers of virgin soils, but may characterize the subsoil horizons as well.

   Although most structural features are usually a product of soil-forming forces, the platy type is often inherited from the parent materials, especially those laid down by water or ice.

3. Prism-like (*columnar* and *prismatic* subtypes). These subtypes are characterized by vertically oriented aggregates or pillars that vary in length with different soils and may reach a diameter of 15 cm or more. They commonly occur in subsoil horizons in arid and semiarid regions and when well developed are a very striking feature of the profile. They also occur in some poorly drained soils of humid regions.

   When the tops of the prisms are rounded, the term *columnar* is used. This rounding may occur when the profile is changing and certain horizons are degrading. When the tops of the prisms are still plane, level, and clean cut, the structural pattern is designated *prismatic*.

4. Block-like (*blocky* and *subangular blocky* subtypes). In this case the original aggregates have been reduced to blocks, irregularly six-faced, with their three dimensions more or less equal. In size these fragments range from about 1 to 10 cm in thickness. In general, the design is so individualistic that identification is easy.

   When the edges of the cubes are sharp and the rectangular faces distinct, the subtype is designated *blocky*. When subrounding has occurred, the aggregates are referred to as *subangular blocky*. These types usually are confined to the subsoil, and their stage of development and other characteristics have much to do with soil drainage, aeration, and root penetration.

As already emphasized, two or more of the structural conditions listed usually occur in the same soil profile. In humid temperate regions, a granular aggregation in the surface horizon with a blocky, subangular blocky, or platy type of some kind in the subsoil is usual, although granular subhorizons are not uncommon. In soils of regions of low rainfall the blocky type in the subsoil may be replaced by a columnar or prismatic arrangement.

Soil Structure Classes.   The peds in each structural type and subtype are further classified according to their size into soil structure *classes* as follows: (a) very fine or very thin, (b) fine or thin, (c) medium, (d) coarse or thick, and (e) very coarse or very thick. While the exact dimension for each class varies from one type or subtype to another, the class designation assures accurate description of the nature of the soil structural units.

Soil Structure Grades.   Soil structure *grades* relate to the degree of interaggregate adhesion and to aggregate stability. Four grades are recognized.

1. *Structureless.* Particles not arranged into peds or aggregates. If separates are not bound together (not coherent), as in a coarse sand, the term *single grain* is used. If they are tightly bound (coherent), as in a very compact subsoil or in a puddled surface soil, *massive* is used.
2. *Weak.* Poorly formed peds or aggregates barely observable in place.
3. *Moderate.* Well-formed and moderately durable peds that are not very distinct in undisturbed soil.
4. *Strong.* Durable peds that are quite evident in undisturbed soil and become separated when the soil is disturbed.

Genesis of Soil Structure.   The mechanics of structure formation is exceedingly complicated and rather obscure. The nature and origin of the parent material play significant roles, as do the physical and biochemical processes of soil formation. Climate is also a prime consideration. Soluble salts influence the development of structural units, particularly in the soils of arid regions. In more humid areas the downward migration of clay, iron oxides, and lime is a factor. Undoubtedly, the accumulation of organic matter and its type of decay are significant, too, especially in the development of the crumb structure so common in the surface soils of grasslands. The need to preserve and encourage this particular structural type is becoming critical in cultivated lands.

2.7   Particle Density of Mineral Soils

One means of expressing soil weight is in terms of the density of the solid particles making up the soil. It is usually defined as the mass (or weight) of a unit volume of soil *solids* and is called *particle density* (D_p). In the metric system, particle density is usually expressed in terms of megagrams per cubic meter (Mg/m^3). Thus, if 1 m^3 of soil solids weighs 2.6 Mg, the particle density is 2.6 Mg/m^3.[1]

---

[1] Since 1 Mg = 1 million grams and 1 m^3 = 1 million cubic centimeters, this particle density can also be expressed as 2.6 g/cm^3.

Although considerable range may be observed in the density of the individual soil minerals, the figures for most mineral soils usually vary between the narrow limits of 2.60 and 2.75 Mg/m³. This occurs because quartz, feldspar, and the colloidal silicates with densities within this range usually make up the major portion of mineral soils. When unusual amounts of minerals with high particle density such as magnetite, garnet, epidote, zircon, tourmaline, and hornblende are present, the particle density may exceed 2.75 Mg/m³. It should be emphasized that *the size of the particles of a given mineral and the arrangement of the soil solids have nothing to do with the particle density.* Particle density depends on the chemical composition and crystal structure of the mineral particle.

Organic matter weighs much less than an equal volume of mineral solids, having a particle density of 1.1–1.4 Mg/m³. Consequently, the amount of this constituent in a soil markedly affects the particle density. This accounts for the fact that mineral surface soils (which almost always have higher organic matter content than the subsoils) usually possess lower particle densities than do subsoils. Some mineral topsoils high in organic matter (say, 15–20%) may have particle densities as low as 2.4 Mg/m³, or even below. Nevertheless, for general calculations, the average arable mineral surface soil (3–5% organic matter) may be considered to have a particle density of about 2.65 Mg/m³.

2.8 Bulk Density of Mineral Soils

Bulk Density. A second important weight measurement of soils is *bulk density* (D_b). It is defined as the mass (weight) of a unit volume of dry soil. This volume includes *both solids and pores.* The comparative calculations of bulk density and particle density are shown in Figure 2.8. A careful study of this figure should make clear the distinction between these two methods of expressing soil weight.

Factors Affecting Bulk Density. Unlike particle density, which is a characteristic of solid particles only, bulk density is determined by the volume of pore spaces as well as soil solids. Thus, soils with a high proportion of pore space to solids have lower bulk densities than those that are more compact and have less pore space. Fine-textured surface soils such as silt loams, clays, and clay loams generally have lower bulk densities than sandy soils. The solid particles of the fine-textured soils tend to be organized in porous grains or granules, especially if adequate organic matter is present. This condition assures high total pore space and a low bulk density. In sandy soils, however, organic matter contents are generally low, the solid particles lie quite closely together, and the bulk densities are commonly higher than in the finer-textured soils.

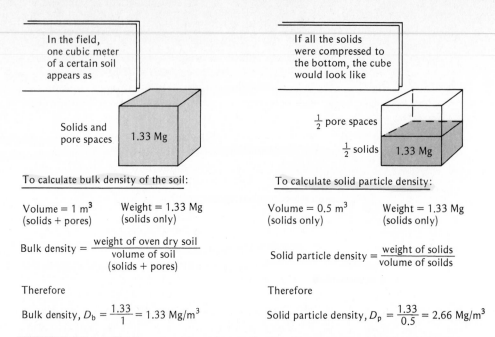

To calculate bulk density of the soil:

Volume = 1 m³   Weight = 1.33 Mg
(solids + pores)   (solids only)

$$\text{Bulk density} = \frac{\text{weight of oven dry soil}}{\text{volume of soil}}$$
(solids + pores)

Therefore

$$\text{Bulk density, } D_b = \frac{1.33}{1} = 1.33 \text{ Mg/m}^3$$

To calculate solid particle density:

Volume = 0.5 m³   Weight = 1.33 Mg
(solids only)   (solids only)

$$\text{Solid particle density} = \frac{\text{weight of solids}}{\text{volume of soilds}}$$

Therefore

$$\text{Solid particle density, } D_p = \frac{1.33}{0.5} = 2.66 \text{ Mg/m}^3$$

FIGURE 2.8 Bulk density, D_b, and particle density, D_p, of soil. Bulk density is the weight of the solid particles in a standard volume of field soil (solids plus pore space occupied by air and water). Particle density is the weight of solid particles in a standard volume of those solid particles. Follow the calculations through carefully and the terminology should be clear. In this particular case the bulk density is one-half the particle density, and the percent pore space is 50.

The bulk densities of clay, clay loam, and silt loam surface soils normally may range from 1.00 to as high as 1.60 Mg/m³, depending on their condition. A variation from 1.20 to 1.80 Mg/m³ may be found in sands and sandy loams. Very compact subsoils may have bulk densities of 2.0 Mg/m³ or even greater. The relationship among texture, compactness, and bulk density is illustrated in Figure 2.9.

Even in soils of the same surface textural class, great differences in bulk density are found, as is clearly shown by data on soils from several locations in Table 2.3. Moreover, there is a distinct tendency for the bulk density to rise with profile depth. This apparently results from a lower content of organic matter, less aggregation and root penetration, and a compaction caused by the weight of the overlying layers.

The system of crop and soil management employed on a given soil is likely to influence its bulk density, especially of the surface layers. The addition of crop residues or farm manure in large amounts tends to lower the weight figure of surface soils, as does a bluegrass sod. Intensive cultivation operates in the opposite direction. Data presented in Table 2.4 show this relationship

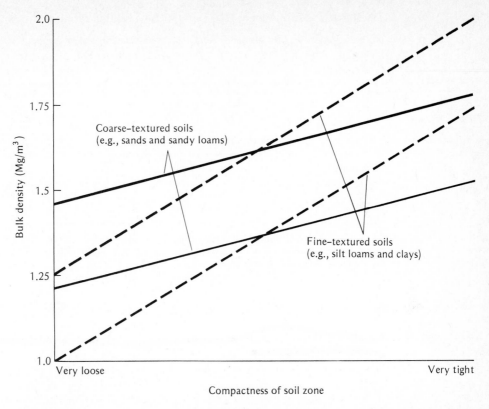

FIGURE 2.9 Generalized relationship between compactness and the range of bulk densities common in sandy soils and in those of finer texture. Sandy soils generally are less variable in their degree of compactness than are the finer-textured soils. For all soils, the surface layers are more likely to be medium to loose in compactness than are the subsoils.

very well. These data are from longtime experiments in different locations, the soils having been under cultivation for from 40 to 150 years. Cropping increased the bulk density of the topsoils in all cases.

Other Weight Figures. Densities expressed in metric units can be converted to the English system. On a dry weight basis clayey and silty surface soils may vary from 65 to 100 lb/ft³; sands and sandy loams may show a range of 75–110 lb/ft³. The greater the organic content, the less is this weight. Very compact subsoils, regardless of texture, may weigh as much as 125 lb/ft³.

Another figure of interest is the weight of soil in one hectare to a depth of normal plowing (15 cm). Such a *hectare–furrow slice* of a typical surface soil weighs about 2.2 million kg. A comparable figure in the English system is 2 million lb per *acre–furrow slice*.

TABLE 2.3 Bulk Density (D_b) Data for Several Soil Profiles[a]

Note that sandy soils generally have higher bulk densities than those high in silt and clay. The Erie soil has a very dense subsoil, while the Oxisol from Brazil is loose and open.

Horizon	Grandfield fine sandy loam (Oklahoma)	Miami silt loam (Wisconsin)	Erie channery silt loam (New York)	Houston clay (Texas)	Oxisol clay (Brazil)
Plow layer	1.72	1.28	1.33	1.24	0.95
Upper subsoil	1.74	1.41	1.46	1.36	—
Lower subsoil	1.80	1.43	2.02	1.51	1.00
Parent material	1.85	1.49	—	1.61	—

[a] The data for Grandfield soil from Dawud and Gray (1979); Miami from Nelson and Muckenhirn (1941); Houston from Yule and Ritchie (1980); Erie calculated from Fritton and Olson (1972); Oxisol from Larson et al. (1980).

TABLE 2.4 Bulk Density and Pore Space of Certain Cultivated Topsoils and of Nearby Uncropped Areas[a] (One Subsoil Included)

The bulk density was increased by cropping in every case while the pore space decreased proportionately.

Soil	Years cropped	Bulk density (Mg/m³) Cropped soil	Uncropped soil	Pore space (%)[b] Cropped soil	Uncropped soil
Hagerstown loam (Pa)	58	1.25	1.07	50.0	57.2
Marshall silt loam (Iowa)	50+	1.13	0.93	56.2	62.7
Nappanese silt loam (Ohio)	40	1.31	1.05	50.5	60.3
Ave. 19 Georgia soils	45–150	1.45	1.14	45.1	57.1
Blaine lake silt loam (Canada)	90	1.30	1.04	50.9	60.8
Sutherland clay (Canada)	70	1.28	0.98	51.7	63.0
Sutherland clay, subsoil	70	1.38	1.21	47.9	54.3

[a] Data for Canadian soils from Tiessen et al. (1982); others from Lyon et al. (1952).
[b] Pore space percentages for the Canadian soils were calculated from bulk density data in Tiessen et al. (1982).

2.9 Pore Space of Mineral Soils

The pore space of a soil is that portion of the soil volume occupied by air and water. The amount of this pore space is determined largely by the arrangement of the solid particles. If they lie close together as in sands or compact subsoils, the total porosity is low. If they are arranged in porous aggregates, as is often the case in medium-textured soils high in organic matter, the pore space per unit volume will be high.

The validity of the above generalizations may be substantiated readily by the use of a very simple formula involving particle density and bulk density figures. The derivation of the formula used to calculate the percentage of total pore space in soil follows.

$$\text{Let} \quad D_b = \text{bulk density} \qquad V_s = \text{volume of solids}$$
$$D_p = \text{particle density} \qquad V_p = \text{volume of pores}$$
$$W_s = \text{Weight of soil (solids)} \qquad V_s + V_p = \text{total soil volume}$$

By definition,

$$\frac{W_s}{V_s} = D_p \quad \text{and} \quad \frac{W_s}{V_s + V_p} = D_b$$

Solving for W_s gives

$$W_s = D_p \times V_s \quad \text{and} \quad W_s = D_b(V_s + V_p)$$

Therefore

$$D_p \times V_s = D_b(V_s + V_p)$$

and

$$\frac{V_s}{V_s + V_p} = \frac{D_b}{D_p}$$

Since

$$\frac{V_s}{V_s + V_p} \times 100 = \% \text{ solid space}$$

$$\% \text{ solid space} = \frac{D_b}{D_p} \times 100$$

Since % pore space + % solid space = 100, and % pore space = 100 − % solid space, then

$$\% \text{ pore space} = 100 - \left(\frac{D_b}{D_p} \times 100\right)$$

Using this formula, a sandy soil having a bulk density of 1.50 and a particle density of 2.65 will be found to have 43.4% pore space. A silt loam for which the corresponding values are 1.30 and 2.65 possesses 50.9% air and water space.

The latter value is close to the pore capacity of a normally granulated silt loam or clay loam surface soil.

Factors Influencing Total Pore Space.   Considerable difference in the total pore space of various soils occurs depending upon conditions. Sandy surface soils show a range of 35–50%, whereas medium- to fine-textured soils vary from 40 to 60%—even more in cases of high organic matter and marked granulation (Figure 2.10). Pore space also varies with depth; some compact subsoils have as little as 25–30%. This accounts in part for the inadequate aeration of such horizons.

Past cropping exerts a decided influence upon pore space of the furrow slice. For instance, the continuous bluegrass sod of the Hagerstown loam of Pennsylvania cited in Table 2.4 had a total porosity of 57.2%, whereas the comparable rotation plot showed only 50%. Additional data presented in this table from three other states show that cropping tends to lower the total pore space to less than that of the virgin or uncropped soils. This reduction usually is associated with a decrease in organic matter content and a consequent lowering of granulation. Pore space in the subsoil also has been found to decrease with cropping, although to a lesser degree.

Size of Pores.   Two types of individual pore space occur in soils—*macro* and *micro*. Although there is no sharp line of demarcation, pores less than about 0.06 mm in diameter are considered as micropores and those larger as macropores. The macropores characteristically allow the ready movement of air and percolating water. In contrast, the micropores are mostly filled with water in a normal field soil and do not permit much air movement into or out of the soil. The water movement is restricted primarily to slow capillary movement. Thus, in a sandy soil, in spite of the relatively low total porosity, the movement of air and water is surprisingly rapid because of the dominance of the macrospaces.

Fine-textured soils, especially those without a stable granular structure, allow relatively slow gas and water movement despite the unusually large volume of total pore space. Here the dominating micropores often stay full of water. Aeration, especially in the subsoil, frequently is inadequate for satisfactory root development and desirable microbial activity. Therefore, the size of the individual pore spaces rather than their combined volume is the important consideration. The loosening and granulating of fine-textured soils promotes aeration not so much by increasing the total pore space as by raising the proportion of the macrospaces.

It has already been indicated (Section 1.6 and Figure 1.4) that in a well-granulated silt loam surface soil at optimum moisture for plant growth the total pore space will be near 50% and is likely to be shared equally by air and water. Soil aeration under such a condition is satisfactory, especially if a similar ratio of air to water extends well into the subsoil.

FIGURE 2.10 Volume distribution of organic matter, sand, silt, clay and pores of macro and micro sizes in a representative sandy loam (a) and two representative silt loams, one with good soil structure (b) and the other with poor structure (c). Both silt loam soils have more total pore space than the sandy loam, but the silt loam with poor structure has a smaller volume of larger (macro) pores than either of the other two soils.

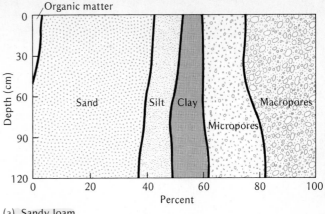

(a) Sandy loam

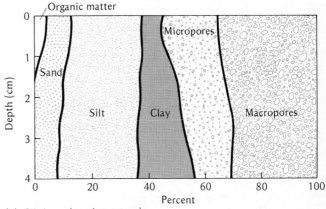

(b) Silt loam (good structure)

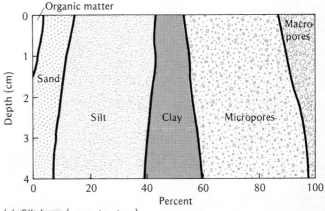

(c) Silt loam (poor structure)

Cropping and Size of Pores. Continuous cropping, particularly of soils originally very high in organic matter, often results in a reduction of large or macropore spaces. Data from a fine-textured soil in Texas (Figure 2.11) show this effect very strikingly. Cropping significantly reduced soil organic matter content and the total pore space. But the most striking effect of cropping was on the size of the soil pores. The amount of macropore space so necessary for ready air movement was reduced about one half by cropping the soil. This severe reduction in pore size also extended into the 15–30 cm layer. In fact, samples taken as deep as 107 cm showed the same trend.

In recent years, so called conservation tillage practices, which minimize plowing and associated soil manipulations, have been widely adopted in the United States (Sections 2.14 and 16.12). Such practices are remarkably effective in reducing soil erosion losses but unfortunately do not seem to increase soil porosity. In fact, in some trials pore space is less with conservation tillage than with conventional tillage.

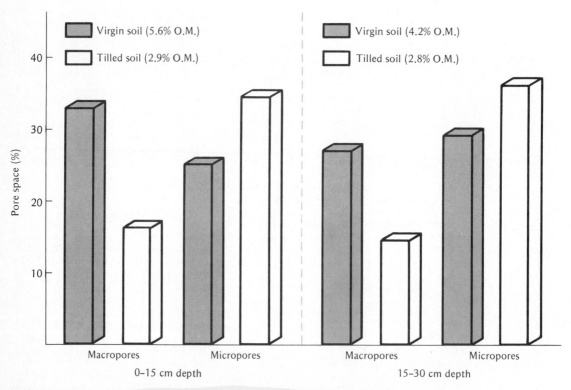

FIGURE 2.11 Effect of continuous cropping for at least 40–50 years (tilled soil) on the macro- and micropore spaces of a Houston Black clay in Texas. The virgin soil is not only higher in total pore space but has much more macropore space to permit good aeration. Note that the cultivated (tilled) soil also is much lower in organic matter than the virgin untilled soil. [*Data from Laws and Evans* (*1949*).]

2.10 Aggregation and Its Promotion in Arable Soils

In a practical sense we are concerned with two sets of factors in dealing with soil aggregation: (a) those responsible for aggregate formation and (b) those that give the aggregates stability once they are formed. Since both sets of factors are operating simultaneously, it is sometimes difficult to distinguish their relative effects on stable aggregate development in soils.

Genesis of Granules and Crumbs. Since the spheroid-type structure (granular and crumb) is the most common in surface soils, its formation and stability are of great practical significance. Although there is some uncertainty about the exact mechanism by which granules and crumbs form, several specific factors influence their genesis. These include (a) wetting and drying, (b) freezing and thawing, (c) the physical activity of roots and soil animals, (d) the influence of decaying organic matter and of the slimes and other products from the microorganisms and other forms of life, (e) the modifying effects of adsorbed cations, and (f) soil tillage.

Any action that will develop lines of weakness, shift the particles back and forth, and force contacts that otherwise might not occur will encourage aggregation. Alternate wetting and drying, freezing and thawing, the physical effects of root extension, and the mixing action of soil organisms and of tillage implements encourage such contacts and, therefore, aggregate formation. The benefits of fall plowing to certain types of soil and the slaking of clods (breaking into smaller aggregates) under the influence of a gentle rain have long been known and utilized in seedbed preparation. And the aggregating influences of earthworms and other soil organisms should not be overlooked.

Influence of Organic Matter. The major agency in the encouragement of granular and crumb type aggregate formation in surface soil horizons is organic matter (Figure 2.12), which not only binds but also lightens and expands, making possible the porosity so characteristic of individual soil aggregates. Plant roots promote this aggregation by their decay in the soil and by the disruptive action of their movement through the soil. The chemical properties of the humus and clay are probably effective in the organization and the later stabilization of the aggregates. Moreover, slime and other viscous microbial products encourage crumb development and exert a stabilizing influence. Granulation thus assumes a highly biological aspect.

Organic matter is of much importance in modifying the effects of clay on soil structure. An actual chemical union may take place between the decaying organic matter and the clay particles. Moreover, the high adsorptive capacity of humus for water intensifies the disruptive effects of temperature changes and moisture fluctuations. The granulation of a clay soil apparently cannot be promoted adequately without the presence of a certain amount of humus.

Effect of Adsorbed Cations. Aggregate formation is definitely influenced by the nature of the cations adsorbed by soil colloids (see Section 5.17). For instance, when sodium is a prominent adsorbed ion, as in some soils of arid

FIGURE 2.12 Puddled soil (left) and well-granulated soil (right). Plant roots and especially humus play the major role in soil granulation. Thus a sod tends to encourage development of a granular structure in the surface horizon of cultivated land. [*Courtesy USDA Soil Conservation Service.*]

and semiarid areas, the particles are dispersed and a very undesirable soil structure results. By contrast, the adsorption of ions such as calcium, magnesium, or aluminum may encourage aggregate formation starting with a process called *flocculation*. These ions encourage the individual colloidal particles to come together in small aggregates called floccules. Flocculation itself, however, is of limited value because it does not provide for the *stabilization* of the aggregates.

Influence of Tillage and Compaction. Tillage has both favorable and unfavorable effects on aggregation. The short-time effect is generally favorable because the implements break up the clods, incorporate the organic matter into the soil, and make a more favorable seedbed. Some tillage is thus considered necessary in normal soil management.

Over longer periods, tillage operations have detrimental effects on surface soil granules. In the first place, by mixing and stirring the soil, tillage generally hastens the oxidation of organic matter from soils. Second, tillage operations, especially those involving heavy equipment, tend to break down the stable soil aggregates. Compaction occurs from repeatedly running over fields with heavy farm equipment. An indication of the effect of such traffic upon bulk density is given in Figure 2.13. These data explain the increased interest in farming systems that drastically reduce the number of tillage operations (Sections 2.14 and 16.12).

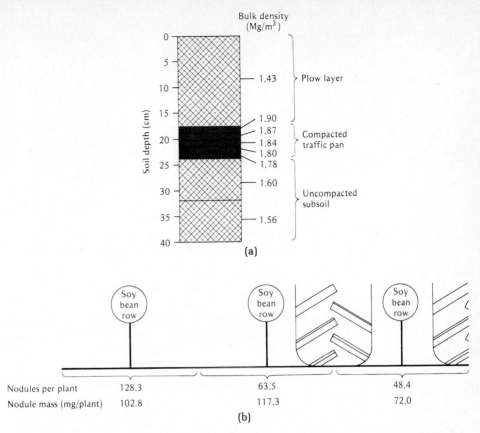

Bulk density (Mg/m³)

Soil depth (cm)

— 1.43   } Plow layer

— 1.90
— 1.87   } Compacted
— 1.84   traffic pan
— 1.80
— 1.78

— 1.60

} Uncompacted
  subsoil

— 1.56

(a)

	Soy bean row		Soy bean row		Soy bean row
Nodules per plant	128.3		63.5		48.4
Nodule mass (mg/plant)	102.8		117.3		72.0

(b)

FIGURE 2.13 Heavy equipment used for tillage and other purposes compacts the soil, increases bulk density, and reduces crop growth. (a) Such equipment compacted the zone just below the plow layer on a Norfolk soil and increased bulk density to more than 1.8 Mg/m³, the limit of penetration for cotton roots. (b) Wheel traffic on a Nicollet silty clay loam reduced the number of nitrogen-fixing nodules on a soybean field in Minnesota and brought about decreased yields. [*Data (a) from Camp and Lund (1964); (b) from Voorhees et al. (1976); used with permission of the American Society of Agronomy.*]

Aggregate Stability.   The stability of aggregates is of great practical importance. Some aggregates readily succumb to the beating of rain and the rough and tumble of plowing and fitting of the land. Others resist disintegration, thus making the maintenance of a suitable soil structure comparatively easy (Figure 2.14). Apparently there are three major factors influencing aggregate stability.

1. The temporary mechanical binding action of microorganisms, especially fungal mycelia. These effects are pronounced when fresh organic matter is

Before wetting   After wetting

High O.M.   Low O.M.   High O.M.   Low O.M.

FIGURE 2.14 The aggregates of soils high in organic matter are much more stable than are those low in this constituent. The low organic matter soil aggregates fall apart when they are wetted; those high in organic matter maintain their stability.

added to soils and are at a maximum a few weeks or months after this application.

2. The cementing action of the intermediate products of microbial synthesis and decay, such as microbially produced gums and certain polysaccharides. These compounds are sometimes called "pre-humus constituents" and are clearly effective as aggregate stabilizers for at least several months.

3. The cementing action of the more resistant stable humus components aided by similar action of certain inorganic compounds, such as oxides of iron. These materials provide most of the long-term aggregate stability.

It should be emphasized that aggregate stability is not entirely an organic phenomenon. There is continual interaction between organic and inorganic components. Polyvalent inorganic cations that cause flocculation (e.g., Ca^{2+}, Mg^{2+}, Fe^{2+}, and Al^{3+}) are thought also to act as bridges between the organic matter and soil clays, encouraging the development of clay–organic matter complexes. In addition, films of clay called "clay skins" often surround the soil peds and help provide stability. The noted stability of aggregates in red and yellow soils of tropical and semitropical areas is due to the hydrated oxides of iron they contain.

As a general rule, the larger the aggregates present in any particular soil, the lower is their stability. This is why it is difficult to build up soil aggregation beyond a certain size of granule or crumb in cultivated land.

2.11 Structural Management of Soils

Coarse-Textured Soils. Looseness, good aeration and drainage, and easy tillage are characteristics of sandy soils. On the other hand, such soils are commonly too loose and open; they lack the capacity to adsorb and hold sufficient moisture and nutrients. They are likely to be droughty and lacking in fertility. They need granulation. There is only one practical method of improving the structure of such a soil—the addition of organic matter. Organic material will not only act as a binding agent for the particles but will also increase the water-holding capacity. The addition of farm manures and the growth of sod crops are practices usually followed to improve the structural condition of sandy soils.

Fine-Textured Soils. The structural management of a clay soil is not such a simple problem as that of a sandy one. In clays and similar soils of temperate regions the potential plasticity and cohesion are always high because of the presence of a large amount of colloidal clay. If such a soil is tilled when wet, its pore space becomes much reduced and it becomes nearly impervious to air and water. The aggregates are broken down, the individual particles tend to act independently, and the soil is said to be *puddled*. When a soil in this condition dries, it usually becomes hard and dense. The tillage of clay soils must be carefully timed. If these soils are plowed too wet, their structural aggregates are broken down and an unfavorable structure results. On the other hand, if they are plowed too dry, large clods, which are difficult to work into a good seedbed, are turned up. In sandy soils and the hydrous oxide clays of the tropics such difficulties usually are at a minimum.

The granulation of fine-textured surface soils should be encouraged by the incorporation of organic matter. In this respect sod crops should be utilized fully and the crop rotation planned to attain their maximum benefits. The data in Table 2.5, from an experiment in Iowa, show the degranulating influence commonly attributed to row crops such as corn. The detrimental effect is less rapid, of course, when the crop is grown in a suitable rotation, and it may be offset by modern conservation tillage practices, which minimize plowing (see Section 2.14). The aggregating tendency of sod, whether it is a meadow mixture or a bluegrass sward, is likewise obvious. The data also suggest the

TABLE 2.5 Water-Stable Aggregation of a Marshall Silt Loam near Clarinda, Iowa, under Different Cropping Systems[a]

Crop	Water-stable aggregates (%)	
	Large (1 mm and above)	Small (less than 1 mm)
Corn continuously	8.8	91.2
Corn in rotation	23.3	76.7
Meadow in rotation	42.2	57.8
Bluegrass continuously	57.0	43.0

[a] From Wilson et al. (1947).

degrees of soil granulation normally expected and disclose the rapidity with which aggregation may decline.

Rice Soils. The detrimental effects of puddling the soil on most upland crops are not evident where wetland rice is grown. In fact the opposite is true—that is, puddling the soil is generally beneficial to the production of rice. In preparation for the planting of rice, the soil is flooded with water, either by irrigation or heavy rains, and is then puddled by intensive tillage, which essentially destroys the structural aggregates. Rice seeds are then sown or seedlings transplanted into the freshly prepared mud. Such soil management helps control weeds and also reduces the rate of water movement down through the soil (Figure 2.15). The second effect is important since it is the common practice to maintain standing water in the rice through much of the growing season. By reducing water percolation, puddled soil markedly decreases the amount of water needed to produce a rice crop.

Unique characteristics of the rice plant account for its positive response to a type of soil management that destroys aggregate stability. Rice survives flooded conditions because oxygen moves downward inside the stem of the plant to supply the roots. This characteristic permits rice to compete well with all but a few aquatic weeds and grasses. The advantage possessed by the rice plant is its response, not to puddling per se, but to a flooded soil

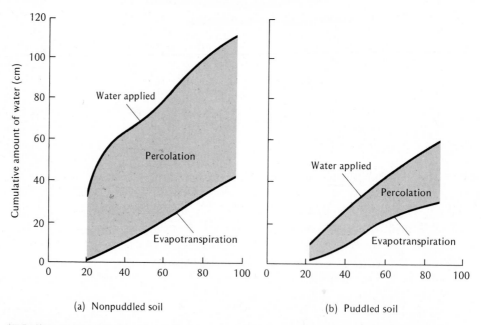

(a) Nonpuddled soil   (b) Puddled soil

FIGURE 2.15 Effects of puddling a rice soil on the loss of water by percolation and evapotranspiration. Note that in this case puddling greatly reduced percolation. [*From De Datta and Kerim (1974).*]

condition. Rice can also be grown successfully on unpuddled but flooded soil. The unpuddled soil maintains at least some structural aggregation. After water is withdrawn and the rice harvested, the soil condition is quite favorable for the growth of a crop such as corn or an edible legume, which may follow rice in a cropping sequence.

The response of rice to soil structural management is important in two ways. It calls attention to a soil management system commonly followed in the tropics and subtropics where the majority of the world's population lives. Also, it illustrates the interaction between plants and soil in determining the type of soil structure required.

2.12   Soil Consistence

Soil *consistence* is a term used to describe the resistance of a soil at various moisture contents to mechanical stresses or manipulations. It is a composite expression of those cohesive and adhesive forces that determine the ease with which a soil can be reshaped or ruptured. It is commonly measured by feeling and manipulating the soil by hand, or by pulling a tillage instrument through it. The consistence of soils is generally described at three soil moisture levels: wet, moist, and dry.

Wet Soils.   For wet soils, consistence is described in terms of *stickiness* and *plasticity*. The degree of stickiness is indicated by *nonsticky, slightly sticky, sticky,* and *very sticky.*

Plasticity is the capacity of soil to be molded—that is, to change shape in response to stress and to keep that shape when the stress is removed. It is evidenced when a thin rod is formed by rolling soil between the thumb and forefinger. Terms used to describe the degree of plasticity are *nonplastic, slightly plastic, plastic,* and *very plastic.*

Moist Soils.   The consistence of moist soils is most important since it best describes the condition of soils when they are tilled in the field. In practice it is a general measure of the resistance of the soil to crushing between the thumb and forefinger. Consistence of a moist soil is described in the following terms, going from the material with least coherence to that which adheres so strongly as to resist crushing between the thumb and forefinger.

1. *Loose:* noncoherent.
2. *Very friable:* coherent but very easily crushed.
3. *Friable:* easily crushed.
4. *Firm:* crushable under moderate pressure.
5. *Very firm:* crushable only under strong pressure.
6. *Extremely firm:* resistant to crushing between thumb and forefinger.

Since the consistence of moist soils is quite dependent on the soil moisture level, the accuracy of field measurement of this soil characteristic is most dependent on the estimate of the soil moisture level. Therefore, coarse sands would be expected to have a loose consistence. Well-granulated loams and silt loams would be very friable, friable, or perhaps firm. Clays, silty clays, and silty clay loams are more likely to be firm or very firm, especially if they are low in organic matter. Such generalizations must be used with caution, however, since soil consistence is influenced by factors such as type of clay and kind and amount of humus present.

Dry Soils. When dry, soils resist crushing or other manipulation. The degree of this resistance is related to the attraction of the particles for each other and is expressed in such terms as rigidity and brittleness. In describing the consistence of dry soils, the following terminology is used, starting with a condition of little interparticle attraction and moving to a state of high cohesive forces.

1. *Loose:* noncoherent.
2. *Soft:* breaks under slight pressure between thumb and forefinger to a powdery mass.
3. *Slightly hard:* breaks under moderate pressure.
4. *Hard:* breaks with difficulty under pressure.
5. *Very hard:* very resistant to pressure; cannot be broken between thumb and forefinger.
6. *Extremely hard:* extreme resistance to pressure; cannot be broken in the hand.

Cementation. Some soil horizons exhibit cementation quite independent of soil moisture level. The cementing agents are compounds such as oxides of iron and aluminum, calcium carbonate, and silica. The consistence of the horizons is expressed in terms of the degree of cementation.

1. *Weakly cemented;* cemented units can be broken in the hand.
2. *Strongly cemented:* units cannot be broken in the hand but can be broken easily with a hammer.
3. *Indurated:* units breakable only with sharp blows of a hammer.

Field Example. How soil consistence varies with moisture content can be explained by using the above terms to describe the variation of this soil property in the field. Fine-textured clay soon after a rain is quite high in soil moisture and is *sticky* to feel. It obviously is too wet to work and if manipulated will tend to puddle or run together. Even when this soil is allowed to dry somewhat, it may still be *slightly sticky* and will be *plastic.* It can be molded into various forms by applying pressure. Although it can be plowed at this moisture content, the furrow slice thus turned will likely form clods when the soil is allowed to dry. The consistence of these clods would then be considered as *hard* or *very hard*.

At a moisture content slightly below that required for plastic consistence, a soil is in optimum condition for working. If it has a high organic matter content, it probably has the appropriate properties to be termed *friable*. The exact moisture range over which this condition occurs will vary for different soils. In general, this range is much wider for medium-textured soils such as loams and some silt loams than for finer-textured clays.

All soils should not be expected to behave like the clay soil just described. Sandy soils, for example, do not become plastic or sticky when wet, or hard when dry. They have a tendency to stay quite *loose* throughout their normal field moisture range. Loams and silt loams will be intermediate in their behavior between the clays and sands.

Consistence has importance for the practical utilization of soils. The terms used to describe this soil property are relevant to soil tillage, compaction by farm machinery, and so on. These subjects are covered in the next section.

2.13 Tilth and Tillage

Although frequent mention has been made of plowing and cultivation in relation to soil structure, attention must be given to seedbed preparation and to the maintenance of stable soil structure throughout the crop-growing season. A convenient term, tilth, will greatly facilitate such a discussion.

Tilth Defined. Simply defined, *tilth* refers to "the physical condition of the soil in relation to plant growth" and hence must include all soil physical conditions that influence crop development.

Tilth depends not only on granule formation and stability but also on such factors as moisture content, degree of aeration, rate of water infiltration, drainage, and capillary-water capacity. As might be expected, tilth often changes rapidly and markedly. For instance, the working properties of fine-textured soils may be altered abruptly by a slight change in moisture and a concomitant change in consistence.

Tillage and Crop Production. *Tillage* involves the manipulation of the soil mechanically with the objective of promoting good tilth and in turn high crop production. For centuries farmers have tilled the soil for three primary reasons: (1) to control weeds, (2) to present a suitable seedbed for crop plants, and (3) to incorporate organic residues into the soil.

Since the Middle Ages the moldboard plow has been the *primary tillage* implement most used in the Western world. Its purpose is to lift, twist, and turn the soil while incorporating crop residues and animal wastes into the plow layer. In more recent times, it has been supplemented by the disk plow, which is used to cut up the residues and partially incorporate them into the soil. In *conventional* practice such primary tillage has been followed by a number of *secondary* tillage operations, such as harrowing, to kill weeds and to break up clods and thereby prepare a suitable seedbed.

After the crop is planted, the soil may receive further secondary tillage to control weeds and to break up crusting of the immediate soil surface. In modern agriculture all these tillage operations are performed with tractors and other heavy equipment that pass over the land many times before the crop is finally harvested.

Short-Term Versus Long-Term Effects on Soil Tilth. The immediate effects of most tillage operations are beneficial. Crop residues are more quickly broken down if they are cut up and incorporated into the soil by tillage implements. Immediately after plowing, the soil is loosened and the total pore space is higher than before plowing. Tillage can provide excellent seedbeds and is a good means of weed control. However, in the short term conventional tillage leaves the soil surface bare and subject to wind and water erosion.

The long-term effects of tillage, especially plowing, are generally undesirable. Rapid breakdown of organic residues can hasten the reduction of soil organic matter content and, in turn, of aggregate stability (see Figure 2.11). Tractors and other heavy implements compact the soil not only through their own weight but by the development of a dense zone (plow pan) immediately below the plowed layer (Figure 2.16).

FIGURE 2.16 Root distribution of a cotton plant. On the right, interrow tractor traffic and plowing have caused a plowpan that restricts root growth. Roots are more prolific on the left where there had been no recent tractor traffic and the plowpan had been broken up by a subsoiling implement. [*Courtesy USDA National Tillage Machinery Laboratory.*]

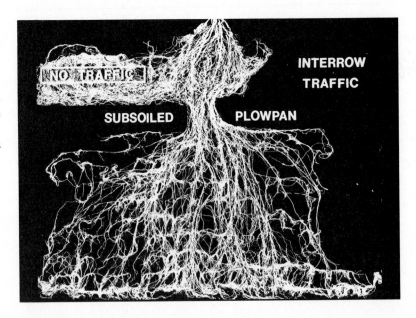

2.14 Conservation Tillage[2]

The weaknesses of conventional tillage practices were emphasized in 1943 by E. H. Faulkner (1943) in his famous book *Plowman's Folly*. However, the minimum tillage practices he advocated were not widely adopted because of the accompanying weed problems. When in the late 1950s and early 1960s effective herbicides became widely available, most weeds were controllable without tillage. This permitted the adoption over wide areas of tillage systems that minimized or even eliminated the use of the plow. These new tillage practices are collectively known as *conservation tillage*. Some involve plowing but little if any secondary tillage, the crop being planted directly on the plowed

FIGURE 2.17 An illustration of one conservation tillage system. Wheat is being harvested and soybeans are planted with no intervening tillage operation. This no-tillage system permits double cropping, saves fuel costs and time, and helps conserve the soil. [*Courtesy Allis-Chalmers Corporation.*]

[2] See Unger and McCalla (1980) for an excellent review of this subject. Also see Phillips and Young (1973).

land or in the wheel tracks of tractors pulling the planting equipment over the plowed land. Others involve no plowing. Instead, the total land area is tilled lightly before planting with implements that stir the soil but do not completely invert it (disks, chisels, sweeps). The extreme in minimum tillage involves essentially no primary or secondary tillage (Figure 2.17). The crop is planted on previously untilled soil. Only in the seedbed area does any mechanical stirring or movement take place. A narrow slot, trench, or band is prepared to permit seed placement and cover. This "no-till" system is growing in popularity especially on well-drained, friable soils.

A key element of all conservation tillage practices is the judicious use of herbicides to control weeds. In the spring, a previous year's sod crop or a winter cover crop can be treated before the new crop is planted. This stops the growth of the sod, leaves the soil covered to prevent erosion and to encourage water penetration, and permits the new crop to grow with little competition.

In the United States, conservation tillage is becoming increasingly popular, the area so tilled having reached more than 33 million hectares (82 million acres) by 1982. The annual rate of increase since 1972 has averaged about 1.7 million hectares.

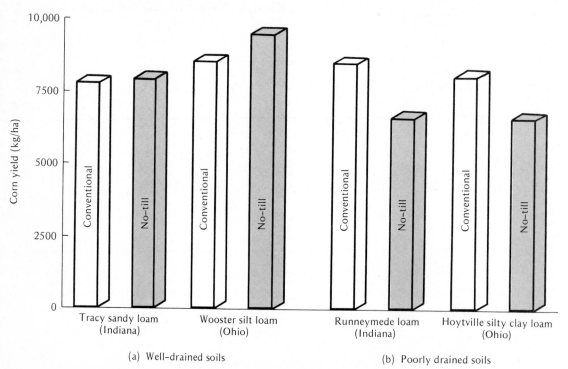

FIGURE 2.18 In well-drained soils no-till systems usually give equal or higher corn yields than conventional tillage does. On poorly drained soils, however, conventional systems—where the land is plowed and then a seedbed prepared—are superior. [*Indiana data from Griffith et al. (1976); Ohio data from Van Doren et al. (1976).*]

Assessment of Conservation Tillage.   In areas where they have been adopted, conservation tillage systems have some advantages over conventional tillage systems. Conservation tillage generally makes possible greater coverage of the soil with crop residues. This assures better water and wind erosion control, increases infiltration of water into the soil, and reduces evaporation from it (see Section 16.12). In addition, the reduction in time required for land preparation and planting permits double cropping in some situations. These matters will all be given more detailed consideration in Chapter 16.

Among the disadvantages of conservation tillage is the cost of the herbicides to keep weeds under control. However, this must be weighed against savings in fuel from the lower energy requirements of the conservation tillage systems. Insect and pathogen damage may be increased by conservation tillage, probably due to the favorable effects of crop residues on these pests.

Crop yields from the two systems are about the same on well-drained soils (Figure 2.18). On poorly drained soils, however, the conventional system, which requires plowing, has given higher yields. This may be due to the fact

FIGURE 2.19   The effect of tillage on the bulk density of a loam soil that had been cropped for 6 years with barley. The no-till plot had higher bulk densities at all depths. Measurements made just before harvest. [*Data selected from Soane and Pidgeon* (1975). *Copyright* © 1975, *The Williams & Wilkins Co., Baltimore, MD; used with permission.*]

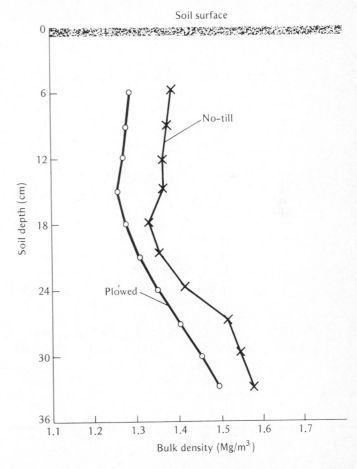

that bulk density is generally higher (Figure 2.19) and pore space is somewhat less in the soils with minimum or no tillage. Also, pore size tends to be smaller. Although reduced porosity of well-drained soils apparently has no adverse effects on crop yields, such is not the case in poorly drained areas.

The use of conservation tillage systems probably will continue to spread, and this trend should have decided benefits for erosion control. It may also have long-term benefits on soil tilth.

REFERENCES

Camp, C. R., and J. F. Lund. 1964. "Effects of soil compaction on cotton roots," *Crops and Soils,* 17:13–14.

Dawud, A. Y., and F. Gray. 1979. "Establishment of the lower boundary of the sola of weakly developed soils that occur in Oklahoma," *Soil Sci. Soc. Amer. J.,* 43:1201–1207.

De Datta, S. K. and A. A. A. Kerim. 1974. "Water and nitrogen economy in rain-fed rice on puddled and nonpuddled soils," *Soil Sci. Soc. Amer. Proc.,* 38:515–518.

Failyer, G. H., et al. 1908. *The Mineral Composition of Soil Particles,* Bull. 54 (Washington, DC: U.S. Department of Agriculture).

Faulkner, E. H. 1943. *Plowman's Folly* (Norman, OK: Univ. of Oklahoma Press).

Fritton, D. D. and G. W. Olson. 1972. "Bulk density of a fragipan soil in natural and disturbed profiles," *Soil Sci. Soc. Amer. Proc.,* 36:686–689.

Griffith, D. R., J. V. Mannering, and C. B. Richey. 1976. "Energy requirements and areas of adaptation for eight tillage-planting systems for corn," *Proc. Conf. on Energy and Agr.,* Center for the Biol. Natural Systems, Washington University, St. Louis, MO.

Joffe, J. S., and R. Kunin. 1942. "Mechanical separates and their fraction in the soil profile: I. Variability in chemical composition and its pedogenic and agropedogenic implications," *Soil Sci. Soc. Amer. Proc.,* 7:187–193.

Larson, W. E., S. C. Gupta, and R. A. Useche. 1980. "Compression of agricultural soils from eight soil orders," *Soil Sci. Soc. Amer. J.,* 44:450–457.

Laws, W. D., and D. D. Evans. 1949. "The effects of long-time cultivation on some physical and chemical properties of two rendzina soils," *Soil Sci. Soc. Amer. Proc.,* 14:15–19.

Lyon, T. L., H. O. Buckman, and N. C. Brady. 1952. *The Nature and Properties of Soils,* 5th ed. (New York: Macmillan), p. 60.

Nelson, L. B., and R. J. Muckenhirn. 1941. "Field percolation rates of four Wisconsin soils having different drainage characteristics," *J. Amer. Soc. Agron.,* 33:1028–36.

Phillips, S. H., and H. M. Young, Jr. 1973. *No Tillage Farming* (Milwaukee, WI: Reiman Associates).

Soane, B. D., and J. C. Pigeon. 1975. "Tillage requirement in relation to soil physical properties," *Soil Sci.,* 119:376–384.

Tiessen, H., J. W. B. Stewart, and J. R. Bettany. 1982. "Cultivation effects on the amounts and concentration of carbon, nitrogen, and phosphorus in grassland soils," *Agron. J.,* 74:831–835.

Unger, P. W., and T. M. McCalla. 1980. "Conservation tillage systems," *Adv. in Agron.,* 33:1–58.

Van Doren, D. M. Jr., G. B. Triplett, and J. E. Henry. 1976. "Influence of long-term tillage crop rotation and soil type combinations on corn yield," *Soil Sci. Soc. Amer. J.,* 40:100–105.

Voorhees, W. B., V. A. Carlson, and C. G. Senst. 1976. "Soybean nodulation as affected by wheel traffic," *Agron. J.,* 68:976–979.

Wilson, H. A., R. Gish, and G. M. Browning. 1947. "Cropping systems and season as factors affecting aggregate stability," *Soil Sci. Soc. Amer. Proc.,* 12:36–43.

Yule, D. F., and J. T. Ritchie. 1980. "Soil shrinkage relationships of Texas Vertisols: I. Small cores," *Soil Sci. Soc. Amer. J.,* 44:1285–1291.

Soil Water: Characteristics and Behavior

We are interested in soil–water relationships for several reasons. First, large quantities of water must be supplied to satisfy the requirements of growing plants as the water is lost by evaporation from leaf surfaces (evapotranspiration). This water must be available when the plants need it, and most of it must come from the soil. Second, water is the solvent that, together with the dissolved nutrients, makes up the soil solution. The significance of this soil component has already been adequately stressed. Third, soil moisture helps control two other important components so essential to normal plant growth— soil air and soil temperature. And last but not least, the control of the disposition of water as it strikes the soil determines to a large extent the incidence of soil erosion—that devastating menace which constantly threatens to impair or even destroy our soils.

3.1 Structure and Related Properties of Water

Water participates directly in dozens of soil and plant reactions and indirectly affects many others. Its ability to do so is determined primarily by its structure. Water is a simple compound, its individual molecules containing one oxygen atom and two much smaller hydrogen atoms. The elements are bonded together covalently, each hydrogen proton sharing its single electron with the oxygen. The resulting molecule is not symmetrical, however. Instead of the atoms being arranged linearly (H—O—H) the hydrogen atoms are attached to the oxygen in sort of a V arrangement at an angle of only 105°. As shown in Figure 3.1, this results in an asymmetric molecule with the shared electrons being closer to the oxygen than to the hydrogen. Consequently, the side on which the hydrogen atoms are located tends to be electropositive and the opposite side electronegative. This accounts for the *polarity* of water and in turn for many reactions so important in soil and plant science.

Polarity.   The property of polarity helps explain how water molecules interact with each other. Each water molecule does not act completely independently but rather is coupled with other neighboring molecules. The hydrogen or positive end of one molecule attracts the oxygen end of another, resulting in a polymer-like grouping. Polarity also accounts for a number of other important properties of water. For example, it explains why water molecules are attracted to electrostatically charged ions. Cations such as Na^+, K^+, and Ca^{2+} become hydrated through their attraction to the oxygen, or negative, end of water molecules. Likewise, negatively charged clay surfaces attract water, this time through the hydrogen, or positive, end of the molecule. Polarity of water mole-

FIGURE 3.1 Two-dimensional representation of a water molecule showing a large oxygen atom and two much smaller hydrogen atoms. The HOH angle of 105° results in an asymmetrical arrangement. One side of the water molecule (with the two hydrogens) is electropositive (P) and the other electronegative (N). This accounts for the polarity of water.

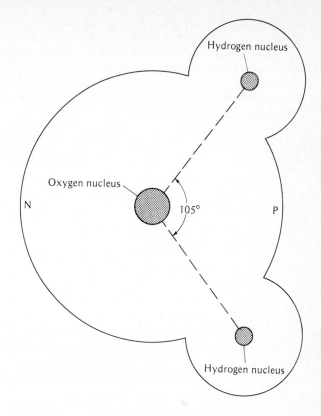

cules also encourages the dissolution of salts in water since the ionic components have greater attraction for water molecules than for each other.

When water molecules become attracted to electrostatically charged ions or clay surfaces, they do so in closely packed clusters. In this state their free energy is lower than in pure water. Thus, when ions or clay particles become hydrated, energy must be released. The released energy is evidenced as *heat of solution* when ions hydrate, or as *heat of wetting* in the case of hydrating clay particles (see Section 2.2).

Hydrogen Bonding. The phenomenon by which hydrogen atoms act as links between water molecules is called *hydrogen bonding*. This is a relatively low energy coupling in which a hydrogen atom is shared between two small electronegative atoms such as O, F, and N. Because of its high electronegativity, an O atom in one water molecule can exert some attraction for the H atom in a neighboring water molecule. This type of bonding accounts for the polymerization and structure of water and for the relatively high boiling point, specific heat, and viscosity of water compared to the same properties of other hydrogen-containing compounds, such as H_2S, which has a similar molecular weight but no hydrogen bonding. It is also responsible for the structural rigidity of

FIGURE 3.2 Comparative forces acting on water molecules at the surface and beneath the surface. Forces acting below the surface are equal in all directions since each water molecule is attracted equally by neighboring water molecules. At the surface, however, the attraction of the air for the water molecules is much less than that of water molecules for each other. Consequently, there is a net downward force on the surface molecules, and the result is something like a compressed film or membrane at the surface. This phenomenon is called *surface tension*.

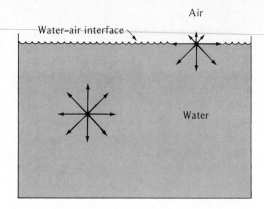

kaolinite crystals and for the structure of some organic compounds, such as proteins.

Cohesion Versus Adhesion.   Hydrogen bonding accounts for two basic forces responsible for water retention and movement in soils. One is the attraction of molecules for each other (cohesion). The other is the attraction of water molecules for solid surfaces (adhesion). By adhesion, solids hold water molecules rigidly at their soil–water surfaces. These molecules in turn hold by cohesion other water molecules further removed from the solid surfaces. Together, these forces make it possible for the soil solids to retain water and control its movement and utilization. They also make possible the property of plasticity possessed by clays (see Section 2.2).

Surface Tension.   One other important property of water that markedly influences its behavior in soils is that of surface tension. This phenomenon is commonly evidenced at liquid–air interfaces and results from the greater attraction of water molecules for each other (cohesion) than for the air above (Figure 3.2). The net effect is an inward force at the surface that causes water to behave as if its surface were covered with a stretched elastic membrane. Because of the relatively high attraction of water molecules for each other, water has a high surface tension compared to that of most other liquids. As we shall see, surface tension is an important property, especially as a factor in the phenomenon of capillarity.

3.2   Capillary Fundamentals and Soil Water

The phenomenon of capillarity is a common one, the classic example being the movement of water up a wick when the lower end is immersed in water. Capillarity is due to two forces: (a) the attractive force of water for the solids on the walls of channels through which it moves (adhesion) and (b) the surface tension of water, which is due largely to the attraction of water molecules for each other (cohesion).

Capillary Mechanism. Capillarity can be demonstrated by placing one end of a fine glass tube in water. The water rises in the tube, and the smaller the tube bore, the higher the water rises (Figure 3.3). The water molecules are attracted to the sides of the tube and start moving up the tube in response to this attraction. The cohesive force between individual water molecules assures that water not directly in contact with the side walls is also pulled up the tube. This continues until the weight of water in the tube counterbalances the cohesive and adhesive forces.

The height of rise in a capillary tube is inversely proportional to the tube diameter and directly proportional to the surface tension, which in turn is determined largely by cohesion between water molecules. The capillary rise can be approximated as

$$h = \frac{2T}{rdg}$$

where h is the height of capillary rise in the tube, T is the surface tension, r is the radius of the tube, d is the density of the liquid, and g is the force of gravity. For water, this equation reduces to the simple expression

$$h = \frac{0.15}{r}$$

FIGURE 3.3 Diagrams illustrating the phenomenon of capillarity. (a) The situation just before lowering a fine glass tube to the water surface. (b) When the tube is inserted in the liquid, water moves up the tube owing to (c) the attractive forces between the water molecules and the wall of the tube (adhesion) and to the mutual attraction of the water molecules for each other (cohesion). Water will move up the tube until the downward pull of gravity equals the attractive forces of cohesion and adhesion.

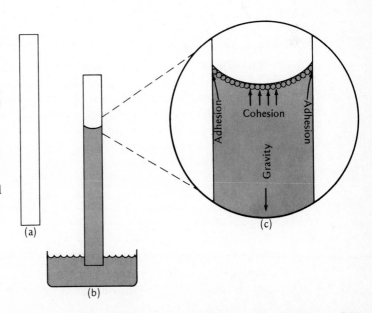

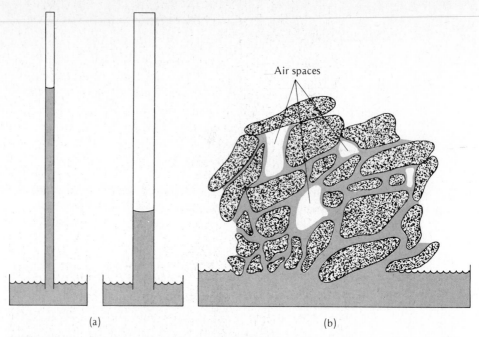

Air spaces

(a)   (b)

FIGURE 3.4 Upward movement by capillarity (a) in glass tubes of different sizes and (b) in soils. While the mechanism is the same in the tubes and in the soil, adjustments are extremely irregular in soil because of the tortuous nature and variability in size of the soil pores and because of entrapped air.

This emphasizes the inverse relation between height of rise and the size of the tube through which the water rises.

Height of Rise in Soils.   Capillary forces are at work in all moist soils. However, the rate of movement and the rise in height are less than one would expect on the basis of soil pore size. One reason is that soil pores are not straight, uniform openings like glass tubes. Furthermore, some soil pores are filled with air, which may be entrapped, slowing down or preventing the movement of water by capillarity (Figure 3.4).

The upward movement due to capillarity in soils is illustrated in Figure 3.5. Usually the height of rise resulting from capillarity is greater with fine-textured soils if sufficient time is allowed and the pores are not too small. This is readily explained on the basis of the capillary size and the continuity of the pores. With sandy soils the adjustment is rapid, but so many of the pores are noncapillary that the height of rise cannot be great.

Although the principle of capillarity is traditionally illustrated as an upward adjustment, movement in any direction takes place in response to capillarity. This is expected since the attractions between soil pores and water are as effective with horizontal pores as with vertical ones. The significance of

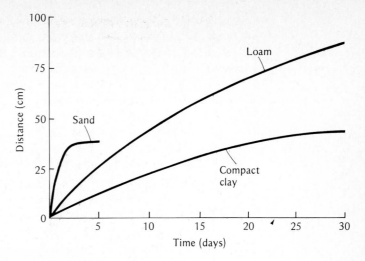

FIGURE 3.5 Upward movement of moisture from a water table through soils of different textures and structures. Note the very rapid rise in the sand but the moderate height attained. Apparently, the pores of the loam are more favorable for movement than those in the compact clay. The rate of movement is thus seen to be of greater significance than the total height.

capillarity in controlling water movement in small pores will become evident as we turn to soil water energy concepts.

3.3 Soil Water Energy Concepts[1]

The retention and movement of water in soils, its uptake and translocation in plants, and its loss to the atmosphere are all energy-related phenomena. Different kinds of energy are involved, including potential, kinetic, and electrical. In the discussions that follow, however, *free energy* is the term we shall use to characterize the energy status of water. This is appropriate since free energy is sort of a summation of all other forms of energy available to do work. Also, its level in a substance is a general measure of the tendency of that substance to change. As we consider energy, we should keep in mind that all substances, including water, have a tendency to move or change from a state of higher to one of lower free energy. Water movement in soils is from a zone where the free energy of the water is high to one where its free energy is low. Water will move readily from a soil saturated with water (high free energy) to a dry soil (low free energy). Therefore, knowledge of the energy levels at various points in a soil makes possible predictions of the direction of water movement and gives some idea of the forces to which the water is subjected. As one might expect, there is great variability in the free energy levels of water in soils. However, the absolute level of free energy of water is not so critical as are *differences* in energy levels from one contiguous site to another.

[1] For a more comprehensive coverage of this subject see Papendick and Campbell (1981) and Hillel (1982).

Forces Affecting Free Energy.   The discussion of properties of water in the previous section suggests three important forces affecting the free energy of soil water. Adhesion, or the attraction of the soil solids (matrix) for water, provides a *matric* force (responsible for capillarity) that markedly reduces the free energy of the adsorbed water molecules and even those held by cohesion. Likewise, the attraction of ions and other solutes for water resulting in *osmotic* forces tends to reduce the free energy of soil solution. Osmotic movement of pure water across a semipermeable membrane into a solution is evidence of the lower free energy state of the solution.

The third major force acting on soil water is *gravity,* which tends to pull the water downward. The free energy of soil water at a given elevation in the profile is thus higher than that of pure water at some lower elevation. Such a difference in free energy level causes water to flow.

Total Soil Water Potential.   While free energy levels are important, the *difference* in free energy from one contiguous site to another is of greater practical significance. This difference, termed the *total soil water potential,* ultimately determines soil water behavior. Technically, the total soil water potential is defined as "the amount of work that must be done per unit quantity of pure water in order to transport reversibly and isothermally an infinitesimal quantity of water from a pool of pure water at a specified elevation at atmospheric pressure to the soil water (at the point under consideration)." While it is impractical to make the measurements specified in this formal definition, the definition stresses that soil water potential is the *difference* between the energy state of soil water and that of pure free water.

Total soil water potential is in effect the sum of the potentials resulting from various forces acting on soil water. Thus, the *gravitational, matric,* and *osmotic* potentials are the differences in free energy resulting from the gravitational, matric, and osmotic forces, respectively. These relationships can be expressed by

$$\psi_t = \psi_g + \psi_m + \psi_o + \ldots$$

where ψ_t is the total soil water potential, ψ_g is the *gravitational potential,* ψ_m is the *matric potential,* and ψ_o is the *osmotic potential.* (Other less significant potentials are indicated by the dots.) The relationship of soil-water potential to free energy is shown in Figure 3.6.

Gravitational Potential.   The force of gravity acts on soil water the same as it does on any other body, the attraction being toward the earth's center. The gravitational potential (ψ_g) of soil water may be expressed mathematically as

$$\psi_g = gh$$

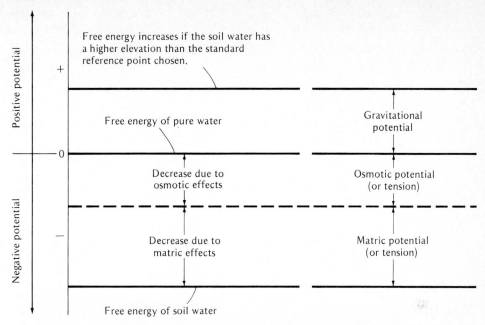

FIGURE 3.6 Relationship between the free energies of pure water and of soil water and the effect of elevation on free energy to illustrate the gravitational potential. Note that osmotic effects and the effects of attraction of the soil solids (matrix) for water both decrease the free energy of soil water. The extent of this decrease represents the osmotic and matric potentials, respectively. The effect of gravity is to increase the free energy if the standard reference point assigned to free water is at a lower elevation than the soil water in the profile. Note that both osmotic and matric potentials are negative, explaining why they are sometimes referred to as *suction* or *tension*. The gravitational potential is generally positive. The behavior of soil water at any one time will be affected by each of these three potentials.

where g is the acceleration of gravity and h is the height of the soil water above a reference elevation. The reference elevation is usually chosen within the soil profile or at its lower edge, thus assuring that the gravitational potential of soil water above the reference point will always be positive. The reference point could be selected at some elevation above the soil water under consideration, in which case the gravitational potential would be negative.

Gravity plays an important role in removing excess water from the upper rooting zones following heavy precipitation or irrigation. It will be given further attention when the movement of soil water is discussed (Section 3.7).

Matric and Osmotic Potentials. The relationship between these two components of total soil water potential is shown in Figure 3.7. Figures 3.6 and 3.7 should be studied carefully to be certain the meaning of these two potentials is clear.

Matric potential is the result of two phenomena, adsorption and capillarity

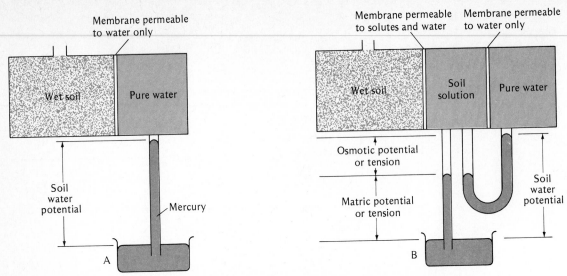

FIGURE 3.7 Relationship among osmotic, matric, and combined soil water potentials. Assume a container of soil separated from pure water by a membrane permeable only to the water (left). Water will move into the soil in response to the attractive forces associated with soil solids (matric) and with solutes (osmotic). At equilibrium, the height of rise in the mercury in the tube above the vessel A is a measure of this combined soil water potential (matric plus osmotic). If a second container were to be placed between the pure water and the soil and if it were separated from the soil by a membrane permeable to *both* water and solutes (right), ions would move from the soil into this container, eventually giving a concentration not too different from the soil solution. The difference between the free energies of the pure water and of the soil solution gives a measure of the osmotic potential. The matric potential is the difference between the combined and osmotic potentials and is measured by the height of rise of the mercury in vessel B. The gravity potential is not shown in this diagram. [*Modified from Richards* (1965).]

(Figure 3.8). The attraction of soil solids and their exchangeable ions for water (adsorption) was emphasized in a previous section, as was the loss of energy (heat of wetting) when the water is adsorbed. This attraction, along with the surface tension of water, also accounts for the capillary force (see Section 3.2). The net effect of these phenomena is to reduce the free energy of soil water as compared to that of unadsorbed pure water. Consequently, matric potentials are always negative.

The matric potential exerts its effect not only on soil moisture retention but on soil water movement as well. The adsorptive and capillary forces tend to resist soil water movement when the soil pores are mostly filled with water. The opposite is true, however, at lower moisture contents where these forces encourage water movement from a moist zone (high free energy) to a dry one (low free energy). Although this latter movement is slow, it is extremely important, especially in supplying water to plant roots.

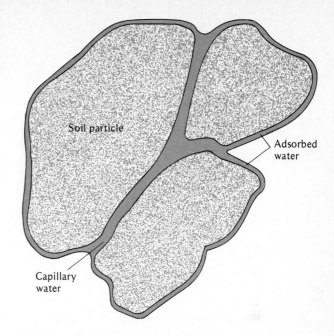

FIGURE 3.8 Two "forms" of water that together give rise to matric potential. The soil solids tightly absorb water, whereas capillary forces are responsible for water's being held in the capillary pores.

Soil particle

Adsorbed water

Capillary water

The osmotic potential is attributable to the presence of solutes in the soil—in other words, to the soil solution. The solutes may be ionic or nonionic, but their net effect is to reduce the free energy of water, primarily because the solute ions or molecules attract the water molecules. The process of osmosis is illustrated in Figure 3.9. This figure should be studied carefully.

Unlike the matric potential, the osmotic potential has little effect on the mass movement of water in soils. Its major effect is on the uptake of water by plant roots. The root membrane, which transmits water more freely than solutes, permits the osmotic effects to be exerted, a matter of considerable importance if the solute content of soils is high. The osmotic potential also affects the movement of water vapor since water vapor pressure is lowered by the presence of solutes.

Suctions and Tensions.   Both the matric and osmotic potentials are negative because the attractive and osmotic forces responsible for these potentials both reduce the free energy level of the soil water (Figure 3.6). Consequently, these negative potentials are sometimes referred to as *tensions* or *suctions*, indicating that they are responsible for the soil's ability to attract and adsorb pure water. The terms "tension" and "suction" have the advantage over "potential" in that they are expressed in positive rather than negative units. Thus, soil solids are responsible for a *negative* potential, which can be expressed as a *positive* tension or suction. For that reason the terms "suction" and "tension" will be used from here on to refer to the matric and osmotic negative potentials.

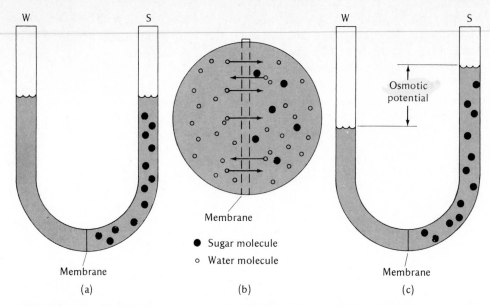

W   S   W   S

Osmotic potential

Membrane

● Sugar molecule
○ Water molecule

Membrane   Membrane

(a)   (b)   (c)

FIGURE 3.9 Illustration of the process of osmosis and of osmotic pressure. (a) A U-tube containing water (W) in the left arm and a solution (S) of sugar in water in the right arm. These are separated by a membrane that is permeable to water molecules but not the dissolved sugar. (b) Enlarged portion of the membrane with H_2O molecules moving freely from the water side to the solution side and vice versa. The sugar molecules, in contrast, are unable to penetrate the membrane. Since the effect of sugar is to decrease the free energy of the water on the solution side, more water passes from left to right than from right to left. (c) At equilibrium sufficient water has passed through the membrane to bring about significant differences in the heights of liquid in the two arms. The difference in the levels in the W and S arms represents the osmotic potential. [*Modified from Keeton (1972).*]

Methods of Expressing Energy Levels. Several units have been used to express differences in energy levels of soil water. A common means of expressing tension (negative potential) is in terms of the height in centimeters of a unit water column whose weight just equals the tension under consideration. The greater the centimeter height, the greater is the suction or tension measured. We may thus express the tenacity with which water is held in soils in centimeters of water or we may convert such readings into other units. For example, the logarithm of the centimeter height is sometimes used.

Another common means of expressing tension is that of the *bar,* which is the pressure exerted by a column of water 1023 cm in height. This approximates the standard atmosphere, which is the average air pressure at sea level, 14.7 lb/in.² or 760 mm Hg. The term *millibar* (mbar) identifies $\frac{1}{1000}$ bar. The suction of water 10 cm high is about $\frac{1}{100}$ bar (10 mbars), that of a column 100 cm high about $\frac{1}{10}$ bar (100 mbars), and that of a 1,000 cm column about 1 bar (1000 mbars). In Table 3.1 the moisture tension in bars is compared with equivalent tension resulting from different heights of a column of water. The bar and millibar units are used in most cases in this text.

TABLE 3.1 Approximate Equivalents of Common Means of Expressing Differences in Energy Levels of Soil Water

Height of unit column of water (cm)	Soil water potential (bars)	Soil water tension (bars)[a]
0	0	0
10.2	−0.01	0.01
102	−0.1	0.1
204	−0.2	0.2
306	−0.3	0.3
1,023	−1.0	1.0
10,230	−10	10
15,345	−15	15
31,713	−31	31
102,300	−100	100

[a] To obtain an expression of soil water tension in the SI unit megapascal (MPa), multiply the number of bars by 0.1.

3.4. Soil Moisture Content Versus Tension

The previous discussions suggest an inverse relation between the water content of soils and the suction or tension with which the water is held. Water is more apt to flow out of a wet soil than from one low in moisture. As we might expect, many factors affect the relationship between soil water tension and moisture content. A few examples will illustrate this point.

Soil Moisture–Energy Curves. The relationship between soil water tension and moisture content of three soils of different texture is shown in Figure 3.10. The absence of sharp breaks in the curves indicates a gradual decrease in tension with increased soil water and vice versa. The clay soil holds much more water at a given tension level than does loam or sand. Likewise, at a given moisture content the water is held much more tenaciously in the clay as compared to the other two soils. As we shall see, much of the water held by clay soils in the field is held so tightly that it cannot be removed by growing plants. In any case, the influence of texture on soil moisture retention is obvious.

The structure of a soil also influences its soil moisture–energy relationships. A well-granulated soil has more total pore space than has a similar soil where the granulation has been destroyed and the soil has become compacted. The reduced pore space may be reflected in a lower water-holding capacity. The compacted soil may also have a higher proportion of small and medium-sized

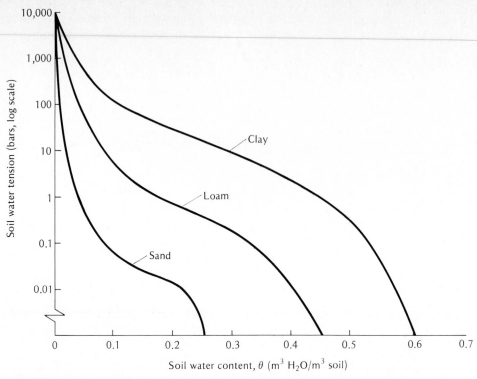

FIGURE 3.10 Soil moisture tension curves for three representative mineral soils. The curves show the relationship obtained by slowly drying completely saturated soils.

pores, which tend to hold the water with a greater suction than do the larger pores.

The soil moisture–suction curves in Figure 3.10 have marked practical significance. They illustrate retention–energy relationships, which influence various field processes, the two most important of which are the movement of water in soils and the uptake and utilization of water by plants. The curves should be referred to frequently as the applied aspects of soil water behavior are considered, in the following sections.

3.5   Measuring Soil Moisture

Two general types of measurements are applied to soil water. The moisture content may be measured, directly or indirectly, or the soil moisture potential (tension or suction) may be determined.

Moisture Content.   A common method of expressing soil moisture percentage is in terms of wet weight percentage or the kilograms of water associated with 1 kg of dry soil. For example, if 1 kg of moist soil (soil and water) loses

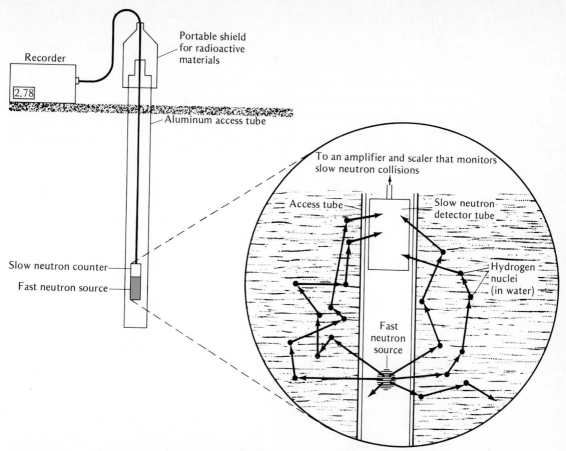

FIGURE 3.11 How a neutron moisture meter operates. The probe, containing a source of fast neutrons and a slow neutron detector, is lowered into the soil through an access tube. Neutrons are emitted by the source (e.g., radium or americium–beryllium) at a very high speed (fast neutrons). When these neutrons collide with a small atom such as hydrogen contained in soil water, their direction of movement is changed and they lose part of their energy. These "slowed" neutrons are measured by a detector tube and a scalar. The reading is related to the soil moisture content.

0.2 kg of water when dried, the 0.8 kg of dry matter is used as the base for the percentage calculation, $0.2/0.8 \times 100 = 25\%$. The weight of the wet soil is an unsatisfactory basis for calculation since it changes with every moisture fluctuation.

Moisture content is also commonly expressed in *volume percentage,* that is, the volume of soil water as a percentage of the volume of the soil sample. Other volume measures, such as cubic meters of H_2O per cubic meter of soil, are employed. These measures have the advantage of giving a better picture of the moisture available to roots in a given volume of soil.

The *gravimetric* method for measuring soil moisture is most commonly used to measure weight percentage. A known weight of a sample of moist soil, usually taken in cores from the field, is dried in an oven at a temperature of 100–110°C and weighed again. The moisture lost by heating represents the soil moisture in the moist sample.

The *resistance* method takes advantage of the fact that the electrical resistance of certain porous materials such as gypsum, nylon, and fiberglass is related to their water content. When blocks of one of these materials, with suitably embedded electrodes, are placed in moist soil, they absorb soil moisture until equilibrium is reached. The electrical resistance in the blocks is determined by their moisture content and in turn by the tension or suction of water in the nearby soil. The relationship between the resistance reading and the soil moisture tension (and the moisture percentage) can be determined by calibration. The blocks are used for measuring the moisture content in selected field locations over a period of time. They give reasonably accurate moisture readings in the range of 1–15 bars tension.

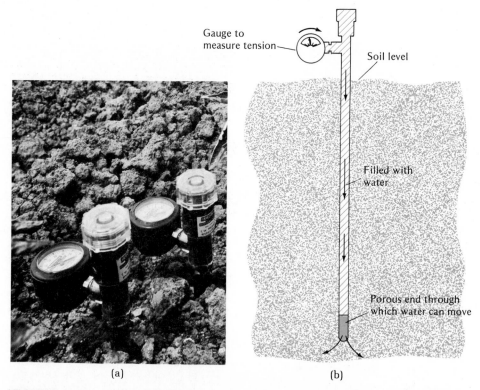

(a)   (b)

FIGURE 3.12 Tensiometer method of determining moisture stress. (a) Tensiometers in place in the field. (b) Cross section showing essential components of a tensiometer. Water will move through the porous end of the instrument in response to the pull of the soil, and the resulting tension is measured by the gauge. [*Photo courtesy T. W. Prosser Co., Arlington, CA.*]

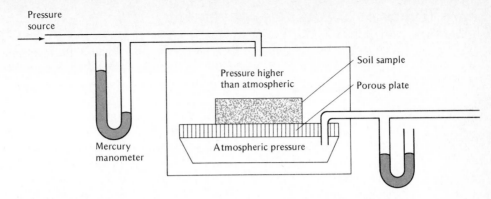

FIGURE 3.13 Pressure membrane apparatus used to determine moisture content–matric tension relations in soils. An outside source of gas creates a pressure inside the cell. Water is forced out of the soil through a porous plate into a cell at atmospheric pressure. This apparatus will measure much higher soil suction values than will tensiometers or tension plates.

Another method of determining soil moisture in the field involves *neutron scattering*. The principle of the neutron moisture meter is based on the ability of hydrogen to reduce drastically the speed of fast-moving neutrons and to scatter them (Figure 3.11). These meters are versatile and give quite accurate results in mineral soils, where water is the primary source of combined hydrogen. In organic soils, however, the method has limited value because much hydrogen in these soils is combined in substances other than water.

Tension Methods. *Field tensiometers,* such as the one shown in Figure 3.12, measure the tension with which water is held in soils. Their effectiveness is based on the principle that water in the tensiometer equilibrates through a porous cup with adjacent soil water and that the suction in the soil is the same as the suction in the potentiometer. They are used successfully in determining the need for irrigation when the moisture is being kept near the field capacity. Their range of usefulness is between 0 and 0.8 bar tension.

A *tension plate* apparatus is a form of tensiometer used under laboratory conditions. A core of soil is placed firmly on a porous plate to which a suction is applied. The soil core eventually reaches equilibrium with the porous plate. The sample is weighed and the relationship between tension and soil moisture content determined. The range of suitability of this apparatus is from 0 to 1 bar only.

A *pressure membrane* apparatus (Figure 3.13) is used to measure matric tension–moisture content relations at tension values as high as 100 bars. This important laboratory tool makes possible the accurate measurement of energy–soil moisture relations of a number of soil samples over a wide energy range in a relatively short time.

3.6　Types of Soil Water Movement

Water is a notably dynamic soil constituent, and three types of movement within the soil are recognized—*unsaturated flow, saturated flow,* and *vapor equilizations.* Both saturated and unsaturated flow involve liquid water in contrast to vapor flow. We shall consider liquid water flow first.

The flow of liquid water is due to a *gradient* in soil water potential from one soil zone to another. The direction of flow is from a zone of higher to one of lower moisture potential. Saturated flow takes place when the soil pores are completely filled (or saturated) with water. Unsaturated flow occurs when the pores in even the wettest soil zones are only partially filled with water. In each case, moisture flow is due to energy–soil water relations. This will be evident as we consider the three types of movement.

3.7　Saturated Flow Through Soils

In most soils, at least some soil pores contain air as well as water; that is, they are unsaturated. Under some conditions, however, at least part of a soil profile may be completely saturated; that is, all pores, large and small, are filled with water. The lower horizons of poorly drained soils are often saturated with water. Even portions of well-drained soils are sometimes saturated. Above stratified layers of clay, for example, the soil pores may all be saturated at times. During and immediately following a heavy rain or irrigation application, pores in the upper soil zones are often filled entirely with water.

The flow of water under saturated conditions is determined by two major factors, the hydraulic force driving the water through the soil and the ease with which the soil pores permit water movement. This can be expressed mathematically as

$$V = Kf$$

where V is the total volume of water moved per unit time, f is the water moving force, and K is the hydraulic conductivity of the soil. It should be noted that the hydraulic conductivity of a saturated soil is essentially constant, being dependent on the size and configuration of the soil pores. This is in contrast to the situation in an unsaturated soil, where hydraulic conductivity decreases with the moisture content.

An illustration of vertical saturated flow is shown in Figure 3.14. The driving force, known as the *hydraulic gradient,* is the difference in height of water above and below the soil column. The volume of water moving down the column will depend upon this force as well as the hydraulic conductivity of the soil.

It should not be inferred from Figure 3.14 that saturated flow occurs only down the profile. The hydraulic force will also cause horizontal and upward

FIGURE 3.14 Saturated (percolation) flow in a column of soil. All soil pores are filled with water. The force drawing the water through the soil is Δh, the difference in the heights of water above and below the soil layer. This same force could be applied horizontally. The water is shown running off into a side container to illustrate that water is actually moving down the profile.

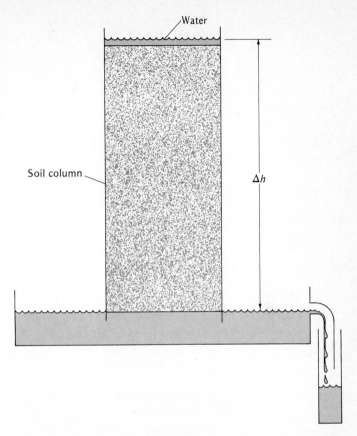

flow. The rate of flow is usually not quite as rapid, however, since the force of gravity does not assist horizontal flow and hinders upward flow. Downward and horizontal flow is illustrated in Figure 3.15, which records the flow of irrigation water into two soils, a sandy loam and a clay loam. Most of the water movement was likely by saturated flow. The water moved down much more rapidly in the sandy loam than in the clay loam. On the other hand, horizontal movement was much more evident in the clay loam.

Factors Influencing the Hydraulic Conductivity of Saturated Soils. Any factor affecting the size and configuration of soil pores will influence hydraulic conductivity. The total flow rate in soil pores is proportional to the fourth power of the radius. Thus, flow through a pore 1 mm in radius is equivalent to that in 10,000 pores with a radius of 0.1 mm even though it takes only 100 pores of radius 0.1 mm to give the same cross-sectional area as a 1 mm pore. Obviously, the macropore spaces will account for most of the saturated water movement.

The texture and structure of soils are the properties to which hydraulic conductivity is most directly related. Sandy soils generally have higher saturated conductivities than finer-textured soils. Likewise, soils with stable granu-

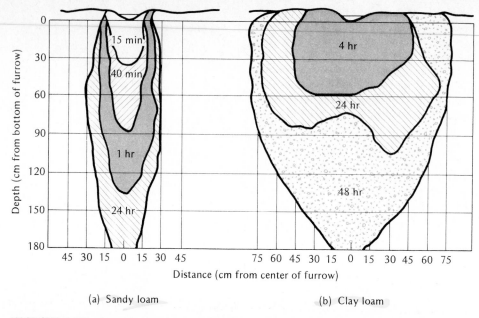

(a) Sandy loam   (b) Clay loam

FIGURE 3.15 Comparative rates of irrigation water movement into a sandy loam and a clay loam. Note the much more rapid rate of movement in the sandy loam, especially in a downward direction. [*Redrawn from Cooney and Peterson (1955).*]

lar structure conduct water much more rapidly than do those with unstable structural units, which break down upon being wetted. Fine clay and silt can clog the small connecting channels of even the larger pores. Fine-textured soils that crack during dry weather at first allow rapid water movement. Later, these cracks swell shut, thereby drastically reducing water movement.

FIGURE 3.16 Relationship between soil water content and hydraulic conductivity for a clay loam soil. Note that the conductivity is about 1000 times greater when the soil is saturated with water ($\theta = 0.52$) than when θ is only 0.3. [*Redrawn from Sisson et al. (1980); used with permission of the Soil Science Society of America.*]

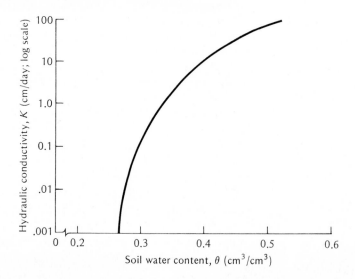

Hydraulic conductivity is influenced markedly by the moisture content of the soil (Figure 3.16). At high moisture levels, especially near saturation, a large proportion of the soil water is in the macropores and saturated flow is relatively rapid. As the moisture content decreases, the soil water is held mostly in the micropores and hydraulic conductivity is greatly reduced.

From a practical point of view saturated flow is very important, especially with poorly drained soils. We shall discuss these aspects in later chapters where percolation and soil drainage are considered.

3.8 Unsaturated Flow in Soils

Under field conditions most soil water movement occurs where the soil pores are not completely saturated with water. The soil macropores are mostly filled with air, and the micropores (capillary pores) with water and some air. Furthermore, the irregularity of soil pores results in isolated pockets of water not in contact with each other. Water movement under these conditions is very slow compared to that occurring when the soil is saturated. This fact is illustrated in Figure 3.17, which shows the generalized relation between matric tension (suction) and conductivity. At or near zero tension, the tension at which saturated flow occurs, the hydraulic conductivity is orders of magnitude greater than at tensions of 0.1 bar and above, which characterize unsaturated flow.

At low tension levels hydraulic conductivity is higher in the sandy soil than in the clay. The opposite is true at higher tension values. This relationship is to be expected since the dominance of large pores in the coarse-textured

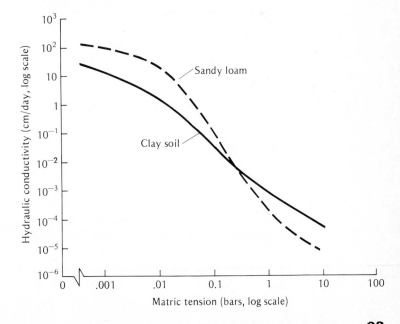

FIGURE 3.17 Generalized relationship between matric tension and hydraulic conductivity for a sandy soil and a clay soil. Saturation flow takes place at or near zero tension, while most of the unsaturated flow occurs at a tension of 0.1 bar or above.

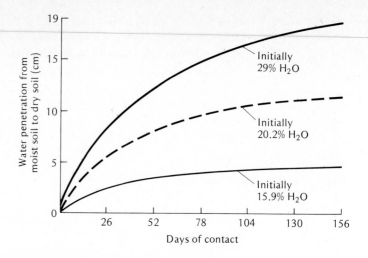

FIGURE 3.18 Rate of water movement from moist soils at three moisture levels to a drier one. The higher the water content of the moist soil, the greater will be the tension gradient and the more rapid will be the delivery. Water adjustment between two slightly moist soils at about the same water content will be exceedingly slow. [*After Gardner and Widtsoe* (*1921*).]

soil encourages saturated flow. Likewise, the prominence of finer (capillary) pores in the clay soil encourages more unsaturated flow than in the sand.

Factors Affecting Unsaturated Flow.   Unsaturated flow is governed by the same general principles affecting saturated flow—that is, its direction and rate are related to the hydraulic conductivity and to a driving force, which in this case is *moisture tension gradient* or *moisture suction gradient*. This gradient is the difference in tension between two adjoining soil zones. Movement will be from a zone of low tension (high matric potential) to one of high tension (low matric potential) or from a zone of thick moisture films to one where the films are thin. The force responsible for this tension is the attraction of soil solids for water.

The influence of tension gradient is illustrated by the moisture curves shown in Figure 3.18, where the rate of water movement from a moist soil into a drier one is shown. The higher the percentage of water in the moist soil, the greater is the tension gradient and the more rapid is the flow. In this case the rate of movement obviously is a function of the tension gradient.

3.9   Water Movement in Stratified Soils

The discussion up to now has dealt almost entirely with soils that are assumed to be quite uniform in texture and structure. In the field, layers differing in physical makeup from the overlying horizons are common. These have a profound influence on water movement and deserve specific attention.

Various kinds of stratification are found in many soils. Impervious silt or clay pans are common, as are sand and gravel lenses or other subsurface layers. In all these cases, the effect on water movement is similar—that is, the downward movement is impeded. The influence of layering is represented

FIGURE 3.19 Downward water movement in soils having a stratified layer of coarse material. (a) Water is applied to the surface of a medium-textured topsoil. Note that after 40 min, downward movement is no greater than movement to the sides, indicating that in this case the gravitational force is insignificant compared to the tension gradient between dry and wet soil. (b) The *downward* movement stops when a coarse-textured layer is encountered. After 110 min no movement into the sandy layer has occurred. (c) After 400 min the moisture content of the overlying layer becomes sufficiently high to give a moisture tension of 0.5 atm or less, and downward movement into the coarse material takes place. Thus, sandy layers, as well as compact silt and clay, influence downward moisture movement in soils. [*Courtesy W. H. Gardner, Washington State University.*]

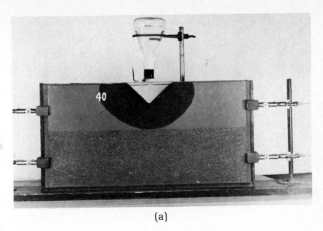

(a)

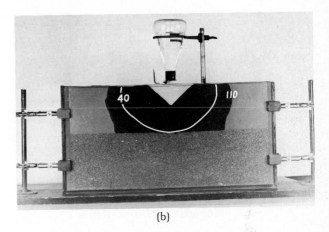

(b)

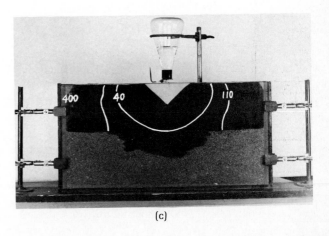

(c)

in Figure 3.19. The change in texture from that of the overlying material results in conductivity differences that prevent rapid downward movement.

The significance of this effect of stratification is obvious. For example, it definitely influences the amount of water the upper part of the soil holds in the field. The layer acts as a moisture barrier until a relatively high moisture level is built up. This gives a much higher field moisture level than that normally encountered in freely drained soils.

3.10 Water Vapor Movement in Soils

Water vaporization as it relates to soils may be distinguished for convenience of discussion as *internal* and *external*. In the one case, the change from the liquid to the vapor state takes place within the soil, that is, in the soil pores. In the second case, the phenomenon occurs at the land surface, and the resulting vapor is lost to the atmosphere by diffusion and convection. The latter, commonly called *surface evaporation,* will be considered later (see Section 15.4). For the present, only vaporization and vapor adjustment tendencies *within* the soil are pertinent.

Relative Humidity of the Soil Air.   The soil air is maintained essentially saturated with water vapor so long as the moisture tension is not above about 31 bars. At this tension and less, water seems to be free enough to maintain the air at nearly 100% relative humidity. But when the moisture is held with a greater tenacity, water vaporizes with great difficulty and its vapor pressure is reduced.

This maintenance of the soil air at or very near a relative humidity of 100% is of tremendous importance, especially in respect to biological activities. Nevertheless, the actual amount of water present in vapor form in a soil at optimum moisture is surprisingly small, being at any one time perhaps not over 11 kg in the upper 15 cm of a hectare of soil (10 lb per acre–furrow slice).

Mechanics of Water Vapor Movement.   The diffusion of water vapor from one area to another in soils occurs in response to differences in free energy. The moving force in this case is the *vapor pressure gradient.* This gradient is simply the difference in vapor pressure of two points a unit distance apart. The greater this difference, the more rapid is the diffusion and the greater is the transfer of vapor water during a unit period. Thus, if a moist soil where the vapor pressure is high is in contact with a drier layer where the vapor pressure is lower, a diffusion of water vapor into the drier area will tend to occur. Likewise, if the temperature of one part of a uniformly moist soil mass is lowered, the vapor pressure will be decreased and water vapor will tend to move toward the cooler part. Heating will have the opposite effect.

The two soil conditions mentioned above—differences in relative humidity

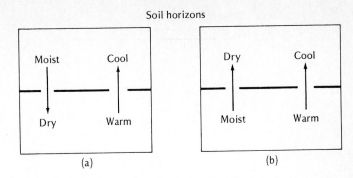

FIGURE 3.20 Vapor movement tendencies that may be expected between soil horizons differing as to temperature and moisture. In (a) the tendencies more or less negate each other, but in (b) they are coordinated and considerable vapor transfer might be possible if the liquid water in the soil capillaries does not interfere.

and in temperature—seem to set the stage for the movement of water vapor under ordinary field conditions. However, they may work at cross purposes and reduce vapor transfer tendencies to a minimum, or they may be so coordinated as to raise them to a maximum. The possible situations are diagrammed in Figure 3.20.

Undoubtedly, some vapor transfer does occur within soils. The extent of the movement by this means, however, probably is not great if the soil water is within the range optimum for higher plants. In dry soils some moisture movement may take place in the vapor form. Such movement may be of some significance in supplying moisture to drought-resistant desert plants, many of which can exist at extremely low soil-moisture levels.

3.11 Retention of Soil Moisture in the Field

With the energy–soil moisture relations covered in previous sections in mind, we now turn to some more practical considerations. We shall start by following the moisture and energy relations of a soil during and following a very heavy rain or the application of irrigation water.

Maximum Retentive Capacity.   During a heavy rain or while being irrigated, a soil may become saturated with water and ready downward drainage will occur. At this point the soil is said to be *saturated* with respect to water (Figure 3.21), and at its *maximum retentive capacity*. The matric tension is essentially zero.

Field Capacity.   Following the rain or irrigation there will be a continued relatively rapid downward movement of some of the water, in response to the hydraulic gradient. After 2 or 3 days this rapid downward movement will become negligible. The soil is then said to be at its *field capacity*. At this time water has moved out of the *macropores*, and its place has been taken by air. The *micropores* or *capillary pores* are still filled with water and will supply the plants with needed moisture. The matric tension will vary slightly

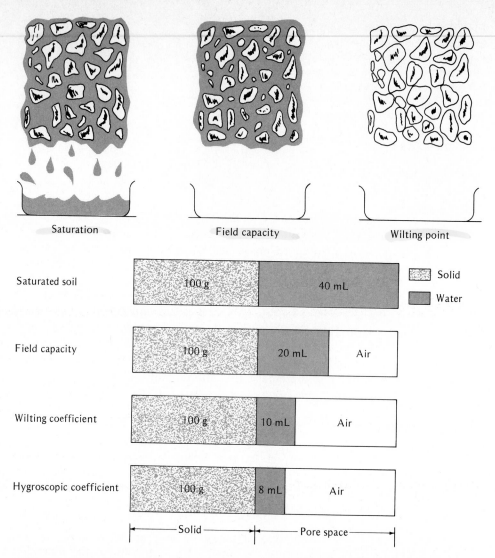

FIGURE 3.21 Volumes of water and air associated with 100 g of a well-granulated silt loam at different moisture levels. The top bar shows the situation when a representative soil is completely saturated with moisture. This situation will usually occur for short periods of time during a rain or when the soil is being irrigated. Water will soon drain out of the larger or *macropores*. The soil is then said to be at the *field capacity*. Plants will remove moisture from the soil quite rapidly until the *wilting coefficient* is approached. Permanent wilting of the plants occurs at this point even though there is still considerable moisture in the soil (wilting coefficient). A further reduction in moisture content to the hygroscopic coefficient is illustrated in the bottom bar. At this point, the water is held very tightly, mostly by the soil colloids. [*Upper drawings modified from Irrigation on Western Farms published by the U.S. Departments of Agriculture and Interior.*]

from soil to soil but is generally in the range of 0.1–0.3 bar, assuming drainage into a less moist zone of similar porosity. Moisture movement will continue to take place, but the rate of movement (unsaturated flow) is quite slow since it is now due primarily to capillary forces, which are effective only in the micropores (Figure 3.20).

Permanent Wilting Percentage. As plants absorb water from a soil, they lose most of it through evapotranspiration at the leaf surfaces. Some water is also lost by evaporation directly from the soil surface. Both of these losses occur simultaneously.

As the soil dries, plants begin to wilt during the daytime, especially if temperatures are high and there is some wind movement. At first the plants will regain their vigor at night. Ultimately, however, the rate of the supply of water to the plants will be so slow that the plant will remain wilted night and day. Although not dead, the plants are now in a permanently wilted condition and will die if water is not added. A measure of soil matric tension would show a value of about 15 bars for most crop plants. Some xerophytes can continue to remove water at this and even higher tensions.

The soil moisture content of the soil at this stage is called the *permanent wilting percentage.* The water remaining in the soil is found in the smallest of the micropores and around individual soil particles (Figure 3.20). Obviously, a considerable amount of the water in soils is not available to higher plants.

Hygroscopic Coefficient. If a soil is kept in an atmosphere that is essentially completely saturated with water vapor (98% relative humidity), it will lose

TABLE 3.2 Hygroscopic and Capillary Water Capacities of Various Soils[a]

Soils	1 Organic matter (%)	2 Hygroscopic coefficient (%)	3 Field capacity (moisture equiv.) (%)	4 Capillary water (col. 3 − 2) (%)	5 Maximum retentive capacity (tension near 0) (%)
Western soils					
Sandy soil (Nebraska)	1.22	3.3	7.9	4.6	34.2
Red loam (New Mexico)	1.07	10.0	19.2	9.2	49.0
Silt loam (Nebraska)	4.93	10.2	27.8	17.8	60.9
Black adobe (Arizona)	2.22	12.9	25.8	12.9	60.3
Iowa soils					
Dickinson fine sand	2.13	3.4	7.6	4.2	44.5
Clarion sandy loam	3.01	6.9	15.5	8.6	58.0
Marshall silt loam	3.58	10.4	24.0	13.6	76.5
Wabash silty clay	5.91	16.1	30.4	14.3	87.0

[a] Data on Western soils from Alway and McDole (1917), on Iowa soils from Russell (1939).

the liquid water held in even the smallest of the micropores. The remaining water will be associated with the surfaces of the soil particles, particularly the colloids, as adsorbed moisture. It is held so tightly that much of it is considered nonliquid and can move only in the vapor phase. The matric tension is about 31 bars. The moisture content of the soil at this point is termed the *hygroscopic coefficient.* As might be expected, soils high in colloidal materials will hold more water under these conditions than will sandy soils and those low in clay and humus (see Figure 3.21 and Table 3.2).

Tension and Moisture Content.   The simple case of water retention described in this section is related directly to the tension–moisture content curves discussed in Section 3.4. This relationship is illustrated in Figure 3.22, which shows the moisture content–matric tension relationship for a loam soil and identifies the ranges in tension for each of the field soil conditions just described. The diagram at the upper part of this figure also suggests physical and biological classification schemes for soil water. However, this diagram is coupled intentionally with the moisture–tension curve to emphasize the fact that there are no clearly identifiable "forms" of water in soil. There is only a gradual change in tension with moisture content. This should be kept in mind in the next section as we discuss some commonly used soil moisture classification schemes.

3.12   Conventional Soil Moisture Classification Schemes

On the basis of observations of the drying of wet soils and of plants growing on these soils, two types of soil water classification schemes have been developed: (a) physical and (b) biological. These schemes are useful in a practical way even though they lack the scientific bases that characterize the preceding moisture–energy discussions.

Physical Classification.   From a physical point of view the terms *gravitational, capillary,* and *hygroscopic* waters are identified in Figure 3.22. Water in excess of the field capacity (0.1–0.3 bar tension) is termed gravitational. Even though energy of retention is low, gravitational water is of little use to plants because it is present in the soil for only a short time and, while in the soil, it occupies the larger pores, thereby reducing soil aeration. Its removal from the soil in drainage is generally a requisite for optimum plant growth.

As the name suggests, capillary water is held in the pores of capillary size and behaves according to laws governing capillarity. Such water includes most of the water taken up by growing plants and exerts tensions between 0.1 and 31 bars.

Hygroscopic water is that bound tightly by the soil solids at tension values greater than 31 bars. It is essentially nonliquid and moves primarily in the vapor form. Higher plants cannot absorb hygroscopic water, but some microbial activity has been observed in soils containing only hygroscopic water.

FIGURE 3.22 Tension–moisture curve of a loam soil as related to different terms used to describe water in soils. The wavy lines in the upper diagram suggest that measurements such as field capacity are not very quantitative. The gradual change in tension with soil moisture change discourages the concepts of different "forms" of water in soil. At the same time, such terms as *gravitational* and *availability* assist in the qualitative description of moisture utilization in soils.

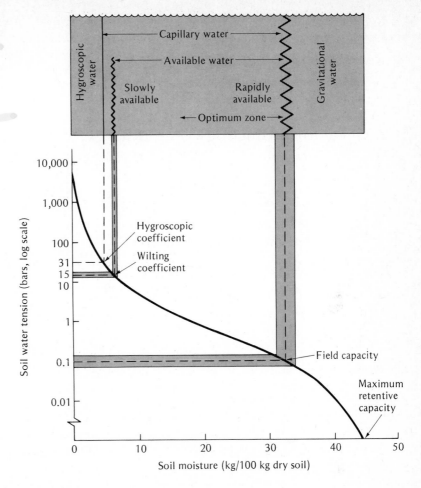

Biological Classification. As one would expect, there is a definite relationship between moisture retention and its utilization by plants. Gravitational water is obviously of little use to plants and may be harmful. In contrast, moisture retained in the soil between the field capacity (0.1–0.3 bar) and the permanent wilting percentage (15 bars) is said to be usable by plants and as such is *available water*. Water held at tensions greater than 15 bars is said to be *unavailable* to most plants (Figure 3.22).

In most soils, optimum growth of plants takes place when the soil moisture content is kept near the field capacity with a moisture tension of 1 bar or less. Thus, the moisture zone for optimum plant growth does not extend over the complete range of moisture availability.

While the various terms employed to describe soil water physically and biologically are useful in a practical way, at best they are only semiquantitative. For example, measurement of field capacity tends to be somewhat arbitrary since the value obtained is affected by such factors as the initial soil moisture in the profile before wetting and the removal of water by plants and surface evaporation during the period of downward flow. Also, the determination as to when the downward movement of water due to gravity has "essentially ceased" is rather arbitrary. These facts stress once again that there is no clear line of demarcation between different "forms" of soil water.

3.13 Factors Affecting Amount of Available Soil Moisture

The amount of water plants absorb from soils is determined by a number of plant, climatic, and soil variables. Rooting habits, basic drought tolerance, and stage and rate of growth are important plant factors. Significant climatic variables include air temperature and humidity and wind velocity and turbulence. These will be considered further in Chapter 15, which is concerned with vapor losses of moisture.

Among the important soil characteristics influencing water uptake by plants are (a) moisture tension relations (matric and osmotic), (b) soil depth, and (c) soil stratification or layering. Each will be discussed briefly.

Matric Tension. Matric tension will influence the amount of available moisture in a soil as it influences the amount of water in a soil at the field capacity and at the permanent wilting percentage. The texture, structure, and organic-matter content all influence these two characteristics and in turn the quality of water a given soil can supply to growing plants. The general influence of texture is shown in Figure 3.23. Note that as fineness of texture increases, there is a general increase in available-moisture storage, although clay soils frequently have a smaller capacity than do well-granulated silt loams. The comparative available water–holding capacities are also shown by this graph.

Ths influence of organic matter deserves special attention. A well-drained mineral soil containing 5% organic matter will probably have a higher available moisture capacity than a comparable soil with 3% organic matter. One might erroneously assume that this favorable effect is all due directly to the moisture-holding capacity of the organic matter. Such is not the case. Rather, most of the benefit of organic matter is attributable to its favorable influence on soil structure and in turn on soil porosity. Although humus does have high moisture content at field capacity, its permanent wilting percentage is proportionately high. Thus, the net direct contribution toward available moisture is less than one would suppose.

Osmotic Tension. The presence of salts in soils, either from applied fertilizers or as naturally occurring compounds, can influence soil water uptake. Osmotic tension effects in the soil solution will tend to reduce the range of available moisture in such soils by increasing the permanent wilting percentage. The

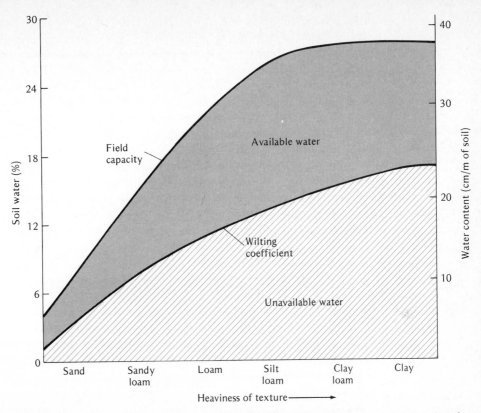

FIGURE 3.23 General relationship between soil moisture characteristics and soil texture. Note that the wilting coefficient increases as the texture becomes heavier. The field capacity increases until we reach the silt loams, then levels off. Remember, these are representative curves. Individual soils would probably have values different from these shown.

total moisture stress in such soils at this point is the matric tension plus the osmotic tension of the soil solution. Although in most humid region soils these osmotic tension effects are insignificant, they become of considerable importance in some saline soils of arid and semiarid regions.

Soil Depth and Layering.   All other factors being equal, deep soils will have greater available moisture–holding capacities than will shallow ones. For deep-rooted plants this is of practical significance, especially in those subhumid and semiarid regions where supplemental irrigation is not possible. Soil moisture measurements to depths as great as 1.5–2 m are sometimes used as bases for predicting wheat yields in the Great Plains area of the United States. Shallow soils are obviously not well suited to these climatic conditions.

Soil stratification or layering will influence markedly the available water and its movement in the soil. Hardpans or impervious layers, for example, slow down drastically the rate of movement of water and also influence unfavorably the penetration of plant roots. They sometimes restrict root growth and effectively reduce the soil depth from which moisture is drawn. Sandy layers also act as barriers to soil-moisture movement from the finer-textured layers above. Movement through a sandy layer is very slow at intermediate and high tensions. The moisture tension in the overlying layers must be less than about 0.5–0.3 bar before movement into the sand will take place. The explanation for this unusual situation was given in Section 3.9 and Figure 3.19.

The capacity of soils to store available moisture determines to a great extent their usefulness in practical agriculture. This capacity is often the buffer between an adverse climate and crop production. It becomes more significant as the utilization of water for all purposes—industrial and domestic as well as agricultural—begins to tax the supply of this all-important natural resource.

3.14   How Plants Are Supplied with Water—Capillarity and Root Extension

At any one time, only a small proportion of the soil water lies in the immediate neighborhood of the absorptive surfaces of plant root systems. Moreover, an immense amount of water (see Section 15.5) is needed to offset transpiration by vigorously growing crops. Two phenomena seem to account for the acquisition of this water: (a) the capillary movement of the soil water to plant roots and (b) the growth of the roots into moist soil.

Rate of Capillary Movement.   When plant rootlets absorb water, they reduce the moisture content of and increase the tension in the soil immediately surrounding them. In response to this higher tension water tends to move toward the plant roots. The rate of movement depends on the magnitude of the tension gradients developed and the conductivity of the soil pores. With some soils, the adjustment may be comparatively rapid and the flow appreciable; in others, especially fine-textured and poorly granulated clays, the movement will be sluggish and only a meager amount of water will be delivered. Thus, a root hair, by absorbing some of the moisture with which it is in contact, automatically creates a tension gradient, and a flow of water is initiated toward its active surface.

How effective this flow may be under field conditions is questionable. Water must be supplied to plant roots at a rapid rate, especially during periods of hot, dry weather when evaporative demand is high, and the rate of capillary movement (capillary flow) is generally too slow to meet this demand. The influence of capillarity is exerted through only a few centimeters so far as the hour-by-hour needs of plants are concerned.

Nevertheless capillary adjustments in the aggregate may be significant.

It is not always necessary for capillary water to move great distances in the soil to be of value to plants. As roots absorb moisture, capillary movement of no more than a few millimeters (if occurring throughout the soil volume) may be of practical importance. Capillary adjustment along with vapor movement is a major factor in supplying water for plants, especially those growing on soils with low moisture contents. Since there is little root extension at moisture tensions approaching the permanent wilting percentage, the water must move to the plants.

Rate of Root Extension. Limitations in rates of capillary movement of water are in part compensated for by rapid rates of root extension, which assure that new root–soil contacts are constantly being established. During favorable growing periods such root elongation may be rapid enough to take care of most of the water needs of a plant growing in a soil at optimum moisture. The mat of roots, rootlets, and root hairs in a meadow is an example of the viable root systems of plants. Table 3.3 provides information on roots of several common crop plants.

TABLE 3.3 Approximate Numbers of Roots and Root Hairs in 1 cm³ of Soil[a]

Plant	Number of roots	Number of root hairs (thousands)	Combined length (m)	Combined surface (cm²)
Oats	7	9	12	6
Rye	9	18	24	12
Kentucky bluegrass	122	61	74	26

[a] Calculated from Ditmer (1938).

The primary limitation of root extension as a means of providing water to plants is the small proportion of the soil with which roots are in contact at any one time. Commonly, root–soil contacts account for less than 1% of the soil surface area. This suggests that much water must move from soil to the root even though the distance of movement may be no more than a few millimeters. It also suggests the complementarity of capillarity and root extension as means of providing soil water for plants.

Root Distribution. The distribution of roots in the soil profile determines to a considerable degree the plant's ability to absorb soil water. Some plants, such as corn and soybeans, have most of their roots in the upper 25–30 cm of the profile (Table 3.4). In contrast, perennial crops such as alfalfa and fruit trees have deep root systems and are able to absorb a high proportion of their moisture from subsoil layers. Even in these cases, however, it is likely that a high proportion of root absorption is in the upper layers of the soil.

TABLE 3.4 Dry Weight of Roots of Soybeans, Corn, and Sorghum in the Surface 0–30 cm of a Silt Loam Soil in Kansas Compared to That in the 30–180 cm Layers (6.7 cm Diameter Cores, Irrigated Plots)[a]

| Crop | Root dry weight (g) | |
	Upper 30 cm	30–180 cm
Soybeans	1.90	0.76
Corn	1.22	0.68
Sorghum	3.60	0.59

[a] From Mayaki et al. (1976).

Root–Soil Contact.   As roots grow into the soil, they move into pores of sufficient size to accommodate them. Contact between the outer cells of the root and the soil is such as to permit ready movement of water from the soil into the plant in response to differences in energy levels (Figure 3.24). When the plant is under moisture stress, however, the roots tend to shrink in size in

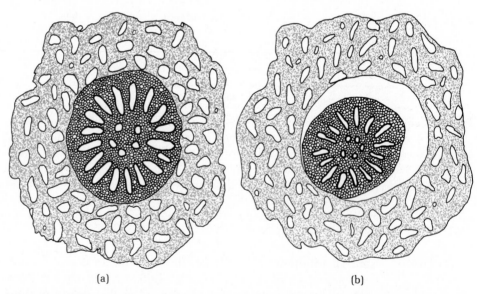

(a)   (b)

FIGURE 3.24   Cross section of a corn root surrounded by soil. (a) During periods of little soil moisture stress on the plant, the root completely fills the soil pore. (b) When the plant is under severe moisture stress such as during a hot dry period, the root shrinks, significantly reducing root–soil contact. Such shrinkage of roots may occur on a hot summer day, even when the soil moisture content is high.

response to this stress. Such conditions exist during a hot, dry spell and are most severe during the daytime when transpiration from plant leaves is at a maximum. The diameter of roots under these conditions may shrink by 30–50%. This reduces considerably the direct root–soil contact as well as the movement of liquid water and nutrients into the plants. Water vapor can still be absorbed by the plant, but its rate of absorption would be much lower than that of liquid water.

3.15   Soil Water and Soil Microorganisms[2]

As is the case for plant roots, soil microbes largely complete their life cycle in the soil and consequently are influenced significantly by soil–water relations. By exerting major control on soil aeration, soil water influences the types of microorganisms present in soils, their numbers and activities. Although many soil microorganisms can survive at moisture levels below the wilting coefficient, their activities are affected even at higher moisture levels. Unfortunately, however, only in recent years has research been focused on the quantitative influences of moisture tension on microbial activity.

The influence of soil moisture tension on selected soil microorganisms is shown in Table 3.5. The movement of bacteria, protozoa, and certain spores

TABLE 3.5   Soil Moisture Tolerance Levels for Different Microbial Activities[a]

Tension tolerance (bars)	Examples of microbial activity in soil
5	Protozoa, zoospores, and bacteria move
15–40	Nitrification, sulfur oxidation
100	Fusarium fungal growth
400	Aspergillus fungal growth

[a] Selected data from several sources compiled by Harris (1981).

(zoospores) ceases at a moisture tension level of about 5 bars. Nitrification and sulfur oxidation may continue until a tension of up to 40 bars is reached, while the growth of some fungi may continue to moisture tensions of up to 400. It is obvious that there are significant differences among organisms in their ability to function at high moisture tension levels. Even so, most soil organisms are able to function at moisture tensions far above those that higher plants can tolerate. This is of considerable significance in the microbial release of nutrients in dry soils.

[2] For review of this subject see SSSA (1981).

3.16 Conclusion

The characteristics and behavior of soil water appear to be complex. As we have gained more knowledge about them, however, it has become apparent that they are governed by relatively simple basic physical principles. Furthermore, researchers are discovering the similarity between these principles and those governing the uptake and utilization of soil moisture by plants. These principles are the subject of Chapter 15.

REFERENCES

Alway, F. J., and G. R. McDole. 1917. "The relation of movement of water in a soil to its hygroscopicity and initial moisture," *J. Agr. Res.,* 10:391–428.

Cooney, J. J., and J. E. Peterson. 1955. *Avocado Irrigation,* Leaflet 50, California Agr. Ext. Serv.

Ditmer, H. J. 1938. "A comparative study of the subterranean members of three field grasses," *Science,* 88:482.

Gardner, W., and J. A. Widtsoe. 1921. "The movement of soil moisture," *Soil Sci.,* 11:230.

Harris, R. F. 1981. "Effect of water potential on microbial growth and activity," *Water Potential Relations in Soil Microbiology,* SSSA Special Publication 9 (Madison, WI: Soil Sci. Soc. of Amer., pp. 23–95.

Hillel, D. 1982. Introduction to Soil Physics. (New York: Academic Press).

Huck, M. G., B. Klepper, and H. M. Taylor. 1970. "Diurnal variations in root diameter," *Plant Physiol.,* 45:529.

Keeton, W. T. 1972. *Biological Science,* 2nd ed. (New York: Norton).

Mayaki, W. C., L. R. Stone, and I. D. Teare. 1976. "Irrigated and nonirrigated soybean, corn and grain sorghum root systems," *Agron. J.,* 68:532–534.

Papendick, R. I., and G. S. Campbell. 1981. "Theory and measurement of water potential," *Water Potential Relations in Soil Microbiology,* SSSA Special Publication 9 (Madison, WI: Soil Sci. Soc. Amer.).

Richards, L. A. 1965. "Physical condition of water in soil," in *Agron. 9: Methods of Soil Analysis,* Part I (Madison, WI: American Society of Agronomy).

Russell, M. B. 1939. "Soil moisture sorption curves for four Iowa soils," *Soil Sci. Soc. Amer. Proc.,* 4:51–54.

Sisson, J. B., H. H. Ferguson, and M. Th. van Genuchten. 1980. "Simple method for predicting drainage from field plots," *Soil Sci. Soc. Amer. J.,* 44:1147–1158.

SSSA. 1981. *Water Potential Relations in Soil Microbiology,* SSSA Special Publication 9 (Madison, WI: Soil Sci. Soc. Amer.).

Soil Air and
Soil Temperature

[*Preceding page*] Differential frost heaving has disrupted the bituminous concrete pavement in this roadway in Pennsylvania. [*Courtesy G. L. Hoffman, Pennsylvania Department of Transportation.*]

As indicated in Chapters 1 and 2, approximately half of the total volume of a representative mineral surface soil is occupied by solid materials. The remaining nonsolid or pore space is occupied by water and gases. Chapter 3 placed major emphasis on the soil moisture phase, but constantly emphasized the interrelationship of soil air and soil water; one cannot be affected without changing the other. We now focus on soil aeration, and on soil temperature, a second physical characteristic significantly affected by soil water.

4.1 Soil Aeration Defined

The two most important biological reactions involving gases that take place in soils are (a) the respiration of higher plants and (b) the decomposition by microorganisms of incorporated organic residues. Although differing in many respects, these processes are similar in that they involve the oxidation of organic compounds. The generalized reaction that illustrates this process may be expressed as follows.

$$\underset{\substack{\text{Organic} \\ \text{compounds}}}{[C]} + O_2 \rightarrow CO_2$$

Thus each of these processes utilizes O_2 and generates CO_2. If there were no mechanism to replenish the O_2 and permit the escape of the CO_2, each process would be curtailed and plant growth would suffer. Soil aeration is the mechanism of gas exchange in soils that prevents oxygen deficiency and CO_2 toxicity. Thus, *a well-aerated soil is one in which gas exchange between the soil air and the atmosphere is sufficiently rapid to prevent a deficiency of oxygen or a toxicity of* CO_2 *and thereby permits normal functioning of plant roots and of aerobic microorganisms.*

A soil in which aeration is considered satisfactory must have at least two characteristics. First, sufficient spaces free of solids and water should be present. Second, there must be ample opportunity for the ready movement of essential gases into and out of these spaces. Water relations largely control the *amount of air space* available, but the problem of adequate *air exchange* is probably a much more complicated feature. The supply of oxygen, a gas constantly being used in biological reactions, must be continually renewed. The activities of the roots of many plants are hampered when the oxygen content of soil air is less than 10%. At the same time, the concentration of CO_2, the major product of these or similar reactions, must not be allowed to build up excessively in the air spaces.

4.2 Soil Aeration Problems in the Field

Under actual field conditions there are generally two situations that may result in poor aeration in soils: (a) when the moisture content is excessively high, leaving little or no room for gases, and (b) when the exchange of gases with the atmosphere is not sufficiently rapid to keep the concentration of soil gases at desirable levels. The latter may occur even when sufficient *total air space* is available.

Excess Moisture. The first case is characterized in the extreme by a water-logged condition in the soil. This may be quite temporary; yet it often seriously affects plant growth. Waterlogging is frequently found in poorly drained, fine-textured soils having a minimum of macropores through which water can move rapidly. It also occurs in soils normally well drained if the rate of water supply to the soil surface is sufficiently rapid or if the soil has been compacted. A low spot in a field, a flat area where water tends to stand for a short while, or an area that has been compacted by heavy machinery or the pressure of plowing (Section 2.10) are good examples of this condition.

Such complete saturation of the soil with water can be disastrous for certain plants in a short time, a matter of a few hours being critical in some cases. Plants previously growing under conditions of good soil aeration are actually more susceptible to damage from flooding than are plants growing on soils where poor aeration has prevailed from the very start.

Prevention of this type of poor aeration requires the rapid removal of excess water either by land drainage or by controlled runoff. Because of the large volume of small capillary pores in some soils, even these precautions leave only a small portion of the soil volume filled with air soon after a rain. This makes the artificial drainage of heavy soils surprisingly effective.

Gaseous Interchange. The seriousness of inadequate *interchanges* of gases between the soil and the free atmosphere above it is dependent primarily on two factors: (a) the rate of biochemical reactions influencing the soil gases and (b) the actual rate at which each gas is moving into or out of the soil. Obviously, the more rapid the usage of O_2 and the corresponding release of CO_2, the greater will be the necessity for the exchange of gases. Factors markedly affecting these biological reactions—temperature, organic residues, etc.—undoubtedly are of considerable importance in determining the air status of any particular soil (see Section 4.4).

The exchange of gases between the soil and the atmosphere above it is facilitated by two mechanisms: (a) *mass flow* and (b) *diffusion*. Mass flow of air is apparently due to pressure differences between the atmosphere and the soil air and is relatively unimportant in determining the total exchange that occurs. However, in the upper few inches of soil diurnal changes in soil temperature may result in mass flow of some significance. The extent of mass flow is determined by such factors as soil and air temperatures, barometric pressure, and wind movements.

Apparently most of the gaseous interchange in soils occurs by diffusion. Through this process each gas tends to move in a direction determined by its own partial pressure (see Figure 4.1). The partial pressure of a gas in a mixture is simply the pressure this gas would exert if it alone were present in the volume occupied by the mixture. Thus, if the pressure of air is 1 atmosphere, the partial pressure of oxygen, which makes up about 21% of the air by volume, is approximately 0.21 bar.

Diffusion allows extensive movement from one area to another even though there is no overall pressure gradient. There is, however, a concentration gradient for individual gases which may be expressed as a partial pressure gradient. Thus, even though the total soil air pressure and that of the atmosphere may be the same, a higher concentration of O_2 in the atmosphere will result in a net movement of this particular gas into the soil. An opposite movement of CO_2 and water vapor is simultaneously taking place since the partial pressures of these two gases are generally higher in the soil air than in the atmosphere. A representation of the principles involved in diffusion is given in Figure 4.1.

In addition to being affected by partial pressure differences, diffusion seems to be directly related to the volume of pore spaces filled with air. On heavy-textured topsoils, especially if the structure is poor, and in compact subsoils, the rate of gaseous movement is seriously slow. Moreover, such soils allow very slow penetration of water into the surface layer. This prevents the rapid escape of CO_2 and the subsequent inward movement of atmospheric O_2 after a heavy rain or irrigation.

FIGURE 4.1 How the process of diffusion takes place. The total gas pressure is the same on both sides of boundary *A–A*. The partial pressure of oxygen is greater, however, in the top portion of the container. Therefore this gas tends to diffuse into the lower portion, where fewer oxygen molecules are found. The carbon dioxide molecules, on the other hand, move in the opposite direction, owing to the higher partial pressure of this gas in the lower half. Eventually equilibrium will be established when the partial pressures of O_2 and CO_2 are the same on both sides of the boundary.

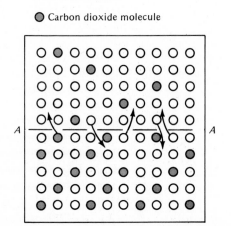

○ Oxygen molecule

● Carbon dioxide molecule

4.3   Means of Characterizing Soil Aeration

The aeration status of a soil can be characterized conveniently in three ways: (a) the content of oxygen and other gases in the soil atmosphere, (b) the oxygen diffusion rate (ODR), and (c) the oxidation–reduction (redox) potential. Each will receive brief attention.

Gaseous Oxygen of the Soil.   The amount of O_2 gas in a soil is determined by the quantity of air-filled pore space and the proportion of that space that is filled with oxygen. These two parameters are usually well correlated since, if the total soil air space is limited, the small amount of O_2 in that space is quickly utilized by plant roots and soil microbes, and CO_2 is released.

The atmosphere above the soil contains nearly 21% O_2, 0.03% CO_2, and nearly 79% N_2. In comparison, soil air may have about the same percent, or somewhat higher, N_2 content, but is consistently lower in O_2 and higher in CO_2. The O_2 content may be only slightly below 20% in the upper layers of a soil with a stable structure and an ample quantity of macropores. It may drop to less than 5% or even to zero in the subsoil of a poorly drained soil with few macropores. Low O_2 contents characterize soils in low-lying areas where water tends to accumulate, and even in well-drained soils after a heavy rain (Figure 4.2). Marked reductions in the O_2 content of soil air may follow

FIGURE 4.2   Oxygen content of soil air before and after heavy rains in a soil on which cotton was being grown. Oxygen was consumed in the respiration of crop roots. The carbon dioxide content probably increased accordingly. [*Redrawn from Patrick* (*1977*); *used with permission of the Soil Science Society of America.*]

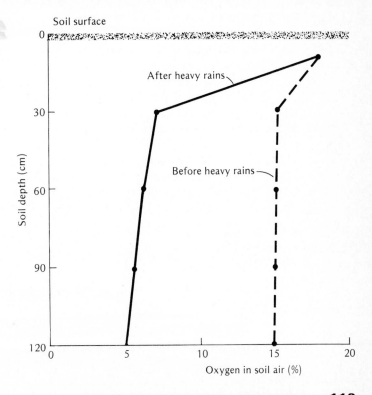

a rain, especially if plants are growing rapidly or if large quantities of manure or other decomposing organic residues are present.

It is well to note that small but significant quantities of O_2 are dissolved in the soil water. Even when all the soil pores are filled with water, soil microorganisms can extract some of the O_2 dissolved in the water for metabolic purposes. This small amount of dissolved O_2 is soon used up, however, and if the excess water is not removed to permit soil air to return to the macropores, even the microbial activities are jeopardized.

There is a general reverse relationship between the O_2 and CO_2 contents of soil air, that of O_2 decreasing as the CO_2 increases. Although the actual differences in CO_2 contents are not impressive, comparatively speaking they are significant. Thus, when the soil air contains only 0.25% CO_2, this gas is more than 8 times as concentrated as it is in the atmosphere. In cases where the CO_2 content becomes as high as 10%, there is more than 300 times as much present as is found in the air above.

Some of the CO_2 gas in soil air dissolves in soil water forming carbonic acid (H_2CO_3). This acid is of universal importance, especially in relation to pH and the solubility of soil minerals.

Oxygen Diffusion Rates.   Perhaps the best measurement of the aeration status of a soil is the *oxygen diffusion rate* (ODR). This determines the rate at which oxygen can be replenished if it is used by respiring plant roots or by microorganisms, or is replaced by water.

Figure 4.3 is a graph showing how ODR decreases with soil depth. Even when atmospheric air with 21% O_2 was used, the ODR rate at 97 cm (38 in.) was less than half that at 11 cm (4.5 in.). When a lower oxygen concentration was used, the ODR decreased even more rapidly with depth. Note that root growth ceased when the ODR dropped to about 20×10^{-8} g/cm² per minute.

Researchers have found that the ODR is of critical importance to growing plants. For example, the growth of roots of most plants cease when the ODR drops to about 20×10^{-8} g/cm² per minute. Top growth is normally satisfactory so long as the ODR remains above 30–40×10^{-8} g/cm² per minute. In Table 4.1 are found some field measurements of ODR along with comments about the condition of the plants. Note the sensitivity of sugar beets to low ODR and the general tendency for difficulty if the ODR gets below the critical level.

Naturally, some plants are affected more by low ODR than are others. Grasses tend to be more tolerant of low diffusion rates than legumes. Sugar beets and alfalfa require higher rates than ladino clover.

Oxidation–Reduction (Redox) Potential (E_h).   One important chemical characteristic of soils that is related to soil aeration is the reduction and oxidation states of the chemical elements in these soils. If a soil is well aerated, oxidized states such as that of ferric iron (Fe^{3+}), manganic manganese (Mn^{4+}), nitrate (NO_3^-), and sulfate (SO_4^{2-}) dominate. In poorly drained and poorly aerated soils the reduced forms of such elements are found, for example, ferrous iron

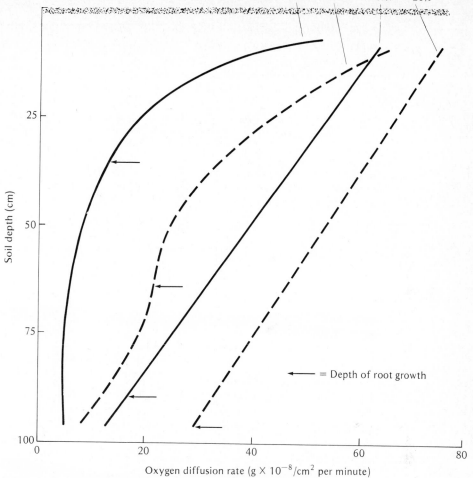

Soil depth (cm)

Oxygen diffusion rate (g × 10^{-8}/cm² per minute)

← = Depth of root growth

FIGURE 4.3 Effect of soil depth and oxygen concentration at the surface on the oxygen diffusion rate (ODR). Arrows indicate snapdragon root penetration depth. Even with a 20% O_2 level at the surface, the diffusion rate at ~ 95 cm is less than half that at the surface. Note that when the ODR drops to ~ 20 × 10^{-8} g/cm² per minute, root growth ceases. [*Redrawn from Stolzy et al. (1961).*]

(Fe^{2+}), manganous manganese (Mn^{2+}), ammonium (NH_4^+), and sulfides (S^{2-}). The presence of these reduced forms is an indication of restricted drainage and poor aeration.

An indication of the oxidation and reduction states of systems (including those in soils) is given by the *oxidation–reduction* or *redox potential* (E_h). It provides a measure of the tendency of a system to reduce or oxidize chemicals and is usually measured in volts or millivolts. If it is positive and high, strong

TABLE 4.1 Relationship Between Oxygen Diffusion Rates (ODR) and the Condition of Different Plants[a]

When the ODR drop below about 40×10^{-8} g/cm² per minute, the plants appear to suffer. Sugar beets require high ODR even at 30 cm depth. ODR values in 10^{-8} g/cm² per minute.

Plant	Soil type	ODR at three soil depths			Remarks
		10 cm	20 cm	30 cm	
Broccoli	Loam	53	31	38	Very good growth
Lettuce	Silt loam	49	26	32	Good growth
Beans	Loam	27	27	25	Chlorotic plants
Sugar beets	Loam	58	60	16	Stunted tap root
Strawberries	Sandy loam	36	32	34	Chlorotic plants
Cotton	Clay loam	7	9	—	Chlorotic plants
Citrus	Sandy loam	64	45	39	Rapid root growth

[a] From Stolzy and Letey (1961).

oxidizing conditions exist. If it is low and even negative, elements are found in reduced forms.

The relationship between O_2 content of soil air in one soil and E_h redox potential is shown in Figure 4.4. In a well-drained soil the E_h is in the 0.4–0.7 volt (V) range. As aeration is reduced, the E_h declines to a level of about 0.3–0.35 V when gaseous oxygen is depleted. At even lower E_h levels, oxygen dissolved in soil water is used by the microbes and finally combined oxygen in nitrates, sulfates, and ferric oxides is utilized for microbial metabolism. Under drastic waterlogged conditions, the E_h may be lowered to an extreme of −0.4 V.

The E_h value at which oxidation–reduction reactions occur varies with the specific chemical to be oxidized or reduced. In Table 4.2 are listed oxidized and reduced forms of several elements of importance in soils along with the approximate redox potentials at which the oxidation–reduction reactions occur. These data emphasize the very close relationship between soil aeration, which is basically a physical process, and the change in forms of chemical elements present in soils, a chemical process. As a consequence, soil aeration has a profound influence on the specific chemical species present in soils and in turn on nutrient availability as well as chemical toxicities in them.

Compared to the atmosphere, soil air usually is much higher in water vapor, being essentially saturated except at or very near the surface of the soil. This fact has already been stressed in connection with the movement of water. Also, under waterlogged conditions, the concentration of gases such as methane and hydrogen sulfide, which are formed by organic matter decomposition, is notably higher in soil air.

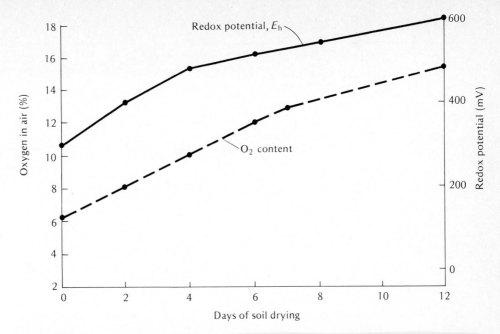

FIGURE 4.4 Relationship between oxygen content of soil air and the redox potential (E_h). Measurements were taken at a 28 cm depth in a soil that had been irrigated continuously for 14 days previous to the drying. Note the general relationship between these two parameters. [*Data selected from Meek and Grass (1975); used with permission of the Soil Science Society of America.*]

TABLE 4.2 The Oxidized and Reduced Forms of Certain Elements in Soils and the Redox Potentials (E_h) at Which Change in Forms Commonly Occurs[a]

Note that gaseous oxygen is depleted at E_h levels of 0.38–0.32 V. At lower E_h levels microorganisms utilize combined oxygen for their metabolism and thereby reduce the elements.

Oxidized forms	Reduced forms	E_h of change of forms (V)
O_2	H_2O	0.38 to 0.32
NO_3^-	N_2	0.28 to 0.22
Mn^{4+}	Mn^{2+}	0.28 to 0.22
Fe^{3+}	Fe^{2+}	0.18 to 0.15
SO_4^{2-}	S^{2-}	−0.12 to −0.18
CO_2	CH_4	−0.2 to −0.28

[a] From Patrick and Reddy (1978).

4.4   Factors Affecting Soil Aeration

The composition of soil air is largely dependent upon the amount of *air space available* together with the rates of *biochemical reactions* and *gaseous interchange.*

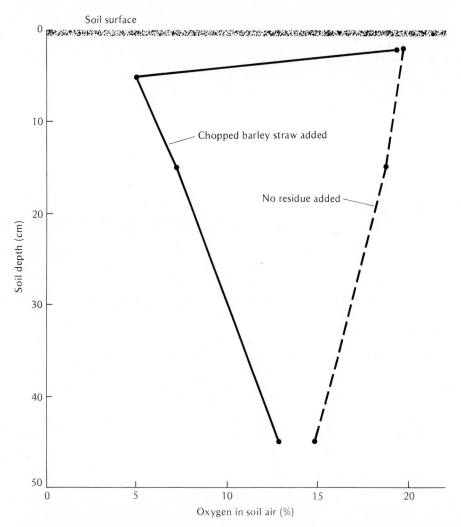

FIGURE 4.5   The effect of soil depth and straw residue application on the oxygen content of a Yolo loam soil. Ten tons of chopped barley straw was mixed in the upper 10 cm of soil 2 months before the measurements were made. Note that the oxygen content at the 5 and 15 cm depths of soil was greatly reduced, the gas having been used in the decomposition of the straw. [*From Rolston et al.* (*1982*); *used with permission of the Soil Science Society of America.*]

FIGURE 4.6 Average oxygen content of two orchard soils during the months of May and June. Normal root growth and functions would occur at much deeper levels on the sandy soil. [*From Boynton* (*1938*).]

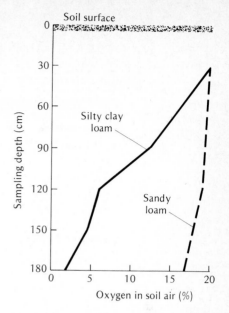

The total porosity of soils is determined chiefly by the bulk density (Section 2.8). This, in turn, is related to factors such as soil texture and structure, and soil organic matter. Not all the pore space in field soils is filled with air, some being occupied by water. In both poorly drained and well-drained soils, a high proportion of the pore space is filled with water immediately after a heavy rain or irrigation. If this situation persists for a period of time, only small amounts of oxygen will be available for plants.

The concentrations of both O_2 and CO_2 in soil air are definitely related to the biological activity in the soil. Microbial composition of organic residues apparently accounts for the major portion of the CO_2 evolved. Incorporation of large quantities of manure, crop residues, or heavy rates of sewage sludge, especially if moisture and temperature are optimum, will alter the soil-air composition appreciably. Respiration by higher plants and the continuous contribution of their roots to the organic mass by sloughage are also significant processes. These influences are well illustrated by data in Figure 4.5. The effects of organic residues are clearly shown.

Subsoil Versus Topsoil. As might be expected, subsoils are usually more deficient in oxygen than are top soils. The total pore space as well as the average size of the pores is generally much less in the deeper horizons. Average oxygen contents at different soil depths at two orchard sites in the months of May and June are shown in Figure 4.6.

In both cases the oxygen percentage in the soil air decreased with depth, the rate of decrease being much more rapid with the heavier soil. Throughout the depth sampled, the lighter-textured sandy loam contained a much higher oxygen percentage. In both cases the decrease in O_2 content could be largely, but not entirely, explained by a corresponding increase in CO_2 content. For example, CO_2 percentages as high as 14.6 were recorded for the lower horizons of the heavier-textured soil.

Soil Heterogeneity.   It should be emphasized that considerable variation exists in the aeration status of different components of a given soil. Thus, poorly aerated zones or pockets may be found in an otherwise well-drained and well-aerated soil. While the poorly aerated component may be a heavy-textured or compacted soil layer, it may also be located inside a soil structural unit where tiny pores may limit ready air exchange. For these reasons oxidation reactions may be taking place within a few centimeters or so from another location where reducing conditions exist. This heterogeneity of soil aeration should be kept in mind.

Seasonal Differences.   As would be expected, there is marked seasonal variation in the composition of soil air. Most of this variation can be accounted

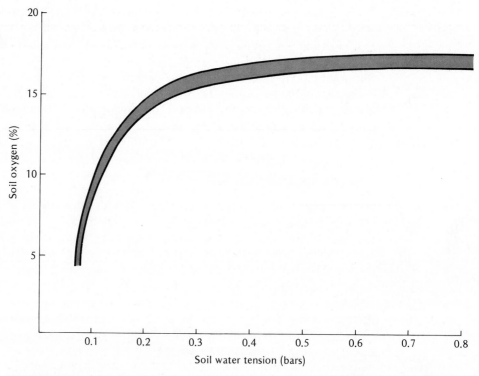

FIGURE 4.7   Relationship between soil matric tension and soil oxygen content in a silty clay loam soil under summer irrigation. Note that at very low soil matric tension levels (high moisture contents) the soil oxygen content is reduced to levels that discourage the growth of a crop such as alfalfa. [*Redrawn from Meek et al.* (*1980*); *used with permission of the Soil Science Society of America.*]

for by soil moisture and soil temperature differences. High soil moisture (low soil tension) tends to favor low oxygen and high CO_2 levels in the soil air. (Figure 4.7). In temperate regions, this situation often prevails in the winter and late spring when the soil moisture is generally highest.

Because soils are normally drier during the summer months, opportunity for gaseous exchange is greatest during this period. This results in relatively high O_2 and low CO_2 levels. Some exceptions to this rule may be found, however. Since high summer temperatures also encourage rapid microbiological release of CO_2, a given soil containing easily decomposable organic matter may have higher CO_2 levels in the summer than in the winter. The dependence of soil air composition on soil moisture and soil temperature is of vital significance.

4.5 Effects of Soil Aeration on Biological Activities

Effects on Activities of Higher Plants. Higher plants are adversely affected in at least four ways by conditions of poor aeration: (a) the growth of the plant, particularly the roots, is curtailed; (b) the absorption of nutrients is decreased, as is (c) the absorption of water; and (d) the formation of certain inorganic compounds toxic to plant growth is favored by poor aeration.

Plant growth. The ability of different plant species to grow in soils with low soil air porosity varies greatly (Table 4.3). Sugar beets, for example, require high air porosities for best growth. In contrast, ladino clover can grow with

TABLE 4.3 Comparative Tolerance by Different Crop Plants of a High Water Table and in Turn of Conditions of Restricted Aeration[a]

Ladino clover grows well with a water table near the soil surface, but wheat and other cereals grow best when the water table is at least 100 cm below the surface.

Plants tolerant to constant table at each depth

15–30 cm	40–60 cm	75–90 cm	100+ cm
Ladino clover	Alfalfa	Corn	Wheat
Orchard grass	Potatoes	Peas	Barley
Fescue	Sorghum	Tomato	Oats
	Mustard	Millet	Peas
		Cabbage	Beans
		Snap beans	Horse beans
			Sugar beets
			Colza

[a] From several sources reviewed by Williamson and Kriz (1970).

very low air porosity, and rice grows normally submerged in water. Furthermore, the tolerance of a given plant to low porosity may be different for seedlings than for rapidly growing plants. The tolerance of red pine to restricted drainage during its early development, and its poor growth or even death on the same site at later stages is a case in point.

In spite of these wide variations in soil air porosity limitation, soil physicists generally consider that if the soil air porosity is reduced below 10–12%, soil oxygen renewal is extremely slow and most plants are likely to suffer.

The abnormal effect of insufficient aeration on root development has been

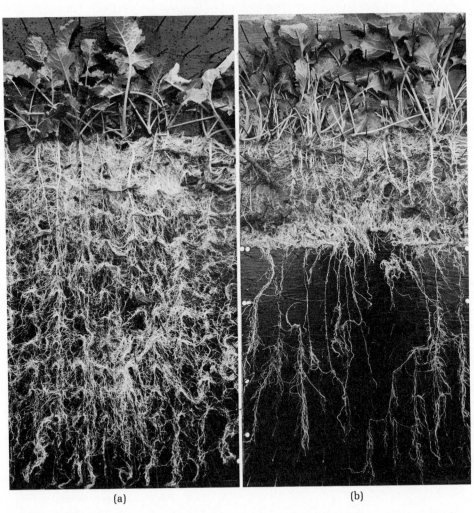

(a)   (b)

FIGURE 4.8 Effect of soil compaction on the development of the root system of the rape plant. (a) Subsoil loosened before planting. (b) Compacted layer at plow depth not loosened. [*From work of H. C. de Roo, Connecticut Agricultural Experiment Station.*]

observed often. With root crops such as carrots and sugar beets the influence is most noticeable. Abnormally shaped roots of these plants are common on compact, poorly aerated soils. Even with sod crops the presence of an impervious soil layer generally results in restricted growth, particularly of the smaller roots. An example of this is found in Figure 4.8b, which shows the root system of rape plants growing on a compacted soil. The lack of extensive root penetration within the impervious layer is noteworthy. Such restricted root systems can hardly be expected to effectively absorb sufficient moisture and nutrients for normal plant growth.

Apparently, different levels of soil oxygen are required for the various functions of roots. Boynton and co-workers (1938) found that apple tree roots required at least 3% oxygen in the soil air to subsist, while 5–10% was sufficient for the growth of existing roots. At least 12% O_2 was required for new root growth. Their results clearly emphasize the complexity of soil aeration problems.

Nutrients and water. A deficiency of oxygen has been found to curtail nutrient and water absorption by plants, two processes that are known to be influenced quite markedly by the rate of root respiration. Apparently, the *energy* of respiration is utilized in bringing about at least part of the nutrient and water absorption. Since a supply of oxygen must be available if roots are to respire normally, a deficiency of this gas results in sluggish nutrient and water uptake. It is surprising as well as ironical that an oversupply of water in the soil tends to reduce the amount of water absorbed by plants.

The effect of aeration on nutrient absorption is of considerable practical significance (Table 4.4). Under poor aeration conditions, for example, plants exhibit nutrient-deficiency symptoms on soils fairly well supplied with available nutrient elements. Also, on certain soils, improper tillage may destroy the granulation, leaving conditions that lead to *inefficient nutrient* utilization. The

TABLE 4.4 The Effect of Water Table Level on the Oxygen Content of Soil Air in a Hoytville Silty Clay, the Yield of Cotton, and the Plant Uptake of N, P, and K[a]

Oxygen content at a depth of 23 cm in this heavy soil was sharply reduced as the water table was raised from 90 to 30 cm depth, and cotton yields and nutrient uptake were decreased accordingly.

Depth of water table (cm)	Oxygen content at 23 cm[b] (%)	Yield of cotton (g)	Nutrient uptake by five plants (mg)		
			N	P	K
30	1.6	57	724	85	1091
60	8.3	108	1414	120	2069
90	13.2	157	2292	156	3174

[a] From Meek et al. (1980).
[b] On June 23, 1976.

desirability of cultivating heavy-textured poorly drained soils when planted to row crops is undoubtedly related to this problem of nutrient uptake.

Effect on Microorganisms. In most soils microbial metabolism accounts for most of the respiration, even when crops are growing vigorously. Since this respiration requires O_2 and releases CO_2, soil microorganisms are significantly affected by soil aeration.

The most apparent evidence of the effect of soil aeration on microbial activity is the slow decay of plant residues in swampy areas. Even in cultivated land, poorly drained soils have high organic matter contents primarily because of the negative influence of poor aeration on plant residue decay.

Soil aeration also affects microorganisms responsible for other critical biochemical reactions. For example, an ample supply of oxygen is required for those bacteria responsible for the oxidation of nitrogen and sulfur compounds to nitrates and sulfates, respectively, which are available to plants.

Soil aeration determines the class of microorganisms that flourishes a soil. Where gaseous O_2 is available, *aerobic* organisms exist and use the O_2 to oxidize organic matter. In the absence of gaseous O_2, alternate microbes (*anaerobic* organisms) take over, utilizing combined oxygen in compounds such as nitrates, sulfates, and iron (ferric) oxides. The reduced forms of some of these elements, such as iron and manganese, may be present in sufficiently high quantities to be toxic to higher plants. While both aerobic and anaerobic organisms play important roles in soils, in general, conditions favoring the aerobic forms likewise favor the growth of most crop plants.

Soil Compaction and Aeration. All negative effects of soil compaction are not due to poor aeration. Soil layers can become so dense as to impede the growth of roots even if an adequate O_2 supply is available. For example, some compacted soil layers adversely affect cotton more by preventing root penetration than by lowering available oxygen content.

4.6 Other Effects of Soil Aeration

As previously indicated, anaerobic decomposition of organic materials is much slower than that occurring when ample gaseous oxygen is available. Moreover, the products of decomposition are entirely different. For example, the complete anaerobic decomposition of sugar occurs as follows.

$$\underset{\text{Sugar}}{C_6H_{12}O_6} \longrightarrow 3\,CO_2 + \underset{\text{Methane}}{3\,CH_4}$$

Less complete decomposition yields other products, such as organic acids, which under extreme conditions may accumulate in toxic quantities. Examples of acids occurring by anaerobic decay are lactic, butyric, and citric acids.

The formation of ethylene (C_2H_4), a hydrocarbon that can adversely affect some plant roots, also occurs in poorly drained soils.

When decomposed anaerobically, organic nitrogen compounds usually yield amines and gaseous nitrogen compounds in addition to ammonia. Of course, these nitrogen compounds remain in reduced forms, not being subject to nitrification under anaerobic conditions. The absence of oxygen thus completely changes the nature of the decay processes and the rates at which they occur.

The chemical forms of certain elements found in soils in the oxidized and reduced states are shown in Table 4.5. In general, the oxidized forms are much more desirable for most of our common crops in humid regions. The oxidized states of the nitrogen and sulfur compounds are readily available to higher plants (Sections 8.7 and 8.8). The desirability of the higher-valent forms of iron and manganese is associated with the solubility of these elements at least in humid region soils. The reduced states, being more soluble than the oxidized forms, are often present in such quantities as to be toxic. This is true especially if the poor aeration occurs in an acid soil. Lack of lime thus intensifies the adverse effects of poor aeration on iron and manganese toxicity.

TABLE 4.5 Oxidized and Reduced Forms of Several Important Elements

Element	Normal form in well-oxidized soils	Reduced form found in waterlogged soils
Carbon	CO_2	CH_4
Nitrogen	NO_3^-	N_2, NH_4^+
Sulfur	SO_4^{2-}	H_2S, S^{2-}
Iron	Fe^{3+} (ferric oxides)	Fe^{2+} (ferrous oxides)
Manganese	Mn^{4+} (manganic oxides)	Mn^{2+} (manganous oxides)

The insolubility of the oxidized forms of iron and manganese under alkaline conditions may result in a deficiency of these elements. Consequently, the interrelation of soil pH and aeration is very important in determining whether an excess or a deficiency of iron and manganese is likely to occur.

In addition to the chemical aspect of these differences, *soil color* is markedly influenced by the oxidation status of iron and manganese. Colors such as red, yellow, and reddish brown are encouraged by well-oxidized conditions. More subdued shades such as grays and blues predominate if insufficient O_2 is present. These facts are utilized in field methods of determining the need for drainage. Imperfectly drained soils usually show a condition wherein alternate streaks of oxidized and reduced materials occur. The *mottled* condition indicates a zone of alternate good and poor aeration, a condition not conducive to proper plant growth.

4.7   Aeration in Relation to Soil and Crop Management

In view of the overwhelming importance of soil oxygen to the growth of most of the common crop plants, the question naturally arises as to the practical means of facilitating its supply. Interestingly enough, almost all methods employed for aeration control are involved directly or indirectly with the management of soil water.

Measures encouraging soil aeration logically fall into two categories: (a) those designed to remove excess soil moisture, and (b) those concerned with the aggregation and cultivation of the soil.

Both surface drainage and underdrainage are essential if an aerobic soil environment is expected. Since the nonsolid spaces in the soil are shared by air and water, the removal of excess quantities of water must take place if sufficient oxygen is to be supplied. The importance of surface runoff and tile drainage will be discussed later (Sections 15.13 and 16.1).

The maintenance of a stable soil structure is an important means of augmenting good aeration. Pores of macrosize, usually greatly encouraged by large stable aggregates, are soon freed of water following a rain, thus allowing gases to move into the soil from the atmosphere. Organic matter maintenance by addition of farm manure and growth of legumes is perhaps the most practical means of encouraging aggregate stability, which in turn encourages good drainage and better aeration.

In poorly drained heavy soils, however, it is often impossible to maintain optimum aeration without resorting to the mechanical cultivation of the soil. Thus, in addition to controlling weeds, cultivation in many cases has a second very important function—that of aiding soil aeration. Yields of row crops, especially those with large tap roots such as sugar beets and rutabagas, are often increased by frequent light cultivations that do not injure the fibrous roots. Part of this increase, undoubtedly, is due to aeration.

Crop–Soil Adaptation.   In addition to the direct methods of controlling soil aeration, one more phase of soil and crop management is of practical importance—that of crop–soil adaptation. The seriousness of oxygen deficiency depends to a large degree on the crop to be grown. Alfalfa, fruit and forest trees, and other deep-rooted plants require deep, well-aerated soils and are quite sensitive to a deficiency of oxygen, even in the lower soil horizons. Shallow-rooted plants such as grasses and alsike and ladino clovers, conversely, do very well on soils which tend to be poorly aerated, especially in the subsoil. The rice plant flourishes even when the soil is submerged in water. These facts are significant in deciding what crops should be grown and how they are to be managed in areas where aeration problems are acute.

The dominating influence of moisture on the aeration of soils is universally apparent. And the control of the moisture means at least a partial control of the aeration. This leads to a second important physical property at times influenced by soil water—that of soil temperature.

4.8 Soil Temperature

The temperature of a soil greatly affects the physical, biological, and chemical processes occurring in that soil. In cold soils chemical and biological rates are slow. Biological decomposition is at a near standstill, thereby limiting the rate at which nutrients such as nitrogen, phosphorus, sulfur, and calcium are

FIGURE 4.9 The influence of soil temperatures on the early growth of corn tops and roots when the air temperature was kept optimum for plant growth. Obviously, corn is quite sensitive to soil temperature differences.

made available. For example, the oxidation of nitrogen to the nitrate form (nitrification) does not begin in the spring until the soil temperature reaches about 5°C (40°F), the most favorable limits being 27–32°C (80–90°F). Also, plant processes such as seed germination and root growth occur only above certain critical soil temperatures. Likewise, absorption and transport of water and nutrient ions by higher plants are adversely affected by low temperature.

Plants vary widely in the soil temperature at which they grow best. There is variability in the optimum temperature for different plant processes. For example, corn germination requires a soil temperature of 7–10°C (45–50°F) and is optimum near 38°C (80–85°F), although this apparently varies under different conditions of air temperature and soil moisture. Potato tubers develop best when the soil temperature is 16–21°C (60–70°F). Oats also grow best at about 21°C, although the roots of this plant apparently do better when the soil temperature is about 15° (Figure 4.9). Optimum vegetative growth of apples and peaches is obtained when the soil temperature is about 18°C. A comparable figure for citrus is 25°C. In cool temperate regions, the yields of some vegetable and small fruit crops are increased by warming the soil (Figure 4.10). The

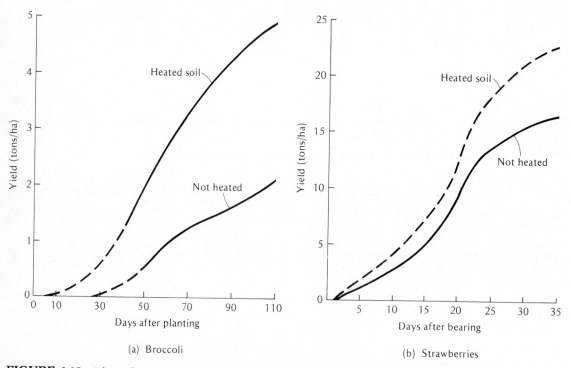

(a) Broccoli (b) Strawberries

FIGURE 4.10 The influence of heating the soil in an experiment in Oregon on the yield of broccoli and strawberries. Heating cables were buried about 92 cm deep. They increased the average temperature of the 0–100 cm layer by about 10°C, although the upper 10 cm of soil was warmed by only about 3°C. [*From Rykbost et al.* (*1975*); *used with permission of the American Society of Agronomy.*]

life cycle of some flowers and ornamentals also is influenced greatly by soil temperature. For example, the tulip bulb requires "chilling" to develop flower buds in early winter although flower development is suppressed until the soil warms up the following spring.

In addition to the direct influence of temperature on plant and animal life the effect of freezing and thawing must be considered. Frost action along with later thawing is responsible for some of the physical weathering that takes place in soils. As ice forms in rock crevices, it forces the rocks apart, causing them to disintegrate. Similarly, alternate freezing and thawing subject the soil aggregates and clumps and rocks to pressures and thus alter the physical setup in the soil. Freezing and thawing of the upper layers of soil can also

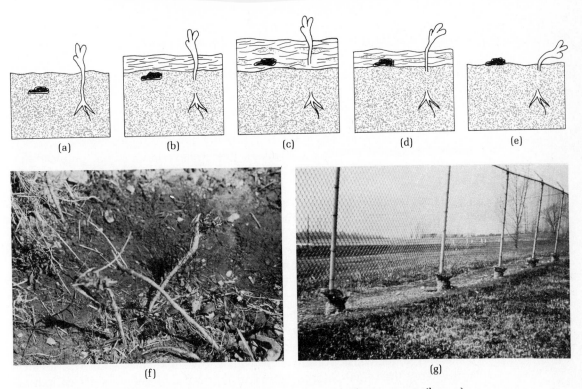

FIGURE 4.11 The diagrams illustrate how frost action in soils can move (heave) plants and stones upward. (a) Position of stone and plant before the soil freezes. (b) Frost action has started. Thin lenses of ice form in the upper few centimeters of soil. Since water expands when it freezes, the lenses of ice expand and move the soil as well as the rock and plant upward, breaking the plant roots. (c) Maximum soil freezing with rock and upper plant parts moved farther up. (d) Partial thawing. (e) Final thawing, which leaves the stone on or near the soil surface and the plant upper roots exposed. (f) Alfalfa plants pushed out of the ground by frost action. (g) Fence posts encased in concrete "jacked" out of the ground by frost action. [Courtesy Bureau of Reclamation, Engineering, and Research Center, U.S. Department of the Interior.]

result in *heaving* of perennial forage crops such as alfalfa (Figure 4.11). This action, which is most severe on bare, imperfectly drained soils, can drastically reduce the stand of alfalfa, some clovers, and trefoil. Changes in soil temperature have the same effect on shallow house foundations and roads with fine material as a base. They show the effects of heaving in the spring.

The temperature of soils in the fields is dependent directly or indirectly upon at least three factors: (a) the net amount of heat the soil absorbs, (b) the heat energy required to bring about a given change in the temperature of a soil, and (c) the energy required for changes such as evaporation, which are constantly occurring at or near the surface of soils. These phases will now be considered.

4.9   Absorption and Loss of Solar Energy

The amount of heat absorbed by soils is determined primarily by the quantity of effective radiation reaching the earth. Only a part of the solar radiation reaches the soil. The remainder, before it reaches the earth, is reflected back into the atmosphere by clouds, is absorbed by atmospheric gases, or is scattered into the atmosphere (Figure 4.12). In relatively cloud-free arid regions as much as 75% of the solar radiation reaches the ground. In contrast, only 35–40% may reach the ground in cloudy humid areas. An average global figure is about 50%.

Some 30–45% of the radiation energy reaching the earth is reflected back to the atmosphere or is lost by thermal radiation. Of that which is not returned to the atmosphere (termed net radiation) about 3% is used to energize photosynthesis and metabolic reactions in plants. Most of the remainder is utilized to evaporate water from soil and plant surfaces (evapotranspiration). Unless the soil is dry and little energy for vaporization is required, only about 5–15% of the net radiation is commonly stored as heat in the soil and plant cover.

Solar radiation in any particular locality depends fundamentally upon climate. But the amount of energy entering the soil is in addition affected by other factors such as (a) the color, (b) the slope, and (c) the vegetative cover of the site under consideration. It is well known that dark soils will absorb more energy than light-colored ones and that red and yellow soils will show a more rapid temperature rise than will those that are white. This does not imply that dark soils are always warmer. Actually the opposite may be true since dark soils usually are high in organic matter and consequently hold large amounts of water, which must also be warmed and evaporated.

Observation has shown that the nearer the angle of incidence of the sun's rays approaches the perpendicular, the greater will be the absorption (Figure 4.13). As an example, a southerly slope of 20°, a level soil, and a northerly slope of 20° receive energy on June 21 at the 42nd parallel north in the proportion 106:100:81.

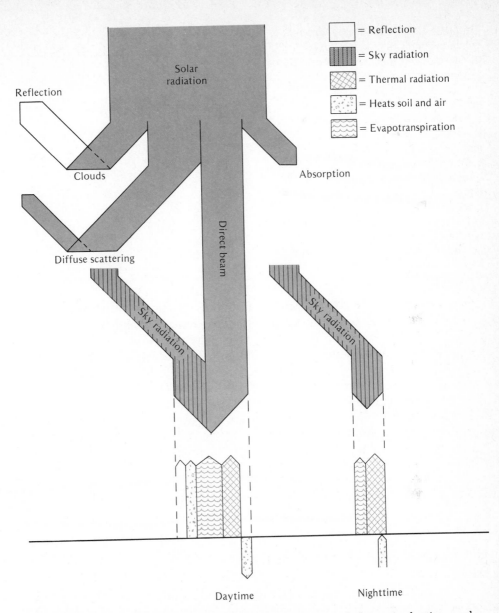

FIGURE 4.12 Schematic representation of the radiation balance in daytime and nighttime in the spring or early summer in a temperate region. About half the solar radiation reaches the earth, either directly or indirectly, from sky radiation. Most radiation that strikes the earth in the daytime is used as energy for evapotranspiration or is radiated back to the atmosphere. Only a small portion, perhaps 10%, is used to heat the soil. At night the soil loses some heat, and some evaporation and thermal radiation take place.

FIGURE 4.13 Effect of the angle at which the sun's rays strike the soil on the area of soil that is warmed. (a) If a given amount of radiation from the sun strikes the soil at right angles, the radiation is concentrated in a relatively small area and the soil warms quite rapidly. (b) If the same amount of radiation strikes the soil at a 45° angle, the area affected is larger by about 40%, the radiation is not so concentrated, and the soil warms up more slowly. This is one of the reasons why north slopes tend to have cooler soils than south slopes. It also accounts for the colder soils in winter as compared to summers.

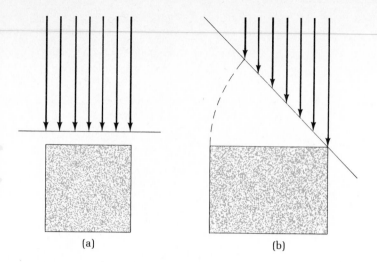

(a) (b)

The temperature of southward slopes varies with the time of year. For instance in the northern hemisphere the southeasterly inclination is generally warmest in the early season, the southerly slope during midseason, and the southwesterly slope in the fall. Southern or southeasterly slopes are often preferred by gardeners. Orchardists and foresters consider exposure an important factor regarding not only the species or variety of tree to be grown but also sunscald and certain plant diseases.

Whether the soil is bare or is covered with vegetation or a mulch is another factor markedly influencing the amount of solar radiation reaching the soil. The effect of a dense forest is universally recognized. Even an ordinary field crop such as bluegrass has a very noticeable influence, especially upon temperature fluctuations. Bare soils warm up more quickly and cool off more rapidly than those covered with vegetation or with artificial mulches. Frost penetration during the winter is considerably greater in bare noninsulated land.

4.10   Specific Heat of Soils

Another major factor affecting the temperature relations of a soil is its specific heat or its thermal capacity compared with that of water. Specific heat may be expressed as a ratio of the quantity of heat required to raise the temperature of a given substance from 15 to 16°C compared to that required for the same temperature rise of an equal weight of water. The importance of this property in soil temperature control undoubtedly is great. The mere absorption of a given amount of heat by a soil does not necessarily assure a rapid rise in temperature. Everything else being equal, a soil with a high specific heat exhibits much less rapid temperature change than does one having a low specific heat. The moisture tends to buffer the soil from too rapid change in temperature.

Under actual field conditions, the soil moisture content determines more than any other factor the energy required to raise the temperature of soils. For instance, the dry weight specific heat of mineral soils, in spite of variations in texture and organic matter, is about 0.20 cal/g or 200 cal/kg. But if the moisture is advanced to 20%, the specific heat of the wet mass becomes 330 cal/kg, while an increase of moisture to 30% raises the wet weight specific heat to 385 cal/kg.[1] Obviously, since moisture is one of the major factors determining the heat capacity of a soil, it has much to do with the rates of warming up and cooling off of soils.

4.11   Heat of Vaporization

Soil moisture is also of major importance in determining the amount of heat used in the process of evaporation of soil water. Vaporization results from an increased activity of the soil water molecules, which is made possible by the absorption of heat from the surrounding environment. This results in a cooling effect, especially at the surface, where most of the evaporation occurs. To evaporate 1 g of water at 20°C (68°F) requires about 540 cal. If evaporation of water from a representative soil lowered the soil moisture content from 25 to 24% (only a 1% decrease), and if all the thermal energy needed to evaporate the water came only from the moist soil, the soil would be cooled by about 12°C. Such a figure is hypothetical because only a part of the heat of vaporization comes from the soil itself. Nevertheless, it indicates the tremendous cooling influence of evaporation.

The low temperature of a wet soil is due partially to evaporation and partially to high specific heat. The temperature of the upper few centimeters of wet soil is commonly 3–6°C (6–12°F) lower than that of a moist or dry soil. This is a significant factor in the spring in a temperate zone when a

---

[1] These figures can be easily verified. For example, consider the above soil at 20% moisture content. Since soil moisture is expressed as kilograms of water per kilogram of soil solids, there is in this case 0.2 kg of H_2O with each kilogram of dry soil. The number of calories required to raise the temperature of 0.2 kg of water by 1°C is

$$0.2 \text{ kg} \times 1000 \text{ cal/kg} = 200 \text{ cal}$$

The corresponding figure for the kilogram of soil solids is

$$1 \text{ kg} \times 200 \text{ cal/kg} = 200 \text{ cal}$$

Thus a total of 400 cal is required to raise the temperature of 1.2 kg of the moist soil by 1°C. Since the *specific heat* is the number of calories required to raise the temperature of 1 kg of wet soil by 1°C, in this case it is

$$400/1.2 = 330 \text{ cal/kg}$$

few degrees will make the difference between the germination or lack of germination of crop seeds. Stand failures are commonly due to wet cold soils.

Soil Color and Temperature.   At this point it is appropriate again to emphasize the interrelationship of soil color and soil temperature. Although dark-colored soils absorb heat readily, such soils, because of their usually high organic matter status, often have high moisture contents. Under such soil conditions, the evaporation and specific heat relationships become especially important. Consequently, a dark-colored soil, particularly if it happens to be somewhat poorly drained, may not warm up as quickly in the spring as will a well-drained light-colored soil nearby.

4.12   Movement of Heat in Soils

As already mentioned (Section 4.9), much of the solar radiation is dissipated into the atmosphere. Some, however, slowly penetrates the profile largely by *conduction*. While this type of movement is influenced by a number of factors, the most important is probably the moisture content of the soil layers. Heat passes from soil to water about 150 times more easily than from soil to air. As the water increases in a soil, the air decreases and the transfer resistance is lowered decidedly. When sufficient water is present to join most of the soil particles, further additions will have little effect on heat conduction. Here again the major role of soil moisture is obvious.

The significance of conduction in respect to field temperatures is not difficult to comprehend. It provides a means of temperature adjustments but, because it is slow, the subsoil changes tend to lag behind those to which the surface layers are subjected. Moreover, the changes occurring are always less in the subsoil. In temperate regions, surface soils in general are expected to be warmer in summer and cooler in winter than the subsoil, especially the lower horizons of the subsoil.

Mention should also be made of the effect of rain or irrigation water on soil temperature. For example, in temperate zones spring rains definitely warm the surface soil as the water moves into it. Conversely, in the summer the rainfall is often cooler than the soil into which it penetrates and thus cools the soil.

4.13   Soil Temperature Data

The temperature of the soil at any time depends on the ratio of the energy absorbed and that being lost. The constant change in this relationship is reflected in the *seasonal, monthly,* and *daily* temperatures. The accompanying data (Figure 4.14) from College Station, Texas, are representative of average seasonal temperatures in relation to soil depth.

FIGURE 4.14 Average monthly soil temperatures for six of the twelve months of the year at different soil depths at College Station, TX (1951–1955). Note the lag in soil temperature change at the lower depths. [*From Fluker (1958)*.]

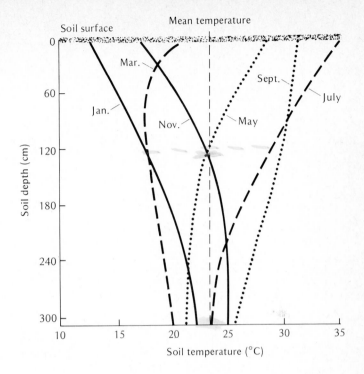

It is apparent from these figures that the seasonal variations of soil temperature are considerable even at the lower depths. The surface layers vary more or less according to air temperature and therefore exhibit a greater fluctuation than the subsoil. On the average, the surface 15-cm layer of soil is warmer than the air at every season of the year, while the subsoil is warmer in autumn and winter but cooler in the spring and summer because of its protected position and the lag in conduction.

Monthly Variations. The lag in temperature change in the subsoil is especially noticeable when the average monthly soil temperatures at Lincoln, Nebraska, are considered. These averages plotted in Figure 4.15 show a greater annual range in temperature for the surface soil than for the air. Note that monthly average soil temperature at a 7.5 cm depth is higher than the air temperature throughout the year. In contrast, the temperature at the 90 cm depth is higher in the winter but lower in the summer than air temperature. Consequently, temperature variation in the subsoil is much less than in the topsoil. This is an advantage not only from the standpoint of agriculture but for other sectors as well. Thus, the deep subsoils are less apt to freeze in the winter months, thereby protecting the roots of perennial crop plants and trees. Likewise, frost action on houses (Figure 4.8) or roads is not as apt to occur if the house or road foundation is anchored in the subsoil.

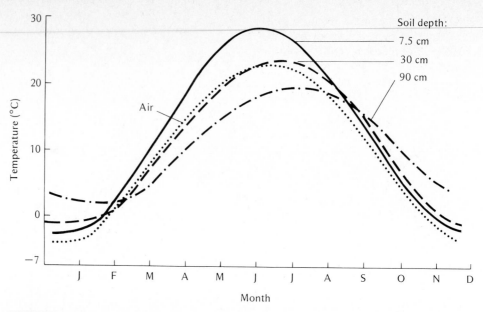

FIGURE 4.15 Average monthly air and soil temperatures at Lincoln, NE (12 years). Note that the 7.5 cm soil layer is consistently warmer than the air above and that the 90 cm soil horizon is cooler in spring and summer, but warmer in the fall and winter, than surface soil.

The temperature lag of subsoils is also of practical importance in temperature regulation in houses that use a so-called heat pump. These pumps utilize water circulating in a subsoil field surrounding a house and take advantage of the fact that subsoils are cooler in summer and warmer in winter to help reduce temperature extremes in the house.

Daily Variations. The daily and hourly temperatures of the atmospheric air and the soil in temperate zones may show considerable agreement or marked divergence according to conditions. Fluctuations are naturally more rapid in the case of air temperatures and usually are greater in temperate regions (Figure 4.16). However, the maximum temperature of a dry surface soil may definitely exceed that of the air, possibly approaching in some cases 50–55°C (122–131°F). But in winter even surface soils do not fall greatly below freezing.

With a clear sky, the air temperature in temperate regions rises from morning to a maximum at about two o'clock. The surface soil, however, does not reach its maximum until later in the afternoon, owing to the usual lag (Figure 4.15). This retardation is greater and the temperature change is less as the depth increases. The lower subsoil shows little daily or weekly fluctuation, the variation, as already emphasized, being a slow monthly or seasonal change.

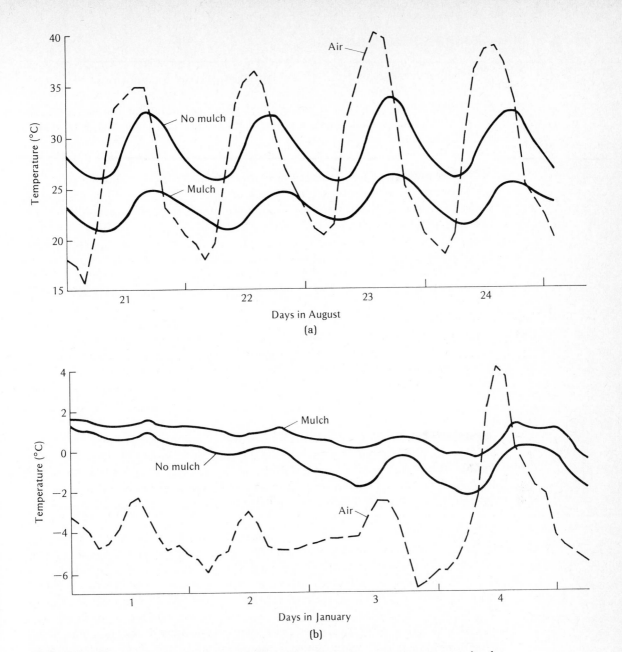

FIGURE 4.16 (a) Influence of straw mulch (8 tons/ha) on air temperature at a depth of 10 cm during a hot spell in August in Bushland, TX. Notice the soil temperatures in the mulched area are consistently lower than when no mulch was applied. (b) During a cold period in January the soil temperature was higher in the mulched than in the no-mulch area. [*Redrawn from Unger (1978); used with permission of the American Society of Agronomy.*]

4.14 Soil Temperature Control

The temperature of field soils is subject to no radical human regulation. However, two kinds of management practice have significant effects on soil temperature: (a) those which keep some type of cover or mulch on the soil and (b) those which reduce excess soil moisture. These effects have meaningful biological implications.

Mulch and Tillage. Soil temperatures are influenced by soil cover and especially by organic residues or other types of mulch placed on the soil surface. Figure 4.16 shows that mulches are definitely soil temperature modifiers. In periods of hot weather they keep the surface soil cooler than where no cover is used. In contrast, during cold spells in the winter, they moderate rapid temperature declines. They tend to buffer extremes in soil temperatures.

Until fairly recently the significant use of mulches in modifying soil temperature extremes was limited mostly to home gardens and flower beds. While these uses are still very important, the effects of mulches have extended to field crop culture in areas where "conservation" tillage practices are followed (Section 2.14). These innovative practices leave much if not all the crop residues at or near the soil surface. The influence of conservation tillage on soil temperature is illustrated by data shown in Figure 4.17. Soil temperatures were consistently lower during July and August where the "no tillage" practice was followed. Crop residues left on the surface undoubtedly accounted for this effect.

Moisture Control. A second means of exercising some control over soil temperature is through management practices that influence soil moisture. The temperature regimes of a well-drained and a poorly drained soil in a humid, temperate region will illustrate this point. The poorly drained soil has a high specific heat. Therefore, large amounts of radiant energy must be absorbed to raise the soil temperature by those few extra degrees so badly needed in early spring. And since the excess water will not percolate through this soil, much of it must be removed by evaporation, another costly process in terms of energy utilization. The low temperatures of poorly drained soils, especially in the spring, are well known. A range from 3 to 6°C lower in the surface layer than in comparable well-drained areas is expected. Drainage is, of course, the only practicable measure.

In addition to the control relations already mentioned, there is the influence of downward-moving water on soil temperature. In the early spring it can have a warming effect, but during the remainder of the year it generally has a cooling effect. Precipitation is usually cooler than the soil in temperate regions in the summer, but even if rainwater should be 5°C warmer than the soil—an improbable assumption—an average rain would raise the temperature of the top 15 cm only slightly, and this would be quickly offset by the cooling effects of surface evaporation.

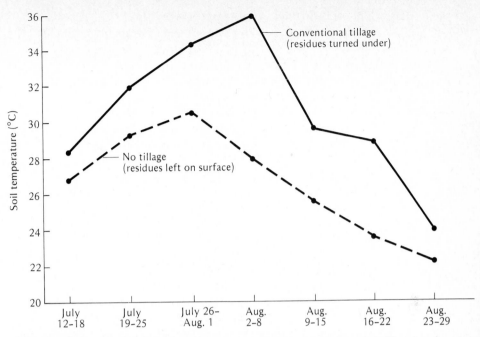

FIGURE 4.17 Effect of plowing under crop residues (conventional tillage) or leaving them on the surface (no tillage) on the temperature of a fine sandy loam in West Virginia. [*Data from Bennett et al.* (*1976*).]

As was the case with soil air, the controlling influence of soil water on soil temperature is apparent everywhere. Whether a question of acquisition of insolation, loss of energy to the atmosphere, or the movement of heat back and forth within the soil, the percentage of water present is always important. Water regulation seems to be the key to what little practical temperature control it is possible to exert on field soils.

REFERENCES

Bennett, O. L., E. L. Mathias, and C. B. Sperow. 1976. "Double cropping for hay and no-tillage corn production as affected by sod species with rates of atrazine and nitrogen," *Agron. J.,* 68:250–254.

Boynton, D. 1941. *Soils in Relation to Fruit Growing in New York. Part XV. Seasonal and Soil Influences on Oxygen and Carbon Dioxide Levels of New York Orchard Soils*, Bull. 763, Cornell Univ. Agr. Exp. Sta.

Boynton, D., et al. 1938. "Are there different critical oxygen concentrations for the different phases of root activity?" *Science*, 88:569–70.

Fluker, B. J. 1958. "Soil temperature," *Soil Sci.,* 86:35–46.

Meek, B. D., and L. B. Grass. 1975. "Redox potential in irrigated desert soils as an indicator of aeration status," *Soil Sci. Soc. Amer. J.,* 39:870–875.

Meek, B. D., E. C. Owen-Bartlett, L. H. Stolzy, and C. K. Labanauskas. 1980. "Cotton yield and nutrient uptake in relation to water table depth," *Soil Sci. Soc. Amer. J.,* 44:301–305.

Meek, B. D., T. J. Donovan, and L. E. Graham. 1980. "Summertime flooding effects on alfalfa mortality, soil oxygen concentrations, and matric potential in a silty clay loam soil," *Soil Sci. Soc. Amer. J.,* 44:433–435.

Patrick, W. H., Jr. 1977. "Oxygen content of soil air by a field method," *Soil Sci. Soc. Amer. J.,* 41:651–652.

Patrick, W. H. Jr., and C. N. Reddy. 1978. "Chemical changes in rice soils," *Soils and Rice* (Los Banos, Laguna, Philippines: The Int. Rice Res. Inst.), pp. 361–379.

Rolston, D. E., A. N. Sharpley, D. W. Toy, and F. E. Broadbent. 1982. "Field measurement of denitrification: III. Rates during irrigation cycles," *Soil Sci. Soc. Amer. J.,* 46:289–296.

Rykbost, K. A., L. Boersma, J. J. Mack, and W. E. Schmisseur. 1975. "Yield response to soil warming: vegetable crops," *Agron. J.,* 67:738–743.

Stolzy, L. H., et al. 1961. "Root growth and diffusion rates as functions of oxygen concentration," *Soil Sci. Soc. Amer. Proc.,* 2:463–67.

Stolzy, L. H., and J. Letey. 1964. "Characterizing soil oxygen conditions with a platinum microelectrode," *Adv. in Agron.,* 16:249–79.

Unger, P. W. 1978. "Straw mulch effects on soil temperatures and sorghum germination and growth," *Agron. J.,* 70:858–864.

Williamson, R. E., and G. J. Kriz. 1970. "Response of agricultural crops to flooding, depth of water table and soil gaseous composition," *Amer. Soc. Agr. Eng. Trans.,* 13:216–220.

Soil Colloids:
Their Nature and
Practical Significance

The colloidal state refers to a two-phase system in which one phase in a very finely divided state is dispersed through a second. In nature, colloids are found as *emulsions* where a liquid is dispersed in a liquid, for example, milk (fat globules in water); as *aerosols* where a solid or liquid is dispersed in a gas, for example, smoke (solid in a gas), and fog (liquid in a gas); or as *gels* where a solid is dispersed in a liquid, for example gelatin and the finer fractions of soil (solid in water). Colloidal particles are generally smaller than 1 micrometer (μm) in diameter. Since the *clay* fraction of soil is 2 μm and smaller, not all clay is strictly colloidal, but even the larger clay particles have colloid-like properties.

The most active portions of the soil are those in the colloidal state, and the two distinct types of colloidal matter, inorganic and organic, exist in intimate intermixture.[1] The inorganic is present almost exclusively as clay minerals of various kinds; the organic is represented by humus. Attention will be focused on both the inorganic and organic fractions.

In a broad way, two groups of clays are recognized: (a) the *silicate clays,* which are so characteristic of temperate regions and (b) the *iron* and *aluminum oxide clays,* which are more prominent in well-weathered soils of the tropics and semitropics. The silicates will be discussed first because they are dominant in the most developed agricultural regions of the world.

5.1 General Constitution of Layer Silicate Clays

Shape.   Early students of colloidal clays visualized the individual particles as more or less spherical and amorphous or noncrystalline internally. It is now definitely established, however, that the particles are laminated, that is, made up of layers of plates or flakes and that they are crystalline, having an ordered internal arrangement (Figure 5.1). Their individual sizes and shapes depend upon their mineralogical organizations and the conditions under which they have developed. Some of these particles are mica-like and definitely hexagonal; some are irregularly plate- or flake-like; some seem to be lath-shaped blades or even rods; and some are spheroidal. The edges of some particles seem to be clean-cut, while the appearance of others is indistinctly frayed or fluffy. In all cases, the horizontal extension of the individual particles greatly exceeds their vertical dimension.

[1] For reviews on soil colloids see ASA/SSSA (1981), Dixon and Weed (1977), and Gieseking (1975).

Surface Area. All clay particles, merely because of their small size, expose a large amount of *external* surface. The external surface area of 1 g of colloidal clay is at least 1000 times that of 1 g of coarse sand. In some clays there are extensive *internal* surfaces as well. This internal interface occurs between the plate-like crystal units that make up each particle (Figure 5.2), and commonly greatly exceeds the external surface area. The surface area of layer silicate clays will range from 10 m^2/g for clays with only external surfaces to more than 800 m^2/g for those with extensive internal surfaces. As a conservative estimate, it is suggested that the active interface due to the clay fraction of a hectare–furrow slice of a slit or clay loam soil probably exceeds the land area of Illinois or Florida at least 100–125 times.

Electronegative Charge and Adsorbed Cations. The minute silicate clay colloid particles, referred to as *micelles* (microcells), ordinarily carry *negative* charges.[2] Consequently, hundred of thousands of positively charged ions or *cations* are attracted to each colloidal crystal (e.g., H^+, Al^{3+}, Ca^{2+}, and Mg^{2+}). This gives rise to what is known as an *ionic double layer* (Figure 5.2). The colloidal particle constitutes the *inner* ionic layer, being essentially a huge *anion,* the external and internal surfaces of which are highly negative in charge. The *outer* ionic layer is made up of a swarm of rather loosely held cations that are attracted to the negatively charged surfaces. Thus, a clay particle is accompanied by an enormous number of cations that are adsorbed[3] or held on the particle surfaces.

Associated with the layer of cations that throng the adsorptive surfaces of clay particles are a large number of water molecules. Some of these water molecules are carried by the adsorbed cations, since most of them are definitely hydrated. In addition, some silicate clays hold numerous water molecules as well as cations packed between the plates (internal surface area) that make up the clay micelle. These various types of water are referred to in toto when the hydration of clays is under consideration.

5.2   Mineral Colloids Other Than Layer Silicates

Oxides and Hydroxides of Iron and Aluminum. Clays comprised of irons and aluminum oxides or hydroxides deserve attention for at least two reasons: (a) they occur in temperature regions intermixed with silicate clays, and (b) they are commonly dominant in the most highly weathered soils of the tropics and semitropics. Properties of the red and yellow soils of these regions are controlled in large degree by these iron and aluminum compounds.

Examples of oxides common in soils are gibbsite ($Al_2O_2 \cdot 3H_2O$) and goethite ($Fe_2O_3 \cdot H_2O$). The formulas may also be written in the hydroxide form, that

---

[2] As we shall see later, certain of these particles also carry positive charges.
[3] For a definition of adsorption see the Glossary.

FIGURE 5.1 Crystals of four silicate clay minerals found in soils. (a) Kaolinite from Illinois magnified at 1600 times (note hexagonal crystal at upper right). (b) Dickite (a kaolinite group mineral) from Kansas magnified about 9000 times. (c) Illite from Wisconsin magnified about 15,000 times. (d) Montmorillonite (a smectite group mineral) from Wyoming magnified about 18,000 times. [*Scanning electron micrographs courtesy Dr. Bruce F. Bohor, Illinois State Geological Survey.*]

(a)

(b)

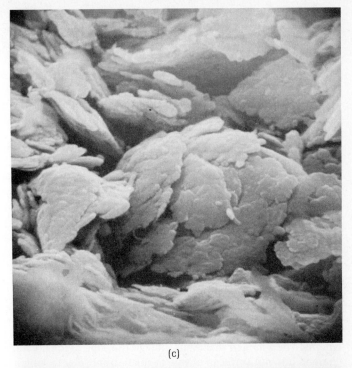

(c)

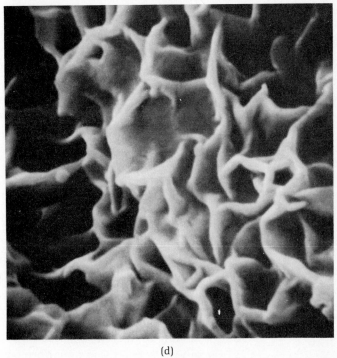

(d)

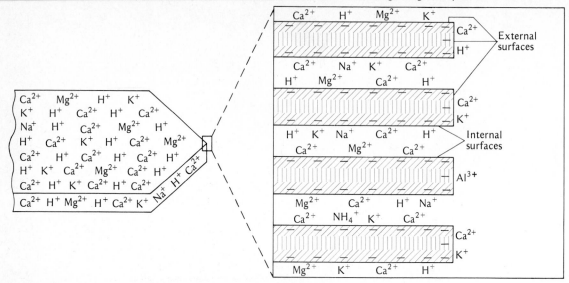

FIGURE 5.2 Diagrammatic representation of a silicate clay crystal (micelle) with its sheet-like structure, its innumerable negative charges, and its swarm of adsorbed cations. The enlarged schematic view of the edge of the crystal illustrates the negatively charged internal surfaces of this particular particle, to which cations and water are attracted. Note that each crystal unit has definite mineralogical structure. (See Figure 5.4 for a more detailed drawing of structure.)

is, gibbsite as $Al(OH)_3$ and goethite as $FeOOH$. For simplicity they will be referred to as the Fe and Al oxide clays.

Although relatively less is known about the Fe and Al oxide clays, they have some properties in common with the silicates. For example, at least some of them are thought to have definite crystalline structure. At high pH values, the small particles may carry a small negative charge and thus serve as a central micelle around which a swarm of cations are attracted. Because of the much smaller number of negative charges per micelle, however, cation adsorption is even lower than for kaolinite.

In acid soils, some Fe,Al oxides are positively charged (see Section 5.11). They tend to counteract the electronegativity of the larger silicate clays and thereby reduce the capacity to adsorb cations. Also, most Fe,Al oxides are not as sticky, plastic, and cohesive as are the silicates. This accounts for the much better physical condition of soils dominated by these oxides.

Allophane and Other Amorphous Minerals. In many soils, significant quantities of noncrystalline colloidal matter are found. For example, part of the iron and aluminum oxides in some soils is amorphous. The same is true of part of the silica, especially in soils formed from volcanic ash. In most cases, the properties of the amorphous minerals do not differ greatly from those of the crystalline materials.

Perhaps the most significant amorphous silicate mineral matter in soil is *allophane,* a somewhat poorly defined aluminum silicate. Having a general composition approximating $Al_2O_3 \cdot 2SiO_2H_2O$, this material is found as a constituent in many soils. It is most prevalent in soils developed from volcanic ash. It has a high capacity to adsorb cations and a considerable capacity to adsorb anions, a property that will be considered later (Section 5.16).

5.3   Organic Soil Colloids—Humus

Because surface soils usually have some organic matter, a brief word about organic colloids is necessary at this point. Otherwise, the soil significance of the colloidal state of matter cannot be fully visualized.

Colloidal Organization.   Humus may be considered to have a colloidal organization similar to that of clay. A highly charged anion (micelle) is surrounded by a swarm of adsorbed cations (Figure 5.3). As later sections will show, the reactions of these cations are similar, whether they are adsorbed by clay or by humus.

Some important difference should be noted, however, between humus and the inorganic micelles. First, the complex humus micelle is composed basically of carbon, hydrogen, and oxygen rather than of aluminum, silicon, and oxygen like the silicate clays. Also, the humus micelle is not considered crystalline but the size of the individual particles, although extremely variable, may be at least as small as the silicate clay particles. Last, humus is not as stable as clay and is thus somewhat more dynamic, being formed and destroyed more rapidly than clay.

Because of their complexity, relatively less is known about the specific structure of humus colloids than of the silicate clays. However, it is known

FIGURE 5.3   Adsorption of captions by humus colloids. The phenolic hydroxyl groups ($>$—OH) are attached to aromatic rings; the carboxyl groups (—COOH) are bonded to other carbon atoms in the central unit. Note the general similarity to the adsorption situation in silicate clays.

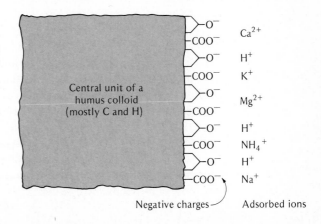

that humus is not a specific compound nor does it have a single structural makeup. The major sources of negative charges are thought to be partially dissociated carboxylic (—COOH), enolic (—OH), and phenolic (⬡—OH) groups associated with central units of varying size and complexity. This relationship is illustrated in a general way in Figure 5.3.

The negative charge on humus colloids is mostly pH dependent, as it is on some of the silicate clays and of hydrous oxides (see Section 5.11). Under strongly acid conditions hydrogen is tightly bound and not easily replaceable by other cations. The colloid therefore exhibits a low negative charge, and the cation adsorptive capacity is low. With a rise in pH, first the hydrogen from the carboxyl groups and then the hydrogen from the enolic and phenolic groups ionize and are replaced by calcium, magnesium, and other cations. Under alkaline conditions the cation adsorptive capacity of humus far exceeds that of most layer silicate clays.

5.4   Adsorbed Cations

Although all cations may be adsorbed by soil colloids, in humid regions the cations of calcium, aluminum, and hydrogen are by far the most numerous, whereas in an arid region soil calcium, magnesium, potassium, and sodium predominate (Table 5.1). A colloidal complex may be represented in the following simple and convenient way for each region.

a Ca e Ca
b Al   ┌─────────┐ f Mg ┌─────────┐
c H   │ Micelle │ g K   │ Micelle │
d M   └─────────┘   h M   └─────────┘
        Humid region   Arid region

The M stands for the small amounts of metallic and other "base-forming" cations adsorbed by the colloids. The a, b, c, d, e, f, g, and h indicate that the numbers of cations are variable.

These examples emphasize that clays and their associated exchangeable ions can be considered in perhaps an oversimplified way as *complex salts.* This can be verified by comparing the formula for a humid region soil with that of $NaHSO_4$, a well-known acid salt. In both cases negatively charged radicals (anions) are associated with metallic cations and hydrogen. The only difference is one of size and charge, the clay anion (micelle) being much larger than the sulfate and, of course, having many more negative charges per anion.

When, by the proper laboratory manipulations, the metallic cations are replaced entirely by hydrogen ions, a colloidal "acid" or H [Micelle] results. In a like manner calcium ions may be given dominance and a calcium "salt" Ca [Micelle] comes into being. Of course, in nature, soil colloids are very seldom wholly acids or wholly salts.

TABLE 5.1 Relative Proportion of Adsorbed Metallic Cations Present in Certain Surface Soils of the United States[a]

The percentage figures in each case are based on the sum of the metallic cations taken as 100. Note the geographic distribution of the soil samples.

Soil	Ca	Mg	K	Na
Penn loam (New Jersey)	60.8	15.8	19.0	4.4
Mardin silt loam (New York)	90.7	5.0	3.1	1.2
Webster series soil (Iowa)	76.8	20.4	1.2	1.6
Sweeney clay loam (California)	76.1	21.3	1.3	1.3
Red River Valley soil (Minnesota)	73.9	21.5	4.2	0.4
Keith silt loam (Nebraska)	77.1	13.3	7.1	2.5
Holdrege silt loam (Nebraska)	66.5	20.9	11.1	1.5

[a] Data compiled from various sources by Lyon et al. (1952).

Why Adsorbed Calcium, Aluminum, and Hydrogen Ions Are So Prominent. In the early stages of clay formation, the solution surrounding the decomposing silicates contains calcium, magnesium, potassium, sodium, iron, and aluminum ions, which have been liberated by weathering. These ions are not all held with equal tightness by the soil colloids. The order of strength of adsorption when they are present in equivalent quantities is Al > H > Ca > Mg > K > Na. Consequently, one would expect the quantity of these ions in the exchangeable form to be in the same order, with aluminum and calcium being the most dominant cations and sodium the least dominant. This is generally the case in most well-drained, moderately acid humid-region soils.

In arid and semiarid regions, the calcium and other metallic cations do not leach from the soil. These metallic cations therefore tend to dominate the adsorptive sites and pH values of 7.0 and above result. Under these conditions the aluminum ions form insoluble compounds and the adsorbed hydrogen ions are replaced by the metallic cations.

When the drainage of an arid region soil is impeded and alkaline salts accumulate, absorbed sodium ions are likely to become prominent and may equal or even exceed those of the adsorbed calcium. Here, then, would be a sodium or a sodium–calcium complex. By the same rule, in humid regions the displacement of the metallic cations by aluminum and hydrogen ions gives an aluminum-hydrogen clay. Since the cation or cations predominant in a colloidal system have much to do with its physical and chemical properties as well as its relationship to plants, this phase of the subject is of much practical importance.

Reference to the displacement of one adsorbed cation by another suggests that these ions are *exchangeable*. The concept of cation exchange and its relationship to plant nutrition will be considered later (Section 5.12).

5.5   Fundamentals of Layer Silicate Clay Structure

Now that the general characteristics of soil colloids and their associated cations are understood, we turn to a more detailed consideration of silicate clays. For many years clays were thought to be amorphous or lacking in orderly internal arrangement of the chemical elements. However, the use of X rays, electron microscopy, and other techniques has shown that the layer silicate clay particles are definitely crystalline. The enlarged schematic view of the edge of a clay mineral crystal presented in Figure 5.2 illustrates that clay particles are comprised of individual layers or crystal units. The mineralogical organization of these layers varies from one type of clay to another and markedly affects the properties of the mineral. For this reason, some attention will be given to the fundamentals of silicate clay structure before consideration of specific silicate clay minerals.

Silica Tetrahedral and Alumina-Magnesia Octahedral Sheets. The most important silicate clays are known as *phyllosilicates* (Gr. *phullon*, leaf), suggesting a leaf-like or platelet structure. They are characterized by alternating *sheets* comprised of *planes* of mineral cations surrounded and linked together by planes of oxygen and hydroxyl anions (Figure 5.4). One type of sheet is dominated by silicon, the other by aluminum and/or magnesium.

The basic building block for the silica-dominated sheet is a unit composed of one silicon cation surrounded by four oxygen anions. It is called the silica *tetrahedron* because of its four-sided configuration. An interlocking array of a series of silica tetrahedra tied together horizontally by shared oxygen anions gives a *tetrahedral sheet*.

Aluminum and/or magnesium are the key cations in the second type of sheet. An aluminum (or magnesium) ion surrounded by oxygens or hydroxyls gives an eight-sided building block termed an *octahedron* (Figure 5.4). Numerous octahedra linked together horizontally comprise the *octahedral* sheet. An aluminum dominated sheet is termed a *dioctahedral* sheet, while one with magnesium dominant is called a *trioctahedral* sheet. The distinction is due to the fact that *two* aluminum ions have the same charge as *three* magnesium ions. As will be seen later, numerous intergrades occur where both cations are present.

The tetrahedral and octahedral sheets are the fundamental structural units of silicate clays. They are associated in different clays in various structural arrangements and combinations, being bound together within the crystals by shared oxygen anions. The specific combinations of sheets, called *layers,* are characteristic of the major groups of silicate clays. The relationship among planes, sheets, and layers shown in Figure 5.4 is important and should be well understood.

Isomorphous Substitution. The structural arrangements just described suggest a very simple relationship among the elements making up silicate clays. In

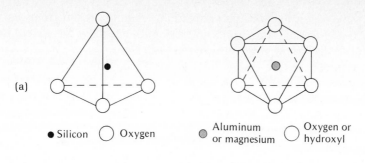

(a)

● Silicon   ◯ Oxygen

● Aluminum or magnesium   ◯ Oxygen or hydroxyl

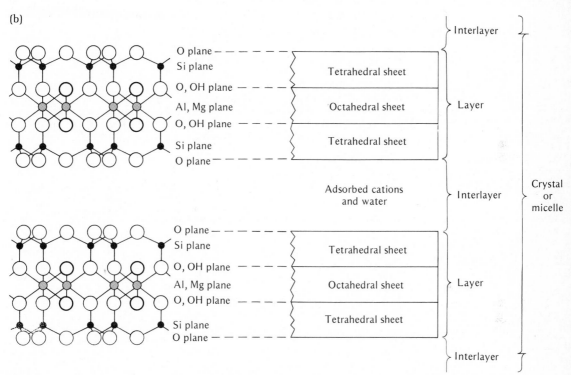

FIGURE 5.4 The basic molecular and structural components of silicate clays. (a) A single *tetrahedron,* a four-sided building block comprised of a silicon ion surrounded by four oxygen ions, and a single eight-sided *octahedron,* in which an aluminum (or magnesium) ion is surrounded by six hydroxyls or oxygens. (b) In clay crystals, thousands of these tetrahedral and octahedral building blocks are connected to give horizontal planes of silicon and of aluminum (or magnesium). These planes alternate with planes of oxygen and hydroxyl ions. The silicon plane and associated oxygen/ hydroxyl planes make up a *tetrahedral sheet.* Similarly, the aluminum/magnesium plane and associated oxygen/hydroxyl planes comprise the *octahedral sheet.* Different combinations of tetrahedral and octahedral sheets are termed *layers.* In some silicate clays, these layers are separated by *interlayers* in which water and adsorbed cations are found. Many layers are found in each *crystal* or *micelle* (micro cell).

nature, however, more complex chemical formulas result from the weathering of a wide variety of rocks and minerals, and cations other than silicon, aluminum, and magnesium enter into the clay lattice.

The silicon in the tetrahedral sheet and the aluminum and magnesium in the octahedral sheet are subject to replacement or substitution by other ions of comparable size. The ionic radii of a number of ions common in clays are listed in Table 5.2 to illustrate this point. Note that aluminum is only slightly larger than silicon. Consequently, aluminum can fit into the center of the tetrahedron in the place of the silicon and does so in some clays. As some silicates form, part of the silicon in the sheet is replaced by aluminum without changing the basic structure of the crystal. The process, called *isomorphous substitution,* is common and accounts for the wide variability in the nature of silicate clays.

TABLE 5.2 Ionic Radii of Elements Common in Silicate Clays and an Indication of Which Are Found in the Tetrahedral and Octahedral Sheets

Note that Al, O, and OH can fit in either.

Ion	Radius (nm)[a]	Found in
Si^{4+}	4.2	
Al^{3+}	5.1	Silica tetrahedra
Fe^{3+}	6.4	
Mg^{2+}	6.6	Alumina octahedra
Zn^{2+}	7.4	
Fe^{2+}	7.0	
Na^+	9.7	Exchange sites
Ca^{2+}	9.9	
K^+	13.3	
O^{2-}	14.0	Both sheets.
OH^-	15.5	

[a] 1 nm = 10^{-9} m.

Isomorphous substitution also occurs in the octahedral sheet. Note from Table 5.2 that ions such as iron and zinc are not too different in size from aluminum and magnesium. As a result, these ions can fit into the position of aluminum or magnesium as the central ion in the octahedral sheet. The isomorphous substitution in a dioctahedral sheet of an ion with two charges such as Mg^{2+} for one with three charges such as Al^{3+} (or Al^{3+} for Si^{4+} in the tetrahedral sheet) leaves unsatisfied negative charges from the oxygen anions in the sheets. As we shall see later (Section 5.11), this type of substitution helps account for the overall negative charge associated with several silicate clays and in turn with the capacity to adsorb cations.

It should be pointed out that substituting a cation such as Al^{3+} or Fe^{3+} with three charges for Mg^{2+} in an otherwise neutral trioctahedral sheet leaves a net positive charge. While such positive charges are usually counterbalanced by negative charges just discussed, they do influence the overall cation adsorptive capacity of silicate clays.

5.6 Mineralogical Organization of Silicate Clays

On the basis of the number and arrangement of tetrahedral (silica) and octahedral (alumina-magnesia) sheets contained in the crystal units or layers, silicate clays are classified into three different groups: (a) 1:1-type minerals—one tetrahedral (Si) to one octahedral (Al) sheet, (b) 2:1-type minerals, and (c) 2:1:1-type minerals. Each of these is discussed briefly in terms of the main member of the group for illustrative purposes.

1:1-Type Minerals. The *layers* of the 1:1-type minerals are made up of one tetrahedral (silica) sheet combined with one octahedral (alumina) sheet; hence the terminology 1:1-type crystal (Figure 5.5). In soils, *kaolinite* is the most prominent member of this group, which includes *halloysite, nacrite,* and *dickite.*

The tetrahedral and octahedral sheets in a given layer of kaolinite are held together by oxygen anions, which are mutually shared by the silicon and aluminum cations in their respective sheets. These layers, in turn, are held together by hydrogen bonding. Consequently, the lattice is *fixed* and no expansion ordinarily occurs between layers when the clay is wetted. Cations and water do not enter between the structural layers of the particle. The effective surface of kaolinite is thus restricted to its outer faces or to its external surface area. Also, there is little isomorphic substitution in this mineral. Along

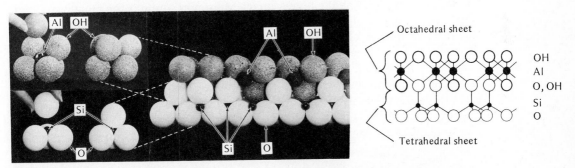

FIGURE 5.5 Models of ions that constitute the 1:1-type clay kaolinite. Note that the mineral is comprised of alternate octahedral (alumina) and tetrahedral (silica) sheets; hence, the designation "1:1." Aluminum ions surrounded by six hydroxyls and oxygens make up the octahedral sheet (upper left). Smaller silicon ions associated with four oxygen ions constitute the tetrahedral sheet. The octahedral and tetrahedral sheets are coupled together (center) to give a layer with hydroxyls on one surface and oxygens on the other. A schematic drawing of the ionic arrangement (right) shows a cross sectional view of the crystal unit or layer.

with the relatively low surface area of kaolinite, this accounts for its low capacity to adsorb cations.[4]

Kaolinite crystals usually are pseudohexagonal in shape (Figure 5.1). In comparison with other clay particles, they are large in size, ranging from 0.10 to 5 μm across, with the majority falling within 0.2–2 μm. Because of the nature of their structural layers kaolinite particles are not readily broken down into extremely thin plates.

In contrast with the other silicate groups, the plasticity (capacity of being molded), cohesion, shrinkage, and swelling properties of kaolinite are very low. Its restricted surface and limited adsorptive capacity for cations and water molecules suggest that kaolinite does not exhibit colloidal properties of a high order of intensity.

Other 1:1-type minerals differ from kaolinite primarily in the stacking of their layers in the crystal. *Halloysite* has sheets of water between these layers, and the crystals have a tubular appearance. This mineral's properties of plasticity, shrinking, and swelling exceed slightly those of kaolinite. *Nacrite* and *dickite,* which are the least common of this group, differ from kaolinite in the orientation of the OH groups involved in the interlayer bonding.

2:1-Type Minerals. The crystal units (layers) of these minerals are characterized by an octahedral sheet sandwiched between two tetrahedral sheets. Three general groups have this basic crystal structure. Two of them, *smectite* and *vermiculite,* are expanding-type minerals, while the third, *fine-grained micas* (*illite*), is nonexpanding.

Expanding minerals. The smectite group, which includes *montmorillonite, beidellite, nontronite,* and *saponite,* is noted for interlayer expansion, which occurs by swelling when the minerals are wetted, the water entering the interlayer space and forcing the layers apart. Montmorillonite is the most prominent member of this group in soils. The flake-like crystals of this mineral (Figure 5.1) are composed of 2:1-type layers, as shown in Figure 5.6. In turn, these layers are loosely held together by very weak oxygen-to-oxygen and cation-to-oxygen linkages. Exchangeable cations and associated water molecules are attracted between layers (the interlayer space), causing *expansion* of the crystal lattice. Consequently, smectite crystals may be easily separated to give particles that approach the size of single crystal layers. Commonly, however, these crystals range in size from 0.01 to 1 μm. They are thus much smaller than the average kaolinite particle (Table 5.3).

The movement of water and cations into the interlayer spaces of the smectite crystals exposes a very large *internal surface,* that, by far, exceeds the external surface area of these minerals. For example, the *specific surface* or total surface area per unit mass (external and internal) of montmorillonite is 700–800 m²/g. A comparable figure for kaolinite is only 15 m²/g.

---

[4] Since the adsorbed cations may be freely exchanged with other cations, the capacity to adsorb cations is usually referred to as *cation exchange capacity* (see Section 5.12).

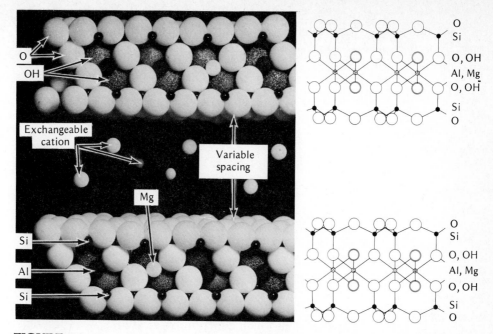

FIGURE 5.6 Model of two crystal units (layers) of montmorillonite, a smectite expanding-lattice 2:1-type clay mineral. Each layer is made up of an octahedral (alumina) sheet sandwiched between two tetrahedral (silica) sheets. There is little attraction between oxygen ions in the bottom sheet of one unit and those in the top sheet of another. This permits a ready and variable space between layers, which is occupied by water and exchangeable cations. This internal surface far exceeds the surface around the outside of the crystal. Note that magnesium has replaced aluminum in some sites of the octahedral sheet. Likewise, some silicon ions in the tetrahedral sheet may be replaced by aluminum (not shown). These substitutions give rise to a negative charge, which accounts for the high cation exchange capacity of this clay mineral.

Isomorphous substitution of magnesium for some of the aluminum in the dioctahedral sheet and of aluminum for silicon in the tetrahedral sheet leaves smectite crystals with a net negative charge. This charge is satisfied by a swarm of cations (H^+, Al^{3+}, Ca^{2+}, K^+, etc.) that are attracted to both the internal and external surfaces (Figure 5.2). The smectites commonly show a high cation exchange capacity, perhaps 10–15 times that of kaolinite.

Smectites also are noted for their high plasticity and cohesion and their marked shrinkage on drying. The shrinkage is due not only to the loss of *interlayer* water, with concomitant reduction in size of each clay particle, but perhaps even more to loss of *interparticle* water, which is present in abundance in wet soils containing smectites because of the extremely small size of the individual clay particles. Wide cracks commonly form as soils dominated

TABLE 5.3 Comparative Properties of Three Major Types of Silicate Clay

Property	Smectite (montmorillonite)	Fine mica (illite)	Kaolinite
Size (μm)	0.01–1.0	0.1–2.0	0.1–5.0
Shape	Irregular flakes	Irregular flakes	Hexagonal crystals
Specific surface (m²/g)[a]	700–800	100–120	5–20
External surface	High	Medium	Low
Internal surface	Very high	Low	None
Cohesion, plasticity	High	Medium	Low
Swelling capacity	High	Medium	Low
Cation exchange capacity (cmol/kg[b]	80–120	15–40	3–10

[a] Square meter of surface per gram.
[b] Centimoles of negative charge per kilogram of clay.

by smectites are dried (see Figure 5.18). The dry aggregates or clods are very hard, making such soils difficult to till.

Montmorillonite, the most common smectite in soils, is characterized by considerable isomorphic substitution of Mg for Al in the octahedral layer.

Beidellite is a 2:1-type clay in the smectite group with its charge determined primarily by ismorphous substitution of aluminum for silicon in the tetrahedral sheet. In *nontronite,* trivalent iron dominates the octahedral sheet, and some of the silicon in the tetrahedral sheet is replaced by aluminum. Hectorite and saponite are trioctahedral minerals, in which magnesium dominates their octahedral sheets. *Saponite* also has some substitution of aluminum for silicon in the tetrahedral sheet.

Vermiculites have structural characteristics similar to those of the smectite group in that an octahedral sheet is found between two tetrahedral sheets. Most soil vermiculties are dioctahedral and have types of isomorphous substitution similar to those for smectites. In the trioctahedral vermiculites, the octahedral sheet is dominated by magnesium rather than aluminum, three magnesium ions rather than two aluminum ions being present. In the tetrahedral sheet of most vermiculites, considerable substitution of aluminum for silicon has taken place. This accounts for most of the very high net negative charge associated with these minerals.

Water molecules, along with magnesium and other ions, are strongly adsorbed in the interlayer space of vermiculite (Figure 5.7). However, they act as bridges holding the units together rather than as wedges driving the units apart. The degree of swelling is, therefore, considerably less for vermiculite than for smectite. For this reason vermiculite is considered a *limited-expansion* clay mineral, expanding more than kaolinite but much less than montmorillonite.

The cation exchange capacity of vermiculites commonly exceeds that of all other silicate clays, including montmorillonite and other smectites. This

FIGURE 5.7 Schematic drawing illustrating the organization of tetrahedral and octahedral sheets in chlorite (a) and vermiculite (b). Magnesium is prominent if not dominant in the tetrahedral sheets of both minerals. In chlorite, Mg-dominated octahedral sheets [brucite, $(Mg(OH)_2]$ alternate with 2:1-type layers giving a 2:1:1- or 2:2-type structure. Water molecules and magnesium and other cations are held rather firmly in the interlayer of vermiculite since they are attracted by the high negative charge in the tetrahedral sheets. Interlayer constituents act as bridges between the 2:1 layers and thereby constrain the interlayer expansion, especially in the case of chlorite.

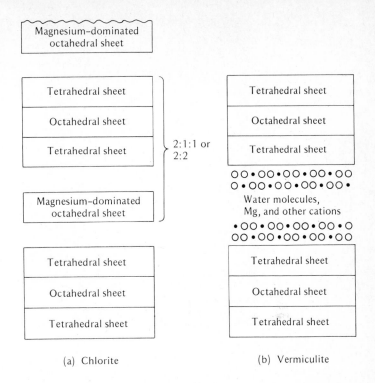

(a) Chlorite   (b) Vermiculite

is due to the very high negative charge in the tetrahedral sheet, which exceeds the positive charge sometimes found in the octahedral sheet. Vermiculite crystals are larger than those of montmorillonite but are much smaller than those of kaolinite.

Nonexpanding minerals.   *Micas* are the type minerals in this group. Muscovite and biotite are examples of micas often found in the sand and silt separates. Minerals similar in structure to these micas are found in the clay fraction of soils. They are called *fine-grained micas* or *illite*.

Like smectites, fine-grained mica has a 2:1-type crystal. However, the particles are much larger than those of the smectites, and the major source of charge is in the tetrahedral rather than the octahedral sheet. About 20% of the tetrahedral silicon sites are occupied by aluminum atoms. This results in a high net negative charge in the tetrahedral sheet, even higher than that found in vermiculite. To satisfy this charge, potassium ions in the interlayer space are strongly attracted and are just the right size to fit snugly into certain spaces in the adjoining tetrahedral sheets (Figure 5.8). The potassium thereby acts as a binding agent, preventing expansion of the crystal. Hence, fine-grained mica is quite nonexpansive.

Such properties as hydration, cation adsorption, swelling, shrinkage, and plasticity are much less intense in fine-grained micas than in smectites. The micas do exceed kaolinite in respect to these characteristics, but this may be due in part to the presence of interstratified layers of smectite or vermiculite.

FIGURE 5.8 Model of a 2:1-type nonexpanding lattice mineral of the fine-grained mica group (illite). The general constitution of the layers is similar to that in the smectites, one octahedral (alumina) sheet between two tetrahedral (silica) sheets. However, potassium ions are tightly held between layers, giving the mineral a more or less rigid type of structure that prevents the movement of water and cations into the space between layers. The internal surface and exchange capacity of illite are thus far below those of the smectites.

In size, too, fine-grained mica crystals are intermediate between the smectites and kaolinites (Table 5.3). Their specific surface area is about 80 m²/g, about 1/10 that for montmorillonite.

2:1:1-Type Minerals. This silicate group is represented by soil *chlorites*, which are common in some soils. Chlorites are basically ferro-magnesium silicates with some aluminum present. In a typical chlorite clay crystal 2:1 layers, such as are found in vermiculite, alternate with a magnesium-dominated trioctahedral sheet, giving a 2:1:1 ratio (Figure 5.7). Magnesium also dominates the trioctahedral position of chlorite in the 2:1 layer. Thus, the crystal unit contains two silica tetrahedral sheets and two magnesium-dominated trioctahedral sheets, giving rise to the term 2:1:1- or 2:2-type structure.

The cation exchange capacity of chlorites is about the same as that of fine-grained mica (illite) and considerably less than that of the smectites or vermiculites. Like fine mica, chlorite may have some vermiculite or smectite interstratified with it in a single crystal. Particle size and surface area for chlorite are also about the same as for illite. There is no water adsorption between the chlorite crystal units, which accounts for the nonexpansive nature of this mineral.

Chlorite-like minerals are formed when aluminum hydroxy polymers are deposited between layers of vermiculites and smectites. Such minerals are common in highly weathered soils and have properties intermediate between those of chlorite and the source mineral, vermiculite or smectite. They are examples of interstratified layers, which will now be discussed briefly.

Mixed and Interstratified Layers. In obtaining a general concept of clay minerals, one should recognize that the specific groups do not occur independently

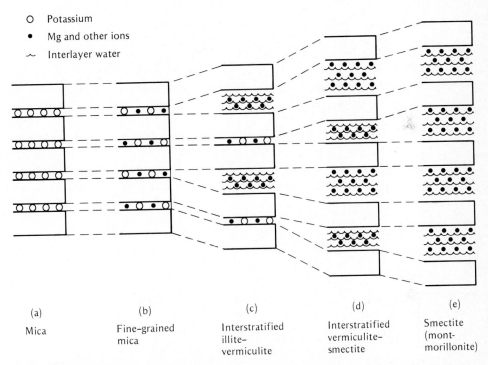

<table>
<tr><td>O</td><td>Potassium</td></tr>
<tr><td>●</td><td>Mg and other ions</td></tr>
<tr><td>~</td><td>Interlayer water</td></tr>
</table>

(a)	(b)	(c)	(d)	(e)
Mica	Fine-grained mica	Interstratified illite-vermiculite	Interstratified vermiculite-smectite	Smectite (mont-morillonite)

FIGURE 5.9 Structural differences among silicate minerals and their mixtures. Potassium-containing micas (a) with rigid crystals lose part of their potassium and weather to fine-grained mica (illite), which is less rigid and attracts exchangeable cations to the interlayer space (b). At a more advanced stage of weathering (c), the potassium is leached from between some of the 2:1 layers; water and magnesium ions bind the layers together and an interstratified illite-vermiculate is present. Further weathering removes more potassium (d), water and exchangeable ions push in between the 2:1 layers and an interstratified vermiculite-smectite is formed. More weathering produces smectite (e), the highly expanded mineral. Smectite in turn is subject to weathering to kaolinite and iron and aluminum oxides. While most smectites actually are formed by other processes, the sequence here illustrates the structural relationships of smectites to the other minerals.

of one another. In a given soil, it is common to find several clay minerals in an intimate mixture. Furthermore, minerals having properties and composition intermediate between those of any two of the well-defined minerals just described will be found. Such minerals are termed *mixed layer* or *interstratified*, because the individual layers within a given crystal may be of more than one type. Terms such as "chlorite-vermiculite" and "illite-smectite" are used to describe mixed-layer minerals (Figure 5.9). In some soils they are more common than the single-structured minerals such as montmorillonite.

5.7   Source of the Negative Charge on Silicate Clays

Exposed Crystal Edges.   There are at least two sources of the negative charges associated with silicate clay particles. The first involves unsatisfied negative charges associated with oxygens and hydroxyl groups exposed at the broken edges and flat external surfaces of minerals such as kaolinite. The O and OH are attached to silicon and aluminum ions within their respective sheets. At pH levels of 7 or higher, the hydrogen of these hydroxyls dissociates slightly and the colloidal surface is left with a negative charge carried by the oxygen. The loosely held hydrogen is readily exchangeable. This situation is illustrated in the diagram in Figure 5.10. The charge sites at crystal edges are at least in part responsible for what has been termed the *pH-dependent* charge of inorganic colloids (see Section 5.11 and Figure 5.16).

The presence of surface and broken-edge OH groups gives the kaolinite clay particles their electronegativity and their capacity to adsorb cations. This phenomenon apparently accounts for most of the cation exchange capacity of 1:1-type colloidal clays and for organic colloids. It is of minor significance with the 2:1-type clays.

FIGURE 5.10   Diagram of a broken edge of a kaolinite crystal, showing oxygens as the source of negative charge. At high pH values, the hydrogen ions tend to be held loosely and can be exchanged for other cations.

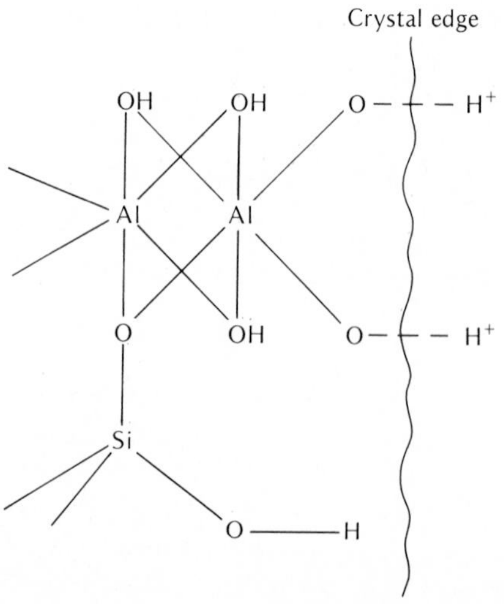

FIGURE 5.11 Atomic configuration in the octahedral sheet of silicate clays (a) without substitution and (b) with a magnesium ion substituted for one aluminum. Where no substitution has occurred, the three positive charges on aluminum are balanced by an equivalent of three negative charges from six oxygens or hydroxyls. With magnesium in the place of aluminum, only two of those negative charges are balanced, leaving one negative charge unsatisfied. This negative charge be satisfied by an adsorbed cation.

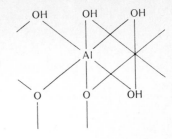

(a) Without substitution

(b) With substitution of Mg^{2+} for Al^{3+}

Isomorphous Substitution.   The phenomenon of the isomorphous substitution of one ion for another in the crystal lattice was mentioned earlier as one of the fundamentals of silicate structure (see Section 5.5). We are now ready to see in a quantitative way the mechanism by which this substitution results in a second source of net negative charge in the clay crystal.

In Figure 5.11 the structural arrangement in a segment of an aluminum-dominated dioctahedral sheet is shown with and without the substitution of a magnesium ion for an aluminum ion. Without substitution, the positive and negative charges are in balance. The three positive charges of aluminum are fully satisfied by an equivalent of three negative charges from the surrounding oxygen or hydroxyl. However, when a magnesium ion replaces one of the aluminum ions by isomorphous substitution, an imbalance occurs. The magnesium ion, with only two positive charges, cannot satisfy the three negative charges associated with the surrounding oxygen and hydroxyl. Consequently, the octahedral sheet assumes one negative charge for each magnesium-for-aluminum substitution. This negative charge must be balanced by a positively charged cation such as Na^+ or K^+, which is adsorbed by the clay surface.

While isomorphous substitution may occur in certain 1:1-type minerals, it is of much greater significance in 2:1-type silicate clays and especially in smectites and dioctahedral vermiculites. The resultant negative charge is far in excess of that resulting from broken crystal edges of these minerals.

Similar reasoning accounts for the net negative charge in the tetrahedral sheet of silicate clays when aluminum, a three-valent ion, has become isomorphically substituted for silicon, which has four positive charges. This can be shown simply as follows.

Tetrahedral sheet (no substitution)	Tetrahedral sheet (Al^{3+} substituted for Si^{4+})
Si_2O_4	$SiAlO_4^-$
No charge	One excess negative charge

Many 2:1-type silicate clays exhibit the substitution phenomenon just explained to a marked degree. Substitution may have occurred in either or both the octahedral and tetrahedral sheets, as discussion of unit formulas in the next section will illustrate. In montmorillonite, most of the substitution has occurred in the octahedral sheet, but some has taken place in the tetrahedral sheet as well. Similarly, substitution in both sheets has occurred in fine micas (illite) and vermiculite, although the substitution and the source of negative charge lie largely in the tetrahedral sheets. In addition to the magnesium and aluminum ions, iron, manganese, and other ions may substitute for aluminum or silicon in the crystal of certain minerals. The ionic diameter largely governs the substitutions that occur (see Table 5.2). Unlike the charge associated with the exposed crystal edges, those resulting from ionic substitution are not dependent on pH.

It is well to note that the isomorphous substitution does not always result in an increase in negative charges. The isomorphous substitution of any cation for one with a lower charge results in an increase in the positive charge. This substitution commonly occurs in trioctahedral layers when two-valent magnesium is replaced by three-valent iron or aluminum. The following is a simple illustration of this effect.

Trioctahedral sheet (no substitution)	Trioctahedral sheet (Al^{3+} substituted for Mg^{2+})
$Mg_3(OH)_6$	$Mg_2Al(OH)_6^+$
No charge	One excess positive charge

The influence of isomorphous substitution on negative and positive charges can be seen in the next section covering the chemical composition of silicate clays.

5.8 Chemical Composition of Silicate Clays

Because of the extensive crystal modifications and ionic substitutions just discussed, simple chemical formulas cannot be used to identify specifically the clays in a given soil. However, "type" formulas can be used to illustrate differ-

TABLE 5.4 Typical Unit Layer Formulas of Several Clay and Other Silicate Minerals Showing Octahedral and Tetrahedral Cations as Well as Coordinating Anions, Charge per Unit Formula, and Fixed and Exchangeable Interlayer Components

Note that the negative charges in the crystal units are counterbalanced by equivalent positive charges in interlayer areas.

Mineral	Octahedral sheet	Tetrahedral sheet	Coordinating anions	Charge per unit formula	Interlayer components	
					Fixed	Exchangeable[a]
Kaolinite	Al_2	Si_2	$O_5(OH)_4$	0	none	none
Serpertine	Mg_3	Si_2	$O_5(OH)_4$	0	none	none
2:1-type dioctahedral minerals						
Pyrophyllite	Al_2	Si_4	$O_{10}(OH)_2$	0	none	none
Montmorillonite	$Al_{1.7}Mg_{0.3}$ -0.3	$Si_{3.9}Al_{0.1}$ -0.1	$O_{10}(OH)_2$	-0.4	none	$M^+_{0.4}$
Beidellite	Al_2	$Si_{3.6}Al_{0.4}$ -0.4	$O_{10}(OH)_2$	-0.4	none	$M^+_{0.4}$
Nontronite	Fe_2	$Si_{3.6}Al_{0.4}$ -0.4	$O_{10}(OH)_2$	-0.4	none	$M^+_{0.4}$
Vermiculite	$Al_{1.7}Mg_{0.3}$ -0.3	$Si_{3.6}Al_{0.4}$ -0.4	$O_{10}(OH)_2$	-0.7	xH_2O	$M^+_{0.7}$
Fine mica (illite)	Al_2	$Si_{3.2}Al_{0.8}$ -0.8	$O_{10}(OH)_2$	-0.8	$K^+_{0.7}$	$M^+_{0.1}$
Muscovite	Al_2	Si_3Al -1.0	$O_{10}(OH)_2$	-1.0	K^+	none
2:1-type trioctahedral minerals						
Talc	Mg_3	Si_4	$O_{10}(OH)_2$	0	none	none
Vermiculite	$Mg_{2.7}Fe^{3+}_{0.3}$ $+0.3$	Si_3Al -1.0	$O_{10}(OH)_2$	-0.7	xH_2O	$M^+_{0.7}$
Chlorite	$Mg_{2.6}Fe^{3+}_{0.4}$ $+0.4$	$Si_{2.5}(Al,Fe)_{1.5}$ -1.5	$O_{10}(OH)_2$	-1.1	$Mg_2Al(OH)_6^+$	$M^+_{0.1}$

[a] Exchangeable cations such as Ca^{2+}, Mg^{2+}, and H^+ are indicated by the singly charged cation M^+.

ences in composition. Examples of these formulas, which are sometimes referred to as *unit layer formulas*, are shown in Table 5.4.

The unit layer formula for kaolinite is quite simple. This is in keeping with the fact that little if any ionic substitution has occurred in the layers of this silicate clay. There are no unit charges in the crystal and no interlayer cations. The small charge associated with kaolinite is from ionized hydroxyl groups on the crystal surfaces and edges described in the previous section.

2:1-Type Dioctahedral Minerals. By comparing the 2:1-type clay mineral formulas with those of minerals having comparable structure but with no charge (e.g., pyrophyllite in dioctahedrals and talc in the trioctahedrals), the relationship between isomorphous substitution, unit charges, and interlayer exchangeable cations becomes obvious. For example, the montmorillonite formula shows the substitution of 0.3 mole of Mg for Al in the octahedral sheet and of 0.1 mole of Al for Si in the tetrahedral sheet. The resulting 0.4 unit negative charge is satisfied by an equivalent positive charge from exchangeable cations in the interlayer space (shown as $M_{0.4}^+$ in Table 5.4). This simple relationship holds for the other expanding-type clays shown in Table 5.4.

In fine mica Al has replaced much of Si in the tetrahedral sheet giving a unit charge of 0.8. Most of this negative charge is counterbalanced by 0.7 mole of K fixed in the interlayer space, leaving 0.1 mole to attract exchangeable cations ($M_{0.1}^+$). An even greater negative charge is found in the tetrahedral sheet of muscovite, but this is satisfied entirely by a K ion firmly fixed in the interlayer space, leaving this mineral with no net charge.

2:1-Type Trioctahedral Minerals. The formulas of 2:1-type trioctahedral minerals illustrates the effect of isomorphous substitution on positive as well as negative charges. For example, compare the formula of trioctahedral vermiculite with that of the comparable mineral having no charge, talc. In vermiculite, 0.3 Mg in the trioctahedral sheet has been replaced by 0.3 Fe^{3+}, a trivalent cation. This results in a 0.3 positive charge in the octahedral sheet, which is more than counterbalanced by the one negative charge in the tetrahedral sheet resulting from the substitution of an aluminum for a silicon ion. The net negative charge (0.7) attracts an equivalent positive charge associated with exchangeable cations in the interlayer space.

In chlorite, a trioctahedral layer is fixed in the interlayer space. One Al ion has replaced one Mg ion in this layer, giving the formula $Mg_2Al(OH)_6$. The positive charge resulting from this substitution, plus the 0.4 positive charge resulting from the substitution of Fe^{3+} for Mg in the trioctahedral sheet of the 2:1 layer, help counterbalance the very strong electronegativity (1.5 negative charges) coming from the high replacement of silicon by aluminum in the tetrahedral sheet. The result is that a net negative charge of only 0.1 remains to attract exchangeable cations to the interlayer in spite of a very high negative charge per unit formula. A study of Table 5.4 should provide an understanding of the relationship between silicate clay structures and practical properties of these clays.

5.9 Genesis of Silicate Clays

The silicate clays are developed most abundantly from such minerals as the feldspars, micas, amphiboles, and pyroxenes. Apparently, the transformation of these minerals into silicate clays has taken place in soils and elsewhere by at least two distinct processes: (a) a comparatively slight physical and chemical *alteration* of the primary minerals and (b) a *decomposition* of the original minerals with the subsequent *recrystallization* of certain of their decomposition products into the silicate clays. These processes will each be given brief consideration.

Alteration. Alteration of the minerals may be encouraged by chemical action involving the removal of certain soluble constituents and the substitution of others within the crystal lattice. The changes that occur as muscovite is altered to fine-grained mica (illite) may be used as an example. Muscovite is a 2:1-type primary mineral with a nonexpanding crystal structure. As the weathering process begins, some potassium is lost from the crystal structure and water molecules enter into the lattice to give a looser and less rigid crystal. Also, there is a relative increase in the silica content as compared to aluminum in the tetrahedral sheet. Some of these changes, perhaps oversimplified, can be shown as

$$KAl_2(Al\,Si_3)O_{10}(OH)_2 + 0.2\,Si^{4+} + 0.1\,M^+ \xrightarrow{\;H_2O\;} M^+_{0.1}(K_{0.7})Al_2(Al_{0.8}Si_{3.2})O_{10}(OH)_2 + 0.3\,K^+ + 0.2\,Al^{3+}$$

Muscovite (rigid crystal)   Soil solution   Fine mica (semirigid crystal)   Soil solution

$M^+_{0.1}$ represents exchangeable cations, and $(K_{0.7})$ represents potassium held semirigidly between crystal units.

There have been a release of potassium and aluminum, a minor change in the chemical makeup, a reduction in rigidity of the crystal, and an initiation of exchangeable properties with little basic change in the crystal structure of the original mineral. It is still a 2:1 type, only having been *altered* in the process of weathering. Continued removal of potassium and substitution of magnesium for some of the aluminum in the alumina layer would result in the formation of montmorillonite (Figure 5.12).

These examples illustrate the structural similarity among some of the silicate clay minerals. They also illustrate the earlier reference to the gradual transition from one mineral to another and to the *intermediate* minerals with properties and characteristics in between those of the distinct groups (see Section 5.6). The presence of "mixed-layer" minerals with such names as "mica-montmorillonite," "chlorite-mica," and "chlorite-vermiculite" (Figure 5.9) suggests that a given colloidal crystal may contain layers of one mineral in between layers of another. Certainly they emphasize the complexity of clay mineralogy and of soils containing these minerals.

Recrystallization. The crystallization of silicate clays from soluble weathering products of other minerals is fully as important in clay genesis as is alteration. A good example is the formation of kaolinite from solutions containing soluble aluminum and silicon that have come from the chemical breakdown of primary

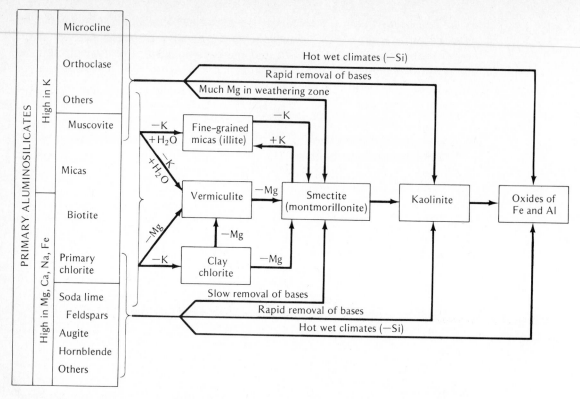

FIGURE 5.12 General conditions for the formation of the various layer silicate clays and oxides of iron and aluminum. Fine-grained micas, chlorite, and vermiculite are formed through rather mild weathering of primary aluminosilicate minerals, whereas kaolinite and oxides of iron and aluminum are products of much more intense weathering. Conditions of intermediate weathering intensity encourage the formation of smectite. In each case silicate-clay genesis is accompanied by the removal of soluble elements such as K, Na, Ca, and Mg.

minerals. This process of recrystallization involves a complete change from the structural makeup of the original minerals and is usually the result of much more intense weathering than that required by the alteration process described above.

Moreover, such crystallization makes possible the formation of more than one kind of clay from a given original mineral. The exact silicate colloid that forms apparently depends upon the condition of the weathering and the ions present in the weathering solution as crystallization occurs.

Relative Stages of Weathering. The more specific conditions resulting in the formation of one or more of the important types of clay are shown in Figure 5.12. Perhaps the first generalization to be drawn from this outline is that there is a difference in the weathering stage of minerals. The magnesium-rich chlorite and the fine mica (illite) apparently represent the earlier weathering stages of the silicates, while kaolinite represents the most advanced. The smectites (e.g., montmorillonite) occupy an intermediate stage of weathering. With Figure 5.12 as a guide, consider briefly the conditions that might yield each of the three groups of clays.

Genesis of Individual Clays.   Fine micas such as illite may be formed by the alteration of primary micas such as muscovite without major changes in the crystal structure, only a comparatively slight alteration being necessary to effect the changes from one to the other. Illite also forms in potassium-rich sediments.

In other cases, illite has apparently been formed from primary minerals such as the potash feldspars by recrystallization under conditions of an abundant potassium supply. In still other instances, illite may be formed from one of the smectites (e.g., montmorillonite) if the latter is in contact with abundant potassium. More common, however, is the reverse reaction by which illite weathers to montmorillonite with the loss of much of its potassium.

Soil chlorite is formed by the alteration of the magnesium- and iron-rich mica, biotite, and of primary chlorite. This change is accompanied by a loss of some magnesium, potassium, and iron. Further alteration and weathering may yield illite (fine mica) or vermiculite, either of which can be altered to form one of the smectites.

Montmorillonite and other smectites may be formed by recrystallization from a variety of minerals provided conditions are appropriate. Apparently, mild weathering conditions (usually slightly acid to alkaline), a relative abundance of magnesium, and an absence of excessive leaching are conducive to the formation of these minerals. Alteration of other silicate clays, such as chlorite, illite, and vermiculite, may also yield montmorillonite or one of the other smectites.

As already stated, kaolinite represents a more advanced stage of weathering than does any of the other major types of silicate clays. It is formed from the decomposition of silicates under conditions of moderately to strongly acid weathering, which results in the removal of the more soluble metal ions, such as Ca, Mg, and Na, and some silica. The soluble aluminum and silicon products that are released may recrystallize under proper conditions to form kaolinite. This mineral, in turn, is subject to decomposition, especially in the tropics, with the formation of iron and aluminum oxides and soluble silica.

As weathering of primary and secondary minerals occurs, ions of several elements are released. The more soluble ions, such as sodium and potassium, are usually removed in leaching waters. Others, such as aluminum, iron, and silicon, either may recrystallize into new silicate-clay minerals or, more commonly, may form insoluble minerals, such as the oxides of iron and aluminum. As shown in Figure 5.12, these compounds represent the most advanced stages of weathering.

The clay of any particular soil is generally made up of a mixture of different colloidal minerals. In a given soil the mixture may vary from horizon to horizon. This occurs because the kind of clay that develops depends not only upon climatic influences and profile conditions but also upon the nature of the parent material. The situation may be further complicated by the presence in the parent material itself of clays that were formed under a preceding and perhaps an entirely different type of climatic regime. Nevertheless, some very general deductions seem possible, taking advantage of the relationships shown in Figure 5.12.

Regional Differences.   The well-drained and well-weathered soils of humid and subhumid tropics tend to be dominated by the oxides of iron and aluminum. These clays are also prominent in the warmer humid regions of the temperate zone such as are found in the southeastern part of the United States. Kaolinite is commonly the dominant silicate mineral in these soils (Table 5.5) and is also found along with the hydrous oxide clays in more tropical areas.

As one might expect, the smectite, vermiculite, and illite (fine mica) groups are more prominent in cooler than in warm climates. Weathering is less intense there than in tropical and subtropical regions. In the northern part of the United States, in Canada, and in similar temperature regions throughout the world,

TABLE 5.5   Dominant Clay Minerals Found in Different Areas of the United States[a]

Soil order[b]	General weathering intensity	Typical location in U.S.	Hydrous oxides	Kaolinite	Montmorillonite	Illite	Vermiculite	Chlorite	Intergrades
Entisol	Low	Variable			x	x			
Inceptisol		Variable				x			x
Aridisols		Desert				x	x	x	x
Vertisols		Ala., Tex.			x				
Mollisols		Central U.S.			x	x	x	x	
Alfisols		Ohio, Pa., N.Y.		x	x	x		x	x
Spodosols		New England	x				x		x
Ultisols		Southeast U.S.	x	x			x		x
Oxisols	High	Tropical zones	x	x					x

[a] Adapted from Jackson (1955).
[b] See Chapter 13 for soil descriptions.

these clay minerals are common. The particular minerals that form depend largely on the parent materials and on the soil-water regime. Where either the parent material or the soil solution surrounding the weathering minerals is high in potassium, illite and related minerals are apt to be formed. Parent materials high in bases, particularly magnesium, or a soil drainage situation that discourages the leaching of these bases, encourages smectite formation. For these reasons, illite and montmorillonite are more likely to be prominent in soils of semiarid and arid regions than in the more humid areas.

The strong influence of parent material on the geographic distribution of clays can be seen in the "black belt" Vertisols of Alabama, Mississippi, and Texas. These soils, which are dark in color, have developed from base-rich marine parent materials and are dominated by montmorillonitic clays. They are surrounded by soils high in kaolinite and hydrous oxides that are more representative of this warm, humid region. Similar situations exist in central India and Sudan.

Data in Table 5.5 show the dominant clay minerals in different soil orders, the descriptions of which are given in Chapter 13. These data tend to substantiate the generalization just discussed. For example, Oxisols and Ultisols are characteristic of areas of intense weathering, and Aridisols are found in desert areas. The dominant clay minerals for these areas are as expected on the basis of Figure 5.12.

Although a few broad generalizations relating to the geographic distribution of clays are possible, these examples suggest that local parent materials and weathering conditions tend to dictate the kinds of clay minerals found in soils.

5.11   The Effect of pH on Surface Charges of Soil Colloids[5]

With the principles affecting the nature and formation of silicate clays in mind, we are now prepared to consider the practical implications of surface charges associated with not only silicate clays but with all soil colloids. As previous sections have suggested, both positive and negative charges are present and at least some of them are influenced by soil pH. These influences will now be discussed.

Negative Charges.   There are two sources of negative charges on soil colloids. First are the *constant* (permanent) charges resulting from the isomorphous substitution with crystals of 2:1-type silicate clays of lower-valent cations for higher valent ones. This process accounts for nearly all of the high negativity associated with these 2:1-type clays.

[5] For a discussion of the theory of surface charge see Uehara and Gillman (1980).

TABLE 5.6 Charge Characteristics of Selected Soil Colloids Showing the Comparative Levels of Their Permanent (Constant) and Variable (pH-Dependent) Negative Charges and Their Positive Charges[a]

| Dominant type of colloid | Negative charge | | | Positive charge (cmol/kg) |
	Total (cmol/kg)	Constant (%)	Variable (%)	
Organic	240	25	75	1
Smectite	118	95	5	1
Vermiculite[b]	85	100	0	0
Illite	19	60	40	3
Allophane	51	20	80	17
Kaolinite	7	43	57	4
Gibbsite	6	0	100	6
Geothite	4	0	100	4

[a] Selected data from Mehlich (1981).
[b] Vermiculites generally have much higher cation exchange capacities than this particular sample.

Permanent negative charges in soils are supplemented by *variable* or pH-dependent charges stemming primarily from the dissociation of H^+ ions from OH^- groups in some 1:1-type clays, organic matter, hydrous oxides of iron and aluminum, and some amorphous materials such as allophane. The relative proportions of permanent and variable charges associated with several soil colloids are shown in Table 5.6.

The variable negative charge is pH dependent, generally being high at high pH values and low under acid conditions. It is controlled by the relative concentrations of H^+ and OH^- ions and their effect on the exposed OH groups in soil colloids. Under conditions of high pH, the OH^- ions in the soil solution bring about the dissociation of H^+ ion from the colloid OH group. This can be shown simply for the inorganic ($>$Al—OH) and organic (—COOH) materials as follows.

$$>Al{-}OH \;+\; OH^- \;\rightleftharpoons\; >Al{-}O^- \;+\; H_2O$$
$$-CO{-}OH \;+\; OH^- \;\rightleftharpoons\; -CO{-}O^- \;+\; H_2O$$

| No charge Soil solids | Soil solution | Negative charge Soil solids | Soil solution |

As the soil pH increases, more OH^- ions are available to force the reactions to the right, and the negative charge on the particle surfaces increases. If the soil pH is lowered, OH^- ions are reduced, the reaction goes back to the left, and negativity is reduced.

Another source of increased negativity as the pH is increased is the removal of complex aluminum hydroxy ions (e.g., $Al(OH)_2^+$), which at low pH levels block negative sites in the silicate clays and make them unavailable for cation exchange. As the pH is raised, these ions react with the OH^- ion to form insoluble $Al(OH)_3$, thereby releasing negatively charged sites.

Since soils normally contain a mixture of soil colloids, both permanent and variable charges are present. However, in soils of temperate climates where 2:1-type clays are common, the permanent negative charges are usually dominant. In highly weathered soils of the tropics where 1:1-type silicate clays and iron and aluminum oxides dominate, and in soils high in organic matter, the variable negative charges are more common.

Positive Charges.   In Sections 5.7 and 5.8, we noted that positive charges result from isomorphous substitution of high valent cations for lower valent ones in trioctahedral clays. In such clays these positive charges are usually countered by much higher negative charges within the crystal units. But in some highly weathered soils of the tropics positive charges may actually exceed the negative ones.

The OH groups associated with the surfaces of iron and aluminum oxides, and allophane, and with the surfaces and edges of 1:1 silicate clays such as kaolinite, are also the sites for the positive charges found in certain acid soils. The source of these positive charges is the protonation or attachment of H^+ ions to the surface OH groups as soils containing these minerals are acidified. The reaction for silicate clays may be shown simply as follows.

$$\rangle AlOH + H^+ \rightarrow\ \rangle AlOH_2^+$$

No charge   Positive charge

Thus, in some cases, the same site on the soil colloid may be responsible for negative charge (high pH), no charge (intermediate pH), or positive charge (low pH). The reaction may occur as follows as H^+ ion is added to a high pH soil.

$$\rangle AlO^- \ + H^+ \rightarrow\ \rangle AlOH \ + H^+ \rightarrow\ \rangle AlOH_2^+$$

Negative charge   No charge   Positive charge
(high pH)   (intermediate pH)   (low pH)

Complex aluminum or iron hydroxy ions may also be the source of positive charges associated with hydrous oxides of iron and aluminum (see Section 5.2). As a soil is acidified these complex ions form by reactions such as the following:

$$Al(OH)_3 + H^+ \rightarrow\ Al(OH)_2^+ \ + H_2O$$

No charge   Positive charge

The positive charge generated attracts anions such as Cl^-, NO_3^-, and HSO_4^-, which are adsorbed by the soil colloid.

Since, in a given soil, mixtures of humus and several inorganic colloids may be found, it is not surprising that positive and negative charges may be exhibited at the same time. In most soils of temperate regions, the negative charges will dominate, but in some acid soils of the tropics high in iron and aluminum oxides, the overall net charge may be positive (Figure 5.13).

Cation and Anion Adsorption.   The charges associated with soil particles attract simple and complex ions of opposite charge. Thus, a given colloidal mixture may exhibit not only a maze of positive and negative surface charges but an equally complex group of simple cations and anions such as Ca^{2+} and SO_4^{2-} as well as complex organic and inorganic charged complexes all of which have been attracted by the particle charges. This is illustrated in Figure 5.14, which shows how soil colloids attract mineral elements so important for plant growth. In neutral to alkaline soils, calcium and magnesium, along with potassium and sodium, are the dominantly adsorbed cations; in acid soils, hydrogen and aluminum ions predominate. The adsorbed anions are commonly present in smaller quantities than the cations since the negative charges on the soil colloid are generally predominant, especially in soils of temperate regions.

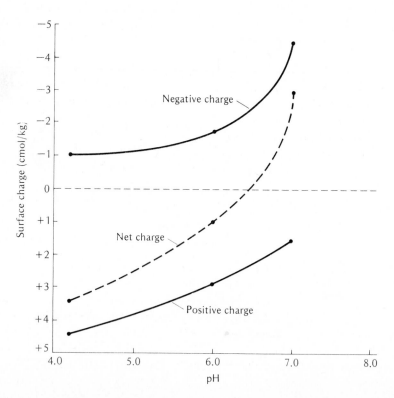

FIGURE 5.13   Surface charge on an Oxisol (Acrorthox) subsoil in relation to pH. At low pH values the positive charges predominate. Only at pH 6.5 was the net charge negative. [*From Van Raij and Peech* (*1972*).]

FIGURE 5.14 Illustration of adsorbed cations and anions on (a) 1:1-type silicate clay and (b) iron/aluminum oxide particles. Note the dominance of the negative charges on the silicate clay and of the positive charges on the aluminum oxide particle.

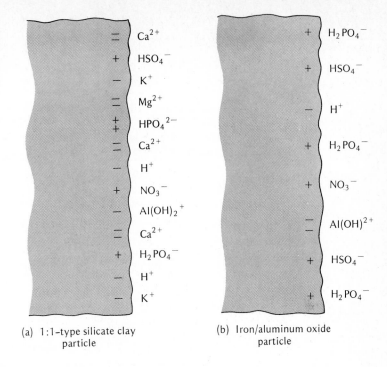

(a) 1:1–type silicate clay particle

(b) Iron/aluminum oxide particle

5.12 Cation Exchange

As previous sections have suggested, the various cations adsorbed by soil colloids are subject to replacement by other cations through a process called *cation exchange*. For example, hydrogen ions generated as organic matter decomposes (see Section 6.3) can displace calcium and other metallic cations from the colloidal complex. This can be shown simply, as follows, where only one adsorbed calcium ion is being replaced.

$$\text{Ca}\boxed{\text{Micelle}} + 2\,\text{H}^+ \rightleftharpoons {}^{\text{H}}_{\text{H}}\boxed{\text{Micelle}} + \text{Ca}^{2+}$$

The reaction takes place rapidly, and the interchange of calcium and hydrogen is chemically equivalent. Furthermore, the reaction is reversible and will go to the left if calcium is added to the system.

Cation Exchange Under Natural Conditions. The reaction as it usually occurs in humid region surface soils is somewhat more complex, but the same principles apply. Assume, for the sake of simplicity, that the numbers of calcium, aluminum, hydrogen, and other metallic cations (M^+) are in the ratio 40, 20, 20, and 20 per micelle, respectively. The other metallic cations are considered monovalent in this case.

$$\begin{matrix} Ca_{40} \\ Al_{20} \\ H_{20} \\ M_{20} \end{matrix} \boxed{Micelle} + 5\,H_2CO_3 \rightleftharpoons \begin{matrix} Ca_{38} \\ Al_{20} \\ H_{25} \\ M_{19} \end{matrix} \boxed{Micelle} + 2\,Ca(HCO_3)_2 + M(HCO_3)$$

Soil solids   Soil solution   Soil solids   Soil solution

Where sufficient rainfall is available to leach the calcium, the reaction tends to go toward the right—that is, hydrogen ions are entering and calcium and other bases (M) are being forced out of the exchange complex into the soil solution. Under these conditions the soils tend to become more acid.

Note that aluminum is not shown as being replaced by the hydrogen. This is because of the tenacity with which the aluminum ion is held. Also, an increase in acidity would likely bring more aluminum ions into solution, which might even increase the adsorption of this element. This situation is discussed in greater detail in Section 5.14.

In regions of low rainfall, the calcium and other salts are not easily leached from the soil, as indicated. This does not permit the reaction to go to completion but instead drives it to the left, keeping the soil neutral or above in reaction. The interaction of climate, biological processes, and cation exchange thus helps determine the properties of soils.

Influence of Lime and Fertilizer. Cation exchange reactions are reversible. Hence, if some form of limestone or other basic calcium compound is applied to an acid soil, the reverse of the exchange just cited occurs. The calcium ions replace the hydrogen and other cations. As a result, the clay becomes higher in exchangeable calcium and lower in adsorbed hydrogen and aluminum, and the soil pH is raised. If, on the other hand, acid-forming chemicals, such as sulfur, are added to an alkaline soil in a dryland area, the H^+ ions would replace the metal cations on the soil colloids and the soil pH would go down.

One more illustration of cation exchange will be offered. If a soil is treated with a liberal application of a water-soluble fertilizer, such as potassium chloride, an exchange may occur (again M is considered to be monovalent).

$$\begin{matrix} Ca_{40} \\ Al_{20} \\ H_{40} \\ M_{20} \end{matrix} \boxed{Micelle} + 7\,KCl \rightleftharpoons \begin{matrix} K_7 \\ Ca_{38} \\ Al_{20} \\ H_{39} \\ M_{18} \end{matrix} \boxed{Micelle} + 2\,CaCl_2 + HCl + 2\,MCl$$

Soil solids   Soil solution   Soil solids   Soil solution

The added potassium is adsorbed on the colloid and replaces an equivalent quantity of calcium, hydrogen, and other elements that appear in the soil solution. The adsorption of the added potassium is considered advantageous because a nutrient so held remains largely in an available condition but is less subject to leaching than are most fertilizer salts. Hence, cation exchange is an important consideration not only for nutrients already present in soils but also for those applied in commercial fertilizers and in other ways.

It should be noted at this point that anionic exchange may take place in soils and in some cases assumes considerable importance. This is especially true with respect to the adsorption of phosphate ions. This phase is discussed in Section 5.16.

5.13 Cation Exchange Capacity

Previous sections have dealt in a qualitative way with the adsorption and exchange of cations on the surface of soil colloids. We now turn to a consideration of the quantitative cation exchange or the *cation exchange capacity*. This property, which is defined simply as "the sum total of the exchangeable cations that a soil can absorb," can be determined rather easily. In commonly used methods, all the adsorbed cations in a soil are replaced by a common ion, such as barium, potassium, or ammonium, and then the amount of adsorbed barium, potassium, or ammonium is determined (Figure 5.15).

Means of Expression.[1]   The cation exchange capacity (CEC) is expressed in terms of moles of positive charge per unit mass. For the convenience of being able to express CEC in whole numbers, we shall use *centimoles of positive charge per kilogram of soil* (cmol/kg). Thus, if a soil has a cation exchange capacity of 10 cmol/kg, 1 kg of this soil is capable of adsorbing 10 cmol of H^+ ion, for example, and of exchanging it with 10 cmol of another uni-charged ion, such as K^+ or Na^+, or with 5 cmol of an ion with two charges, such as Ca^{2+} or Mg^{2+}. In each case, the 10 cmol of negative charge associated with 1 kg of soil is attracting 10 cmol of positive charges, whether they come from H^+, K^+, Na^+, Ca^{2+}, Mg^{2+}, Al^{3+}, or any other cation. This reemphasizes the fact that cations are adsorbed and exchanged on a *chemically equivalent* basis. One mole of charge is provided by 1 mole of H^+, K^+, or any other monovalent cation, by $\frac{1}{2}$ mole of Ca^{2+}, Mg^{2+}, or other divalent cation, and by $\frac{1}{3}$ mole of Al^{3+}, or other trivalent cation.

Keeping this chemical equivalency in mind, it is easy to express cation exchange in practical field terms. For example, when an acid soil is limed and calcium replaces part of the adsorbed hydrogen, the reaction that occurs between the uni-charged H^+ ion and the bi-charged Ca^{2+} ion is

$$\boxed{\text{Micelle}}\ H^+ + \tfrac{1}{2} Ca^{2+} \rightleftharpoons \boxed{\text{Micelle}}\ \tfrac{1}{2} Ca^{2+} + H^+$$

[1] In the past, cation exchange capacity generally has been expressed as milliequivalents per 100 g soil. In this text the International System of Units (SI) will be used. Fortunately, since one milliequivalent per 100 g soil is equal to one centimole of positive or negative charge per kilogram of soil, it is easy to compare soil data using either of these methods of expression.

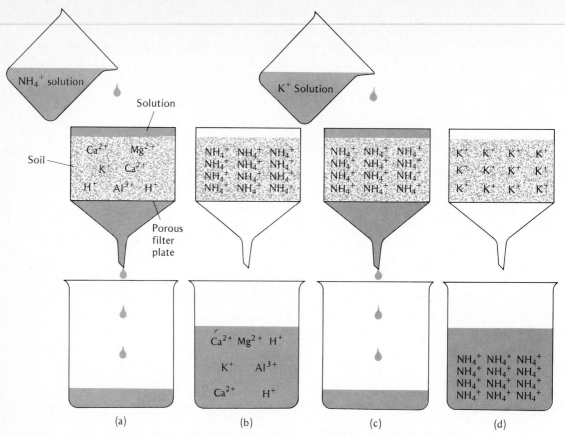

FIGURE 5.15 Illustration of a method of determining the cation exchange capacity of soils. (a) A given mass of soil containing a variety of exchangeable cations is leached with an NH_4^+-containing salt solution. (b) The NH_4^+ ions replace the other adsorbed cations, which are leached into the container below. (c) After the excess NH_4^+ salt solution is removed with an organic solvent, a K^+-containing salt solution is used to replace and leach the adsorbed NH_4^+ ions. (d) The amount of NH_4^+ released and washed into the lower container can be determined, thereby measuring the chemical equivalent of the cation exchange capacity (i.e., the negative charge on the soil colloids).

Note that the 1 mole of charge associated with the H^+ ion is replaced by the equivalent charge associated with $\frac{1}{2}$ Ca^{2+}. In other words, 1 mole of H^+ ion (1 g) would exchange with $\frac{1}{2}$ mole of Ca^{2+} ion ($40 \div 2 = 20$ g). Accordingly, to replace 1 *centi*mole H^+/kg would require $20 \div 100 = 0.2$ g calcium/kg soil. The amount of calcium required for a hectare–furrow slice (2.2 million kg) is $0.2 \times 2.2 \times 10^6 = 440{,}000$ g or 440 kg. This can be expressed in terms of the amount of limestone ($CaCO_3$) to supply the calcium by multiplying by the ratio $CaCO_3/Ca = 100/40 = 2.5$. Thus, $440 \times 2.5 = 1100$ kg limestone per hectare–furrow slice would exchange with 1 cmol H^+/kg soil. A comparable figure in the English system is 1000 pounds per acre–furrow slice. These are practical relationships worth remembering.

Cation Exchange Capacities of Soils. The cation exchange capacities of humus, vermiculite, montmorillonite, illite, chlorite, kaolinite, and Fe,Al oxides at high soil pH values are more or less in the order of 200, 150, 100, 30, 30, 8, and 4 cmol/kg, respectively.

It is easy to see why the clay complex of soils in the southern part of the United States, when dominated by kaolinite, should have a low exchange capacity, ranging perhaps between 5 and 20 cmol/kg of soil. On the other hand, the clay mixtures functioning in the soils of the Midwest, where illite and montmorillonite are more likely to be prominent, have a much higher adsorptive capacity, ranging from 50 to possibly 100 cmol/kg, depending on conditions. Colloids of soils very high in humus (such as mucks) will have cation exchanges in excess of 180 cmol/kg.

These generalizations are verified by the cation exchange capacities (CEC) of a number of soils shown in Table 5.7. Note that the figures range from 2 to nearly 60 cmol/kg. This is to be expected because soils vary so markedly in humus content and in the amounts and kinds of clay present.

TABLE 5.7 Cation Exchange Capacity[a] of a Wide Variety of Surface Soils from Various Parts of the United States[b]

Soil series	Exchange capacity (cmol/kg)	Soil series	Exchange capacity (cmol/kg)
Sand		*Silt loam*	
Sassafras (N.J.)	2.0	Delta (Miss.)	9.4
Plainfield (Wis.)	3.5	Fayette (Minn.)	12.6
		Spencer (Wis.)	14.0
Sandy loam		Dawes (Nebr.)	18.4
Greenville (Ala.)	2.3	Carrington (Minn.)	18.4
Sassafras (N.J.)	2.7	Penn (N.J.)	19.8
Norfolk (Ala.)	3.0	Miami (Wis.)	23.2
Cecil (S.C.)	5.5	Grundy (Ill.)	26.3
Coltneck (N.J.)	9.9		
Colma (Calif.)	17.1	*Clay and clay loam*	
		Cecil clay loam (Ala.)	4.0
Loam		Cecil clay (Ala.)	4.8
Sassafras (N.J.)	7.5	Coyuco sandy clay (Calif.)	20.2
Hoosic (N.J.)	11.4	Gleason clay loam (Calif.)	31.5
Dover (N.J.)	14.0	Susquehanna clay (Ala.)	34.2
Collington (N.J.)	15.9	Sweeney clay (Calif.)	57.5

[a] Centimoles of charge per kilogram of dry soil.
[b] Data compiled from Lyon et al. (1952).

Factors Affecting the Cation Exchange Capacity. Finer-textured soils tend to have higher cation exchange capacities than sandy soils. Further, within a given textural group, organic matter content and the amount and kind of clay influence the CEC. The marked difference between the CECs of the Cecil and Susquehanna clay soils in Alabama (Table 5.7) is due to the dominance of 1:1-type clays in the Cecil and of 2:1-type clays in the Susquehanna.

As was pointed out in previous sections, the cation exchange capacity in most soils increases with pH (Figure 5.16). At a very low pH value, only the "permanent" charges of the clays (see Section 5.11) and a small portion of the charges of organic colloids hold ions that can be replaced by cation exchange. Under these conditions the cation exchange capacity is generally low.

As the pH is raised, the hydrogen held by the organic colloids and silicate clays, such as kaolinite, becomes ionized and is replaceable. Also, the adsorbed aluminum hydroxy ions (see Section 5.11) are removed, forming $Al(OH)_3$,

FIGURE 5.16 Influence of pH on the cation exchange capacity of smectite and humus. Below pH 6.0 the charge for the clay mineral is relatively constant. This charge is considered permanent and is due to ionic substitution in the crystal unit. Above pH 6.0 the charge on the mineral colloid increases slightly because of ionization of hydrogen from exposed O—H groups at crystal edges. In contrast to the clay, essentially all of the charges on the organic colloid are considered pH dependent. [*Smectite data from Coleman and Mehlich (1957); organic colloid data from Helling et al. (1964).*]

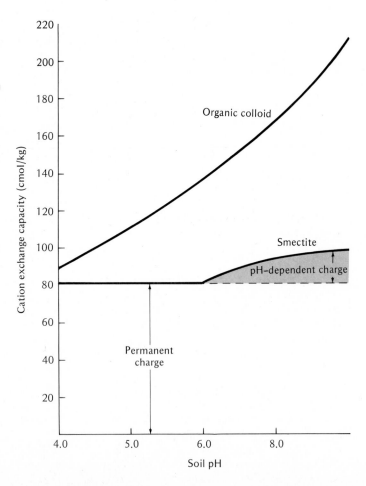

thereby releasing additional exchange sites on the mineral colloids. The net result is an increase in the negative charge on the colloids and in turn an increase in the cation exchange capacity.

In most cases the cation exchange capacity is determined at a pH of 7.0 or above. This means that it includes most of those charges dependent on pH as well as the permanent ones.

Representative Figures. From the preceding discussion, average figures for the cation exchange capacity of silicate clay (0.5 cmol/kg for each 1%) and for well-humidified organic matter (2.0 cmol/kg for each 1%) can be ascertained. By using these figures, it is possible to calculate in a rough way the cation exchange capacity of a humid temperate region surface soil from the percentages of clay and organic matter present. For soils dominated by kaolinite and Fe,Al oxides, comparable figures might be 0.1 for each 1% of clay and 2.0 for each 1% of organic matter.

5.14   Percentage Base Saturation of Soils

Two groups of adsorbed cations have opposing effects on soil acidity and alkalinity. Hydrogen and aluminum tend to dominate very acid soils, both contributing to the concentration of H^+ ions in the soil solution. Adsorbed hydrogen contributes directly to the H^+ ion concentration in the soil solution. Al^{3+} ions do so indirectly through hydrolysis. This may be illustrated as follows.

$$Al^{3+} + H_2O \rightarrow Al(OH)^{2+} + H^+$$
$$Al(OH)^{2+} + H_2O \rightarrow Al(OH)_2^+ + H^+$$
$$Al(OH)_2^+ + H_2O \rightarrow Al(OH)_3 + H^+$$

Most of the other cations, called *exchangeable bases,* neutralize soil acidity and dominate in neutral and alkaline soils. The proportion of the cation exchange capacity occupied by these bases is called the *percentage base saturation.*

$$\% \text{ base saturation} = \frac{\text{exchangeable bases (cmol/kg)}}{\text{CEC (cmol/kg)}}$$

Thus, if the percentage base saturation is 80, four-fifths of the exchange capacity is satisfied by bases, the other by hydrogen and aluminum. The example in Figure 5.17 should be helpful in showing this relationship.

Percentage Base Saturation and pH. A general correlation exists between the percentage base saturation of a given soil and its pH. As the base saturation is reduced owing to the loss in drainage of calcium and other metallic constituents, the pH also is lowered in a more or less definite proportion. This is in

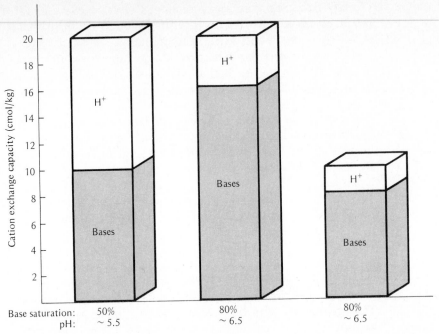

FIGURE 5.17 Three soils with percentage base saturations of 50, 80, and 80, respectively. The first is a clay loam; the second, the same soil satisfactorily limed; and the third, a sandy loam with a cation exchange capacity of only 10 cmol/kg. Note especially that soil pH is correlated more or less closely with percentage base saturation. Also note that the sandy loam (right) has a higher pH than the acid clay loam (left), even though the latter contains more exchangeable bases.

line with the common knowledge that leaching tends ordinarily to increase the acidity of humid region soils. The mechanism by which colloids exert a control on soil pH will be considered later (Section 6.2).

Within the range pH 5–6, a ratio of roughly 5% base saturation change for every 0.10 change in pH is applicable for humid temperate region mineral soils. Thus, if the percentage base saturation is 50% at pH 5.5, it should be 25 and 75% at pH 5.0 and 6.0 respectively. This relationship is worth remembering.

Humid Versus Arid Region Soils. Cation exchange data for representative surface soils from two temperate regions, one humid and the other semiarid, are shown in Table 5.8. The differences are striking. The humid region soil is distinctly acid and about two-thirds base saturated. In contrast, the soil from a semiarid region has a pH near 7, and its percentage base saturation is nearly 100. The hydrogen and aluminum ions are prominent in the exchange complex of the humid region soil; calcium tends to dominate the soil from the semiarid region.

TABLE 5.8 Cation Exchange Data for Representative Mineral Surface Soils in Humid Temperate and Semiarid Temperate Regions

Characteristics	Humid region soil	Semiarid region soil
Exchangeable calcium (cmol/kg)	6–9	13–16
Other exchangeable bases (cmol/kg)	2–3	6–8
Exchangeable hydrogen and/or aluminum (cmol/kg)	4–6	1–2
Cation exchange capacity (cmol/kg)	12–18	20–26
Base saturation (%)	66.6	95 and 92
Probable pH	5.6–5.8	~7

5.15 Cation Exchange and the Availability of Nutrients

Exchangeable cations generally are available to both higher plants and microorganisms. By cation exchange, hydrogen ions from the root hairs and microorganisms replace nutrient cations from the exchange complex. They are forced into the soil solution, where they can be assimilated by the adsorptive surfaces of roots and soil organisms, or they may be removed by drainage water.

Cation Saturation and Nutrient Adsorption by Plants.   Several factors operate to expedite or retard the release of nutrients to plants. First, there is the proportion of the cation exchange capacity of the soil occupied by the nutrient cation in question. For example, if the percentage calcium saturation of a soil is high, the displacement of this cation is comparatively easy and rapid. Thus, 6 cmol/kg of exchangeable calcium in a soil whose exchange capacity is 8 cmol/kg probably would mean ready availability, but 6 cmol/kg when the total exchange capacity of a soil is 30 cmol/kg produces quite the opposite condition. This is one reason why, for a crop that requires abundant calcium, such as alfalfa, the base saturation of at least part of the soil should approach or even exceed 90%.

Influence of Associated Ions.   A second important factor influencing the plant uptake of a given cation is the effect of the ions held in association with it. For example, potassium availability to plants has been shown to be limited by excessive quantities of exchangeable calcium. Likewise, in some cases, high-exchangeable-potassium contents have depressed the availability of magnesium.

Effect of Type of Colloid.   Third, the several types of colloidal micelles differ in the tenacity with which they hold specific cations and the ease of cation exchange. At a given percentage base saturation, smectite holds calcium much more strongly than does kaolinite. As a result, smectite clay must be raised to about 70% base saturation before calcium will exchange easily and rapidly enough to satisfy growing plants. A kaolinite clay, on the other hand, seems to exchange calcium much more readily, serving as a satisfactory source of this constituent at a much lower percentage base saturation. Obviously, the need to add limestone to the two soils will be somewhat different, partly because of this factor.

5.16 Anion Exchange

The positive charges associated with hydrous oxides of iron and aluminum, some 1:1-type clays and amorphous materials, such as allophane, give rise to the adsorption of anions (Figure 5.14). In turn, these anions are subject to replacement by other anions just as cations replace each other. The *anion exchange* which takes place, while it may not approach cation exchange quantitatively, is extremely important as a means of providing higher plants readily available nutrient anions.

The basic principles of cation exchange apply also to anion exchange except that the charges on the colloids are positive and the exchange is among negatively charged anions. A simple example of an anion exchange reaction is

$$\boxed{\text{Micelle}}\, NO_3^- \; + \; Cl^- \; \rightleftharpoons \; \boxed{\text{Micelle}}\, Cl^- \; + \; NO_3^-$$

Soil solid $\qquad$ Soil solution $\qquad$ Soil solid $\qquad$ Soil solution

Just as in cation exchange, *equivalent* quantities of NO_3^- and Cl^- are exchanged, the reaction can be reversed, and plant nutrients can be released for plant adsorption.

While simple reactions such as this are common, it should be pointed out that the adsorption and exchange of some anions including phosphates and sulfates is somewhat more complex. This is due to specific reactions between the anions and soil constituents. For example, the $H_2PO_4^-$ ion may react with the protonated hydroxyl group rather than remain as an easily exchanged anion.

$$>\!Al\!-\!OH_2^+ \; + \; H_2PO_4^- \; \rightarrow \; >\!Al\!-\!H_2PO_4 \; + \; H_2O$$

Soil solid $\qquad$ Soil solution $\qquad$ Soil solid $\qquad$ Soil solution

Note that this reaction actually reduces the net positive charge on the soil colloid.

In spite of these complexities, anion exchange is an important mechanism for interactions in the soil and between the soil and plants. Together with cation exchange it largely determines the ready ability of soils to provide nutrients to plants.

5.17 Physical Properties of Colloids

In addition to the cation exchange property of colloids, certain other characteristics also assume considerable importance. We shall discuss plasticity, cohesion, swelling, shrinkage, dispersion, and flocculation. As might be expected, they are all surface phenomena, and their intensities depend upon the amount and nature of the interfaces presented by the colloids.

Plasticity. Soils containing more than about 15% clay exhibit *plasticity*—that is, pliability and the capacity to be molded. This property is probably due to the plate-like nature of the clay particles and the lubricating though binding influence of the adsorbed water. Thus, the particles easily slide over each other, much like panes of glass with films of water between them.

Plasticity is exhibited only when soils are moist or wet. The range of moisture contents in which plasticity is evident is set by two *plastic limits*. The lower plastic limit (LPL) is the moisture content below which the soil no longer can be molded. The moisture content of a soil when tilled should be no higher than the LPL. At moisture contents above the upper plastic limit (UPL), also termed *liquid limit*, the soil ceases to be plastic but instead tends to flow much as a liquid. While these limits have some usefulness for agricultural purposes, they have special meaning in the classification of soils for engineering purposes, such as the bearing strength for a building or a highway bed.

A comparison of the upper and lower plastic limits of three types of clay is given in Table 5.9. Note that the limits and the range between them are much higher for montmorillonite and others in the smectite group than for kaolinite; illite is intermediate in its properties. Also, the saturating cation affects these limits; sodium gives much higher limits and ranges for the montmorillonite clay than does calcium. In general, soils with wide ranges between these two limits are difficult to handle in the field.

In a practical way, plasticity is extremely important because it encourages such a ready change in soil structure. This must be considered in tillage opera-

TABLE 5.9 Plastic Limits of Three Types of Silicate Clays Saturated with Calcium or Sodium[a]

Type of clay	Calcium saturated (%)		Sodium saturated (%)	
	Lower limit	Upper limit	Lower limit	Upper limit
Montmorillonite	63	177	97	700
Illite	40	90	34	61
Kaolinite	36	73	26	52

[a] From White (1949).

tions. As everyone knows, the cultivation of a fine-textured soil when it is too wet will result in a "puddled" condition detrimental to suitable aeration and drainage. With clayey soils, especially those of the smectite type, plasticity presents a real problem. Suitable granulation therein is often difficult to establish and maintain.

Cohesion.   A second characteristic, somewhat related to plasticity, is *cohesion.* As the water content of a very wet clay is reduced, there is a seeming increase in the attraction of the colloidal particles for each other. This tendency of the clay particles to stick together probably is due primarily to the attraction of the clay particles for the remaining water molecules held between them. Hydrogen bonding between clay surfaces and water, and among the water molecules, is the attractive force responsible for the cohesion. It results in cloddy field conditions which are not conducive to easy tillage.

As one might expect, montmorillonite and illite exhibit cohesion to a much more noticeable degree than does kaolinite or hydrous oxides. Humus, by contrast, tends to reduce the attraction of individual clay particles for each other.

Swelling and Shrinkage.   The third and fourth characteristics of silicate clays to be considered are *swelling* and *shrinkage.* If the clay in question has an expanding crystal and very small particle size, as is the case with the smectites (e.g., montmorillonite), extreme swelling may occur upon wetting. Kaolinite, chlorite, fine mica (illite), and most hydrous oxides with a static crystal do not exhibit the phenomenon to any extent; vermiculite is intermediate in this respect. After a prolonged dry spell, soils high in montmorillonite often are criss-crossed by wide, deep cracks which, at first, allow rain to penetrate rapidly (Figure 5.18). But later, because of swelling, such a soil is likely to close up and become much more impervious than one dominated by kaolinite.

The swelling is due in part to water moving between crystal layers, resulting in intracrystal expansion. But most of the swelling is accounted for by water attracted to ions adsorbed by the clays and by tiny air bubbles entrapped as water moves into the extremely small pores of these soils (Figure 5.19).

Apparently, swelling, shrinkage, cohesion, and plasticity are closely related. They are dependent not only upon the clay mixture present in a soil and the dominant adsorbed cation, but also upon the nature and amount of humus that accompanies the inorganic colloids. These properties of soils are responsible to no small degree for the development and stability of soil structure, as has been previously stressed (see Section 2.10).

Dispersion and Flocculation.   A typical condition of a dilute colloidal suspension in water is that of complete *dispersion;* that is, the particles tend to repel each other, permitting each particle to act completely independent of the others. This condition is encouraged by the smallness of the colloidal particles and by their negative charge and hydration, the latter being enhanced by the swarm

FIGURE 5.18 A field scene showing the cracks that result when a soil high in clay dries out. The type of clay in this case was probably a smectite. [*Courtesy USDA Soil Conservation Service.*]

of hydrated cations around each micelle. Dispersion is encouraged by higher pH values, where the micelle electronegativity is at a maximum (see Figure 5.16). Also, highly hydrated monovalent ions, such as Na^+, which are not very tightly held by the micelles, help to stabilize dispersed colloids. Apparently, these loosely held ions do not effectively reduce the electronegativity of the micelle, permitting the individual micelles to repel each other and to stay in dispersion.

The electronegativity can be reduced in several ways. First, the pH can be lowered. This will have the effect of reducing the negative charge on the micelle (see Section 5.11) and of replacing at least some of the sodium or other monovalent ion with hydrogen. Second, a divalent or trivalent ion such

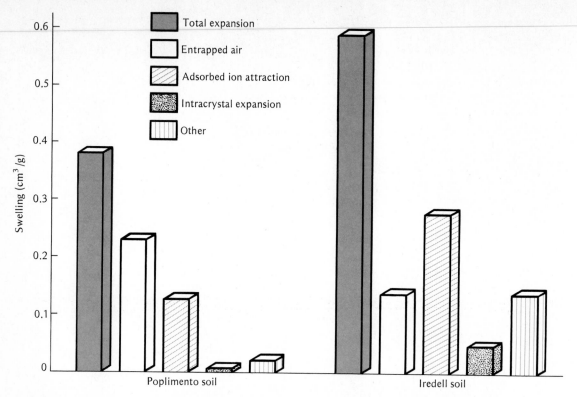

FIGURE 5.19 Total expansion (swelling) of two soils upon wetting and the approximate proportion of the swelling due to various factors. The Iredell soil with 63% smectite plus vermiculite swells more than the Poplimento with only 15% of these two clays. Note that the attraction of adsorbed ions for water and the air entrapped as the soils are wetted are primarily responsible for the swelling. [*Modified from Parker et al.* (*1982*); *used with permission of the Soil Science Society of America.*]

as Ca^{2+}, Mg^{2+}, or Al^{3+}, can be introduced to exchange with the monovalent sodium. These multicharged ions are more tightly adsorbed by the micelle, thereby reducing the electronegativity and enhancing the opportunities for the natural attractive forces between particles to dominate as the colloids approach each other and collide. Finally, adding simple salts will increase the concentration of cation around the micelle, thereby reducing the electronegativity.

Each of these techniques will tend to encourage the opposite of dispersion, *flocculation.* This is a condition that is generally beneficial from the standpoint of agriculture, since it is a first step in the formation of stable aggregates or granules. The ability of common cations to flocculate soil colloids is in the general order of Al > H > Ca > Mg > K > Na. This is fortunate, since the colloidal complexes of humid and subhumid region soils are dominated by aluminum, hydrogen, and calcium; those of semiarid regions are high in calcium. These ions all encourage flocculation.

In limited areas of arid regions, sodium ions have become prominent on the exchange complex. This results in a dispersed condition of the soil colloids, making the soils largely impervious to water penetration. Most plants will not grow under these conditions. The sodium must be replaced before normal growth will occur.

It is clear that the six colloidal properties under discussion are of great importance in the practical management of arable soils. The field control of soil structure must definitely take them into account. With the colloidal viewpoint now provided, it might be worthwhile to review the discussion already offered (Section 2.11) relating to structural management of cultivated lands.

5.18   Conclusion

No attempt will be made to summarize this chapter except to reemphasize three things: (a) the unique and somewhat complicated physical and chemical organization of soil colloids, (b) their capacity to expedite certain phenomena vital to plant and animal life, and (c) the bearing of these phenomena on soil management and crop production.

REFERENCES

ASA/SSSA. 1981. *Chemistry in the Soil Environment,* ASA Publication No. 40 (Madison, WI: Amer. Soc. Agron. and Soil Sci. Soc. Amer.).

Buseck, P. R. 1983. "Electron microscopy of minerals," *Amer. Scientist,* 71:175–185.

Coleman, N. T., and A. Mehlich. 1957. "The chemistry of soil pH," *The Yearbook of Agriculture (Soil)* (Washington, DC: U.S. Department of Agriculture).

Dixon, J. B., and S. B. Weed (Eds.). 1977. *Minerals in Soil Environments* (Madison, WI: Soil Sci. Soc. Amer.).

Gieseking, J. E. (Ed.). 1975. *Soil Components,* Vol. 2: *Inorganic Components* (New York: Springer-Verlag).

Helling, C. S., et al. 1964. "Contribution of organic matter and clay to soil cation exchange capacity as affected by the pH of the saturated solution," *Soil Sci. Soc. Amer. Proc.,* 28:517–20.

Jackson, J. L. 1955. "Chemical composition of soils," in F. E. Bear (Ed.), *Chemistry of the Soil,* 2nd ed. (New York: Reinhold).

Lyon, T. L., H. O. Buckman, and N. C. Brady. 1952. *The Nature and Properties of Soils,* 5th ed. (New York: Macmillan).

Mehlich, A. 1981. "Charge Properties in Relation to Sorption and Desorption of Selected Cation and Anions," in ASA/SSSA (1981).

Parker, J. C., D. F. Amos, and L. W. Zelanzny. 1982. "Water adsorption and swelling of clay minerals in soil systems," *Soil Sci. Soc. Amer. J.*, 46:450–456.

Uehara, G., and G. P. Gillman. 1980. "Charge characteristics of soils with variable and permanent charge minerals: I. Theory," *Soil Sci. Soc. Amer. J.*, 44:250–252.

Van Raij, B., and M. Peech. 1972. "Electrochemical properties of some Oxisols and Alfisols of the tropics," *Soil Sci. Soc. Amer. Proc.*, 36:587–593.

White, W. A. 1949. "Atterberg plastic limits of clay minerals," *Amer. Mineralogist,* 34:508–12.

Soil Reaction: Acidity and Alkalinity

One of the outstanding physiological characteristics of the soil solution is its reaction—that is, whether it is acid, alkaline, or neutral. Since microorganisms and higher plants respond so markedly to their chemical environment, the importance of soil reaction and the factors associated with it have long been recognized.

Soil acidity is common in all regions where precipitation is high enough to leach appreciable amounts of exchangeable bases from the surface layers of soils. So widespread is its occurrence and so pronounced is its influence on plants that acidity has become one of the most discussed properties of soils.

Alkalinity occurs when there is a comparatively high degree of base saturation. The presence of soluble salts and calcium, magnesium, and sodium carbonates, also gives a preponderance of hydroxyl ions over hydrogen ions in the soil solution.[1] Under such conditions the soil is alkaline, sometimes very strongly so, especially if sodium carbonate is present, a pH of 9 or 10 being common. Alkaline soils are, of course, characteristic of most arid and semiarid regions. Their discussion will follow that of acid soils.

6.1 Source of Hydrogen and Hydroxyl Ions

As previously indicated (Section 5.14), two adsorbed cations are largely responsible for soil acidity—hydrogen and aluminum. The mechanisms by which these two ions exert their influence differ, however, depending on the level of soil acidity and on the source and nature of the negative charge on soil colloids.

Effects of Type of Negative Charges. The two types of negative charges discussed in Section 5.7, *permanent* and *variable* or *pH dependent,* each influence soil acidity and the association of hydrogen and aluminum with soil colloids. The permanent charges, which are due to isomorphous substitutions in silicate clay crystals, offer exchange sites throughout the pH range. Hydrogen, aluminum, and the base-forming cations are exchangeable at all pH levels at these permanent charge sites.

In contrast, the magnitude of negative charge and, in turn, of cation ex-

---

[1] When salts of strong bases and weak acids, such as Na_2CO_3, K_2CO_3, and $MgCO_3$, go into solution, they undergo hydrolysis and develop alkalinity. For Na_2CO_3 the reaction is

$$2\,Na^+ + CO_3^{2-} + 2\,HOH \rightleftharpoons 2\,Na^+ + 2\,OH^- + H_2CO_3$$

Since the dissociation of the NaOH is greater than that of the weak H_2CO_3, a domination of OH^- ions results.

change at the variable charge sites depends on the soil pH. Under strongly acid conditions, the variable charges and cation exchange are negligible. The covalently bonded hydrogen of surface OH groups of 1:1-type clays and certain organic compounds is tightly bound and therefore does not exchange with other cations. Only when the pH is increased does the hydrogen dissociate, leaving negative charges on the soil colloids that serve as exchange sites for the other cations.

Also, as the pH rises, the complex aluminum and iron hydroxy ions (e.g., $Al(OH)_2^+$) that block some of the negative charge sites of certain 2:1-type clays are removed, forming insoluble $Al(OH)_3$ and $Fe(OH)_3$. The pH-dependent charge is thereby further increased, and additional exchange sites are made available. These permit increased cation exchange involving hydrogen and base-forming cations. Thus both the pH-dependent and permanent negative charges play an important role in determining the nature and degree of soil acidity.

Strongly Acid Soils.   Under very acid soil conditions much aluminum becomes soluble (Figure 6.1) and is present in the form of aluminum or aluminum hydroxy cations. These become adsorbed even in preference to hydrogen by the permanent charges of soil colloids.

FIGURE 6.1   The effect of soil pH on exchangeable aluminum in a soil–sand mixture (0% O.M.) and a soil–peat mixture (10% O.M.).
Apparently, the organic matter binds the aluminum in an unexchangeable form. This organic matter/aluminum interaction helps account for better plant growth at low pH values on soils high in organic matter.
[*Modified from Hargrove and Thomas (1981) and used with permission of the American Society of Agronomy.*]

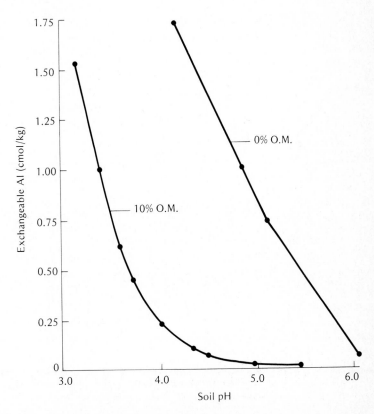

The adsorbed aluminum is in equilibrium with aluminum ions in the soil solution. The latter contribute to soil acidity through their tendency to hydrolyze. Two simplified reactions illustrate how adsorbed aluminum can increase acidity in the soil solution.

$$\boxed{\text{Micelle}}\,\text{Al} \quad \rightleftharpoons \quad \text{Al}^{3+} \tag{6.1}$$

Adsorbed   Soil solution
aluminum   aluminum

The aluminum[2] ions in the soil solution are then hydrolyzed in a manner such as the following.

$$\text{Al}^{3+} + \text{H}_2\text{O} \rightarrow \text{Al(OH)}^{2+} + \text{H}^+ \tag{6.2}$$

The hydrogen ions thus released give a very low pH value in the soil solution and are the major source of hydrogen in most very acid soils.

Adsorbed hydrogen is a second but minor source of hydrogen ions in very acid soils. Under these conditions, much of the hydrogen, along with some iron and aluminum, is held so tightly by covalent bonds in the organic matter and on clay crystal edges that it contributes little to the soil solution (see Section 5.7). On only the strong acid groups of humus and some of the permanent charge exchange sites of the clays is the hydrogen held in an exchangeable form. This hydrogen is in equilibrium with the soil solution. A simple equation to show the release of adsorbed hydrogen to the soil solution is

$$\boxed{\text{Micelle}}\,\text{H} \quad \rightleftharpoons \quad \text{H}^+ \tag{6.3}$$

Adsorbed   Soil solution
hydrogen   hydrogen

Thus, it can be seen that the effect of both adsorbed hydrogen and aluminum is to increase the hydrogen ion concentration in the soil solution.

Moderately Acid Soils. Aluminum and hydrogen compounds also account for soil solution hydrogen ions in these soils, but again by different mechanisms. These soils have somewhat higher percentage base saturations and pH values. The aluminum can no longer exist as Al^{3+} ions but is converted to aluminum hydroxy ions by reactions such as

$$\text{Al}^{3+} + \text{OH}^- \rightarrow \text{Al(OH)}^{2+}$$
$$\text{Al(OH)}^{2+} + \text{OH}^- \rightarrow \text{Al(OH)}_2{}^+ \tag{6.4}$$

Aluminum
hydroxy
ions

[2] The Al^{3+} ion is actually highly hydrated, being present in forms such as $\text{Al(H}_2\text{O)}_6{}^{3+}$. For simplicity, however, it will be shown as the simple Al^{3+} ion.

The actual aluminum hydroxy ions are much more complex than those shown. Formulas such as $[Al_6(OH)_{12}]^{6+}$ and $[Al_{10}(OH)_{22}]^{8+}$ are examples of the more complex ions.

Some of the aluminum hydroxy ions are adsorbed and act as exchangeable cations. As such, they are in equilibrium with the soil solution just as is the Al^{3+} ion in very acid soils. In the soil solution, they are able to produce hydrogen ions by the following hydrolysis reactions, using again as examples the most simple of the aluminum hydroxy ions:

$$Al(OH)^{2+} + H_2O \rightarrow Al(OH)_2^+ + H^+$$
$$Al(OH)_2^+ + H_2O \rightarrow Al(OH)_3 + H^+ \tag{6.5}$$

In some 2:1-type clays, particularly vermiculite, the aluminum hydroxy ions (as well as iron hydroxy ions) play another role. They move into the interlayer space of the crystal units and become very tightly adsorbed. In this form they prevent intracrystal expansion and block some of the exchange sites. Their removal, which can be accomplished by raising the soil pH, results in the release of these exchange sites. In this way they are partly responsible for the "pH-dependent" charge of soil colloids.

In moderately acid soils adsorbed hydrogen also makes a contribution to the soil solution hydrogen. The readily exchangeable hydrogen held by the permanent charges contributes in the same manner shown for very acid soils. In addition, with the rise in pH, some hydrogen ions which have been held tenaciously through covalent bonding by the organic matter and clay are now subject to release. These are associated with the pH-dependent sites previously mentioned. Their contribution to the soil solution might be illustrated as follows.

$$\boxed{Micelle}\begin{matrix}H\\H\end{matrix} \quad + \quad Ca^{2+} \quad \rightarrow \quad \boxed{Micelle}\,Ca^{2+} \quad + \quad 2\,H^+ \tag{6.6}$$

| Bound hydrogen (not dissociated) | Soil solution calcium | | Exchangeable calcium | Soil solution hydrogen |

Again, the colloidal control of soil solution pH has been demonstrated, as has the dominant rôle of the calcium and aluminum ions.

Neutral to Alkaline Soils. Soils that are neutral to alkaline in reaction are no longer dominated by either hydrogen or aluminum ions. The permanent charge exchange sites are now occupied primarily by exchangeable bases, both the hydrogen and aluminum hydroxy ions having been largely replaced. The aluminum hydroxy ions have been converted to gibbsite by reactions such as

$$Al(OH)_2^+ + OH^- \rightarrow Al(OH)_3 \tag{6.7}$$

Insoluble gibbsite

More of the pH-dependent charges have become available for cation exchange, and the hydrogen released therefrom moves into the soil solution and reacts with OH^- ions to form water. The place of hydrogen on the exchange complex is taken by calcium, magnesium, and other bases. The reaction is the same as that shown for the moderately acid soils (reaction 6.6).

Soil pH and Cation Associations. Figure 6.2 presents the distribution of ions in a hypothetical soil as affected by pH. Study it carefully, keeping in mind that for any particular soil the distribution of ions might be quite different.

The effect of pH on the distribution of bases and of hydrogen and aluminum in a muck and in a soil dominated by 2:1 clays is shown in Figure 6.3. Note that permanent charges dominate the exchange complex of the mineral soil, whereas the pH-dependent charges account for most of the adsorption in the muck soil. Kaolinite and related clays have a distribution intermediate between that of the two soils shown.

In Figure 6.3 two forms of hydrogen are shown. That tightly held by the pH-dependent sites (covalent bonding) is termed *bound* hydrogen. The hydrogen ions associated with permanent electrostatic charges is exchangeable. Note

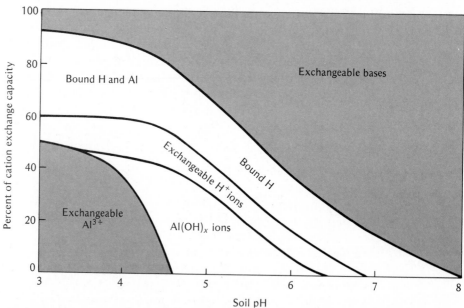

FIGURE 6.2 General relationship between soil pH and the cations held by soil colloids. Under very acid conditions exchangeable aluminum ions and bound H and Al dominate. At higher pH values the exchangeable bases predominate, while at intermediate values aluminum hydroxy ions such as $Al(OH)^{2+}$ and $Al(OH)_2^+$ are prominent. This diagram is for average conditions. Any particular soil would likely give a modified distribution.

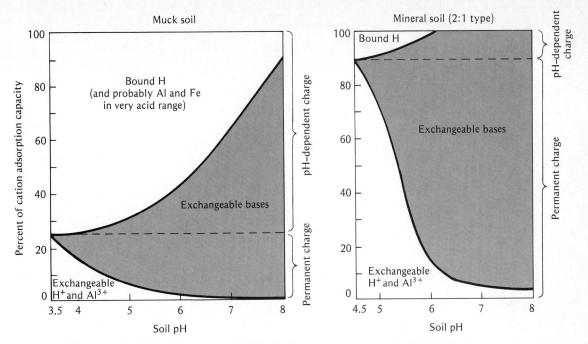

FIGURE 6.3. Relationship between soil pH and the proportion of the adsorptive complex satisfied by bases and by hydrogen and aluminum. The mineral soil (right) is of the White Store series in North Carolina. Its properties tend to be dominated by 2:1-type clays. The muck soil (left) is also from North Carolina. Note the dominance of the permanent charge in the mineral soil and the large pH-dependent charge in the muck. Soils dominated by 1:1-type colloids have distributions intermediate between these two extremes. [*Redrawn from Mehlich* (*1964*).]

that aluminum also exists in both the bound and exchangeable forms (Figure 6.1).

It is obvious that the factors responsible for soil acidity are far from simple. At the same time, there are two dominant groups of elements in control. Aluminum and hydrogen generate acidity, and most of the other cations combat it. This simple statement is worth remembering.

Source of Hydroxyl Ions. If adsorbed hydrogen and aluminum are replaced from acid soils by cations such as calcium, magnesium, and potassium, the hydrogen ion concentration in the soil solution will decrease. The concentration of hydroxyl ions will simultaneously increase since there is an inverse relationship between the hydrogen and hydroxyl ions. Thus, the "base-forming" cations become sources of hydroxyl ions merely by replacing the adsorbed hydrogen.

The metallic cations such as calcium, magnesium, and potassium also have a more direct effect on the hydroxyl ion concentration of the soil solution. A definite alkaline reaction results from the hydrolysis of colloids saturated with these cations, for example,

$$Ca \boxed{Micelle} + 2\,H_2O \rightleftharpoons \begin{smallmatrix}H\\H\end{smallmatrix}\boxed{Micelle} + Ca^{2+} + 2\,OH^- \qquad (6.8)$$

Soil solid   Soil solution   Soil solid   Soil solution

In a calcium-saturated soil, the tendency for the metallic cations to encourage hydroxyl ion formation is the factor controlling soil pH. In a soil containing hydrogen and aluminum as well as calcium ions the same tendency is there, but its effect is not so obvious because it is countered by the effect of the adsorbed hydrogen and aluminum ions. Under natural conditions the reactions to furnish hydrogen and hydroxyl ions to the soil solution occur simultaneously; that is, hydrogen and aluminum ions and the basic cations are held at one time by the same micelle (see Figure 6.2). The pH of the soil solution therefore will depend upon the relative amounts of adsorbed hydrogen and aluminum compared to adsorbed metallic cations. Where the effect of the hydrogen and aluminum is dominant, acidity results. Excess bases yield alkalinity, whereas at just the right balance the pH of the soil solution will be 7 (see Figure 1.13).

6.2 Colloidal Control of Soil Reaction

Percentage Base Saturation. The relative proportions of the adsorbed hydrogen and aluminum and the exchangeable bases of a colloidal complex are shown by the *percentage base saturation* (see Section 5.14). Obviously, a low percentage base saturation means acidity, whereas a percentage base saturation approaching 100 will result in neutrality or alkalinity. In general, humid region soils dominated by *silicate clays* and *humus* are acid if their percentage base saturation is much below 80. When such soils have a percentage base saturation of 80 or above, they usually are neutral or alkaline. The exact pH value in any case is determined by at least three other factors in addition to the percentage base saturation: (a) the nature of the colloidal micelles, (b) the kind of adsorbed bases, and (c) the concentration of soluble salts in solution.

Nature of the Micelle. At the same percentage base saturation, different types of colloids will have different pH values. This is because the various colloidal materials differ in their ability to furnish hydrogen ions to the soil solution. For example, the organic complex contains enough strong acid exchange sites to give very low pH values when the degree of base saturation is low. Even as bases are added, the degree of hydrogen ionization at the pH-dependent sites is sufficiently rapid to give lower pH values than are found commonly among mineral soils of comparable base saturation.

In contrast, the dissociation of the adsorbed hydrogen from the iron and aluminum hydrous oxides is relatively low. Consequently, soils dominated by this type of colloid have relatively high pH values for a given percentage base saturation. The dissociation of adsorbed hydrogen from silicate clays is intermediate between that from humus and from the hydrous oxides.

The relative abilities to supply soil solution hydrogen can be seen by comparing the pH values found when the various colloids are about 50% saturated with bases. The organic colloids would have pH values of 4.5–5.0, the silicate clays 5.2–5.8, and the hydrous oxides 6.0–7.0. These figures verify the importance of type of colloid in determining the pH of a soil.

The diverse acid silicate clays—the kaolinite, smectite, and hydrous mica types—evidently supply hydrogen ions in somewhat different degrees, the kaolinite least and the smectite greatest (see Figure 6.9). Likewise, the organic colloids exhibit considerable variety among themselves. In spite of this, however, the organic group apparently has a lower pH value than any of the clays when at corresponding percentage base saturations.

Kind of Adsorbed Bases. Another factor that influences the pH of a soil is the comparative amounts of the *particular* bases present in the colloidal complex. Soils with high sodium saturation have much higher pH values than those dominated by calcium and magnesium. Thus, at a percentage base saturation of 90, the presence of calcium, magnesium, potassium, and sodium ions in the ratio 10:3:1:1 would certainly result in a lower pH than if the ratio were 4:1:1:9. In the one instance, calcium is dominant—in the other, a sodium–calcium complex is dominated by sodium.

Neutral Salts in Solution. The presence in the soil solution of neutral salts, such as the sulfates and the chlorides of sodium, potassium, calcium, and magnesium, has a tendency to increase the activity of the H^+ ion and consequently decrease the soil pH. For example, if $CaCl_2$ were added to a slightly acid soil, the following reaction would occur.

$$\begin{array}{c} H \\ H \end{array} \boxed{Micelle} + Ca^{2+} + 2\,Cl^- \; \rightleftharpoons \; Ca\,\boxed{Micelle} + 2\,H^+ + 2\,Cl^- \qquad (6.9)$$

$$\underset{\text{Soil solid}}{} \quad \underset{\text{Soil solution}}{} \qquad \underset{\text{Soil solid}}{} \quad \underset{\text{Soil solution}}{}$$

Obviously, the amount of H^+ ion in the soil solution has increased.

The addition of neutral salts to an alkaline soil will also reduce pH, but by a different mechanism. The salts depress the hydrolysis of colloids saturated with metallic cations. Thus, the presence of NaCl in a soil with high sodium saturation will tend to reverse the following reaction.

$$Na\,\boxed{Micelle} + H_2O \; \rightleftharpoons \; H\,\boxed{Micelle} + Na^+ + OH^- \qquad (6.10)$$

$$\underset{\text{Soil solid}}{} \quad \underset{\text{Soil solution}}{} \qquad \underset{\text{Soil solid}}{} \quad \underset{\text{Soil solution}}{}$$

The Na^+ ion in the added NaCl salt will drive the reaction to the left, thereby reducing the concentration of OH^- ion in the soil solution and increasing that of the H^+ ion. This is a reaction of considerable practical importance in certain alkaline soils of arid regions.

Since the reaction of the soil solution is influenced by four distinct and uncoordinated factors—percentage base saturation, nature of the micelle, the ratio of the exchangeable bases, and the presence of neutral salts—a close correlation would not be expected between percentage base saturation and pH when comparing soils at random. Yet with soils of similar origin, texture, and organic content, a rough correlation does exist.

6.3 Major Changes in Soil pH

Two major groups of factors cause large changes in soil pH: (a) those resulting in increased adsorbed hydrogen and aluminum and (b) those which increase the content of adsorbed bases.

Acid-Forming Factors. In the process of organic matter decomposition, both organic and inorganic acids are formed. The simplest and perhaps the most widely found is carbonic acid (H_2CO_3), which results from the reaction of carbon dioxide and water. The solvent action of H_2CO_3 on the mineral constituents of the soil is exemplified by its dissolution of limestone or calcium carbonate (Section 12.4). The long-time effects of this acid have been responsible for the removal of large quantities of bases by solution and leaching. Because carbonic acid is relatively weak, however, it cannot account for the low pH values found in many soils.

Inorganic acids such as H_2SO_4 and HNO_3 are potent suppliers of hydrogen ions in the soil. In fact, these acids, along with strong organic acids, are responsible for the development of moderately and strongly acid conditions (Figure 6.4). Sulfuric and nitric acids are formed not only by the organic decay processes but also from the microbial action on certain fertilizer materials such as sulfur

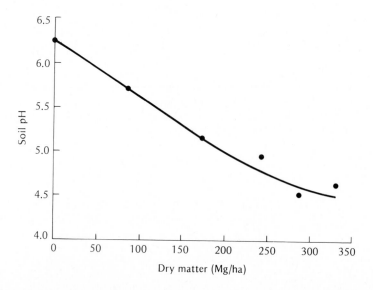

FIGURE 6.4 Effect of large applications of sewage sludge over a period of 6 years on the pH of a paleudult soil (Orangeburg fine sandy loam). The reduction was due to the organic and inorganic acids such as HNO_3 formed during decomposition of the organic matter. [*Data from Robertson et al.* (*1982*).]

and ammonium sulfate. In the latter case, both nitric and sulfuric acids are formed (see Section 18.11).

In recent years a significant source of soil acidity, especially near large cities or around large industrial complexes, has been acid rainwater. The acidity in the atmosphere is generated from oxides of nitrogen and sulfur coming mostly from the combustion of coal, gasoline, and other fossil fuels. When combined with water vapor in the atmosphere, these gaseous oxides form strong acids such as HNO_3 and H_2SO_4. The resulting rainwater commonly has a pH value of 4.0–4.5, but in extreme cases may be as low as 2.0 (see Section 20.13). Although the total quantity of hydrogen thus added to the soil is not sufficient to bring about significant pH changes at once, over a period of time the addition has a decided acidifying effect.

Leaching also encourages acidity. Consequently, bases that have been replaced from the colloidal complex or dissolved by percolating acids are removed in the drainage waters. This process encourages the development of acidity in an indirect way by removing those metallic cations that might compete with hydrogen and aluminum on the exchange complex. The effect of leaching on acidity of soils developed under grassland and forests is shown in Figure 6.5.

Base-Forming Factors. Any process that will encourage high levels of the exchangeable bases such as calcium, magnesium, potassium, and sodium will contribute toward a reduction in acidity and an increase in alkalinity. Of great significance are the weathering processes, which release these exchangeable cations from minerals and make them available for absorption. The addition of limestone is a common procedure for augmenting nature's supply of metallic cations. Irrigation waters also frequently contain various kinds of salts whose

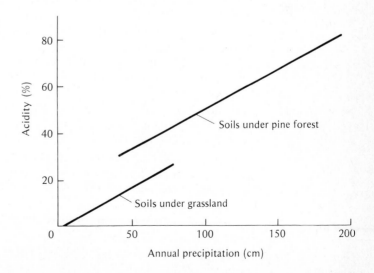

FIGURE 6.5 Effect of annual precipitation on the percent acidity of untilled California soils under grassland and pine forests. Note that the degree of acidity goes up as the precipitation increases. Also note that the forest produced a higher degree of soil acidity than did the grassland. [*From Jenny et al.* (*1968*).]

cations are adsorbed by soil colloids, thus increasing soil alkalinity, sometimes excessively.

Conditions that permit the exchangeable bases to remain in the soil will encourage high pH values. This accounts for the relatively high pH of soils of the semiarid and arid regions. Leaching waters do not remove most of the metallic cations as they are weathered from soil minerals. Consequently, the percentage base saturation of these soils remains high. In general, this situation is favorable for crop production. Only when the pH is too high or when sodium is the dominant cation is plant growth unfavorably affected.

6.4   Minor Fluctuations in Soil pH

Not only do soil solutions suffer major changes in hydrogen ion concentration but they also exhibit minor fluctuations. For instance, the drying of soils, especially above field temperatures, will often cause a noticeable increase in acidity.

The pH of mineral soils declines during the summer, especially if under cultivation, as a result of the acids produced by microorganisms. The activities of the roots of higher plants, particularly with regard to acidic exudates, may also be a factor. In winter and spring an increase in pH often is noted because biotic activities during this time are considerably slower.

6.5   Hydrogen Ion Variability in the Soil Solution

In considering the hydrogen concentration of the soil solution, it must not be inferred that this is an ordinary homogeneous solution. Differences in pH are noted from one portion of soil to that only a few inches away. Such variation results from local microbial action and the uneven distribution of organic residues in the soil.

The variability of the soil solution is important in many respects. For example, it affords microorganisms and plant roots a great variety of solution environments. Organisms unfavorably influenced by a given hydrogen ion concentration may find, at an infinitesimal distance away, another that is more satisfactory. This may account in part for the many different microbial species present in normal soils.

Even at a given location in the soil—in fact, around a given micelle—there are marked differences in the distribution of hydrogen and aluminum ions (Figure 6.6). These cations are concentrated near the surfaces of the colloid and become less numerous as the distance from the micelle is increased. They are least concentrated in the soil solution. Furthermore, equilibrium conditions exist between the adsorbed and soil solution ions, permitting the ready movement from one form to another. This fact is of great practical importance since

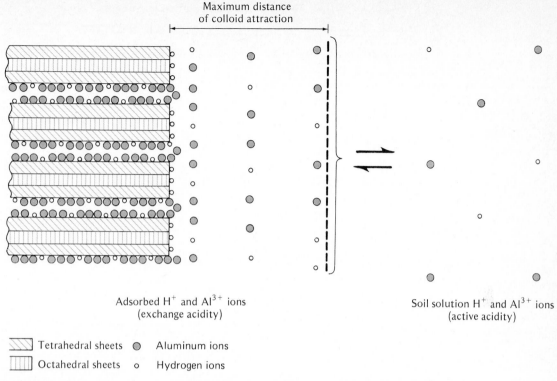

Maximum distance
of colloid attraction

Adsorbed H^+ and Al^{3+} ions
(exchange acidity)

Soil solution H^+ and Al^{3+} ions
(active acidity)

Tetrahedral sheets   ● Aluminum ions
Octahedral sheets   ○ Hydrogen ions

FIGURE 6.6 Equilibrium relationship between exchange (reserve) and soil solution
(active) acidity on 2:1-type colloid. Note that the adsorbed ions are much more numerous
than those in the soil solution even when only a small portion of the clay crystal is
shown. Remember that the aluminum ions, by hydrolysis, also supply hydrogen ions
to the soil solution [see (6.2), p. 192]. It is obvious that neutralizing only the hydrogen
and aluminum ions in the soil solution will be of little consequence. They will be
quickly replaced by ions adsorbed by the colloid. This means high buffering capacity.

it provides the basis for the soil's buffering capacity or its resistance to change
in pH (Section 6.7). The mechanisms by which adsorbed aluminum and hydrogen ions supply H^+ ions to the soil solution have been discussed (see Section
6.1).

6.6 Active Versus Exchange Acidity

Figure 6.6 illustrates the presence in acid soils of two kinds of acidity: (a)
active acidity due to the hydrogen ion concentration of the soil solution and
(b) *exchange acidity*, which refers to those hydrogen and aluminum ions adsorbed on the soil colloids. This situation is shown graphically by Figure 6.6
and by the following equation.

$$\text{Adsorbed } H^+ \text{ (and } Al^{3+}) \text{ ions} \rightleftharpoons \text{Soil solution } H^+ \text{ (and } Al^{3+}) \text{ ions} \qquad (6.11)$$

(Exchange acidity)   (Active acidity)

Since the adsorbed hydrogen and aluminum move into the soil solution when its acidity is reduced, the exchange acidity is sometimes referred to as "reserve" acidity, which suggests the role it plays. Distinct as the active and exchange acidity apparently are, however, they exist in an equilibrium relationship in the soil.

Amounts of Active and Exchange Acidity. Even though the active acidity is what determines the pH of the soil solution at any given time, the quantity of H^+ ions contributing to active acidity is very small compared to that in the exchange acidity. For example, only about 0.02 kg of calcium carbonate would be required to neutralize the active acidity in a hectare–furrow slice of an average mineral soil at pH 6 and 20% moisture. If the soil were at pH 4.0, about 2 kg would be required. Thus, the *active* acidity is extremely small even at its maximum.

Since limestone at the rate of 2, 4, or even 8 metric tons per hectare is often recommended, this neutralizing agent is obviously applied in amounts thousands of times in excess of the active acidity. The reason for such heavy applications is the magnitude of the exchange acidity. These adsorbed hydrogen and aluminum ions move into the soil solution when the hydrogen ion concentration becomes depleted. The reserve acidity thus must be depleted before the pH of the soil solution can be changed appreciably.

Conservative calculations indicate that the exchange acidity may be perhaps 1000 times greater than the active acidity in a sandy soil, and 50,000 or even 100,000 times greater for a clayey soil high in organic matter. The figure for a peat soil is even greater. The practical significance of this tremendous difference in magnitude of the active and reserve acidities of soils becomes apparent in the next section.

6.7   Buffering of Soils

As previously indicated, there is a distinct resistance to a change in the pH of the soil solution. This resistance, called *buffering,* can be explained very simply in terms of the equilibrium that exists between the active and exchange acidities (Figure 6.7). Removal of hydrogen ions from the soil solution results in their being largely replenished from the exchange acidity. The resistance to change in hydrogen ion concentration (pH) of the soil solution is thus established (see equation 6.11).

If just enough liming material was added to neutralize the hydrogen ions in the soil solution, equation 6.11 would immediately be shifted to the right, resulting in more hydrogen ions moving out into the soil solution. Consequently,

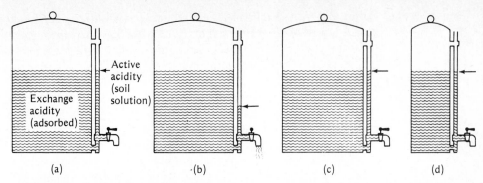

(a)   ·(b)   (c)   (d)

FIGURE 6.7 The buffering action of a soil can be likened to that of a coffee dispenser. (a) The active acidity, which is represented by the coffee in the indicator tube on the outside of the urn, is small in quantity. (b) When hydrogen ions are removed, this active acidity falls rapidly. (c) The active acidity is quickly restored to near the original level by movement from the exchange or adsorbed acidity. By this process there is considerable resistance to the change of active acidity. A second soil with the same active acidity (pH) level but a much smaller exchange acidity (d) would have a lower buffering capacity.

the resulting pH rise would be negligibly small and remain so until enough lime had been added to deplete appreciably the reserve acidity.

This resistance to pH change is equally important in preventing a rapid lowering of the pH of soils. When hydrogen ions are added to a soil or they result from certain biochemical changes, a temporary increase in the hydrogen ions in the soil solution occurs. In this case the equilibrium reaction above would immediately shift to the left and more hydrogen ions would become adsorbed on the micelle. Again, the resultant pH change, this time a lowering, in the soil solution would be very small.

These two examples indicate clearly the principles involved in buffering. In addition, they show that the basis of buffer capacity lies in the adsorbed cations of the complex. Hydrogen and aluminum ions, together with the adsorbed metallic cations, not only indirectly control the pH of the soil solution but also determine the quantity of lime or acidic constituents necessary to bring about a given pH change.

6.8 Buffer Capacity of Soils and Related Phases

The higher the exchange capacity of a soil, the greater will be its buffer capacity, other factors being equal. This is because more reserve acidity must be neutralized to effect a given rise or lowering of the percentage base saturation. In practice this is fully recognized in that the heavier the texture of a soil and the higher its organic content, the larger must be the application of lime to force a given change in pH.

Buffer Curves.   Another important phase of buffering logically presents itself at this point. Is the buffer capacity of soils the same throughout the percentage base saturation range? This is best answered by reference to an average theoretical titration curve presented by Peech (1941) for a large number of Florida soils (Figure 6.8).

Three things are clearly obvious from the curve. First, there is a correlation between the percentage base saturation of these soils and their pH, as previously suggested (Section 5.14). Second, the generalized curve indicates that the degree of buffering varies, being lowest at the extreme base saturation values. Between these extremes where the curve is flatter, the buffering reaches a maximum. Theoretically, the greatest buffering occurs at about 50% base saturation, a situation extremely important both technically and practically. Third, the buffering, as indicated by the curve, is uniform over the pH range 4.5–6.5. This is vitally significant, indicating that under field conditions approximately the same amount of lime will be required to change the soil pH from 5.0 to 5.5 as from 5.5 to 6.0.

Before leaving this phase, note that the curve of Figure 6.8 is a composite and represents the situation in respect to any particular soil only in a general way. Titration curves for individual soils will deviate widely from this general-

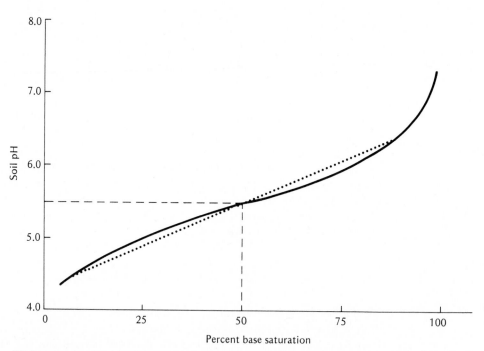

FIGURE 6.8   Theoretical titration curve based on a large number of Florida soils. The dotted line indicates the zone of greatest buffering. The maximum buffering should occur at approximately 50% base saturation. [*From Peech (1941).*]

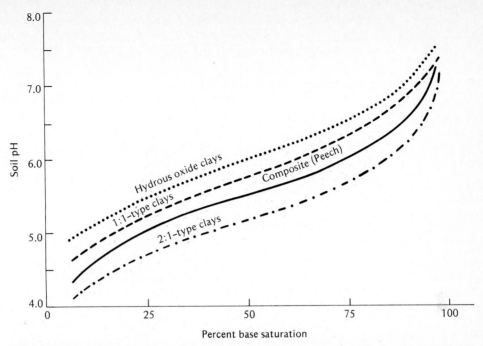

FIGURE 6.9 Theoretical titration curves of different mineral colloids compared with that of the composite of soils used by Peech (Figure 6.8). The hydrous oxide clays generally exhibit the highest pH values at a given percent base saturation, the 2:1-type clays the lowest, and the 1:1 type in between. Keep in mind that these are only comparative curves and would not represent any given soil.

ized curve since the colloidal complex of different soils varies with the kinds and amounts of clay and humus present.

Variation in Titration Curves.   Significant differences are found in the relationship of pH to percent base solution of different mineral colloids. The theoretical titration curves shown in Figure 6.9 illustrate this point. At a given percent base saturation the pH is normally highest in hydrous oxide clay materials, intermediate in kaolinitic materials and lowest in 2:1-type minerals. Obviously, the type of clay definitely affects the pH–base saturation percentage relationship.

6.9   Importance of Buffering

Stabilization of Soil pH.   A marked change in pH affects the availability of several plant nutrients and the levels of other elements that may be toxic to higher plants and microorganisms. Thus, higher plants and microorganisms may be affected by rapid changes in pH either directly due to the change in hydrogen ion concentration, or indirectly due to nutrient deficiency or chemical toxicities (Figures 6.10 and 6.11). The stabilization of soil pH through buffering seems to be an effective guard against these difficulties.

Quantities of Amendments Required. Obviously, the greater the buffering capacity of a soil, the larger the amounts of lime or sulfur necessary to effect a given change in pH. Therefore, in deciding the amount of lime to apply to a soil of known pH, texture and organic content are among the important soil factors to be considered (see Section 17.11). These properties give a rough idea of the adsorptive capacity of a soil and hence of its buffering. Chemical tests are available to measure the buffering capacity directly.

6.10 Soil-Reaction Correlations

Exchangeable Calcium, Magnesium, and Sodium. As the exchangeable calcium and magnesium are lost by leaching, the acidity of the soil gradually increases. Consequently, in humid region soils there is a fairly definite correlation between the pH and the amounts of these two constituents present in exchangeable form (Figure 6.11). In arid regions, the same general relationships hold except under conditions where a substantial amount of sodium is adsorbed (see Section 6.16).

Aluminum and the Micronutrients. The relationships between soil reaction and the activity of aluminum, iron, and manganese are shown in Figure 6.10. When the pH of a mineral soil is low, appreciable amounts of these three constituents are soluble, so much that they may become extremely toxic to certain plants (Figure 6.11). However, as the pH is increased, precipitation takes place and the amounts of these ions in solution become less and less until at neutrality or thereabout certain plants may suffer from a lack of available manganese and iron. This is especially true if a decidedly acid sandy soil is suddenly brought to a neutral or alkaline condition by an excessive application of lime.

While deficiencies of manganese and iron are not widespread, they do occur in certain areas, particularly on overlimed sandy soils or alkaline arid region soils. If the soil reaction is held within a soil pH range of 6.0–7.0, the toxicity of the aluminum, iron, and manganese, and the deficiency of iron and manganese may be avoided. Copper and zinc are affected in the same way by a rise in pH, the critical point being between pH 6 and 7; above 7 their availability definitely declines (Figure 6.10).

With boron, the situation is somewhat different and more complicated. Although neither the soil untreated nor the lime alone appreciably precipitate boron, the two in combination render it unavailable. Also, the excess of calcium, in spite of its solubility, may hinder in some way the movement of boron into the plant. Too much calcium in the plant cells might even interfere with

FIGURE 6.10 Relationships existing in mineral soils between pH on the one hand and the activity of microorganisms and the availability of plant nutrients on the other. The wide portions of the bands indicate the zones of greatest microbial activity and the most ready availability of nutrients. Considering the correlations as a whole, a pH range of ~6–7 seems to promote the most ready availability of plant nutrients. In short, if soil pH is suitably adjusted for phosphorus, the other plant nutrients, if present in adequate amounts, will be satisfactorily available in most cases.

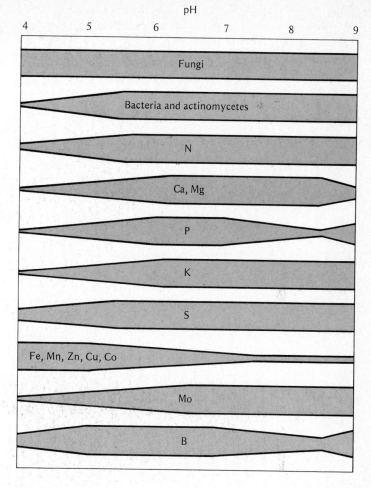

boron metabolism, even if plenty of the latter were present. It has also been suggested that lime may create serious competition for boron by stimulation of soil microorganism activity.

Molybdenum availability is significantly dependent on pH. In strongly acid soils it is quite unavailable. As the pH is raised to 6 and above, its availability increases. The correlation between molybdenum availability and pH is so strong that some researchers believe the main reason for liming is to increase the molybdenum supply.

Available Phosphorus. The kind of phosphate ion present in the soil solution varies with the soil pH. When the soil is neutral to slightly alkaline, the HPO_4^- ion apparently is the most common form. As the pH is lowered and the soil becomes slightly to moderately acid, both the HPO_4^{2-} ion and the $H_2PO_4^-$

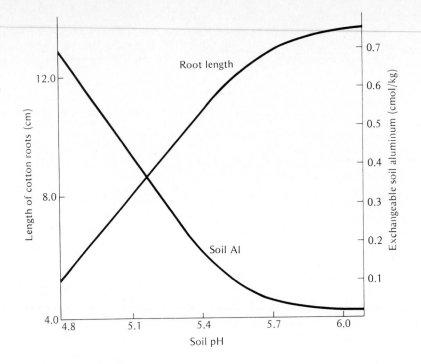

FIGURE 6.11 High concentrations of exchangeable aluminum in acid soils are toxic to cotton roots, thereby restricting their length. [*Data from Adams and Lund* (1966).]

ion prevail. At higher acidities $H_2PO_4^-$ ions tend to dominate. Because of the formation of insoluble compounds, a soil reaction which yields a mixture of HPO_4^- and $H_2PO_4^{2-}$ ions is usually preferred.

The activity of the soil phosphorus is indirectly related to pH in another way. It has already been explained that as soil acidity increases, there is a rise in the activity of the iron, aluminum, and manganese. Under such conditions soluble phosphates are markedly "fixed" as very complex and insoluble compounds of these elements. This fixation is most serious when the soil pH is below 5.0. The situation is shown graphically in Figure 10.5.

If the pH of a mineral soil is raised much above 7, the phosphate nutrition of higher plants is disturbed in other ways. In the first place, at these high pH values, complex insoluble calcium phosphates are formed. Thus, the solubility of both the native and applied phosphorus may be very seriously impaired. Furthermore, at pH values above 7, the excess calcium may hinder phosphorus absorption and utilization of plants.

Between pH 6 and 7, phosphorus fixation is at a *minimum* and availability to higher plants is at a maximum (Figure 6.10). In the regulation of the phosphorus nutrition of crops, it is therefore important that soil pH be kept within, or very near, the conservative limits of 6–7. Even then, higher plants often do not readily absorb one half or even one third of the available phosphorus supplied in fertilizers. (For further consideration of phosphorus fixation and availability in soils, see Chapter 10).

Soil Organisms and pH.   Soil organisms are influenced by fluctuations in the pH of the soil solution. This may be due in extreme cases to the hydrogen ion itself, but in most soils it must be attributed to the factors correlated with soil pH.

Bacteria and actinomycetes generally function better in mineral soils at intermediate and higher pH values, the activity being sharply curtailed when the pH drops below 5.5. Fungi, however, are particularly versatile, flourishing satisfactorily at a wide range of pH. In normal soils, therefore, fungi predominate at the lower pH values, but at intermediate and higher ranges they meet strong competition from the bacteria and actinomycetes.

The oxidation and "fixation" of nitrogen take place vigorously in mineral soils only at pH values well above 5.5, although such reactions occur in organic soils at lower pH values. However, general organic matter degradation, although curtailed, will still proceed with considerable intensity at lower pH values because most fungi are able to effect these enzymic transfers at high acidities. This is fortunate since higher plants growing on very acid soils are provided with at least ammoniacal nitrogen from decaying organic matter.

All in all, a soil in the intermediate pH range presents the most satisfactory biological regime. Nutrient conditions are favorable without being extreme, and phosphorus availability is at a maximum.

One very significant exception to the generalized correlation of bacteria with soil reaction should be mentioned. The organisms that oxidize sulfur to sulfuric acid (Section 9.27) seem to be markedly versatile. Apparently they not only function vigorously in soils at medium to high pH values but under decidedly acid conditions as well. This is extremely important since it is therefore possible to apply sulfur to soils and develop highly acid conditions through the activity of these bacteria. If these organisms were at all sensitive to low pH and its correlated factors, their activity would soon be retarded and finally brought to a halt by their own acidic products. Under such conditions sulfur would be relatively ineffective as a soil acidifier.

6.11   Response of Higher Plants to Soil Reaction

Because of the many physiological factors involved, it is often difficult to correlate the optimum growth of plants on mineral soils with their pH. On the other hand, the general relationships of higher plants to strong acidity or alkalinity can be established and in a practical way are just as significant. With this relationship as a basis, some of the important plants are rated in Figure 6.12.

Crop Ratings.   Legume crops such as alfalfa and sweet clover grow best in near neutral or alkaline soil reaction. Except under especially favorable conditions, humid region mineral soils must be limed in order to grow crops of this group satisfactorily.

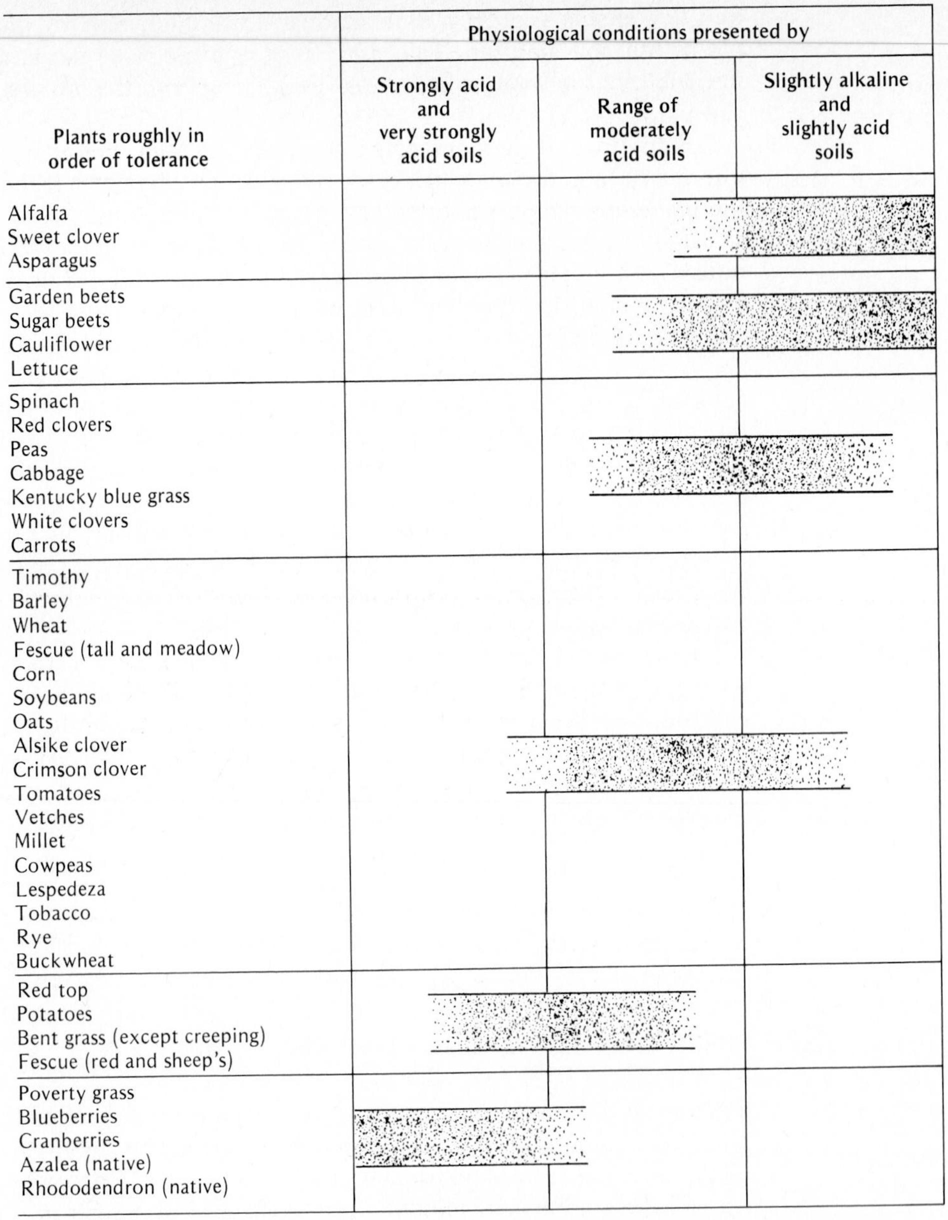

Plants roughly in order of tolerance	Physiological conditions presented by		
	Strongly acid and very strongly acid soils	Range of moderately acid soils	Slightly alkaline and slightly acid soils
Alfalfa Sweet clover Asparagus			▓▓▓
Garden beets Sugar beets Cauliflower Lettuce			▓▓▓
Spinach Red clovers Peas Cabbage Kentucky blue grass White clovers Carrots		▓▓▓	▓
Timothy Barley Wheat Fescue (tall and meadow) Corn Soybeans Oats Alsike clover Crimson clover Tomatoes Vetches Millet Cowpeas Lespedeza Tobacco Rye Buckwheat	▓	▓▓▓	
Red top Potatoes Bent grass (except creeping) Fescue (red and sheep's)	▓▓▓		
Poverty grass Blueberries Cranberries Azalea (native) Rhododendron (native)	▓▓▓		

FIGURE 6.12 Relation of higher plants to the physiological conditions presented by mineral soils of different reactions. Note that the correlations are very broad and are based on pH ranges. The fertility level will have much to do with the actual relationship in any specific case. Such a chart is of great value in deciding whether or not to add lime and the rate of application, if any.

Native rhododendrons and azaleas are at the other end of the scale. They apparently require a considerable amount of iron, which is abundantly available only at low pH values. Undoubtedly, highly acid soils also present other conditions physiologically favorable for this type of plant. Also, soluble aluminum apparently is not detrimental to these plants as it is to certain plants higher up the scale. If the pH and the percentage base saturation are not low enough, these plants will show chlorosis and other symptoms indicative of an unsatisfactory nutritive condition.

Because arable soils in a humid region are usually somewhat acid, it is fortunate that most cultivated crop plants not only grow well on moderately to slightly acid soils but seem to prefer the physiological conditions therein (Figure 6.12). Since pasture grasses, many legumes, small grains, intertilled field crops, and a large number of vegetables are included in this broadly tolerant group, soil acidity is not such a deterrent to their growth. In terms of pH, a range from 5.8–6.0 to slightly above 7.0 is most suitable for this group. (The significance of this as it relates to liming is presented in Section 17.11.) Forest trees seem to grow well over a wide range of soil pH values. Most show at least some tolerance of acid soil. Many species, particularly the conifers, tend to intensify soil acidity. Forests exist as natural vegetation in regions of acid soils. This is not a direct response to soil pH but rather to the climatic environment, which incidentally encourages the development of acid soils. Also, there are some isolated areas of high soil pH in humid regions where the natural vegetation is grass rather than trees. The "black belt" soil areas of Alabama and Mississippi are examples. These areas are surrounded by acid soils on which forests are the natural vegetation. This indicates that the trees are better competitors on the more acid soil areas.

6.12 Determination of Soil pH

The importance of pH measurements as a tool in liming and similar problems is obvious. In fact, pH is a diagnostic figure of unique value and as a result its determination has become one of the routine tests made on soils. Moreover, its determination is easy and rapid.

Electrometric Method. The most accurate method of determining soil pH is by a pH meter. In this electrometric method the hydrogen concentration of the soil solution is balanced against a standard hydrogen electrode or a similarly functioning electrode. Although the instrument gives very consistent results, its optimum use requires a skilled operator since the mechanism is rather complicated.

Dye Methods. A second method, very simple and easy but somewhat less accurate than the electrometric, consists in the use of certain indicators (Figure 6.13). Many dyes change color with an increase or decrease of pH, making it

FIGURE 6.13 The indicator method for determining pH is widely used in the field. It is simple and is accurate enough for most purposes. [*Courtesy New York State College of Agriculture and Life Sciences, Cornell University.*]

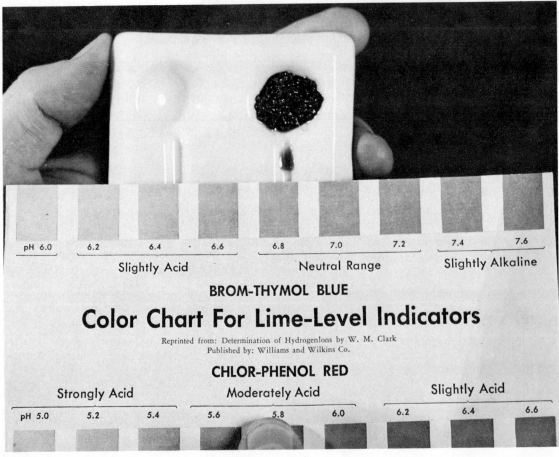

| pH 6.0 | 6.2 | 6.4 | 6.6 | 6.8 | 7.0 | 7.2 | 7.4 | 7.6 |

Slightly Acid   Neutral Range   Slightly Alkaline

BROM-THYMOL BLUE
Color Chart For Lime-Level Indicators

Reprinted from: Determination of HydrogenIons by W. M. Clark
Published by: Williams and Wilkins Co.

CHLOR-PHENOL RED

Strongly Acid   Moderately Acid   Slightly Acid

| pH 5.0 | 5.2 | 5.4 | 5.6 | 5.8 | 6.0 | 6.2 | 6.4 | 6.6 |

possible, within the range of the indicator, to estimate the approximate hydrogen ion concentration of a solution. By using a number of dyes, either separately or mixed, a pH range of 3–8 is easily covered. In making such a pH determination on soil, the sample is saturated with the dye, and after standing in contact a few minutes a drop of the liquid is run out and its color observed in thin layer. By the use of a suitable color chart the approximate pH may be ascertained. When properly manipulated, the indicator method is accurate within about 0.2 pH unit.

Limitations of pH Values. Because of the precision with which pH readings can be duplicated electrometrically, interpretations may be made which the pH measurement cannot justify. There are several reasons for this. First, there is considerable variation in the pH from one spot in a given field to another. Even at a given location there are seasonal variations in pH. Localized effects of fertilizers may give sizable pH variations within the space of a few inches. Last, the pH of a given soil sample will vary depending on the amount of water used in wetting the soil prior to the measurement. Obviously, standardization against field performance must be obtained.

Because of these limitations, it might seem peculiar that so much reliance is placed on soil pH. In the first place, it is easily and quickly determined. More important, however, is its susceptibility to certain broad correlations that are of great practical significance (Figures 6.10 and 6.12). Thus, a great deal may be inferred regarding the physiological condition of a soil from its pH value, much more than from any other single analytical datum. Furthermore, the variations in pH value from one local soil area to another in a given field indicate that great accuracy in estimating soil acidity may not be so necessary from a practical point of view.

6.13 Soil Acidity Problems[3]

Other than the maintenance of fertility in general, two distinct procedures are often necessary on acid soils, especially those at intermediate pH values. One is the intensification of the acidity to encourage such plants as azaleas and rhododendrons. The other is the application of lime, usually in amounts to raise the pH at least to 6.0, 6.5, or 7.0. This modifies the physiological conditions, thereby favoring alfalfa, sweet clover, red clover, and other lime-loving crops.

Since liming is such an important agricultural feature, its consideration will be reserved for a later and fuller discussion (Chapter 17). However, the methods of intensifying the acidity of the soil are briefly discussed next.

---

[3] For an extensive review of soil acidity and liming, see Adams (1984).

Methods of Intensifying Soil Acidity

A reduction of the pH of soils is often desirable to favor such plants as rhododendrons and azaleas as previously suggested and also to discourage certain diseases, especially the actinomycetes that produce potato scab. In arid regions treatments are sometimes made to reduce the high pH of alkali soils sufficiently to allow common field plants to grow and to eliminate deficiencies of iron, manganese, and zinc in other soils (see Section 11.5)

Acid Organic Matter.   When dealing with ornamental plants, acid-forming organic matter may be mixed with the soil already at hand to lower the pH of the latter. Leafmold, pine needles, tanbark, sawdust, and moss peat, if highly acid, are quite satisfactory in preparing such a compost. Farm manure, however, may be alkaline and consequently should be used with caution for this purpose.

Use of Chemicals.   When the above methods are not feasible, chemicals may be used. For rhododendrons, azaleas, and other plants that require considerable iron, such as blueberries and cranberries, ferrous sulfate is sometimes recommended. This salt by hydrolysis develops sulfuric acid, which drastically lowers the pH and liberates some of the iron already present in the soil. At the same time, soluble and available iron is being added. Such a chemical thus serves a double purpose in effecting a change in the physiological condition of a soil.

Other materials that are even better in some respects are flowers of sulfur or, on a field scale, sulfuric acid. Sulfur usually undergoes vigorous microbial oxidation in the soil (see Section 9.27) and sulfuric acid is produced. Under favorable conditions sulfur is four or five times more effective pound for pound in developing acidity than is ferrous sulfate. Moreover, it is comparatively inexpensive and easy to obtain and is a material often used on the farm for other purposes.

No definite recommendation can be made as to the amounts of ferrous sulfate or sulfur that should be applied, since the buffering of soils and their original pH are so variable.

Control of Potato Scab.   Sulfur is also effective in the control of potato scab since the actinomycetes responsible are discouraged by acidity. Ordinarily when the pH is lowered to 5.3, their virulence is much reduced. In using sulfur to increase soil acidity, the management of the land, especially the rotation, should be such that succeeding crops are not unfavorably affected.

The amount of sulfur to apply to control potato scab will vary depending on circumstances. The buffering capacity of the soil and the original soil pH will be the determining factors. The results of a given treatment both on the pH and the crop should be checked and succeeding applications changed to conform with the influence exerted by the previous dosage.

6.15   Reaction of Soils of Arid Regions[4]

Arid region soils occur in areas where the rainfall is seldom more than 50 cm per year. Lack of extensive leaching leaves the base status of these soils high. A fully and normally developed profile usually carries at some point in its profile (usually in the C horizon) a calcium carbonate accumulation greater than that of its parent material. The lower the rainfall, the nearer the surface this layer will be (see Figure 13.12).

As a result, these soils may have alkaline subsoils and alkaline or neutral surface layers. When enough leaching has occurred to free the solum of calcium carbonate, a mild acidity may develop in the surface horizons. The genetic classification of soils of arid regions with a pertinent description of each group are presented in Section 13.7.

6.16   Reaction of Saline and Sodic Soils

When the drainage of arid region soils is impeded and the surface evaporation becomes excessive, soluble salts accumulate in the surface horizon. Such soils have been classified under three headings: saline, saline-sodic, and sodic.

Saline Soils.[5]   These soils contain a concentration of neutral soluble salts sufficient to seriously interfere with the growth of most plants. The electrical conductivity of a saturated extract (EC_e) is greater than 4 decisiemens per meter (dS/m).[6] Less than 15% of the cation exchange capacity of these soils is occupied by sodium ions and the pH usually is below 8.5. This is because the soluble salts present are mostly neutral, and because of their domination, only a small amount of exchangeable sodium is present.

Such soils are sometimes called *white alkali* soils because a surface incrustation, if present, is light in color (Figure 6.14). The excess soluble salts, which are mostly chlorides and sulfates of sodium, calcium, and magnesium, can readily be leached out of these soils with no appreciable rise in pH. This is a very important practical consideration in the management of these soils. Care must be taken to be certain that the leaching water is low in sodium.

Saline-Sodic Soils.   The saline-sodic soils contain appreciable quantities of neutral soluble salts and enough adsorbed sodium ions to seriously affect most plants. Although more than 15% of the total exchange capacity of these soils is occupied by sodium, their pH is likely to be below 8.5. This is because of

---

[4] See Bresler et al. (1982) for an excellent discussion of these soils.

[5] Salinization is the term used in reference to the natural processes that result in the accumulation of neutral soluble salts in soils.

[6] Previously expressed as millimhos per centimeter (mmho/cm). Since 1 S = 1 mho, 1 dS/m = 1 mmho/cm.

(a)

(b)

FIGURE 6.14 [*opposite*] (a) White "alkali" spot in a field of alfalfa under irrigation. Because of upward capillarity and evaporation, salts have been brought to the surface where they have accumulated. (b) White salt crust on a saline soil from Colorado. The white salts are in contrast with the darker-colored soil underneath. (Scale is shown in inches and centimeters.) [(a) *courtesy USDA Soil Conservation Service.*]

the repressive influence of the neutral soluble salts, as in the saline soils previously described. The electrical conductivity of a saturated extract is more than 4 dS/m.

But unlike the saline soils, leaching will markedly raise the pH of saline-sodic soils unless calcium or magnesium salt concentrations are high in the soil or in the irrigation water. This is because the exchangeable sodium, once the neutral soluble salts are removed, readily hydrolyzes and thereby sharply increases the hydroxyl ion concentration of the soil solution. In practice, this is detrimental since the sodium ions disperse the mineral colloids, which then develop a tight, impervious soil structure. At the same time, sodium toxicity to plants is increased.

Sodic Soils. Sodic soils do not contain any great amount of neutral soluble salts, the detrimental effects on plants being largely due to the toxicity of the sodium as well as of the hydroxyl ions. The high pH is largely due to the hydrolysis of sodium carbonate.

$$2\,Na^+ + CO_3{}^{2-} + 2\,H_2O \rightleftharpoons 2\,Na^+ + 2\,OH^- + H_2CO_3 \qquad (6.12)$$

The sodium complex also undergoes hydrolysis.

$$Na\,\boxed{Micelle} + HOH \rightleftharpoons H\,\boxed{Micelle} + Na^+ + OH^- \qquad (6.13)$$

The exchangeable sodium, which occupies decidedly more than 15% of the total exchange capacity of these soils, is free to hydrolyze because the concentration of neutral soluble salts is rather low. The electrical conductivity of a saturated extract is less than 4 dS/m. Consequently, the pH is always above 8.5, often rising to 10.0 or higher. Owing to the deflocculating influence of the sodium, such soils usually are in an unsatisfactory physical condition. As already stated, the leaching of a saline-sodic soil will readily change it to a characteristic sodic soil.

Because of the extreme alkalinity resulting from the Na_2CO_3 present, the surface of sodic soils usually is discolored by the dispersed humus carried upward by the capillary water—hence the name *black alkali* is frequently used. These soils are often located in small areas called *slick spots* surrounded by soils that are relatively productive.

Characterization of Sodic and Salt-Affected Soils. Several means not commonly employed for acid soils are used to characterize the high pH soils of arid regions. Soil salinity (an expression of the presence of soluble salts) is measured by the electrical conductivity (EC_e) of a saturated soil extract is also expressed in decisiemens per meter. It is an important characteristic of saline and saline-sodic soils.

For saline-sodic and sodic soils, parameters relating to sodium content are most important. While the level of sodium salts is of some significance, measurements of sodium associated with the exchange complex are of greater concern. The presence of sodium in an exchangeable form has a deleterious effect on the chemical and physical properties of soils and in turn on plant growth. The sodium hazard is expressed as *exchangeable sodium percentage* (ESP), which is calculated as follows.

$$ESP = \frac{\text{exchangeable sodium}}{\text{cation exchange capacity}} \times 100 \qquad (6.14)$$

Thus ESP is a measure of the degree to which the exchange complex is saturated with sodium.

The adverse influences of exchangeable sodium are moderated by levels of exchangeable calcium and magnesium. Years of study and experimentation have shown that the best measure of potential hazards from high sodium levels is a cation ratio that takes calcium and magnesium as well as sodium into consideration. This ratio, called the *sodium adsorption ratio* (SAR), is defined as

$$SAR = \frac{[Na^+]}{\sqrt{[Ca^{2+}] + [Mg^{2+}]}} \qquad (6.15)$$

Where $[Na^+]$, $[Ca^{2+}]$, and $[Mg^{2+}]$ are the concentrations of these ions in a soil extract or irrigation water in millimoles per liter. The SAR of a soil extract gives an indication of the level of exchangeable sodium in comparison with those of calcium and magnesium in the soil. The SAR of irrigation water warns of possible increases in exchangeable sodium through the use of that water.

6.17   Growth of Plants on Halomorphic Soils

Saline and saline-sodic soils with their relatively low pH (usually less than 8.5) detrimentally influence plants largely because of their high soluble-salt concentration (see Figure 6.14). When a water solution containing a relatively large amount of dissolved salts is brought into contact with a plant cell, it will cause a shrinkage of the protoplasmic lining. This action, called *plasmolysis,* increases with the concentration of the salt solution. The phenomenon is due to the osmotic movement of the water, which passes from the cell toward the more concentrated soil solution. The cell then collapses. The nature of the salt, the species, and even the individuality of the plant, as well as other factors, determine the concentration at which the individual succumbs. The adverse physical condition of the soils, especially the saline-sodic, may also be a factor.

Sodic soils, dominated by active sodium, exert a detrimental effect on plants in three ways: (a) caustic influence of the high alkalinity induced by the sodium carbonate and bicarbonate, (b) toxicity of the bicarbonate and other anions, and (c) the adverse effects of the active sodium ions on plant metabolism and nutrition.

6.18 Tolerance of Higher Plants to Halomorphic Soils

The capacity of higher plants to grow satisfactorily on salty soils depends on a number of interrelated factors. The physiological constitution of the plant, its stage of growth, and its rooting habits certainly are among them. It is interesting to note that old alfalfa is more tolerant than young alfalfa and that deep-rooted legumes show a greater resistance than those with shallow rootage.

Concerning the soil, the nature of the various salts, their proportionate amounts, their total concentration, and their distribution in the solum must be considered. The structure of the soil and its drainage and aeration are also important.

As a result, it is difficult to forecast accurately the tolerance of crops. Perhaps the best comparative data are those presented in Table 6.1.

TABLE 6.1 Relative Tolerance of Certain Plants to Salty Soils[a]

Tolerant	Moderately tolerant	Moderately sensitive	Sensitive
Barley, grain	Barley, forage	Alfalfa	Apple
Bermuda grass	Beet, garden	Broad bean	Apricot
Bougainvillea	Broccoli	Cauliflower	Bean
Cotton	Brome grass	Cabbage	Blackberry
Date	Clover, berseem	Clover, alsike, Ladino, red, strawberry	Carrot
Natal plum	Fig	Corn	Celery
Mutall alkali grass	Orchard grass	Cowpea	Grapefruit
Rescue grass	Oats	Cucumber	Lemon
Rosemary	Rye, hay	Lettuce	Onion
Sugar beat	Rye grass, perennial	Pea	Orange
Salt grass	Sorghum	Peanut	Peach
Wheat grass, crested	Soybean	Potato	Pear
Wheat grass, fairway	Sudan grass	Rice, paddy	Pineapple; Guava
Wheat grass, tall	Trefoil, birdsfoot	Sweet clover	Raspberry
Wild rye, altai	Wheat	Timothy	Rose
Wild rye, Russian	Wheat grass, western	Tomato	Strawberry

[a] Selected from Carter (1981).

6.19 Management of Saline and Sodic Soils

Since saline and sodic soils are found mostly in arid regions, their use for agricultural purposes requires irrigation water. The quality of that water, especially in relation to its salt content and to its SAR, is extremely important in the management of the soils. Water high in sodium salts can bring about harmful effects unless these salts are counterbalanced by soluble calcium and magnesium salts. Knowledge of the quality of irrigation water is a requisite for good management of saline and sodic soils.

Three kinds of general management practices have been used to maintain or improve the productivity of saline and sodic soils. The first is *eradication;* the second is a *conversion* of some of the salts to less injurious forms; the third may be designated *control.* In the first two methods, an attempt is made actually to eliminate by various means some of the salts or to render them less toxic. In the third, soil management procedures are utilized which keep the salts so well distributed throughout the soil solum that there is no toxic concentration within the root zone.

Eradication. The most common methods used to free the soil of excess salts are (a) underdrainage and (b) leaching or flushing. A combination of the two, flooding after tile drains have been installed, is the most thorough and satisfactory. When this method is used in irrigated regions, heavy and repeated applications of water can be made. The salts that become soluble are leached from the solum and drained off through the tile. The irrigation water used must be relatively free of silt and salts, especially those containing sodium.

The leaching method works especially well with pervious saline soils, whose soluble salts are largely neutral and high in calcium and magnesium. Of course, little exchangeable sodium should be present. Leaching saline-sodic soils (and even sodic soils if the water will percolate) with waters very high in salt but low in sodium may be effective. Conversely, treating sodic and saline-sodic soils with water low in salt may intensify their alkalinity because of the removal of the neutral soluble salts. This allows an increase in the percent sodium saturation thereby increasing the concentration of hydroxyl ions in the soil solution. This may be avoided, as explained next, by converting the toxic sodium carbonate and bicarbonate to sodium sulfate by first treating the soil with heavy applications of gypsum or sulfur. Leaching will then render the soil more satisfactory for crops.

Conversion. The use of gypsum on sodic soils is often recommended for the purpose of changing part of the caustic alkali carbonates into sulfates. Several tons of gypsum per acre are usually necessary. The soil must be kept moist to hasten the reaction, and the gypsum should be cultivated into the surface, not plowed under. The treatment may be supplemented later by a thorough leaching of the soil with irrigation water to free it of some of its sodium sulfate. The gypsum reacts with both the Na_2CO_3 and the adsorbed sodium.

$$Na_2CO_3 + CaSO_4 \rightleftharpoons CaCO_3 + \underset{\text{Leachable}}{Na_2SO_4}$$

$$\begin{array}{c} Na \\ Na \end{array}\boxed{Micelle} + CaSO_4 \rightleftharpoons Ca\boxed{Micelle} + \underset{\text{Leachable}}{Na_2SO_4} \qquad (6.16)$$

It is also recognized that sulfur can be used to advantage on salty lands, especially where sodium carbonate abounds. The sulfur upon oxidation yields sulfuric acid, which not only changes the sodium carbonate to the less harmful sulfate but also tends to reduce the intense alkalinity. The reactions of the sulfuric acid with the compounds containing sodium may be shown as follows.

$$Na_2CO_3 + H_2SO_4 \rightleftharpoons CO_2 + H_2O + \underset{\text{Leachable}}{Na_2SO_4}$$

$$\begin{array}{c} Na \\ Na \end{array}\boxed{Micelle} + H_2SO_4 \rightleftharpoons \begin{array}{c} H \\ H \end{array}\boxed{Micelle} + \underset{\text{Leachable}}{Na_2SO_4} \qquad (6.17)$$

Not only is the sodium carbonate changed to sodium sulfate, a mild neutral salt, but the carbonate radical is entirely eliminated. When gypsum is used, however, the carbonate remains as a calcium salt.

Control. The retardation of evaporation is an important feature of the control of salty soils. This will not only save moisture but will also retard the trans-location upward of soluble salts into the root zone. As previously indicated, there are no inexpensive methods of reducing evaporation from large acreages. Consequently, other control practices must be explored.

Where irrigation is practiced, an excess of water should be avoided unless it is needed to free the soil of soluble salts. Frequent light irrigations are often necessary, however, to keep the salts sufficiently dilute to allow normal plant growth.

The timing of irrigation is extremely important on salty soils, particularly during the spring planting season. Since young seedlings are especially sensitive to salts, irrigation often precedes or follows planting to move the salts down-ward. After the plants are well established, their salt tolerance is somewhat greater.

The practice in recent years of adding nitrogen as anhydrous ammonia to irrigation water as it is applied to a field has created soil problems. The high pH brought about by the NH_3 causes some of the calcium in the water to precipitate and thus raises the sodium adsorption ratio (SAR) and the hazard of increased exchangeable sodium percentage. To counteract this, sulfuric acid is sometimes applied to the irrigation water to reduce its pH as well as that of the soil. This practice may well spread where there are economical sources of sulfuric acid.

The use of salt-resistant crops is another important feature of the successful management of saline and alkali lands. Sugar beets, cotton, sorghum, barley, rye, sweet clover, and alfalfa are particularly advisable (Table 6.1). Moreover, a temporary alleviation of alkali will allow less-resistant crops to be established. Farm manure is very useful in such an attempt. A crop such as alfalfa, once it is growing vigorously, may maintain itself in spite of the salt concentrations that may develop later. The root action of tolerant plants is exceptionally helpful in improving sodic soils that are in a poor physical condition.

6.20 Conclusion

Although the discussion closes with pertinent suggestions regarding the management of saline and sodic soils, the major theme of this chapter is soil reaction or soil pH. And it is obvious that as many features of practical concern arise from soil reaction as from any other single soil characteristic. Material in Chapters 17 and 18 emphasizes this point even more decisively.

REFERENCES

Adams, Fred, Ed. *Soil Acidity and Liming* (Madison, WI: American Society of Agronomy, 1983. (In Press).

Adams, F., and Z. F. Lund. 1966. "Effect of chemical activity of soil solution aluminum on cotton root penetration of acid subsoils," *Soil Sci.*, 101:193–198.

Bresler, E., B. L. McNeal, and D. L. Carter. 1982. *Saline and Sodic Soils, Principles—Dynamics—Modeling* (Berlin: Springer-Verlag).

Carter, D. L. 1981. "Salinity and plant productivity," in CRC Handbook Series in Nutrition and Food, Chemical Rubber Co., Cleveland, Ohio.

Hargrove, W. L., and G. W. Thomas. 1981. "Effect of organic matter on exchangeable aluminum and plant growth in acid soils," pp. 151–166 in *Chemistry in the Soil Environment*, ASA Special Publication #40. (Madison, WI: American Society of Agronomy and the Soil Science Society of America).

Jenny, H., et al. 1968. "Interplay of soil organic matter and soil fertility with state factors and soil properties," in *Organic Matter and Soil Fertility*, Pontificiae Acadamiae Scientiarum Scripta Varia 32 (New York: Wiley).

Mehlich, A. 1964. "Influence of sorbed hydroxyl and sulfate on neutralization of soil acidity," *Soil Sci. Soc. Amer. Proc.*, 28:492–96.

Peech, M. 1941. "Availability of ions in light sandy soils as affected by soil reaction," *Soil Sci.*, 51:473–486.

Richards, L. A., Ed. 1947. *Diagnosis and Improvement of Saline and Alkali Soils* (Riverside, CA: U.S. Regional Salinity Lab.).

Robertson, W. K., M. C. Lutrick, and T. L. Yuan. 1982. "Heavy applications of liquid-digested sludge on three ultisols: I. effects on soil chemistry," *J. Environ. Qual.*, 11:278–282.

Organisms of the Soil

7

Bacteria

Mucig

cell wall

Root cells

oil particle

Humus is a product of degradation and synthesis. And the agencies responsible are the living organisms in the soil, both the animals (fauna) and the plants (flora).[1] These organisms stimulate a myriad of biochemical changes as decay takes place. They also physically churn the soil and help stabilize soil structure. Our concern for these organisms is in terms of their activity rather than their scientific classification. Consequently, the grouping employed (Figure 7.1) is very broad and simple.

Vast numbers of organisms live in the soil, and by far the greater proportion of these belong to plant life. Yet the role of animals is not to be minimized, especially in the early stages of organic decomposition. Most soil organisms, both plant and animal, are so minute as to be visible only with the aid of the microscope. The number of the large organisms ranging in size to that of the larger rodents is comparatively small. All are of significance in the intricate biological processes operative in soils.

7.1 Organisms in Action

The activities of soil flora and fauna are so interrelated as to make it rather difficult to study them independently. To illustrate this point, consider how various soil organisms are involved in the degradation of higher plant tissue (Figure 7.2). Even while growing, plants are subject to attack by soil organisms known as *herbivores*. Examples are parasitic nematodes, snails, slugs, and the larvae of some insects that attack plant roots. Likewise, soil-borne termites and beetle larvae devour aboveground woody materials, as do some of the larger mammals, such as woodchucks and mice.

Primary Consumers. As soon as a leaf or a stalk or a piece of bark drops to the ground, it is subject to a coordinated attack by microflora and by *detritivores,* animals that live on dead and decaying plant tissues (Figure 7.2). If a little moisture is present, bacteria and fungi initiate the attack. They are joined by mites, snails, beetles, millipedes, woodlice, collembola, earthworms, and enchytraeid worms. These animals chew or tear holes in the tissue, opening it up to more rapid attack by the microflora. Together with microflora, they utilize the energy stored in the plant residues and are termed *primary consumers.*

While the action of the microflora is mostly chemical, that of the fauna is both physical and chemical. The animals chew the plant parts and move them from one place to another on the soil surface and even into the soil. Earthworms incorporate the plant residues into the mineral soil by literally

---

[1] For reviews of soil organisms, see Alexander (1977) and Wallwork (1976).

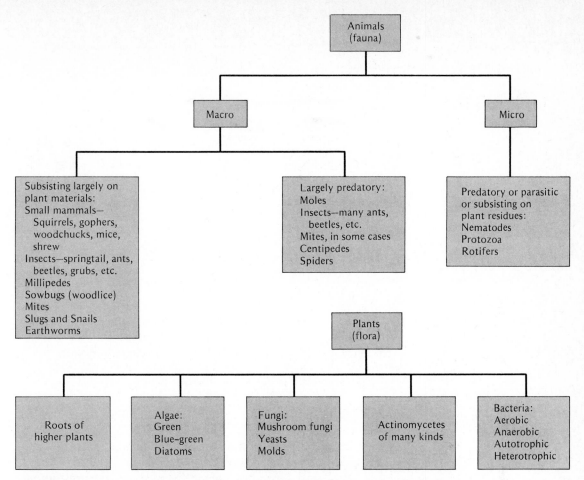

FIGURE 7.1 The more important groups of organisms commonly present in soils. The groupings are very broad and general because the emphasis is to be placed upon biochemical activity rather than on classification.

eating their way through the soil. Larger animals, such as gophers, moles, voles, prairie dogs, and rats, also burrow into the soil and bring about considerable soil mixing and granulation. Detritivores, such as collembola, mites, and enchytraeid worms, edge their way into cracks and crevices in the soil.

Secondary and Tertiary Consumers. The primary consumers are themselves food sources for predators and parasites existing in the soil. These are *secondary consumers* and include carnivores (animal consumers) and microphytic feeders, which consume small plants (bacteria, fungi, algae, and lichens). Examples of carnivores are centipedes, which consume small insects, collembola, spiders, nematodes, slugs, and snails, and the European mole, which feeds primarily on earthworms. Examples of microphytic feeders are some mites,

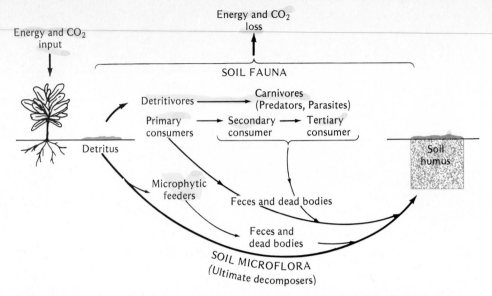

FIGURE 7.2 Diagram of the general pathway for the breakdown of higher plant tissue. Because they capture energy and CO_2, the higher plants are known as *primary producers*. When the debris from dead plants (detritus) falls to the soil surface, it is attacked by the soil fauna and microflora—the primary, secondary, or tertiary *consumers*. These organisms release energy and CO_2 and produce humus. Note that 80–90% of the total soil metabolism is due to the microflora.

springtails, termites, and protozoa, which utilize the microflora as sources of food (Figure 7.2 and Table 7.1).

Moving farther up the food chain, the secondary consumers are prey for still other carnivores, called *tertiary consumers*. For example, ants consume centipedes, spiders, mites, and scorpions, which themselves can prey on primary or other secondary consumers. Again, the microflora are intimately involved in the decomposition of organic material associated with the fauna. In addition to their direct attack on plant tissue, they are active within the digestive tract of some of the animals. They also attack the finely shredded organic material in animal feces and later decompose the bodies of dead animals. For this reason they are referred to in Figure 7.2 as the *ultimate decomposers*.

7.2 Organism Numbers, Biomass, and Metabolic Activity

The specific flora and fauna inhabiting soils are dependent upon many factors. The climate and the resultant vegetation greatly influence which organisms are dominant. The species composition in an arid desert will certainly be differ-

TABLE 7.1 Examples of Microphytic Feeders and of Carnivores That Act as Secondary and Tertiary Consumers Within or on Top of Soil

Note that some animals (e.g., centipedes) are both secondary and tertiary consumers.

| Microphytic feeders | | Carnivores | | | |
| | | Secondary consumers | | Tertiary consumers | |
Organism	Microflora consumed	Predator	Prey	Predator	Prey
Springtails	Algae Bacteria Fungi	Mites	Collembola Nematodes Enchytraeids	Ants	Spider Centipedes Mites Scorpions
Mites	Fungi Algae Lichens	Centipedes	Collembola Nematodes Snails Slugs Aphids Flies	Centipedes	Spiders Mites Centipedes (other)
Protozoa	Bacteria and other microflora	Mole	Earthworm Insects	Beetles	Spiders Mites Beetles (other)
Nematodes	Bacteria Fungi				
Termites	Fungi				

ent from that in a humid forest area, which in turn will be quite different from that in a cultivated field. Soil temperature, acidity, and moisture relations are also factors that govern the activity of both the flora and fauna. For these reasons, it is not easy to predict the number, kinds, and activities of organisms that one might expect to find in a given soil. But there are a few generalizations that might be made. For example, vegetation under forests generally supports a more diverse fauna than do grasslands. However, the grassland fauna are metabolically more active and their total weight per hectare is greater (see Table 7.2).

Compared to virgin areas, cultivated fields are generally lower in numbers and weight of soil organisms, especially the soil fauna. Exceptions may be soils that in the virgin state were very acid and have since been well limed and fertilized. Such cultivated soils may contain a higher microflora population than their untilled counterparts.

Comparative Organism Activity. The activities of specific groups of soil organisms are commonly identified by (a) their numbers in the soil, (b) their weight

TABLE 7.2 Biomass of Groups of Soil Animals under Grassland and Forest Cover[a]

The mass and in turn the metabolism are greatest under grasslands. The spruce with low-base-containing leaves encourages acid conditions and slow organic breakdown.

Group of organisms	Biomass (g/m²)[b]		
	Grassland meadow	Forest	
		Oak	Spruce
Herbivores	17.4	11.2	11.3
Detritivores			
Large	137.5	66.0	1.0
Small	25.0	1.8	1.6
Predators	9.6	0.9	1.2
Total	189.5	79.9	15.1

[a] Data from Macfadyen (1963).
[b] Depth about 15 cm.

per unit volume or area of soil (biomass), and (c) their metabolic activity. The numbers and biomass of groups of organisms that commonly occur in soils are shown in Table 7.3. Although the relative metabolic activities are not shown, they are generally related to the biomass of the organisms.

As might be expected, the numbers are highest among the microorganisms, both plant and animal. So numerous are microflora that they dominate the biomass in spite of the minute size of each individual organism. Together with the earthworms, the microflora monopolize the metabolic activity in soils. It is estimated that 60–80% of the total soil metabolism is due to the microflora. Not only do they destroy plant residues but they function in the digestive tracts of animals and eventually decompose the dead bodies of all organisms. Furthermore, soil humus and nutrients in a form plants can use are the significant end products of their activities. For these reasons major attention will be given to the microflora along with earthworms, protozoa, and other microanimals.

Before leaving the soil fauna, recognition should be given to their importance in soil formation and management. Rodents pulverize, mix, and granulate soil and incorporate organic materials into lower horizons. The medium-sized detritivores translocate and partially digest organic residues and leave their excrement for microfloral degradation. By living in the soil, many animals favorably affect its physical condition. Others utilize the soil as a habitat for destructive action against higher plants. In any case, all these organisms are part of perhaps the most intricate biological cycle in the biosphere.

TABLE 7.3 Relative Number and Biomass of Soil Flora and Fauna Commonly Found in Soils

Since the metabolic activity is generally related to the biomass, it is obvious that the microflora and earthworms dominate the life of soils.

| | Values common in surface soils[a] | | | |
| | Number | | Biomass[b] | |
Organisms	per square meter	per gram	Kg/HFS	(lb/AFS)
Microflora				
Bacteria	10^{13}–10^{14}	10^{8}–10^{9}	450–4500	400–4000
Actinomycetes	10^{12}–10^{13}	10^{7}–10^{8}	450–4500	400–4000
Fungi	10^{10}–10^{11}	10^{5}–10^{6}	1,120–11,200	800–8000
Algae	10^{9}–10^{10}	10^{4}–10^{5}	56–560	50–500
Microfauna				
Protozoa	10^{9}–10^{10}	10^{4}–10^{5}	17–170	15–150
Nematoda	10^{6}–10^{7}	10–10^{2}	11–110	10–100
Other fauna	10^{3}–10^{5}		17–170	15–150
Earthworms	30–300		110–1100	100–1000

[a] Generally considered 15 cm (6 in.) deep, but in some cases (e.g., earthworms) a greater depth is used.

[b] The biomass values are on a live weight basis. Dry weights are about 20–25% of these values.

Source of Energy and Carbon. Soil organisms may be classified on the basis of their source of energy and carbon as either *autotrophic* or *heterotrophic*. The autotrophs obtain their energy from the oxidation of inorganic compounds (those containing, e.g., nitrogen, sulfur and iron) and most of their carbon from carbon dioxide. Some bacteria and blue-green algae are examples. Though small in number, the autotrophic organisms perform essential functions in changing the inorganic forms in soils.

The heterotrophic organisms obtain their energy and carbon from the breakdown of organic materials. These organisms are far more numerous than the autotrophs and are responsible for most of the general-purpose decay. They include the soil fauna, most bacteria, and the fungi and actinomycetes.

7.3 Earthworms

The ordinary earthworm is, probably, the most important soil macroanimal (Figure 7.3). Of the more than 200 species known, *Lumbricus terrestris,* a deep-boring reddish organism, and *Allolobophora caliginosa,* a shallow-boring pale pink organism with a length of up to 25 cm, are the two most common, both in Europe and in the eastern and central United States. In the tropics and

FIGURE 7.3 Earthworms are perhaps the most significant soil macroorganism, especially in relation to their effect on the physical condition of soils.

semitropics, still other types are prevalent, some small and others surprisingly large, up to 3 m long. In respect to species, it is rather interesting that *Lumbricus terrestris* is not native to America. As our forests and prairies were put under cultivation, this European worm rapidly replaced the native types, which could not withstand the change in soil environment. Virgin lands, however, still retain at least part of their native populations.

Influence On Soil Fertility and Productivity. Earthworms are important in many ways, especially in the upper 15–25 cm of soil. The amount of soil these creatures pass through their bodies annually may amount to as much as 34 Mg/ha (15 tons/acre) of dry earth, a startling figure. During the passage through the worms, not only the organic matter that serves the earthworms as food, but also the mineral constituents are subjected to digestive enzymes and to a grinding action within the animals. Earthworm casts on a cultivated field may weigh as much as 18,000 kg/ha (16,000 lb/acre). Compared to the soil itself, the casts are definitely higher in bacteria and organic matter, total and nitrate nitrogen, exchangeable calcium and magnesium, available phosphorus and potassium, pH and percentage base saturation, and cation exchange capacity (Table 7.4). The rank growth of grass around earthworm casts suggests an increased availability of plant nutrients therein. Earthworms are noted for their favorable effect on soil productivity.

Other Effects. Earthworms are important in other ways. The holes left in the soil serve to increase aeration and drainage, an important consideration in soil development. Moreover, the worms mix and granulate the soil by dragging into their burrows quantities of undecomposed organic matter such as

TABLE 7.4 Comparative Characteristics of Earthworm Casts and Soils (Average of Six Nigerian Soils[a]).

Characteristic	Earthworm casts	Soils
Silt and clay (%)	38.8	22.2
Bulk density (Mg/m³)	1.11	1.28
Structural stability[b]	849	65
Cation exchange capacity (cmol/kg)	13.8	3.5
Exchangeable Ca (cmol/kg)	8.9	2.0
Exchangeable K (cmol/kg)	0.6	0.2
Soluble P (ppm)	17.8	6.1
Total N (%)	.33	.12

[a] From de Vleeschauwer and Lal (1981).

[b] Numbers of raindrops required to destroy structural aggregates.

leaves and grass, which they use as food. In some cases, the accumulation is surprisingly large. In uncultivated soils this is more important than in plowed land, where organic matter is normally turned under in quantity. Without a doubt, earthworms increase both the size and stability of the soil aggregates.

Factors Affecting Earthworm Activity. Earthworms prefer a well-aerated but moist habitat. For this reason, they are found mostly in medium-textured upland soils where the moisture capacity is high, rather than in droughty sands or poorly drained lowlands. They must have organic matter as a source of food. Consequently, they thrive where farm manure or plant residues have been added to the soil. A few species are reasonably tolerant to low pH, but most earthworms thrive best where the soil is not too acid.

Soil temperature affects earthworm numbers and their distribution in the soil profile. For example, a temperature of about 10°C (50°F) appears optimum for *Lumbricus terrestris,* earthworm numbers declining above or below this temperature. This relationship, as well as soil moisture requirements, probably accounts for the maximum earthworm activity noted in spring and autumn in temperate regions.

Earthworm counts must take into consideration the ability of some of these organisms to burrow deeply into the profile, thereby avoiding unfavorable moisture and temperature conditions. Although species differ markedly in the depth of soil they normally penetrate, cold weather or dry upper soil conditions will drive them into a more favorable environment deeper in the profile. Penetration as deep as one to two meters is not uncommon. Unfortunately, in barren soils a sudden heavy frost in the fall may kill the organisms before they can move lower in the profile. Soil cover is important in maintaining a high earthworm population under such circumstances.

Because of their sensitivity to soil and other environmental factors, there is a wide variation in the numbers of earthworms in different soils. In very acid soils under conifers, an average of fewer than one organism per square meter is common. In contrast, more than 500 per square meter have been found on rich grassland soils. The numbers commonly found in arable soils range from 30 to 300 per square meter (Table 7.3), equivalent to from 300,000 to 3 million per hectare–furrow slice. The biomass or live weight for this number would range from perhaps 110 to 1100 kg/ha (100–1000 lb/acre).

7.4   Termites

Termites, or "white ants," are major contributors to the breakdown of organic material in or at the surface of soils. They are found in about two-thirds of the land areas of the world (Yost and Fox, 1982) but are most prominent in tropical and subtropical areas. There they supplement and even surpass the activities of earthworms. These insects are found in temperate regions but are most prominent in tropical grasslands or savannahs and in tropical forested areas (Table 7.5). Their mud nests or mounds characterize open savannah areas in Africa, Latin America, and Asia. They carry on a very complex social life in these honeycombed mounds, which serve essentially as their "cities." In doing so they transport soil from lower layers to and above the surface soil level thereby bringing about extensive mixing of soil materials and of the plant residues they use as food. The quantities of materials they deposit often compare favorably with the surface castings of earthworms, amounting to tens or even hundred of tons per hectare at any one time. Their mounds may be 6 m or more in height and may extend to an even greater depth into the soil in some desert areas. Each mound provides the home for one or more millions of termites.

The effect of termites on soil productivity is generally less beneficial than that of earthworms. The termites' digestive processes are generally more efficient than those of earthworms. Microorganisms such as bacteria and protozoa in the gut of the termites readily attack cellulose and other complex organic materials, releasing sugars and other nutrients for use by the termites. While this symbiotic process is helpful in rapidly breaking down woody materials, it contributes less to soil fertility than does the earthworm metabolism.

Termite deposits commonly have a lower organic matter content than the surrounding undisturbed topsoil. This may be due in part to the fact that these deposits contain much low organic subsoil material, which was brought to the surface as the mounds were being formed. Crop growth in soil from areas where these mounds have existed is often poor, not only because of low nutrient content in the surface layers, but because of the greater compact-

TABLE 7.5 Area and Biomass Production in Different Ecological Regions of the World and Estimated Percent of the Biomass Consumed by Termites[a]

Ecological region	Area $(10^{12}\,M^2)$	Biomass Total produced $(10^{15}\,g/yr)$	Biomass % consumed by termites
Tropical Forest			
Wet	4.6	5.5	12
Moist	6.1	9.2	41
Dry	7.8	9.4	36
Temperate	12.0	15.0	7
Wood/shrub land	8.5	6.0	9
Savannah			
Wet	14.2	17.0	51
Dry	4.3	3.9	13
Temperate grassland	9.0	5.4	50
Cultivated land	11.9	7.7	60
Desert scrub	18.0	1.6	38
Clearing burning	6.8	9.6	68
Total	103.2	90.3	37

[a] Data from various sources summarized by Zimmermen et al. (1982).

ness of some of the mound material the particles of which had been cemented together by the termites as the mound was constructed.

Termites are significant factors in the formation of soils of tropical and subtropical areas. They also have both positive and negative effects on current land use in these areas. They accelerate the decay of dead trees and grasses but also disrupt crop production by the rapid development of their nests or mounds.

7.5 Soil Microanimals

Of the abundant microscopic animal life in soils, two groups are of some importance—nematodes and protozoa. These groups will be considered in order.

Nematodes. Nematodes—commonly called threadworms or eelworms—are found in almost all soils, often in surprisingly large numbers (Table 7.3). These organisms are round and spindle-shaped, the caudal end usually being acutely pointed. In size, they are almost wholly microscopic, seldom being large enough to be seen at all readily with the naked eye (Figure 7.4).

Three groups of nematodes may be distinguished on the basis of their food demands: (a) those that live on decaying organic matter (saprophytes);

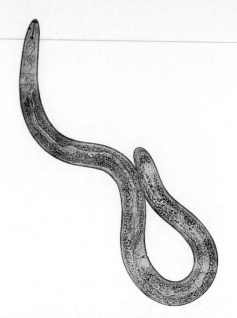

FIGURE 7.4 A nematode commonly found in soil (magnified about 120 times). More than 1000 species of soil nematodes are known. They can cause serious damage to the roots of higher plants on which many of them feed. [*Courtesy William F. Mai, Cornell University.*]

(b) those that are predatory on other nematodes, bacteria, algae, protozoa, and the like; and (c) those that are parasitic, attacking the roots of higher plants to pass at least a part of their life cycle embedded in such tissue. The first and second groups are by far the most numerous in the average soil and most varied. They are important primarily because of their interrelationships with the soil microflora.

The last group, especially those of the genus *Heterodera,* is the most important to the plant specialist. Because of their pointed form and adaptable mouth parts, they find it easy to penetrate plant tissue. The roots of practically all plants are infested even in cool temperature regions and the damage done is often very great, especially to vegetable crops. Even in greenhouses, nematodes may become a serious pest unless care is taken to avoid infestation. Because of the difficulties encountered in control, an appreciable nematode infection is a serious matter.

Protozoa. Protozoa probably are the simplest form of animal life. Although one-celled organisms, they are considerably larger than bacteria (5–100 mm diam.) and of a distinctly higher organization. For convenience of discussion, soil protozoa are divided into groups: (a) amoebae, (b) ciliates or infusoria, and (c) flagellates (Figure 7.5). The flagellates are usually most numerous in soil, followed in order by the amoebae and the ciliates.

The protozoa are the most varied and numerous in the microanimal population of soils although they have only a modest effect on the decay of organic matter. More than 250 species have been isolated, sometimes as many as 40

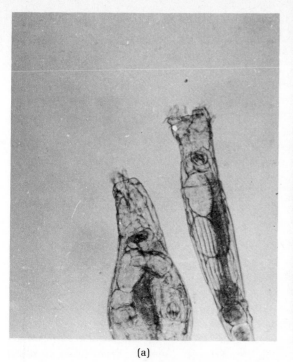
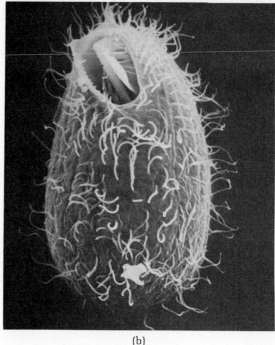

(a)   (b)

FIGURE 7.5 Two microanimals typical of those found in soils. (a) Photomicrograph
of two species of rotifer, *Rotaria rotatoria* (stubby one) and *Philodina acuticornus*
(slender one). (b) Scanning electron micrograph of a ciliated protozoan, *Glaucoma
scintillaus*. [(*a*) *courtesy F. A. Lewis and M. A. Stirewalt, Biomedical Research Institute,
Rockville, MD; (b) courtesy J. O. Corliss, University of Maryland.*]

or 50 of such groups occurring in a single sample of soil. A considerable number
of serious animal and human diseases are attributed to protozoan infections.

The numbers of protozoa in the soil are extremely variable. Aeration, as
well as the available food supply, is probably a very important factor. Most
of the organisms thrive best under conditions of good aeration and are therefore
confined to the surface horizons. Usually, their numbers are highest in the
spring and autumn. Populations of a few thousand to several hundred thousand
per gram of the upper few centimeters of soil have been observed. Perhaps
1–10 billion protozoa of all kinds per square meter 15 cm deep[2] (10,000–100,000
per gram) might be considered a normal range. This might amount to a live
weight of as much as 170 or even 200 kg to the hectare–furrow slice.

As a source of food, protozoa ingest bacteria and, to a lesser extent, other
microflora as well. There is some evidence that this protozoan predation on
bacteria may hasten the turnover of readily available nutrients, especially in

---

[2] A common areal unit for comparative organism numbers is the square meter; the depth of
surface soil is normally 15 cm or 6 in.

the vicinity of plant roots (the rhizosphere). Everything considered, however, the protozoa are generally not sufficiently abundant in soils to be a major factor in organic matter decay and nutrient release.

Rotifers. This third group of soil microanimals, of which about 100 species have been described, thrives under moist conditions, especially in swampy land where their numbers may be great. These animals are mostly microscopic in size. The anterior is modified into a retractile disk bearing circles of cilia that, in motion, give the appearance of moving wheels, hence, the name. These hairs sweep floating food materials into the animal. Just how important rotifers are in soils is unknown, but their activities are likely confined largely to peat bogs and in wet places occurring in mineral soils.

7.6 Roots of Higher Plants

Higher plants are the primary producers of organic matter and storers of the sun's energy (Figure 7.2). Their roots grow and die in the soil and in so doing supply the soil fauna and microflora with food and energy. At the same time the living roots physically modify the soils as they push through cracks and make new openings of their own (Figure 7.6). Tiny initial channels are increased in size as the roots swell and grow. By removing moisture from the soil, plant roots bring about further physical stresses that help soil aggregate initiation. The roots also provide a mass of living organic matter that stabilizes these aggregates. As they later decompose, the roots provide building materials for humus not only in the top few inches, but to greater soil depths as well.

Amounts of Organic Tissue Added. The mass of the root residues of crop plants remaining in the soil after harvest, commonly vary from 15 to 40% of the mass of the aboveground crop. If an average figure of 25% is used, good crops of oats, corn, and sugar cane would be expected to leave about 2500, 4500, and 8500 kg/ha of root residues, respectively. Many farmers do not appreciate the fact that the maintenance of a satisfactory supply of organic matter in arable soils is possible only because of root residues added in this way.

Rhizosphere. The roots of higher plants also function in a more intimate manner than as a source of dead tissue for the nutrition of soil microbes. Live roots not only affect the equilibrium of the soil solution by the withdrawal of soluble nutrients, but they also influence nutrient availability directly. Significant quantities of organic compounds are exuded, secreted, or otherwise released at the surface of young roots (Figure 7.6). Organic acids so excreted can solubilize plant nutrients. Amino acids and other simple carbon-containing compounds stimulate microflora in the *rhizosphere* or zone immediately surrounding these young roots. The number of organisms in the rhizosphere may be as many as 100 times greater than elsewhere in the soil (Figure 7.7). This

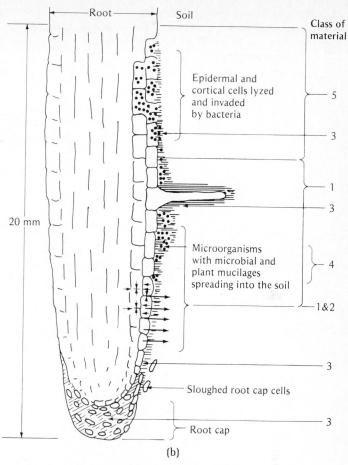

(a)   (b)

FIGURE 7.6   (a) Photograph of a root tip illustrating how roots penetrate soil and emphasizing the root cells through which nutrients and water move into and up the plant. (b) A diagram of a root showing the origins of organic materials in the rhizosphere including (1) simple *exudates,* which leak from plant cells to the soil; (2) *secretions,* simple compounds released by metabolic processes; (3) plant *mucilages* originating in root cells or from bacterial degradation; (4) *mucigel, a* gelatinous layer composed of mucilages and soil particles intermixed; and (5) *lyzates,* compounds released through digestion of cells by bacteria. [(*a) from Chino (1976); used with permission of the Japanese Society of Soil Science and Plant Nutrition. (b) redrawn from Rovira et al. (1979). Copyright Academic Press, Inc. (London) Ltd.; used with permission.]*

means that the absorptive surfaces of the root hairs lie within the zone of enhanced availability.

Also surrounding young roots are compounds called mucilages, which are released or sloughed off from plant roots. These compounds when intermixed with clay particles, form a gelatinous layer called *mucigel,* which surrounds the roots (Figure 7.6). The mucigel layer may facilitate contact with surrounding

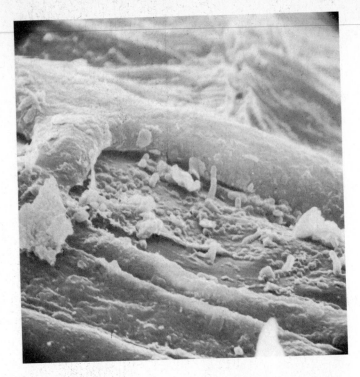

FIGURE 7.7 Soil microorganisms are found closely associated with the rhizosphere of wheat roots. Note the bacteria colonies on the root surface in this scanning electron micrograph (magnified 3900 times). [Courtesy L. F. Elliott, Washington State University.]

soil, especially during periods of moisture stress when the root proper may shrink in size and lose direct contact with the soil.

The chemical and physical characteristics of compounds in the rhizosphere are partially responsible for the plant roots being classified as soil organisms. The influence of these roots on soil properties justifies their classification.

7.7 Soil Algae

Most algae are chlorophyll-bearing organisms and, like higher plants, are capable of performing photosynthesis. To do this, they need light and must live at or very near the surface of the soil. They perform best under moist to wet conditions. However, a few forms obtain their energy largely from organic matter and live within and below the surface horizon. Several hundred species of algae have been isolated from soils, those most prominent being the same the world over. Soil algae are divided into four general groups: (a) blue-green, (b) green, (c) yellow-green, and (d) diatoms. The green algae are most evident in soils, especially if the pH is low.

Soil algae in vegetative form are most numerous in the surface layers. In subsoils, most algae are present as resting spores, or cysts, or in vegetative forms that do not depend upon chlorophyll. Grassland seems especially favorable for the blue-green forms, whereas in old gardens diatoms are often numerous. All of the ordinary types of algae are greatly stimulated by the application of farm manure.

Both the green and the blue-green algae outnumber the diatoms. A common range in algal population is from 1 to 10 billion per square meter 15 cm deep (10,000–100,000 per gram). Blue-green algae are especially numerous in rice soils, and when such lands are flooded, appreciable amounts of atmospheric nitrogen are fixed or changed to a combined form by these organisms. Blue-green algae, growing within the leaves of the aquatic fern, *azolla,* are also able to fix nitrogen in quantities significant for rice production.

7.8 Soil Fungi

Although the influence of fungi is by no means entirely understood, it is known that they play a very important part in the transformations of the soil constituents. Over 690 species have been identified, representing 170 genera. Like the bacteria and actinomycetes, fungi contain no chlorophyll and must depend for their energy and carbon on the organic matter of the soil.

For convenience of discussion, fungi may be divided into three groups: (a) yeasts, (b) molds, and (c) mushroom fungi. Only the last two are considered important in soils; yeasts are rare in such a habitat.

Molds. The distinctly filamentous, microscopic, or semimacroscopic fungi, the molds (Figure 7.8) play a role infinitely more important in the soil than the mushroom fungi, approaching or even excelling at times the influence of bacteria. Molds will develop vigorously in acid, neutral, or alkaline soils, some being favored, rather then harmed, by lowered pH (see Figure 6.10). Consequently, they are noticeably abundant in acid soils, where bacteria and actinomycetes offer only mild competition. This is especially important in decomposing the organic residues in acid forest soils.

The greatest numbers of molds are found in the surface layers, where organic matter is ample and aeration adequate. Many genera are represented, four of the more evident being *Penicillium, Mucor, Fusarium,* and *Aspergillus.* All of the common species occur in most soils, conditions determining which shall dominate. Their numbers fluctuate greatly with soil conditions; perhaps 100,000 to 1 million individuals to 1 g dry soil (10–100 billion per square meter) representing a more or less normal range in population. This would amount to a range in biomass of 800–8000 kg/ha. Molds are an important part of the general-purpose heterotrophic group of soil organisms that fluctuate so greatly in most soils. The fungal population is constantly changing not only in numbers but in respect to the dominant species. The complexity of the organic compounds being attacked seems to determine the particular mold or molds that prevail.

FIGURE 7.8 The three most important plant microorganisms of the soil: (a) fungal mycelium, (b) various types of bacteria cells, and (c) actinomycetes threads. The bacteria and actinomycetes are much more highly magnified than the fungus.

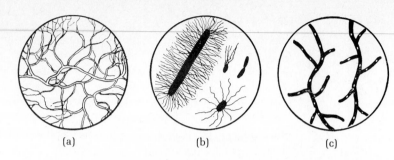

(a)   (b)   (c)

Activities of Fungi. In their ability to decompose organic residues, fungi are the most versatile and perhaps the most persistent of any group. Cellulose, starch, gums, lignin, as well as the more easily affected proteins and sugars, succumb to their attack. In affecting the processes of humus formation and aggregate stabilization, molds are more important than bacteria. They are especially active in acid forest soils, but also play a significant role in all soils.

Moreover, fungi function more efficiently than bacteria in that they transform into their tissues a larger proportion of decaying plant residues and give off as by-products less carbon dioxide and ammonium. Up to 50% of the substances decomposed by molds may become organism tissue, compared to about 20% for bacteria. However, they apparently cannot oxidize ammonium compounds to nitrates as do certain bacteria, nor can they *fix* or bind elemental nitrogen into combined forms. Nevertheless, soil fertility depends in no small degree on molds, since they continue the decomposition process after bacteria and actinomycetes have essentially ceased to function.

Mycorrhizae.[3]   An economically important symbiotic association between numerous fungi and the roots of higher plants is called *mycorrhizae,* a term meaning "fungus root." This association, first noted on certain forest tree species, is now known to be widespread and to affect most plant species, including many agronomic crops. The association is of great practical significance since it markedly increases the availability of several plant nutrients, especially from infertile soils. Apparently the symbiotic association provides the fungi with sugars and other organic exudates for use as food. In return, the fungi provide an enhanced availability of several essential nutrients, including phosphorus, zinc, copper, calcium, magnesium, manganese, and iron.

There are two types of mycorrhizal associations of considerable practical importance, *ectomycorrhiza* and *endomycorrhiza.* The ectomycorrhiza group includes hundreds of different fungal species associated primarily with trees, such as pine, birch, hemlock, beech, oak, spruce, and fir. These fungi, stimulated by root exudates, cover the surface of feeder roots with a fungal mantle. Their hyphae penetrate the roots and develop around the cells of the cortex but do not penetrate these cells (Figure 7.9).

[3] For an excellent but brief presentation of mycorrhizae, see Menge (1981).

FIGURE 7.9 Diagram of ectomycorrhiza and vesicular-arbuscular (VA) mycorrhiza association with plant roots. (a) The ectomycorrhiza fungi association produces short branched rootlets that are covered with a fungal mantle, the hyphae of which extend out into the soil and between the plant cells but do not penetrate the cells. (b) In contrast, the VA mycorrhizae penetrate not only between cells but into certain cells as well. Within these cells, the fungi form structures known as *arbuscules* and *vesicles,* the former to transfer nutrients to the plant and the latter to store nutrients. In both types of association, the host plant provides sugars and other food for the fungi and receives in return essential mineral nutrients that the fungi absorb from the soil. [*Redrawn from Menge* (*1981*).]

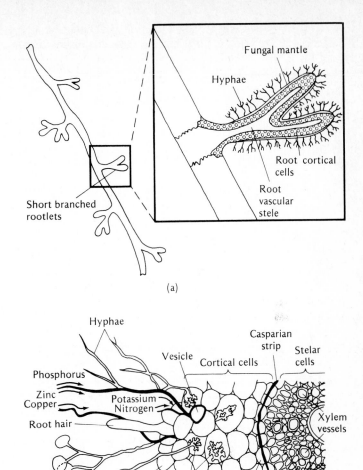

The endomycorrhiza group, also called *vesicular arbuscular* (VA) *mycorrhizae,* are perhaps the most common and most widespread. Some 89 species of fungi found in soils from the tropics to the arctic form VA mycorrhizal associations. The roots of most agronomic crops, including corn, wheat, potatoes, beans, alfalfa, sugar cane, cassava, and dryland rice, have VA mycorrhizal associations. Many trees, including maple, yellow poplar, redwood, and such important tree crops as apple, cocoa, coffee, citrus, and rubber also have VA mycorrhizae.

The root cortical cells of host plants are penetrated by the hyphae of VA mycorrhizae. Inside the plant cells, small structures known as *arbuscules* are formed by the fungi. These structures are considered to be the sites of transfer of nutrients from the fungi to the host plants (Figure 7.9).

TABLE 7.6 Effect of Inoculation with Mycorrhiza and of Added Phosphorus on the Content of Different Elements in the Shoots of Corn, Expressed in μg.[a]

Element in plant	Content of elements in shoots (μg)			
	No phosphorus		25 ppm phosphorus added	
	No mycorrhiza	Mycorrhiza	No mycorrhiza	Mycorrhiza
P	750	1,340	2,970	5,910
K	6,000	9,700	17,500	19,900
Ca	1,200	1,600	2,700	3,500
Mg	430	630	990	1,750
Zn	28	95	48	169
Cu	7	14	12	30
Mn	72	101	159	238
Fe	80	147	161	277

[a] From Lambert et al. (1979).

The increased nutrient availability from mycorrhizae is thought to be due to the nutrient-absorbing surface provided by mycorrhizae fungi, which has been calculated to be as much as 10 times that of the uninfested roots. Also, the soil volume from which nutrients are absorbed is greater for mycorrhizal associations. The very fine fungal hyphae extend up to 8 cm into soil surrounding the roots, thereby increasing the absorption of nutrients, such as phosphorus, zinc, and copper, which do not diffuse readily to the roots (Table 7.6). Likewise, water stress in mycorrhizal-infested plants is less during drought than in uninfested plants. VA mycorrhizae are thought to be responsible for the transfer of phosphorus from one plant to nearby neighboring species, especially in grasslands. Research findings of recent years have brought to light the significant role mycorrhizae play in helping plants absorb nutrients from relatively infertile soils (Figure 7.10).

7.9   Soil Actinomycetes

Actinomycetes resemble molds in that they are filamentous, often profusely branched, and produce fruiting bodies in much the same way. Their mycelial threads are smaller, however, than those of fungi. Actinomycetes are similar to bacteria in that they are unicellular and of about the same diameter. When they break up into spores, they closely resemble bacteria. On the basis of organization, actinomycetes occupy a position between true molds and bacteria. Although they often are classified with the fungi, they are sometimes called *thread bacteria* (Figure 7.8).

Actinomycetes develop best in moist, well-aerated soil. But in times of

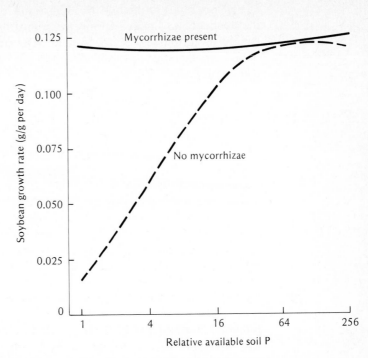

FIGURE 7.10 Growth rates of soybeans with and without mycorrhizae. The mycorrhizae help solubilize soil phosphorus even where available soil phosphorus levels are low. [*Modified from Yost and Fox (1982); used with permission of the American Society of Agronomy.*]

drought, they remain active to a degree not generally exhibited by either bacteria or molds. They are in general rather sensitive to acid soil conditions, their growth being practically prohibited in mineral soils at a pH of 5.0 or below. Their optimum development occurs at pH values between 6.0 and 7.5 (see Figure 6.10). This marked relationship to soil reaction is used for control of potato scab, an actinomycete disease of wide distribution. It is possible by the use of sulfur to effect a reduction in pH sufficient to keep the disease under control (see Section 6.13).

Numbers and Activities of Actinomycetes. Except for bacteria, no other microorganisms are so numerous in the soil as the actinomycetes, their numbers sometimes reaching hundreds of millions, one tenth the number of the bacteria. In actual live weight, they often exceed bacteria. Under especially favorable conditions, a biomass of more than 4,500 kilograms of actinomycete threads and spores might be present in a hectare–furrow slice. These organisms are especially numerous in soils high in humus, such as old meadows or pastures, where the acidity is not too great. There they sometimes exceed in numbers all other microscopic forms of life. The addition of farm manure markedly stimulates their activities. The aroma of freshly plowed land that is so noticeable at certain times of the year is probably due to actinomycetes as well as to certain molds.

Actinomycetes undoubtedly are of great importance in the decomposition of soil organic matter and the liberation of nutrients therefrom. Apparently, they reduce to simpler forms even the more resistant compounds, such as cellulose, chitin, and phospholipids. The presence of actinomycetes in abundance in soils long under sod is an indication of their capacity to attack complex compounds.

7.10 Soil Bacteria

Characteristics. Bacteria are single-cell organisms, one of the simplest and smallest forms of life known. Their almost unlimited capacity to increase in numbers is extremely important in soils. It allows them to quickly adjust their activities in response to changes in their environment.

Bacteria are very small; the larger individuals seldom exceed 4–5 μm (0.004–0.005 mm) in length, and the smaller ones approach the size of an average clay particle. The shape of bacteria is varied in that they may be nearly round, rod-like, or spiral. In the soil, the rod-shaped organisms seem to predominate (Figure 7.8). The surface exposed by soil bacteria is remarkable, amounting to about 2 m²/kg of dry soil or about 460 hectares to the hectare–furrow slice.

Bacterial Populations in Soils. The numbers of bacteria present in soil are variable, as many conditions decidedly affect their growth. In general the greatest population is in the surface horizons since conditions of temperature, moisture, aeration, and food are more favorable. The numbers of bacteria are high, normally ranging from a few billion to 3 trillion in each kilogram of soil. A biomass of from a few hundred kilograms to 5 Mg (metric tons) live weight to the hectare–furrow slice of fertile soils is commonly encountered (2.2 tons/acre–furrow slice). The bacterial flora, as well as the other soil organisms, fluctuates sharply with the season, the numbers in humid temperate regions usually being the greatest in early summer and in the autumn.

In the soil, bacteria exist as mats, clumps, and filaments, called *colonies,* on and around soil particles wherever food and other conditions are favorable. The jelly-like mixture of mineral and organic colloidal matter makes an almost ideal medium for their development.

Many of the soil bacteria are able to produce spores or similar resistant bodies, thus presenting both a vegetative and a resting stage. This latter capacity is important as it allows the organisms more readily to survive unfavorable conditions.

Source of Energy. Soil bacteria are either *autotrophic* or *heterotrophic,* (see Section 7.2). The autotrophs obtain their energy from the oxidation of mineral constituents such as ammonium, sulfur, and iron, and most of their carbon from carbon dioxide. They are not very numerous but are vital to the sustenance of higher plants. However, most soil bacteria are heterotrophic—that is, both their energy and their carbon come directly from organic matter.

Importance of Bacteria. Bacteria as a group, almost without exception, participate vigorously in all of the organic transactions so vital if a soil is to support higher plants. They not only rival but often excel both fungi and actinomycetes in this regard. Also, they hold near monopolies on three basic enzymatic transformation: (a) *nitrogen oxidation* (*nitrification*), (b) *sulfur oxidation*, and (c) *nitrogen fixation*. If these were to fail, life for higher plants and for animals would be endangered. In this respect, bacteria, the simplest and most numerous of all life forms, are perhaps basically the most consequential.

7.11 Conditions Affecting the Growth of Soil Bacteria

Many conditions of the soil affect the growth of bacteria. Among the most important are the supplies of oxygen and moisture, the temperature, the amount and nature of the soil organic matter, the pH, and the amount of exchangeable calcium present. These effects are briefly outlined below.

1. Oxygen requirements:
    a. Some bacteria use mostly oxygen gas (aerobic).
    b. Some use mostly combined oxygen (anaerobic).
    c. Some use either of the above forms (facultative).
    d. All three of the above types usually function in a soil at one time.
2. Moisture relationships:
    a. Optimum moisture level for higher plants usually best for most bacteria.
    b. The moisture content affects oxygen supply (see 1 above).
3. Suitable temperature range:
    a. Bacterial activity generally is greatest at 20–40°C (about 70–100°F).
    b. Ordinary soil temperature extremes seldom kill bacteria.
4. Organic matter requirements:
    a. Used as energy source for majority of bacteria (heterotrophic).
    b. Organic matter not required as energy source for others (autotrophic).
5. Exchangeable calcium and pH relationships:
    a. High calcium concentration, and pH from 6 to 8 generally best for most bacteria.
    b. Calcium and pH values determine the specific bacteria present.
    c. Certain bacteria function at very low pH (<3.0) and others at high pH values.
    d. Exchangeable calcium seems to be more important than pH.

7.12 Injurious Effects of Soil Organisms on Higher Plants

Soil Fauna. It has already been suggested that certain of the soil fauna are injurious to higher plants. For instance, rodents and moles may, in certain cases, greatly damage crops. Snails and slugs in some climates are dreaded pests, especially of vegetables. Ants, as they transfer aphids on certain plants, must be diligently combated by gardeners. Also, most plant roots are infested with nematodes, sometimes so seriously as to make the successful growth of certain crops both difficult and expensive. Tobacco, potatoes, and cotton are examples of commercial crops adversely affected by nematodes. Crop rotation and the development of resistant varieties are the most acceptable means of combating these pests.

Microflora and Plant Diseases. In general, it is the plant forms of soil life that exert the most devastating effects on higher plants. All three groups—bacteria, fungi, and actinomycetes—contribute their quota of plant diseases. However, fungi are responsible for most of the common soil-borne diseases of crop plants. Some of the more common diseases produced by soil flora are wilts, damping off, root rots, clubroot of cabbage and similar crops, and the actinomycetes scab of potatoes. In short, disease infestations occur in great variety, induced by many different organisms (Figure 7.11).

Injurious organisms live for variable periods in the soil. Some of them will disappear within a few years if their host plants are not grown, but others are able to maintain existence on almost any organic substance. Once a soil is infested, it is likely to remain so for a long time. Infection usually occurs easily. Organisms from infested fields may be carried on implements, plants, or rubbish of any kind. Even manure from animals having been fed infected plants may act as disease carriers. Erosion, if soil is washed from one field to another, may also be a means of transfer.

Disease Control by Soil Management. Prevention is one of the best defenses against diseases produced by such soil organisms, hence the strict quarantines often attempted. Once a disease has procured a foothold, it is often very difficult to eradicate it. Rotation of crops is adequate for some diseases, but the absence of the host plant is often necessary.

The regulation of the pH is effective to a certain extent with potato scab (Figure 7.11) and the clubroot of cabbage. If the pH is held somewhat below 5.3 or even 5.5, the former disease, which is due to an actinomycete, is much retarded. With clubroot, a fungus disease, the addition of calcium hydroxide until the pH of the soil is definitely above 7.0 seems fairly effective.

Wet, cold soils favor some seed rots and seedling diseases known as damping off. Good drainage and ridging help control these diseases. Steam or chemical sterilization is a practical method of treating greenhouse soils for a number of diseases. The breeding of plants immune to particular diseases has been successful in the case of a number of other crops.

Competition for Nutrients. Another way in which soil organisms may detrimentally affect higher plants, at least temporarily, is by competition for available nutrients. Nitrogen is the element for which competition is usually greatest, although organisms may also utilize appreciable quantities of phosphorus, po-

FIGURE 7.11 Soil-borne pathogens damage roots as well as other below-ground organs. This potato has been attacked by the potato scab actinomycete, which may be present in soils with pH above ~5.0. [*Courtesy Dr. F. E. Manzer, University of Maine.*]

tassium, and calcium, making them unavailable to the crops growing on the land. Competition for trace elements may even be serious. Soil organisms usually exact their nutrient quota first and higher plants must subsist on what remains available. This subject is considered in greater detail in Chapter 8.

Other Detrimental Effects. Under conditions of somewhat restricted drainage, active soil microflora may deplete the already limited oxygen supply to the soil. This may affect plants adversely in two ways. First, the plant roots require a certain minimum amount of O_2 for normal growth and nutrient uptake. Second, oxidized forms of several elements, including nitrogen, sulfur, iron, and manganese, will be chemically reduced by further microbial action. In the cases of nitrogen and sulfur, some of the reduced forms are gaseous and these may be lost to the atmosphere. Iron and manganese reduction may result in soluble forms of these elements being present in toxic quantities, especially if the soil is quite acid. Thus, nutrient deficiencies and toxicities, both microbiologically induced, can result from the same basic set of conditions.

7.13 Competition Among Soil Microorganisms

In addition to competition between microorganisms and higher plants, there exists in soils an intense intermicrobial rivalry for food. When fresh organic matter is added, the vigorous heterotrophic soil organisms (bacteria, fungi, and actinomycetes) compete with each other for this source of food. The bacteria dominate initially since they can reproduce rapidly and prefer simple com-

FIGURE 7.12 Organisms compete with each other in the soil. The growth of a fungus (*Fusarium*) was rapid when this organism was grown alone in a soil (*left*), but when a certain bacterium (*Agrobacterium*) was also introduced, the fungal growth did not appear. [*Courtesy M. A. Alexander, Cornell University.*]

pounds. As these simple compounds are broken down the fungi, and particularly the actinomycetes, become more competitive. Undoubtedly, such food competition is the rule and not the exception in soils (Figure 7.12).

Antibiotics Produced in Soils. Besides the food competition just stressed, there is another type of microbial rivalry just as intense and even more deadly. Certain bacteria, fungi, and actinomycetes can produce substances that will inhibit, or even actually kill, other microbes. Not only are organisms alien to the soil thus affected, but also many of those that normally flourish vigorously therein.

The discovery that many soil organisms can produce *antibiotics,* as they are called, has in some respects revolutionized the treatment of certain human and animal diseases, greatly minimizing their seriousness. Many preparations carrying specific bactericidal ingredients are now on the market, such as penicillin, streptomycin, and aureomycin. Undoubtedly, many new and valuable substances of this type are yet to be discovered. Although the soil harbors numerous types of disease organisms, at the same time it supports others that are the source of life-saving drugs, the discovery of which marks an epochal advance in medical science.

7.14 Effects of Agricultural Practice on Soil Organisms

Changes in environment affect both the number and kinds of soil organisms. Placing either forested or grassland areas under cultivation drastically changes the soil environment. In the first place, the amount of plant residues (food for the organisms) is markedly reduced. Also, the species of higher plants are changed and generally less numerous. Monoculture or even crop rotation provides a much narrower range of original plant materials than is generally encountered in nature.

Tillage of the soil and applications of lime and fertilizer present a changed environment to the soil inhabitants. Drainage or irrigation likewise affects soil moisture and aeration relations with their concomitant effects on soil organisms. Also, commercial agriculture transports organism species from one location to another, making possible their introduction in new areas.

It is obvious that agricultural practices have different effects on different organisms. There are a few generalizations that can be made, however. Figure 7.13, intended originally to show the ecological effects of agricultural practices on soil microarthropods, illustrates some principles relating to the total soil organism population. For example, agricultural practices generally reduce the species diversity as well as the total organism population. These practices can greatly increase or reduce the numbers of a given species, depending on the situation. Adding lime, fertilizers, and manures to an infertile soil will definitely increase the activities of certain bacteria and actinomycetes. Pesticides (especially the fumigants), on the other hand, can sharply reduce organism

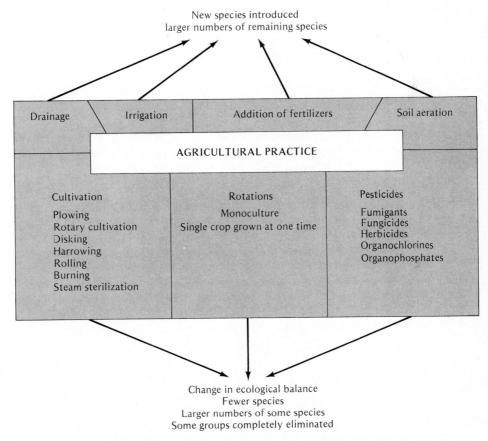

FIGURE 7.13 Ecological effects of agricultural practices on soil microarthropod population. Some agricultural practices (drainage, fertilization, etc.) are beneficial and help to maintain or even increase species numbers. But excessive tillage, monoculture, and pesticides may reduce species numbers and even eliminate some species. [*From Edwards and Lofty* (*1969*).]

numbers, at least on a temporary basis. Likewise, monoculture cropping systems reduce species numbers but may actually increase the organism count of the remaining species.

Most changes in agricultural technology have ecological effects on soil organisms that can affect higher plants and animals, including humans. For this reason, the ecological side effects of modern technology must be carefully scrutinized. The effects of pesticides, both positive and negative, provide evidence of this fact.

7.15 Activities of Soil Organisms Beneficial to Higher Plants

In their influence on crop production, the soil fauna and flora are indispensable. Of their many beneficial effects on higher plants, only the most important can be emphasized here.

Organic Matter Decomposition. Perhaps the most significant contribution of the soil fauna and flora to higher plants is that of organic matter decomposition. By this process, plant residues are broken down, thereby preventing an unwanted accumulation. Furthermore, nutrients held in organic combinations within these residues are released for use by plants. Nitrogen is a prime example. At the same time, the stability of soil aggregates is enhanced not only by the slimy intermediate products of decay, but by the more resistant portion, humus. Plants naturally profit from these beneficial chemical and physical effects.

Inorganic Transformations. The appearance in the soil of ammonium compounds and nitrates is the result of a long series of biochemical transfers beginning with proteins and related compounds (see Section 9.4). These successive changes are of vital importance to higher plants since the plants absorb most of their nitrogen in ammoniacal and nitrate forms.

The production of sulfates is roughly analogous to the biological simplification of nitrogen (see Section 9.26). Here again a complicated chain of enzymatic activities culminates in a simple soluble product—in this case the sulfate—the only important form utilized by higher plants.

Other biologically instigated inorganic changes that may be helpful to plants are those relating to mineral elements such as iron and manganese. In well-drained soils these elements are oxidized by autotrophic organisms to their higher valent states, in which forms their solubilities are very low at intermediate pH values. This keeps the greater portion of iron and manganese, even under fairly acid conditions, in insoluble and nontoxic forms. If such oxidation did not occur, plant growth would be jeopardized because of toxic quantities of these elements in solution.

Nitrogen Fixation. The fixation of elemental nitrogen into compounds usable by plants is one of the most important microbial processes in soils. Nitrogen gas, so plentiful in the atmospheric air, cannot be used directly by higher plants. It must be in combined form before it can satisfy their nutritional needs.

Blue-green algae and certain actinomycetes are significant nitrogen-fixing organisms. But worldwide, bacteria are probably the most important group in the capture of gaseous nitrogen. The *nodule organisms,* especially those of legumes, and *free-fixing bacteria* of several kinds are most noted for their ability to fix nitrogen.

The legume bacteria, as they are often called, use the carbohydrates of their hosts as an energy source, fix the nitrogen, and pass part of it on to the infected host. The nitrogen-fixing soil bacteria acquire their energy from the soil organic matter, fix the free nitrogen, and make it a part of their own tissue. When they die and decay, part of this nitrogen is available to higher plants.

It is obvious that the organisms of the soil must have energy and nutrients if they are to function efficiently. In obtaining them they break down organic matter, aid in the production of humus, and leave behind compounds that are useful to higher plants. These biotic features and their practical significance are considered in the next chapter, which deals with soil organic matter.

REFERENCES

Alexander, M. 1977. *Introduction to Soil Microbiology,* 2nd ed. (New York: Wiley).

Chino, M. 1976. "Electron microprobe analysis of zinc and other elements within and around rice root growth in flooded soil," *Soil Sci. and Plant Nut. J.,* 22:449.

De Vleeschauwer, D., and R. Lal. 1981. "Properties of worm casts under secondary tropical forest regrowth," *Soil Sci.,* 132:175–181.

Edwards, C. A., and J. R. Lofty. 1969. "Agricultural practice and soil microarthropods," in J. G. Sheals, Ed., *The Soil Ecosystem.* (London: The Systematics Association, Publ. 8).

Lambert, D. H., D. E. Baker, and H. Cole, Jr. 1979. "The role of Mycorrhizae in the interactions of phosphorus with zinc, copper, and other elements," *J. Soil Sci. Soc. Amer.,* 43:976–980.

Macfadyen, A. 1963. In J. Doeksen and J. van der Drift, Eds., *Soil Organisms* (Amsterdam: North-Holland Publ. Co.).

Menge, J. A. 1981. "Mycorrhizae agriculture technologies," pp. 383–424 in *Background Papers for Innovative Biological Technologies for Lesser Developed Countries,* paper No. 9., Office of Technology Assessment Workshop, Nov. 24–25, 1980 (Washington, DC: U.S. Government Printing Office).

Rovira, A. D., R. C. Foster, and J. K. Martin. 1979. "Origin, nature and nomenclature of the organic materials in the rhizophere," in J. L. Harley and R. S. Russell, Eds., *The Soil-Root Interface* (New York: Academic Press).

Wallwork, J. A. 1976. The Distribution and Diversity of Soil Fauna (London: Academic Press).

Yost, R. S., and R. L. Fox. 1982. "Influence of mycorrhizae on the mineral contents of cowpea and soybean growth in an oxisol," *Agron. J.,* 74:475–481.

Zimmerman, P. R., J. P. Greenberg, S. O. Wandiga, and P. J. Crutzen. 1982. "Termites: a potentially large source of atmospheric methane, carbon dioxide and molecular hydrogen," *Science,* 218:563–565.

Organic Matter of Mineral Soils

[*Preceding page*] Soil organic matter, which is found mostly in the upper soil layers, supplies plant nutrients and simultaneously promotes soil physical properties favorable for plant growth. [*Courtesy USDA Soil Conservation Service.*]

Organic matter[1] influences physical and chemical properties of soils far out of proportion to the small quantities present. It commonly accounts for at least half the cation exchange capacity of surface soils and is responsible perhaps more than any other single factor for the stability of soil aggregates. Furthermore, it supplies energy and body-building constituents for most of the microorganisms whose general activities have just been considered.

8.1 Sources of Soil Organic Matter

The original source of the soil organic matter is plant tissue. Under natural conditions, the tops and roots of trees, shrubs, grasses, and other native plants annually supply large quantities of organic residues. A good portion of cropped plants is commonly removed from cropped soils, but one-tenth to one-third of the tops and all of the roots are left in the soil. As these materials are decomposed and digested by soil organisms of many kinds, they become part of the underlying horizons by infiltration or by actual physical incorporation. Thus, higher plant tissue is the primary source not only of food for the various soil organisms but of organic matter, which is so essential for soil formation (see Section 13.2).

Animals are usually considered secondary sources of organic matter. As they attack the original plant tissues, they contribute waste products and leave their own bodies as their life cycles are consummated. Certain forms of animal life, especially the earthworms, centipedes, and ants, also play an important role in the translocation of plant residues.

8.2 Composition of Plant Residues

The moisture content of plant residues varies from 60 to 90%, 75% being a representative figure (Figure 8.1). On a weight basis, the dry matter is mostly carbon and oxygen, with less than 10% each of hydrogen and inorganic elements (ash). However, on an elemental basis (number of atoms of the elements), hydrogen predominates. Thus, there are 8 hydrogen atoms for every 3.7 carbon atoms and 2.5 oxygen atoms. These three elements can truly be said to dominate the bulk of organic tissue in the soil.

Even though more than 90% of the dry matter is carbon, hydrogen, and oxygen, the other elements play a vital role in plant nutrition and in meeting microorganism body requirements. Nitrogen, sulfur, phosphorus, potassium,

[1] For a review of the subject see Allison (1973).

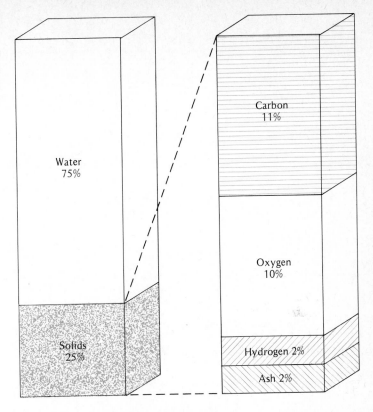

FIGURE 8.1 Elemental composition of representative green plant tissue. Plant tissue solids are made up largely of carbon, hydrogen, and oxygen. The nitrogen is included in the ash.

calcium, and magnesium are particularly significant, as are the micronutrients contained in plant materials. These will be discussed later in more detail.

The actual compounds in plant tissue are many and varied. However, they can be grouped into a small number of classes, as shown in Figure 8.2. Representative percentages as well as ranges common in plant material are shown.

General Composition of Compounds. The carbohydrates, which are made up of carbon, hydrogen, and oxygen, range in complexity from simple sugars to the celluloses. The fats and oils are glycerides of fatty acids such as butyric, stearic, and oleic. These are associated with resins of many kinds and are somewhat more complex than most of the carbohydrates. They, too, are made up mostly of carbon, hydrogen, and oxygen.

Lignins occur in older plant tissue such as stems and other woody tissues. They are complex compounds, some of which may have "ring" structures. The major components of lignins are carbon, hydrogen, and oxygen. They are very resistant to decomposition.

Of the various groups, the crude proteins are among the more complicated. They contain not only carbon, hydrogen, and oxygen, but also nitrogen and

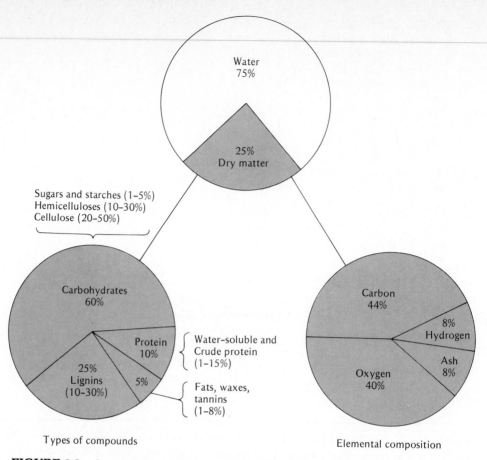

FIGURE 8.2 Composition of representative green plant materials added to soils. All inorganic elements, including nitrogen, are represented in the ash. Common ranges in the percentage of compounds present are shown in parentheses. [*Data from Waksman* (*1948*).]

smaller amounts of such elements as sulfur, iron, and phosphorus. As a consequence, they are compounds of great significance as carriers of essential elements. Their reactions in soils are means by which these nutrients are first conserved and eventually made available for plant uptake.

8.3 Decomposition of Organic Compounds

Rate of Decomposition. Organic compounds vary greatly in their rate of decomposition. They may be listed in terms of ease of decomposition as follows.

1. Sugars, starches, and simple proteins   rapidly decomposed
2. Crude proteins
3. Hemicelluloses
4. Cellulose
5. Lignins, fats, waxes, etc.   very slowly decomposed

It should be remembered that all of these compounds usually begin to decompose simultaneously when fresh plant tissue is added to a soil. The rate at which decomposition occurs, however, decreases as we move from the top to the bottom of the list. Thus, sugars and water-soluble proteins are examples of readily available energy sources for soil organisms. Lignins are a very resistant source of food, although they eventually supply much total energy.

When organic tissue is added to soil, three general reactions take place.

1. The bulk of the material undergoes enzymatic oxidation with carbon dioxide, water, energy, and heat as the major products.
2. The essential elements, nitrogen, phosphorus, and sulfur are released and/ or immobilized by a series of specific reactions relatively unique for each element.
3. Compounds resistant to microbial action are formed either through modification of compounds in the original plant tissue or by microbial synthesis.

Each kind of reaction has great practical significance.

Decomposition—an Oxidation Process. In spite of the differences in composition of the various organic compounds, the similarity of the ultimate end products of decay is quite striking, especially if aerobic organisms are involved. Under such conditions the major portion of all these compounds undergoes essentially a "burning" or oxidation process. The oxidizable fractions of organic materials are composed largely of carbon and hydrogen, which make up more than half of their dry weight (Figure 8.1). Consequently, the complete oxidation of most of the organic compounds in the soil may be represented as:

$$-[C, 4H] \quad + 2\,O_2 \xrightarrow[\text{oxidation}]{\text{enzymatic}} CO_2 + 2\,H_2O + \text{energy}$$

Carbon- and
hydrogen-containing
compounds

It is recognized, of course, that many intermediate steps are involved in this overall reaction. Also, important side reactions that involve elements other than carbon and hydrogen are occurring simultaneously. Neither of these facts, however, detracts from the importance of this basic reaction in accounting for most of the organic matter decomposition in the soil.

Breakdown of Proteins. The plant proteins and related compounds yield other very important products upon decomposition in addition to the carbon dioxide and water previously mentioned. For example, they break down into amides and amino acids[2] of various kinds, the rate of breakdown depending on conditions. Once these compounds are formed, they may be hydrolyzed readily to carbon dioxide, ammonium compounds, and other products. The ammonium compounds may be changed to nitrates, the form in which higher plants take up much of their nitrogen.

Example of Organic Decay. The process of organic decay in time sequence is illustrated in Figure 8.3. First, assume a situation where no readily decomposable materials are present in a soil. The microbial numbers and activity are low. Next, under favorable conditions, introduce an abundance of fresh, decomposable tissue. A marked change occurs immediately as the number of soil microorganisms suddenly increases manyfold. Soon microbial activity is at its peak, at which point energy is being liberated rapidly and carbon dioxide is being formed in large quantities. General-purpose decay bacteria, fungi, and actinomycetes are soon fully active and are decomposing and synthesizing at the same time.

The organic matter at this stage contains a great variety of substances: intermediate products of all kinds, ranging from the more stable compounds, such as modified lignins, to microbial cells, both living and dead. The microbial tissue may even at times account for as much as half of the organic fraction of a soil.

Dead microbial cells soon decay, and the compounds present are devoured by living microbes, with the profuse evolution of carbon dioxide. As the readily available energy is used up and food supplies diminish, microbial activity gradually lessens and the general-purpose soil organisms again sink back into comparative quiescence. This is associated with a release of simple products such as nitrates and sulfates. The organic matter now remaining is a dark, incoherent, and heterogeneous colloidal mass usually referred to as *humus*.

The decomposition of both plant residues and soil organic matter is nothing more than a process of enzymatic digestion. It is just as truly a digestion as though the plant materials entered the stomach of a domestic animal. The products of these enzymatic activities, although numerous and tremendously varied, may be listed for convenience of discussion under three headings: (a) energy appropriated by the microorganisms or liberated as heat, (b) simple end products, and (c) humus. They will be considered in order.

---

[2] Amino acids are produced by replacing one of the alkyl hydrogens in an organic acid with NH_2. Acetic acid (CH_3COOH) thereby becomes aminoacetic acid or glycine (CH_2NH_2COOH). Amides are formed from organic acids by replacing the hydroxyl of the carboxyl group with NH_2. Acetic acid (CH_3COOH) thus becomes acetamide (CH_3CONH_2).

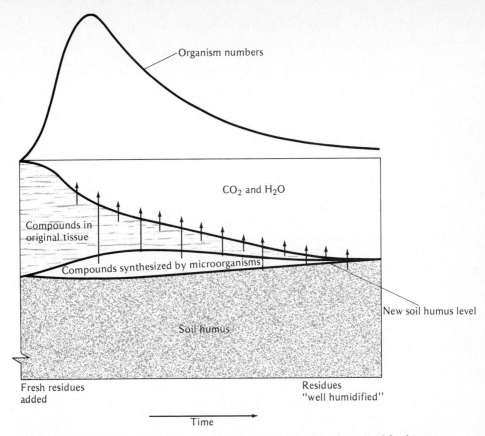

FIGURE 8.3 General changes in the chemical form of carbon- and hydrogen-containing compounds when plant residues are added to soil. Initially, organisms attack easily decomposable compounds such as sugars and celluloses, releasing CO_2 and H_2O and rapidly increasing their own numbers as well as the quantities of new compounds they synthesize. Note that as the initial organism buildup takes place, even the original soil organic matter is subject to some breakdown, CO_2 and H_2O probably being the decomposition products. As soon as the easily decomposed food is exhausted, microorganism numbers decline. The remaining microbes attack the more resistant compounds, both those in the original plant material and those that have been synthesized. In time, modified compounds from the original plant materials and new synthesized compounds, all of which are very slowly decomposed, become indistinguishable from the original soil humus. The time required for the process will depend upon the nature of both the residue and the soil.

8.4 Energy of Soil Organic Matter

The microorganisms of the soil must not only have substance for their tissue synthesis but energy as well. For most of the microorganisms, both of these are obtained from the soil organic matter. All manner of compounds are utilized as energy sources, some freely, others slowly and indifferently.

Potential Energy in Organic Matter. Organic matter contains considerable potential energy, a large proportion of which is readily transferable to other latent forms or is liberated as heat. Plant tissue, such as that entering the soil, has a heat value approximating 5 kcal/g of air-dry substance. For example, the application of 20 Mg (metric tons) of farm manure containing 5000 kg of dry matter would mean an addition in round numbers of 25 million kcal of latent energy. A soil containing 4% of organic matter carries about 400 million kcal of potential energy per hectare–furrow slice. This is equivalent in heat value to about 55 Mg of anthracite coal.

Of this large amount of energy carried by the soil organic matter, only a part is used by soil organisms. The remainder is left in the residues or is dissipated as heat. The natural heating of a compost pile and the spontaneous combustion of hay are practical illustrations of the dissipated heat. This heat loss represents a large and continual removal of energy from the soil.

Rate of Energy Loss from Soils. Certain estimates made at the Rothamsted Experiment Station, England, on two plots of the Broadbalk field give some idea of the rate of energy dissipation from soils (Russell and Russell, 1950). It was calculated that 2.47 million kcal/ha (1 million kcal/A) were lost annually from the untreated, low-producing soil, and about 37 million kcal/ha (15 million kcal/A) were dissipated from the more productive soil, which was receiving liberal supplies of farm manure. The magnitude of such loss is surprising even for the poorer plot.

8.5 Simple Decomposition Products

As the enzymatic changes of the soil organic matter proceed, simple products begin to manifest themselves. Some of these, especially carbon dioxide and water, appear immediately. Others, such as nitrate nitrogen, accumulate only after the peak of the vigorous decomposition has passed and the general purpose decay organisms have diminished in numbers.

The more common simple products resulting from the activity of the soil microorganisms may be listed as follows.

Carbon	CO_2, CO_3^{2-}, HCO_3^-, CH_4, elemental carbon
Nitrogen	NH_4^+, NO_2^-, NO_3^-, gaseous nitrogen

Sulfur	S, H_2S, SO_3^{2-}, SO_4^{2-}, CS_2
Phosphorus	$H_2PO_4^-$, HPO_4^{2-}
Others	H_2O, O_2, H_2, H^+, OH^-, K^+, Ca^{2+}, Mg^{2+}, etc.

Some of the significant relationships of each of the five groups are presented in the following sections. They are considered in the order just listed.

8.6   The Carbon Cycle

Carbon is a common constituent of all organic matter and is involved in essentially all life processes. Consequently, the transformations of this element, termed the *carbon cycle* are in reality a *biocycle* that makes possible the continuity of life on earth. These changes are shown graphically in Figure 8.4. Note that humus and carbon dioxide are relatively stable components of this cycle.

Release of Carbon Dioxide.   Through the process of photosynthesis, carbon dioxide is assimilated by higher plants and converted into numerous organic compounds, classes of which were described in Section 8.2. As these organic compounds reach the soil in plant residues, they are digested and carbon dioxide is given off. Microbial activity is the main soil source of CO_2, although appreciable amounts come from the respiration of rapidly growing plant roots and some is brought down in rain water. Under optimum conditions more than 100 kg/ha of carbon dioxide may be evolved per day, 25–30 kg being more common. Much of the carbon dioxide of the soil ultimately escapes to the atmosphere, where it may again be used by plants, thus completing the cycle.

A much lesser amount of carbon dioxide reacts in the soil, producing carbonic acid (H_2CO_3) and the carbonates and bicarbonates of calcium, potassium, magnesium, and other bases. The bicarbonates are readily soluble and may be removed in drainage or used by higher plants.

Other Carbon Products of Decay.   The simplification of organic matter results in other carbon products. Small quantities of elemental carbon are found in soils. Under highly anaerobic conditions, methane (CH_4) and carbon bisulfide (CS_2) may be produced in small amounts. But of all the simple carbon products, carbon dioxide is by far the most abundant.

It is now obvious that the carbon cycle is all-inclusive since it involves not only the soil and its teeming fauna and flora and higher plants of every description, but also all animal life, including humans. Its failure to function properly would mean disaster to all. It is an energy cycle of such vital import that, with its many ramifications, it might properly be designated the "cycle of life."

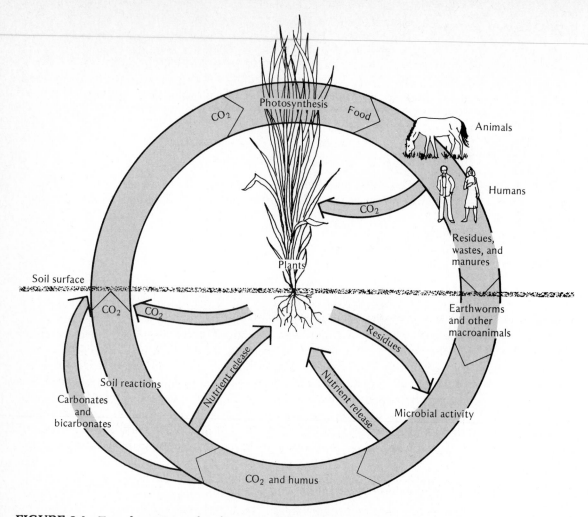

FIGURE 8.4 Transformations of carbon commonly spoken of as the *carbon cycle,* the *biocycle,* or *life cycle.* Plants assimilate CO_2 from the stmosphere into organic compounds using energy from the sun. Man and other higher animals obtain energy and body tissue from plant products and return wastes and residues to the soil. Macro- and microorganisms digest these organic materials, releasing nutrients for plants and leaving CO_2 and humus as relatively stable products. Carbonates and bicarbonates of Ca, Mg, K, etc., are removed in leaching, but eventually the carbon returns to the cycle in the form of CO_2. The total CO_2 is released to the atmosphere where it is again available for plant assimilation. This biocycling illustrates how carbon is the focal point of *energy* transformations and makes possible the continuity of life on earth.

8.7 Simple Products Carrying Nitrogen

Ammonium salts are the first inorganic nitrogen compounds produced by microbial digestion, but other inorganic or mineral forms follow. Hence the process is called *mineralization*. Proteins split up into amino acids and similar nitrogenous materials that readily yield ammonium compounds by enzymatic hydrolysis. These transformations are brought about by a large number of general-purpose heterotrophic organisms—bacteria, fungi, and actinomycetes. The ammonium ion is readily available to microorganisms and most higher plants.

Nitrification. If conditions are now favorable, ammonium ions are subject to ready oxidation, principally by two special-purpose organisms, the nitrite and the nitrate bacteria. The process, *nitrification*, will be discussed in more detail later (see Section 9.7). However, it may be shown simply as follows.

$$2\,NH_4^+ + 3\,O_2 \xrightarrow[\text{oxidation}]{\text{enzymatic}} 2\,NO_2^- + 2\,H_2O + 4\,H^+ + \text{energy}$$

$$2\,NO_2^- + O_2 \xrightarrow[\text{oxidation}]{\text{enzymatic}} 2\,NO_3^- + \text{energy}$$

The autotrophic bacteria obtain energy by these reactions, which produce nitrates that are assimilated by plants and are also subject to leaching loss. The hydrogen ions formed show that nitrification tends to result in an increase in soil acidity. This effect is of even greater importance when dealing with commercial fertilizers such as anhydrous ammonia, urea, and $(NH_4)_2SO_4$ which are sources of ammonium nitrogen. Extra increments of lime are often added to counteract this acidifying effect.

Release of Gaseous Nitrogen. One more group of end products must be considered—gaseous nitrogen compounds. Under certain conditions, the reduction of nitrates and nitrites takes place in soils, and free nitrogen or oxides of nitrogen may be evolved. This transfer is most likely to occur in poorly drained soils or in acid soils containing nitrites. The processes may be chemical as well as biochemical. In any case, the reduction is serious since these nitrogen gases are not readily useful to higher plants. Under conditions of poor soil drainage soils lose considerable nitrogen in this way (see Section 9.10).

8.8 Simple Products Carrying Sulfur

Many organic compounds, including proteins, carry sulfur, which in turn appears in simple forms as decay progresses. General heterotrophic types of organisms apparently simplify the complex organic compounds. The sulfur of these simplified by-products is then subjected to oxidation by special autotrophic bacteria. The final transformation carried to completion by the sulfur-oxidizing organisms may be shown as

$$-[HS] \quad + \; 2\,O_2 \; \xrightarrow[\text{oxidation}]{\text{enzymatic}} \; SO_4{}^{2-} \; + \; H^+ \; + \; \text{energy}$$

Organic
combination

The organisms involved obtain energy by the transfer and leave the sulfur as sulfate, the form in which it is taken up by plants. Note that this process also increases soil acidity.

8.9 Mineralization of Organic Phosphorus

A large proportion of the soil phosphorus is carried in organic combinations (see Section 10.4). Upon attack by microorganisms the organic phosphorus compounds are mineralized; that is, they are changed to inorganic combinations. The particular forms present depend to a considerable degree upon soil pH. As the pH goes up from 5.5 to 7.5, the available phosphorus changes from $H_2PO_4{}^-$ to $HPO_4{}^{2-}$. Both of these forms are available to higher plants. Since the small amount of phosphorus held in complex mineral combinations in soils usually is very slowly available, the organic sources mentioned above become especially important.

It must not be assumed, however, that the maintenance of soil organic matter at normal or even high levels will solve the phosphorus problem. Most field soils require phosphatic fertilizers for best plant growth. Yet strangely enough, the economic use of such phosphorus depends to a considerable degree upon the organic transformation previously described. Since microorganisms utilize phosphorus freely, some of that added in fertilizers quickly becomes part of the soil organic matter. Thus, this phosphorus is held in an organic condition and is later mineralized by microbial activity.

8.10 Humus—Genesis and Definition

The formation of humus, although an exceedingly complicated biochemical process, may be described in general terms rather simply. As organic tissue is incorporated into a moist warm soil it is immediately attacked by a host of different soil organisms. The easily decomposed compounds quickly succumb, first yielding intermediate substances and finally the simple, soluble products already enumerated.

Humus Formation. As this decomposition occurs, two major kinds of organic compounds tend to be stabilized in the soil. First are the microbial-resistant compounds of higher plant origin that have been modified by biochemical reactions, which, in turn, have made the compounds even more resistant to microbial attack. Modified lignin[3] compounds are most abundant, although minor amounts of oils, fats, and waxes may be present. Second are new compounds, such as polysaccharides and polyuronides, which are synthesized by microorganisms and held as part of their tissue. The modified lignin and similar materials are at least partially oxidized during the decomposition, thereby increasing their reactivity. The aromatic groups associated with lignin are bonded with protein substances, thereby protecting the protein nitrogen from further attack.

The compounds of microbial origin are not insignificant in quantity, studies having shown that up to one third of the organic carbon may be in this form. Apparently these two groups of compounds, one modified from the original plant material and one newly synthesized by the microorganisms, provide the basic framework for humus.

As these humic substances form, side reactions occur that bind proteins and other organic nitrogen compounds as integral parts of the humus complex, thereby protecting the nitrogen from degradation to simple inorganic forms. A wide variety of organic materials including lignin, tannins, melanins, and humic acids are known to react with proteins and thereby protect them from microbial attack. Exactly how these reactions occur is not well known. Some nitrogen compounds are thought to react with aromatic and quinone groups as well as polysaccharides. Among the reaction products are amino combinations, in which form about half the bound nitrogen occurs.

Regardless of the mechanism by which nitrogen is bound, the important fact is that the resultant product, newly formed humus, is quite resistant to further microbial attack. The nitrogen and other essential nutrients are thereby protected from ready solubility and dissipation.

Protein–Clay Combinations.[4]   Another means of stabilizing nitrogen in soil is through the reaction between certain clays and proteins and other nitrogen compounds. Clays with expanding crystals such as the smectites (e.g., vermiculite or montmorillonite), seem to have such a faculty; the proteins and other nonionic molecules can function as bases in satisfying the adsorption capacity of the inorganic colloids. The proteins seem to be protected against rapid decomposition either because the protein molecule moves into the interlayer space within a clay crystal and is thereby protected or because the clay inhibits the action of protein-destroying enzymes. The importance of this interaction will vary from soil to soil. Nevertheless, it suggests that intermixed with the nitrogen-containing modified lignin and the polysaccharides there are clay–protein combinations, all of which tend to protect nitrogen from microbial attack.

[3] Lignin is a complex resinous material that impregnates the cell walls of plants as they increase in age. It is exceedingly complex and varies with different plants and different tissues of the same plant. The chemical formulas of the various lignins are in doubt.

[4] For a recent discussion of this subject see Loll and Bollag (1983).

Solubility Groupings. On the basis of resistance to degradation and of solubility in acids and alkalis, humic substances have been classified into three chemical groupings: (a) *fulvic acid,* lowest in molecular weight and lightest in color, soluble in both acid and alkali, and most susceptible to microbial attack; (b) *humic acid,* medium in molecular weight and color, soluble in alkali but insoluble in acid, and intermediate in resistance to degradation; and (c) *humin,* highest in molecular weight, darkest in color, insoluble in both acid and alkali, and most resistant to microbial attack. It should be emphasized, however, that even fulvic acid, the most easily decomposed by microbes, is still quite stable in the soil. Depending on the environment it may take 15–25 years for the destruction in the soil of fulvic acid–type compounds. In general, these materials are all more resistant to microbial attack than most freshly applied crop residues. In spite of differences in chemical and physical makeup, the three groupings of materials have some similarities with regard to such properties as the ability to absorb and release cations. Consequently, they will all be considered under the general term "humus."

Humus Defined. From the previous discussion two facts are obvious: (a) humus is a mixture of complex compounds, not a single material, and (b) these compounds are either resistant materials that have been only slightly modified from the original plant tissue or compounds synthesized by microorganisms in the soil. These two facts lead to the following definition.

> *Humus is a complex and rather resistant mixture of brown or dark brown amorphous and colloidal substances modified from the original tissues or synthesized by the various soil organisms.*

Humus is a naturally occurring material, and although it is exceedingly variable and heterogenous, it possesses properties that distinguish it sharply from the original parent tissues and from the simple products that develop during its synthesis.

8.11 Humus—Nature and Characteristics

In Chapter 5, the colloidal nature of humus was emphasized along with the following major characteristics of this important soil constituent.

1. The tiny colloidal humus particles (micelles) are composed of carbon, hydrogen, and oxygen (probably in the form of modified lignins, polyuronides, and polysaccharides).
2. The surface area of humus colloids is very high, generally exceeding that of silicate clays.
3. The colloidal surfaces of humus are negatively charged, the sources of the charge being carboxylic (—COOH) or phenolic (⟨◯⟩—OH) groups. The extent of the negative charge is pH dependent (high at high pH values).

4. At high pH values the cation exchange capacity of humus far exceeds that of most silicate clays (150–300 cmol/kg).
5. The water-holding capacity of humus on a weight basis is 4–5 times that of the silicate clays (humus will adsorb from a saturated atmosphere water equivalent to 80–90% of its weight).
6. Humus has low plasticity and cohesion, which helps account for its very favorable effect on aggregate formation and stability.
7. The black color of humus tends to distinguish it from most of the other colloidal constituents in soils.
8. Cation exchange reactions with humus are qualitatively similar to those occurring with silicate clays.

The range in complexity, size, and molecular weight of the humic acids has already been discussed (see Section 8.10).

The humic micelles, like the particles of clay, carry a swarm of adsorbed cations (Ca^{2+}, H^+, Mg^{2+}, K^+, Na^+, etc.). Thus, humus colloidally may be represented by the same illustrative formula used for clay.

$$\begin{matrix} Ca \\ Al \\ H \\ M \end{matrix} \boxed{Micelle}$$

and the same reactions will serve to illustrate cation exchange in both (see Section 5.12). M represents other metallic cations, such as potassium, magnesium, and sodium.

Effect of Humus on Nutrient Availability. One particular characteristic of humus merits attention—the capacity of this colloid when saturated with H^+ ions to increase the availability of certain nutrient bases such as calcium, potassium, and magnesium. It seems that an H $\boxed{Micelle}$ of humus, like an H $\boxed{Micelle}$ of clay, acts much like an ordinary acid and can react with soil minerals in such a way as to extract their bases. Acid humus has an unusual capacity to effect such a transfer since the organic acid is comparatively strong. Once the exchange is made, the bases so affected are held in a loosely adsorbed condition and are easily available to higher plants. In the following generalized reaction microcline is used as an example of the various minerals so affected.

$$\underset{\text{Microcline}}{KAlSi_3O_8} + H\boxed{Micelle} \rightarrow \underset{\text{Acid silicate}}{HAlSi_3O_8} + \underset{\text{Adsorbed K}}{K\boxed{Micelle}}$$

The potassium is changed from a molecular to an adsorbed status, which is rated as rather readily available to higher plants.

Humus–Clay Complex in Soils. In considering all the colloidal matter of a mineral soil, we must remember that a mixture of many very different kinds of colloids is involved. The crystalline clayey nucleus is markedly stable under ordinary conditions and is active mainly in respect to cation exchange. Conversely, the amorphous organic micelle is susceptible to slow but continuous microorganic attack and has a twofold activity—cation exchange and nutrient release. These contrasting details are as important practically as the colloidal similarities already noted.

8.12 Direct Influence of Organic Compounds on Higher Plants

One of the early concepts regarding plant nutrition was that organic matter as such is directly absorbed by higher plants. Although this opinion was later discarded, there is some evidence that certain organic nitrogen compounds can be absorbed by higher plants, often rather readily. For example, some amino acids, such as alanine and glycine, can be absorbed directly. Such substances are absorbed in too small quantities to satisfy plant needs for nitrogen, however. The uptake of vanillic acid and other phenol carboxylic acids has been established by the use of radioactive carbon, but the significance of these acids in practical agriculture is not known. The beneficial effects of an exceedingly small absorption of organic compounds might be explained by the presence of growth-promoting substances. In fact, various growth-promoting compounds such as vitamins, amino acids, auxins, and gibberellins are developed as organic matter decays. These may at times stimulate both higher plants and microorganisms.

On the other hand, some soil organic compounds may be harmful. As an example, dihydroxystearic acid, which is toxic to higher plants, has been isolated from soils. However, such "toxic matter" may be a product of unfavorable soil conditions that will disappear when such conditions are corrected. Apparently, good drainage and tillage, lime, and fertilizers reduce the probability of organic toxicity.

8.13 Influence of Soil Organic Matter on Soil Properties

Before discussing the practical maintenance of soil organic matter, a brief review of the most obvious influences of this all-important constituent follows.

1. Effect on soil color—brown to black.
2. Influence on physical properties:
   a. Granulation encouraged.
   b. Plasticity, cohesion, etc., reduced.
   c. Water-holding capacity increased.
3. High cation exchange capacity:
   a. Two to thirty times as great as mineral colloids (weight basis).
   b. Accounts for 30–90% of the adsorbing power of mineral soils.

4. Supply and availability of nutrients:
  a. Easily replaceable cations present.
  b. Nitrogen, phosphorus, sulfur and micronutrients held in organic forms.
  c. Release of elements from minerals by acid humus.

8.14   Carbon/Nitrogen Ratio

Attention has been called several times to the close relationship existing between the organic matter and nitrogen contents of soils. Since carbon makes up a large and rather definite proportion of this organic matter, it is not surprising that the carbon to nitrogen ratio of soils is fairly constant. This fact is important in controlling the available nitrogen, total organic matter, and the rate of organic decay, and in developing sound soil management schemes.

Ratio in Soils.   The ratio of carbon to nitrogen in the organic matter of the furrow slice of arable soils commonly ranges from 8:1 to 15:1, the median being between 10:1 and 12:1. In a given climatic region, little variation is found in this ratio, at least in similarly managed soils. The variations that do occur seem to be correlated in a general way with climatic conditions, especially temperature and the amount and distribution of rainfall. For instance, it is rather well established that the carbon/nitrogen ratio tends to be lower in soils of arid regions than in those of humid regions when annual temperatures are about the same. It is also lower in warmer regions than in cooler ones if the rainfalls are about equal. Also the ratio is narrower for subsoils, in general, than for the corresponding surface layers.

Ratio in Plants and Microbes.   The carbon/nitrogen ratio in plant material is variable, ranging from 20:1 to 30:1 in legumes and farm manure and 100:1 in certain strawy residues and to as high as 400:1 in sawdust. Conversely, the carbon/nitrogen ratio of the bodies of microorganisms is not only more constant but much narrower, ordinarily falling between 4:1 and 9:1. Bacterial tissue in general is somewhat richer in protein than fungi and consequently has a narrower ratio.

It is apparent, therefore, that most organic residues entering the soil carry large amounts of carbon and comparatively small amounts of total nitrogen; that is, their carbon/nitrogen ratio is high, and the C/N values for soils are in between those of higher plants and the microbes.

8.15   Significance of the Carbon/Nitrogen Ratio

The carbon/nitrogen ratio in soil organic matter is important for two major reasons: (a) keen competition among microorganisms for available nitrogen results when residues having a high C/N ratio are added to soils, and (b) because this ratio is relatively constant in soils, the maintenance of carbon—

and hence soil organic matter—depends largely on the soil nitrogen level. The significance of the C/N ratio will become obvious as a practical example of the influence of highly carbonaceous material on the availability of nitrogen is considered.

Practical Example.   Assume that a cultivated soil in a condition favoring vigorous nitrification is examined. Nitrates are present in moderate amounts and the C/N ratio is narrow. The general purpose decay organisms are at a low level of activity, as evidenced by low carbon dioxide production (Figure 8.5).

Now suppose that large quantities of organic residues with a wide C/N ratio (50:1) are incorporated in this soil under conditions supporting vigorous digestion. A change quickly occurs. The heterotrophic flora—bacteria, fungi, and actinomycetes—become active and multiply rapidly, yielding carbon dioxide in large quantities. Under these conditions, nitrate nitrogen practically disappears from the soil because of the insistent microbial demand for this element to build their tissues. And for the time being, little or no mineral nitrogen

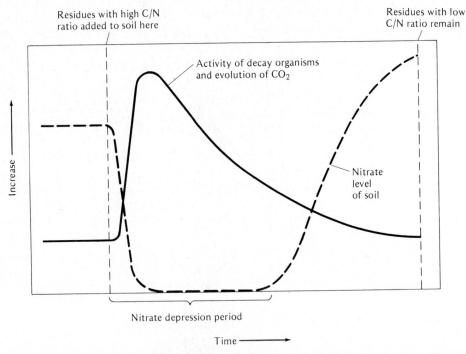

FIGURE 8.5   Cyclical relationship between the stage of decay of organic residues and the presence of nitrate nitrogen in soil. As long as the C/N ratio is large, the general-purpose decay organisms are dominant and the nitrifiers are more or less inactive. During the period of nitrate depression that results, higher plants can obtain little nitrogen from the soil. The length of this period will depend upon a number of factors, of which the C/N ratio is most important.

(NH_4^+ or NO_3^-) is available to higher plants. As decay occurs, the C/N ratio of the remaining plant material decreases since carbon is being lost and nitrogen conserved.

This condition persists until the activities of the decay organisms gradually subside owing to a lack of easily oxidizable carbon. Their numbers decrease, carbon dioxide formation drops off, nitrogen demand by microbes becomes less acute, and nitrification can proceed. Nitrates again appear in quantity and the original conditions again prevail except that, for the time being, the soil is somewhat richer both in nitrogen and humus. This sequence of events, an important phase of the carbon cycle, is shown in Figure 8.5.

Reason for C/N Constancy. As the decomposition processes continue, both carbon and nitrogen are subject to loss—the carbon as carbon dioxide and the nitrogen as nitrates that are leached or absorbed by plants. It is only a question of time until percentage rate of disappearance from the soil becomes approximately the same; that is, the percentage of the total nitrogen being removed equals the percentage of the total carbon being lost. At this point the carbon/nitrogen ratio, whatever it happens to be, becomes more or less constant, always being somewhat greater than the ratios characterizing microbial tissue. As already stated, the carbon/nitrogen ratio in humid temperate region soils, especially those under cultivation, usually stabilizes in the neighborhood of 10:1 to 12:1. The ratio in some cultivated well-weathered tropical soils may be somewhat higher.

Period of Nitrate Depression. The time interval of nitrate depression (Figure 8.5) may be long or short depending on conditions. It may be for only a week or so, or it may last throughout the growing season. The rate of decay will lengthen or shorten the period as the case may be. And the greater the amount of carbonaceous residues applied, the longer will nitrification be blocked. Also, the narrower the C/N ratio of the residues applied, the more rapidly the cycle will run its course. Hence, alfalfa and clover residues interfere less with nitrification and yield their nitrogen more quickly than if oats or wheat straw were plowed under. Also, mature residues, whether legume or nonlegume, have a much higher C/N ratio than do younger succulent materials (Figure 8.6). These facts are of much practical significance and should be considered when organic residues are added to a soil.

Moreover, cultivation by hastening oxidation should encourage nitrification, whereas a vigorous sod crop such as bluegrass should discourage it. This is because the root and top residues of bluegrass maintain a wide C/N ratio, and the small quantities of nitrate and ammonical nitrogen that appear are immediately appropriated by the sod itself. All of this is illustrative of the influence exerted by the C/N ratio on the transfer of nitrogen in the soil and its availability to crop plants.

Carbon/Nitrogen Ratio and Organic Matter Level. In time, carbon and nitrogen in a given surface soil are reduced to a more or less definite ratio (about

FIGURE 8.6 The C/N ratio of organic residues added to soil will depend upon the maturity of the plants turned under. The older the plants, the larger will be the C/N ratio and the longer will be the period of nitrate suppression. Obviously, leguminous tissue has a distinct advantage over nonlegumes since it promotes a more rapid organic turnover in soils.

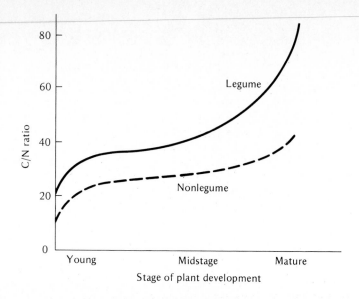

12:1 for many humid temperate region soils). Consequently, the level of soil nitrogen largely determines the level of organic carbon present when stabilization occurs and, in turn, the soil nitrogen content cannot be increased without a corresponding increase in carbon content. Thus, the greater the amount of nitrogen present in the original residue, the greater will be the possibility of an accumulation of organically combined carbon. And since a rather definite ratio (about 1:1.7) exists between the organic carbon and the soil humus, the amount of organic matter that can be maintained in any soil is largely contingent on the amount of organic nitrogen present. The ratio between nitrogen and organic matter is thus rather constant. A value for the organic matter/nitrogen ratio of 20:1 is commonly used for average soils. The practical significance of this relatively constant ratio is that a soil's organic matter content cannot be increased without simultaneously increasing its organic nitrogen content, and vice versa. This should be kept in mind in reading the sections that follow.

8.16   Amount of Organic Matter[5] and Nitrogen in Soils

The amount of organic matter in mineral soils varies widely; mineral surface soils contain from a trace to 20 or 30% of organic matter. The average organic matter and nitrogen contents of large numbers of soils from different areas

[5] No figures are given as to the amount of humus present in mineral soils. This is partly because no very satisfactory method is available for its determination. Unfortunately there is no very satisfactory method of determining directly the exact quantity of total organic matter present in soil. The usual procedure is to find first the amount of organic carbon, which can be done quite accurately. This figure multiplied by the factor 1.7 will give the approximate amount of organic matter present.

TABLE 8.1 Average Nitrogen and Organic Matter Contents and Ranges of Mineral Surface Soils in Several Areas of the United States[a]

Soils	Organic matter (%)		Nitrogen (%)	
	Range	Av.	Range	Av.
240 West Va. soils	0.74–15.1	2.88	0.044–0.54	0.147
15 Pa. soils	1.70–9.9	3.60	—	—
117 Kansas soils	0.11–3.62	3.38	0.017–0.27	0.170
30 Nebraska soils	2.43–5.29	3.83	0.125–0.25	0.185
9 Minn. prairie soils	3.45–7.41	5.15	0.170–0.35	0.266
21 Southern Great Plains soils	1.16–2.16	1.55	0.071–0.14	0.096
21 Utah soils	1.54–4.93	2.69	0.088–0.26	0.146

[a] From Lyon et al. (1952).

of the United States are shown in Table 8.1. The relatively wide ranges in organic matter encountered in these soils even in comparatively localized areas is immediately apparent. Thus, the West Virginia soils have a range from 0.74 to more than 15% of organic matter. The several factors that may account for this wide variability as well as for the differences between averages in Table 8.1 will be considered in Section 8.17.

The data in the table are for surface soils only. The organic matter contents of the subsoils are generally much lower (Figure 8.7). This is readily explained by the fact that most of the organic residues in both cultivated and virgin soils are incorporated in or deposited on the surface. This increases the possibility of organic matter accumulation in the upper layers. Also note that poorly

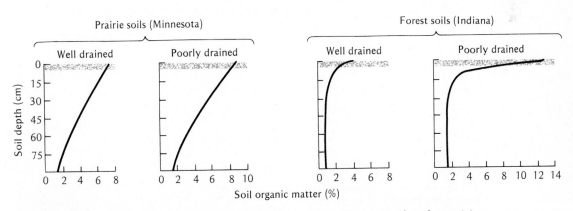

FIGURE 8.7 Distribution of organic matter in four soil profiles. Note that the prairie soils have a higher organic matter content in the profiles as a whole than the soils developed under forest vegetation. Poor drainage results in a higher organic matter content, particularly in the surface horizon.

drained soils, such as those in boggy areas, are always higher in organic matter than their well-drained counterparts.

Organic Matter/Nitrogen Ratio. Data in Table 8.1 illustrate the close correlation between the organic matter and nitrogen contents of soils mentioned in the preceding section. The average organic matter content is about 20 times that of nitrogen. This O.M./N ratio of 20:1 is of considerable value in making rough calculations involving these two constituents.

8.17   Factors Affecting Soil Organic Matter and Nitrogen

Influence of Climate. Climatic conditions, especially *temperature* and *rainfall,* exert a dominant influence on the amounts of nitrogen and organic matter found in soils (see Jenny, 1941). As one moves from a warmer to a cooler climate, the organic matter and nitrogen of comparable soils tend to increase. At the same time, the C/N ratio widens somewhat. In general, the decomposition of organic matter is accelerated in warm climates; a lower loss is the rule in cool regions. Within belts of uniform moisture conditions and comparable vegetation, the average total organic matter and nitrogen increase from two to three times for each 10°C fall in mean annual temperature.

The situation is well illustrated by conditions in the Mississippi Valley region (Figure 8.8). Here the northern Great Plains soils contain considerably greater amounts of total organic matter and nitrogen than those to the south. The influence of temperature on the rapidity of decay and disappearance of organic material is apparent.

Soil moisture also exerts a positive control upon the accumulation of organic matter and nitrogen in soils. Ordinarily, under comparable conditions, the nitrogen and organic matter increase as the effective moisture becomes greater (Figure 8.8). At the same time, the C/N ratio becomes wider, especially in grasslands areas. The explanation lies mostly in the scantier vegetation of the drier regions. In arriving at this rainfall correlation, however, it must be remembered that the level of organic matter in any one soil is influenced by both temperature and precipitation as well as other factors. Climatic influences never work singly.

Influence of Natural Vegetation. It is difficult to differentiate between the effects of climate and vegetation on organic matter and nitrogen contents of soil. Grasslands generally dominate the subhumid and semiarid areas, while trees are dominant in humid regions. In those climatic zones where the natural vegetation includes both forests and grasslands, the organic matter is higher in soils developed under grasslands than under forests (Figure 8.7). Apparently, the nature of the grassland organic residues and their mode of decomposition

FIGURE 8.8 Influence of the average annual temperature and the effective moisture on the organic matter contents of grassland soils of the Midwest. Of course, the soils must be more or less comparable in all respects except for climatic differences. Note that the higher temperatures yield soils lower in organic matter. The effect of increasing moisture is exactly opposite, favoring a higher level of this constituent. These climatic influences affect forest soils in much the same way.

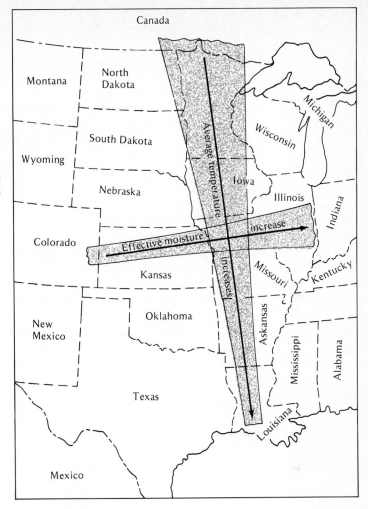

encourage a reduced rate of decay and in turn a higher organic level than is found under forests.

Effect of Texture, Drainage, and Other Factors.   Beside the two broader aspects just discussed, numerous local relationships are involved. In the first place, the *texture* of the soil, other factors being constant, influences the percentage of humus and nitrogen present. A sandy soil, for example, usually carries less organic matter and nitrogen than one of a finer texture (Table 8.2). This is probably because of the lower moisture content and the more ready oxidation occurring in the lighter soils. Also the natural addition of residues normally is less with the lighter soil, and the soils high in clay are able to protect the protein nitrogen and in turn the organic matter from degradation.

TABLE 8.2   Relationship Between Soil Texture
and Approximate Organic Matter Contents of a Number of North Carolina Soils[a]

*Percent of organic matter is calculated by multiply-
ing percent N × 20.*

Soil type	Number of soils	Organic matter (%)	
		Topsoil	Subsoil
Cecil sands	15	0.80	0.50
Cecil clay loams	10	1.32	0.56
Cecil clays	27	1.46	0.64

[a] From Williams et al. (1915).

Again, *poorly drained* soils, because of their high moisture relations and relatively poor aeration, are generally much higher in organic matter and nitrogen than their better drained equivalents (Figure 8.7). For instance, soils lying along streams are often quite high in organic matter. This is due in part to their poor drainage.

The *calcium content* of a soil, its *erosion* status, and its *vegetative cover* are other factors that influence the accumulation and activity of the soil organic matter and its nitrogen.

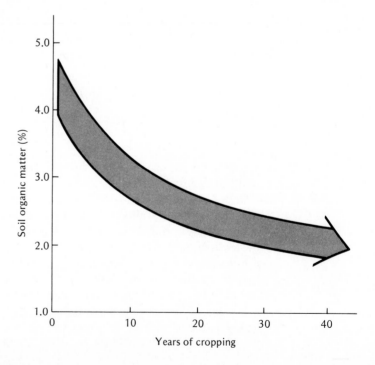

FIGURE 8.9   The general effect of cropping on the organic matter level of soils starting with virgin grasslands. Note the very rapid initial decline and a leveling off with time.

Influence of Cropping. A very marked change in the soil organic matter content occurs when a virgin soil developed under either forest or prairie is brought under cultivation. This is illustrated in Figure 8.9, which shows the general decline in organic matter of grassland soils with time of cultivation. It is common to find cropped land much lower in both nitrogen and organic matter than comparable virgin areas. This is not too surprising, since in nature all the organic matter produced by the vegetation is returned to the soil. By contrast, in cultivated areas much of the plant material is removed for human or animal food and relatively less finds it way back to the land. Also, soil tillage breaks up the organic residues and brings them into easy contact with soil organisms, thereby increasing the rate of decomposition. The depressing effect of cropping on soil organic matter goes far beyond the plow layer, however, as shown by the graph in Figure 8.10. Subsoil layers are also depleted of their organic matter and nitrogen contents.

Soil organic matter and nitrogen contents are also influenced by level of crop production. Field trials in which no fertilizer was used have demonstrated dramatically the marked loss in soil organic matter and nitrogen contents with

FIGURE 8.10 Average organic matter content of three North Dakota soils before and after an average of 43 years of cropping. About 25% of the organic matter was lost from the 0–15 cm layer as a result of cropping. [*From Haas et al. (1957).*]

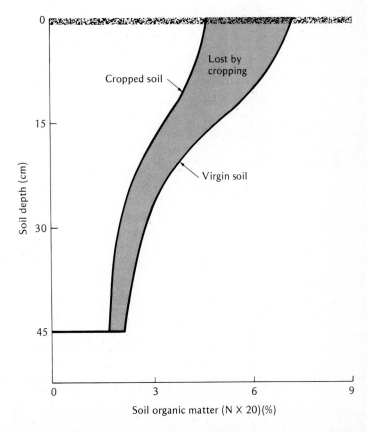

TABLE 8.3 Organic Matter and Nitrogen Contents of Unfertilized Soil Plots at Wooster, Ohio, After Being Cropped in Various Ways for 32 Years[a]

Cropping	Organic matter (Mg/ha)	Nitrogen (kg/ha)
Original cropland	39.2	2439
Continuous corn	14.3	942
Continuous oats	25.6	1597
Continuous wheat	24.7	1474
Corn, oats, wheat, clover, timothy	30.0	1733
Corn, wheat, clover[b]	33.2	1995

[a] Calculated from Salter et al. (1941).
[b] Continued for 29 years instead of 32.

cropping. Crop rotation, especially if legumes are included, helps to maintain soil organic matter (Table 8.3).

Soils that are kept highly productive by supplemental applications of fertilizers, lime, and manures, and by proper choice of disease-free, high-yielding varieties are apt to have higher organic matter contents than comparable less productive soils. The amounts of root and top residues returned to the soil are invariably dependent upon the level of soil productivity. It is well to remember, however, that even the most productive cultivated soil will likely be considerably lower in organic matter than soil in a nearly virgin area.

8.18   Regulation of Soil Organic Matter

The preceding discussion has established two definite conclusions regarding the organic matter and nitrogen of cultivated soils. First, the inherent capacity of soils to produce crops is closely and directly related to their organic matter and nitrogen content. Second, the satisfactory level of these two constituents is difficult to maintain in the majority of farm soils. Consequently, the methods of organic matter additions and upkeep should receive priority consideration in all soil management programs.

Source of Supply.   Organic matter is added to cultivated soils in several ways. One is the plowing under of crops when in an immature, succulent stage. This is called *green-manuring*. Such crops as rye, buckwheat, oats, peas, soybeans, and vetch, as well as others, lend themselves to this method of soil improvement.

A second source of organic matter supply on many farms, especially in dairy sections, is *farm manure*. When applied at the usual rates of 10 or 15 Mg/ha during a five-year rotation, perhaps 500–750 kg/ha of dry matter goes into the soil as a yearly average. Such additions greatly aid in organic matter maintenance.

In the same category with farm manure but of increasing concern today are *artificial farm manures*, and nonfarm organic wastes such as *sludge*, and *composts*. Composts are used mostly in the management of nursery, garden, and greenhouse soils, while land application of sewage sludge is becoming more common in areas near cities and towns (see Section 19.8).

The third and probably the most important source of organic residue is the *current crop* (Figure 8.11). Stubble, aftermath, and especially root residues of various kinds left in or on top of the soil to decay make up the bulk of such contributions. Few farmers realize how much residual root systems aid in the conditioning of their soils. Without them, the practical maintenance of the humus in most cases would be impossible.

In this connection it should also be remembered that the maintaining of high crop yields through proper *liming* and *fertilizer practices* may be as effective as adding manure or other organic residues. In some cases organic matter can be maintained by simply raising bumper crops. Optimum yields usually

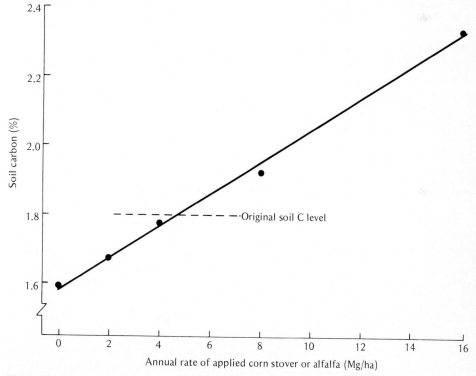

FIGURE 8.11 The effect of applying different rates of residues for 11 consecutive years on the carbon content of a typic Hapludoll soil. The Crop was corn, and conventional tillage was used. [*From Larson et al.* (*1972*); *used with permission of The American Society of Agronomy.*]

mean more residues to return to the soil and certainly increase the amount of roots remaining after harvest.

Crops and Crop Sequence. It has already been suggested that certain sod crops, such as meadow and pasture grasses and legumes, tend to facilitate the maintenance of humus. This is partly due to their liberal contributions of organic residues, the slow decay of these materials, and their wide C/N ratios. Under these conditions, little nitrogen can appear in the soil in the nitrate form and therefore only slight losses of this constituent will occur. In legumes even these losses may be partially offset by nitrogen fixation. Because the amount of humus depends to a considerable extent upon the amount of organic nitrogen, sod crops by their nitrogen economy promote the highest possible yields of humus.

Cultivated crops, on the other hand, generally are not humus conservers. Tillage and other features of their management encourage an extremely rapid

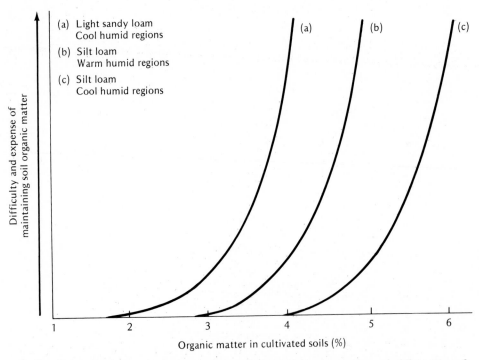

FIGURE 8.12 How the difficulty and expense of maintaining the organic matter of cultivated soils increase as the support level is raised. While the curves are quite similar, their position in relation to the percentage of organic matter possible varies with *texture* [compare (a) with (c)] and *climate* [compare (b) with (c)]. Other factors also are involved, especially the type of crop rotation employed. In practice the average amount of organic matter in a cultivated soil should be maintained as high as is economically feasible.

rate of decay and dissipation of organic matter. Thus, intertilled crops are associated with humus reduction instead of humus accumulation. Small grains, such as oats and wheat, are in the same category; they are generally humus wasters but to a lesser degree.

In practice, crop sequences may be so arranged as to offset depreciation with humus conservation. Much can be accomplished by using a suitable rotation involving some sod crops (Table 8.3). Also, adequate liming and fertilization to provide bumper crops encourages residues that help maintain the organic level.

The growing adoption of "no tillage" practices in the United States may influence future soil organic matter levels (see Sections 2.14 and 16.12). Crop residues are left mostly on the soil surface where these practices are used. Consequently, the organic residues are more slowly broken down. In time this could lead to higher soil organic matter levels.

Economy in Humus Maintenance. Since the rate at which carbon is lost from the soil increases very rapidly as the organic content is raised, maintenance of the humus at a high level is not only difficult but also expensive (Figure 8.12). It is therefore unwise to hold the organic matter above a level consistent with crop yields that pay best. Just what this level should be will depend on climatic environments, soil conditions, and the particular crops grown and their sequence. Obviously, it should be higher in North Dakota than in central Kansas, where the temperature is higher, or in northern Montana, where the effective rainfall is lower. In any event, the soil organic matter and nitrogen should always be maintained at as high a level as economically feasible.

REFERENCES

Allison, F. E. 1973. *Soil Organic Matter and Its Role in Crop Production* (Amsterdam: Elsevier).

Haas, H. J., et al. 1957. *Nitrogen and Carbon Changes in Great Plains Soils as Influenced by Cropping and Soil Treatments,* Tech. Bull. 1167 (Washington, DC: U.S. Department of Agriculture).

Jenny, H. 1941. *Factors of Soil Formation* (New York: McGraw-Hill).

Larson, W. E., C. E. Clapp, W. H. Pierre, and Y. B. Morachan. 1972. "Effect of increasing amounts of organic residues on continuous corn: organic carbon, nitrogen, phosphorus, and sulfur," *Agron. J.,* **64**:204–208.

Loll, M. J., and J. M. Bollag. 1983. "Protein transformation in soil," *Adv. in Agron.,* **36**:352–382.

Lyon, T. L., H. O. Buckman, and N. C. Brady. 1952. *The Nature and Properties of Soils,* 5th ed. (New York: Macmillan), p. 171.

Russell, E. J., and E. W. Russell. 1950. *Soil Conditions and Plant Growth* (London: Longmans, Green), p. 194.

Salter, R. M., R. D. Lewis, and J. A. Slipher. 1941. *Our Heritage—The Soil,* Extension Bull. 175, Ohio Agr. Exp. Sta., p. 9.

Stevenson, F. J. 1982. "Origin and distribution of nitrogen in soil," in F. J. Stevenson (Ed.), *Nitrogen in Agricultural Soils* (Madison, WI: Amer. Soc. Agron., Crop Sci. Soc. Amer., and Soil Sci. Soc. Amer.), pp. 1–42.

Waksman, S. A. 1948. *Humus* (Baltimore: Williams & Wilkins), p. 95.

Williams, C. B., et al. 1915. *Report on the Piedmont Soils Particularly with Reference to Their Nature, Plant-Food Requirements and Adaptability for Different Crops,* Bull. 36, No. 2, North Carolina Dept. Agr.

Nitrogen and Sulfur
Economy of Soils

9

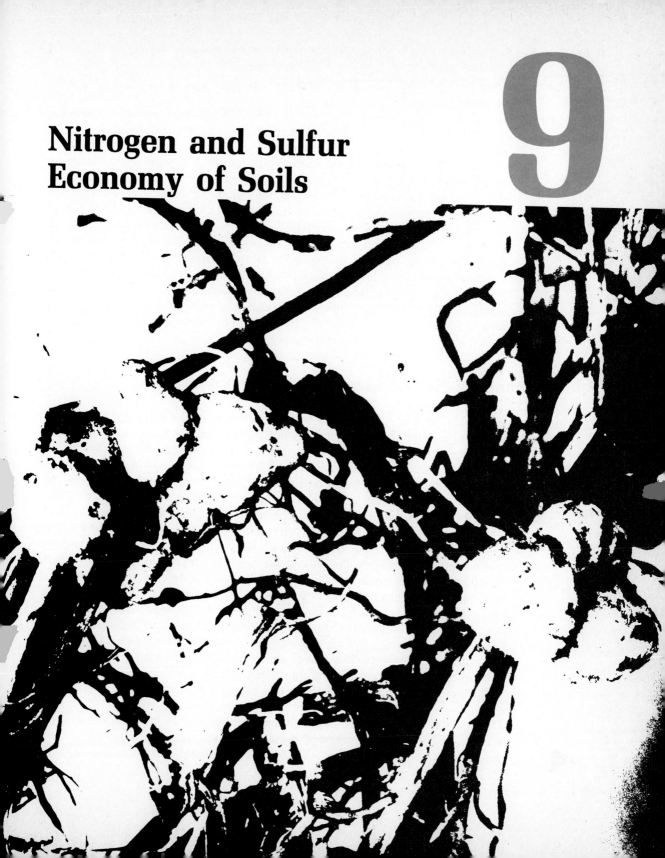

Of the various essential elements, nitrogen probably has been subjected to the most study and still receives much attention. And there are very good reasons. The amount of this element in available forms in the soil is small, while the quantity withdrawn annually by crops is comparatively large. At times, there is too much nitrogen in readily soluble forms, and it is lost in drainage, and may even become a water pollutant; at other times, it is lost from the soil by volatilization; at all times, most of the soil nitrogen is unavailable to higher plants. Moreover, its effects on plants usually are very marked and rapid. Thus overapplication, which may be harmful, sometimes occurs.

Some plants, such as the legumes, have associated with their roots soil organisms that "fix" atmospheric nitrogen into forms that they can use. Others, such as the grasses, are largely dependent upon outside sources—either through nonsymbiotic fixation or through the addition of fertilizers or other combined forms of nitrogen. All in all, nitrogen is a potent nutrient element that should be not only conserved but carefully regulated.

9.1 Influence of Nitrogen on Plant Development

Favorable Effects. Nitrogen is an integral component of many compounds essential for plant growth processes including chlorophyll and many enzymes. It is an essential component of the proteins and related amino acids, which are critical not only as building blocks for plant tissue but in the cell nuclei and protoplasm in which the hereditary control is vested. It is essential for carbohydrate utilization within plants and stimulates root growth and development as well as the uptake of other nutrients. Of the macronutrients usually applied in commercial fertilizers, nitrogen seems to have the quickest and most pronounced effect. It encourages aboveground vegetative growth and imparts deep green color to the leaves. With cereals, it increases the plumpness of the grain and the percentage of protein. With all plants, nitrogen is a regulator that governs to a considerable degree the utilization of potassium, phosphorus, and other constituents. Moreover, its application tends to produce succulence, a quality particularly desirable in such crops as lettuce and radishes.

Plants receiving insufficient nitrogen are stunted in growth and possess restricted root systems. The leaves turn yellow or yellowish green and tend to drop off. The addition of available nitrogen will cause a remarkable change, indicative of the unusual activity of this element within the plant.

Oversupply. When higher applications of nitrogen than necessary are made, the leaves become dark green in color and excess vegetative growth occurs. As a result, the stems are not able to hold the plants upright and they lodge or fall over with the slightest wind. Also, crop maturity is delayed and the plants are more susceptible to diseases and insect pests. An oversupply of nitrogen may adversely affect grain and fruit quality, as in barley, apples, and peaches. It also adversely affects sugar levels in sugar cane or sugar beets.

It must not be inferred, however, that all plants are detrimentally affected by large amounts of nitrogen. Many crops, such as the grasses and vegetables, should have plenty of this element for their best and most normal development. Detrimental effects are not to be expected with such crops unless exceedingly large quantities of nitrogen are applied. Nitrogenous fertilizers may be used freely in such cases, the cost of the materials in respect to the value of the crop increases being the major consideration.

9.2 Forms of Soil Nitrogen

There are three major forms of nitrogen in mineral soils: (a) organic nitrogen associated with the soil humus, (b) ammonium nitrogen fixed by certain clay minerals, and (c) soluble inorganic ammonium and nitrate compounds.

Most of the nitrogen in surface soils is associated with the organic matter. In this form it is protected from rapid microbial release, only 2–3% a year being mineralized under normal conditions. About half the organic nitrogen is in the form of amino compounds. The form of the remainder is uncertain.

Some of the clay minerals (e.g., vermiculites and some smectites) have the ability to fix ammonium nitrogen between their crystal units. The amount fixed varies depending on the nature and amount of clay present. Up to 8% of the total nitrogen in surface soils and 40% of that in subsoils may be in the "clay-fixed" form. In most cases, however, both these figures would be considerably lower. Even so, the nitrogen so fixed is only slowly available to plants and microorganisms.

The amount of nitrogen in the form of soluble ammonium and nitrate compounds is seldom more than 1–2% of the total present, except where large applications of inorganic nitrogen fertilizers have been made. This is fortunate since inorganic nitrogen is subject to loss from soils by leaching and volatilization. Only enough is needed to supply the daily requirements of the growing crops.

9.3 The Nitrogen Cycle[1]

In all soils the considerable intake and loss of nitrogen in the course of a year are accompanied by many complex transformations. Some of these changes may be controlled more or less by man, whereas others are beyond

[1] For this and subsequent sections on nitrogen, the reader is referred to the monograph edited by Stevenson (1982).

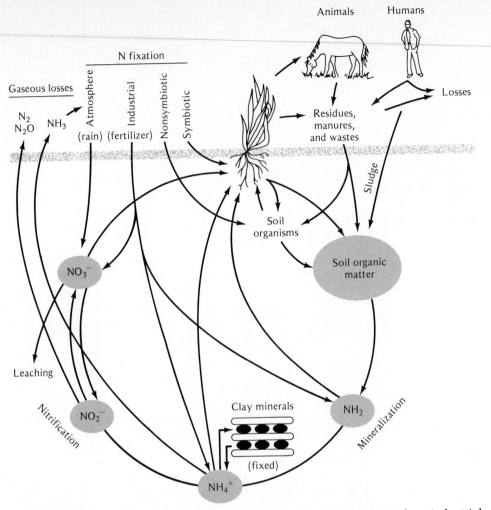

FIGURE 9.1 Main portions of the nitrogen cycle. Chemical fertilizers from industrial nitrogen fixation are an increasingly important source of this element.

his command. This interlocking succession of largely biochemical reactions constitutes what is known as the *nitrogen cycle* (Figure 9.1). It has attracted scientific study for years, and its practical significance is beyond question.

The nitrogen income of arable soils is derived from such materials as commercial fertilizers, crop residues, green and farm manures, and ammonium and nitrate salts brought down by precipitation. In addition, there is the fixation of atmospheric nitrogen accomplished by certain microorganisms. The depletion is due to crop removal, drainage, erosion, and to loss in a gaseous form.

Much of the nitrogen added to the soil undergoes many transformations before it is removed. The nitrogen in organic combination is subjected to espe-

FIGURE 9.2 Two major pools of soil nitrogen, their primary sources and processes of transfer.

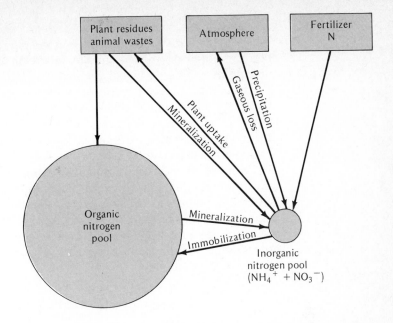

cially complex changes. Proteins are converted into various decomposition products, and finally some of the nitrogen appears in the nitrate form. Even then it is allowed no rest since it is either appropriated by microorganisms and higher plants, removed in drainage, or lost by volatilization. And so the cyclic transfer goes on and on. The mobility of nitrogen is remarkable, rivaling carbon in its ease of movement.

Major Divisions of the Nitrogen Cycle.   At any one time, the great bulk of the nitrogen in a soil is in organic combinations protected from loss but largely unavailable to higher plants. For this reason much scientific effort has been devoted to the study of organic nitrogen—how it is stabilized and how it may be released to forms usable by plants. The process of tying up nitrogen in organic forms is called *immobilization;* its slow release—specifically, organic to inorganic conversion—is called *mineralization* (Figure 9.2).

9.4   Immobilization and Mineralization

During the process of microbial decomposition of plant and animal residues, especially those low in nitrogen, the plant inorganic nitrogen as well as that in the soil is converted to organic forms (see Section 8.15) primarily as microbial tissue. As the rate of microbial activity subsides, some of this *immobilized* nitrogen will be mineralized and ammonium and nitrate ions will again appear in the solution. However, most of the immobilized nitrogen remains in the organic form.

The mechanism by which simple nitrogen compounds are changed to organic combinations that resist breakdown is still obscure (see Section 8.10). Somehow the immobilized nitrogen in the microbial tissue becomes an integral part of the soil organic matter. In this form it is only slowly mineralized to compounds usable by higher plants.

Isotopically tagged nitrogen experiments have demonstrated that only about 2–3% of the immobilized nitrogen is mineralized annually. Even so, this release of nitrogen to inorganic forms has long supplied a significant portion of crop needs and may be about 60 kg/ha of nitrogen per year for a representative mineral surface soil.

Although the general term *mineralization* covers a whole series of reactions, the net effect can be rather simply visualized. Heterogenous soil organisms, both plant and animal, attack the organic nitrogen compounds. As a result of enzymatic digestion, the more complex proteins and allied compounds are simplified and hydrolyzed. The end product is ammonia. The enzymatic process may be indicated as follows, using an amino compound as an example of the nitrogen source.

$$\underset{\substack{\text{Amino} \\ \text{combination}}}{R\text{—}NH_2} + HOH \xrightarrow[\text{hydrolysis}]{\text{enzymatic}} R\text{—}OH + NH_3 + \text{energy}$$

$$2\,NH_3 + H_2CO_3 \longrightarrow (NH_4)_2CO_3 \rightleftharpoons 2\,NH_4^+ + CO_3^{2-}$$

Mineralization seems to proceed to the best advantage in well-drained aerated soils with plenty of basic cations present. It will take place to some extent under almost any conditions, however, because of the great number of different organisms capable of accomplishing such a change. This is one of the advantages of a general-purpose flora and fauna.

9.5   Utilization of Ammonium Compounds

The fate of the ammonium nitrogen (as shown in Figure 9.1) is fourfold. First, considerable amounts are appropriated by organisms capable of using this type of compound, for example, mycorrhizal fungi (see Section 7.8) undoubtedly are able to absorb ammoniacal nitrogen and pass it on in some form to their host.

Second, higher plants are able to use this form of nitrogen, often very readily. Young plants of almost all kinds are especially capable in this respect, although they seem to grow better if some nitrate nitrogen is also available. Azaleas, laurel, and other plants requiring low-lime soil are additional examples. Still other plants, such as blueberries and some lowland rices, even prefer ammonium nitrogen to the nitrate.

Third, ammonium ions are subject to fixation by vermiculite and to a certain extent by some smectites and organic matter. In this fixed form, the nitrogen is not subject to rapid oxidation, although in time it may become available. This will receive attention in the next section.

Finally, when plant and animal syntheses temporarily are satisfied, the remaining ammonium nitrogen may go in a third direction. It is readily oxidized by certain special purpose forms of bacteria, which use it not only as a source of nitrogen but also as a source of energy. Thus, a much discussed and perhaps comparatively overemphasized phase of biochemistry—*nitrification*—is reached. It is so named because its end product is nitrate nitrogen, and it will receive attention following a consideration of ammonia fixation.

9.6   Ammonia Fixation

Both the organic and inorganic soil fractions have the ability to bind or "fix" ammonia in forms relatively unavailable to higher plants or even microorganisms. Since different mechanisms and compounds are involved in these two types of fixation, they will be considered separately.

Fixation by Clay Minerals.   Several clay minerals with a 2:1-type structure have the capacity to "fix" ammonium and potassium ions. Vermiculite has the greatest capacity, followed by the smectites.

It will be remembered that these minerals have internal negative charges that attract cations to internal surfaces between crystal units (see Section 5.6). Most cations that satisfy these charges can move freely into and out of the crystal. In other words, they are exchangeable. Ammonium and potassium ions, however, are apparently just the right size to fit into the "cavities" between crystal units, thereby becoming fixed as a rigid part of the crystal (see Figure 10.16). They prevent the normal expansion of the crystal and in turn are held in a nonexchangeable form, from which they are only slowly released to higher plants and microorganisms. The relationship of the various forms of ammonium might be represented as follows.

$$
\underset{\text{(soil solution)}}{NH_4^+} \quad \rightleftharpoons \quad \underset{\text{(exchangeable)}}{NH_4^+} \quad \rightleftharpoons \quad \underset{\text{(fixed)}}{NH_4^+}
$$

Ammonium fixation by clay minerals is generally greater in subsoils than in topsoil because of the higher clay content of subsoils (Table 9.1). In some cases this fixation may be considered to be an advantage since it is a means of conserving soil nitrogen. In others, the rate of release of the fixed ammonium is too slow to be of much practical value.

Fixation by Organic Matter.   Anhydrous ammonia or other fertilizers that contain free ammonia or that form it when added to the soil can react with soil organic matter to form compounds that resist decomposition. In this sense the ammonia can be said to be "fixed" by the organic matter. The exact mecha-

TABLE 9.1   Percentage of Total Nitrogen Fixed as NH_4^+ Ion in Clay Minerals at Different Depths in Three Trinidad Soils[a]

These soils had significant levels of 2:1-type clays that are able to fix large quantities of NH_4^+ ion.

Soil	Composition of clay fraction	Soil depth (cm)	Fixed NH_4^+ nitrogen as % of total N
Maracas	Mica, kaolinite, geothite	0–10	28.5
		10–30	43.4
		50–70	72.9
McBean	Illite, kaolinite, iron oxides	0–10	24.2
		10–30	33.2
		50–70	68.9
Tarouba	Montmorillonite	0–10	18.9
		10–30	48.1
		50–70	63.5

[a] From Dalal (1977).

nism by which the fixation occur is not known, although reactions with phenols and quinones as well as with carbohydrates are suspected (Nommik and Vahtras, 1981). The reaction takes place most readily in the presence of oxygen and at high pH values.

The practical significance of organic fixation depends upon the circumstances. In organic soils with a high fixing capacity it could be serious and would dictate the use of fertilizers other than those which supply free ammonia. In normal practice on mineral soils, however, organic fixation should not be too disadvantageous. In the first place, the fertilizers is often banded, thereby contacting a relatively small portion of the entire soil mass and minimizing opportunities for fixation. Furthermore, the fixed ammonia is subject to subsequent slow release by mineralization.

9.7   Nitrification

Nitrification is a process of enzymatic oxidation of ammonia to nitrates brought about by certain microorganisms in the soil. It takes place in two coordinated steps. The first step is the production of nitrite ions by one group of organisms apparently followed immediately by their oxidation to the nitrate form by another. The enzymatic oxidation is represented very simply as follows.

$$NH_4^+ \xrightarrow{[O]} HONH_2 \xrightarrow{-2H} \tfrac{1}{2}\,HONNOH \xrightarrow{[O]} NO_2^- + H^+ + energy$$

Ammonium   Hydroxylamine   Hyponitrite   Nitrite

$$\downarrow [O]$$

$$NO_3^- + energy$$

Nitrate

Under most conditions favoring the two reactions, the second transformation is thought to follow the first so closely as to prevent any great accumulation of the nitrite. This is fortunate since this ion in any concentration is toxic to higher plants. In very alkaline soils or where large fertilizer ammonia applications have been made, there is some evidence that the second reaction may be delayed until after the ammonium ion concentration is reduced to a relatively low level. This may result in nitrite accumulation of sufficient magnitude to have an adverse effect on plant growth or to encourage gaseous losses of nitrogen (see Section 9.10).

Organisms Concerned. Although some heterotrophic bacteria, fungi, and actinomycetes are able to convert ammonia to nitrites, the bulk of this oxidation is due to a group of autotrophic bacteria including *Nitrosomonas, Nitrosolobus,* and *Nitrosospira.* Organisms responsible for the oxidation of nitrites to nitrates are the autotrophic bacteria, *Nitrobacter.* In common usage these groups are referred to as nitrite and nitrate organisms.

Rate of Nitrification. Under ideal temperature, soil, and moisture conditions, nitrification occurs at a very rapid rate, especially where adequate ammonium ion is available. Under ideal conditions the nitrifying organisms can supply nitrates at a rate that more than meets the needs of crop plants and may result in excess nitrates, some of which are lost in drainage or by volatilization (see Section 9.9).

9.8   Soil Conditions Affecting Nitrification

The nitrifying bacteria are quite sensitive to their environment, much more so than most heterotrophic organisms. Consequently, soil conditions that influence the vigor of nitrification deserve practical consideration. They are (a) a presence of adequate ammonium ion for oxidation, (b) aeration, (c) temperature, (d) moisture, (e) active lime, (f) fertilizer salts, (g) the carbon/nitrogen ratio, and (h) pesticides.

Ammonia Level. Nitrication can take place only if there is a source of ammonia to be oxidized. Factors such as a high C/N ratio of residues, which prevents the release of ammonia, also prevent nitrification. However, if the ammonia is present at too high a level, it constrains nitrification. Heavy localized applications to alkaline soils of anhydrous ammonia fertilizer or urea, which by hydrolysis forms ammonia, appear to be toxic to *Nitrobacter,* resulting in the accumulation of toxic levels of nitrite. Obviously, the level of ammonia is an important regulator of nitrification.

Aeration. Since nitrification is a process of oxidation, any practice that increases the soil aeration should, up to a point, encourage it. Plowing and cultivation, especially if granulation is not impaired, are recognized means of promoting nitrification. Rates of nitrification are generally somewhat slower under minimum tillage than where plowing and cultivation are practiced.

Temperature.   The temperature most favorable for the process of nitrification is from 27 to 32°C (80–90°F). In temperate regions the process proceeds slowly at temperatures slightly above freezing and increases until the optimum temperature is reached. Nitrification declines at higher temperatures, and essentially ceases at temperatures above 50°C (122°F). The lateness with which nitrification attains its full vigor in the spring is well known, application of fertilizer nitrogen being used to offset the delay.

Moisture.   Nitrification is markedly influenced by soil water content, the process being retarded by both very low and very high moisture conditions. In practice, it is safe to assume that the optimum moisture as recognized for higher plants is also optimum for nitrification. One reservation must be made, however. Nitrification will progress appreciably at soil moisture contents at or even below the wilting coefficient.

Exchangeable Bases and pH.   Nitrification proceeds most rapidly where there is an abundance of exchangeable bases. This accounts in part for the weak nitrification in acid mineral soils and the seeming sensitiveness of the organisms to a low pH. However, within reasonable limits, acidity itself seems to have little influence on nitrification when adequate bases are present. This is especially true of peat soils. Even at pH values below 5 these soils may show remarkable accumulations of nitrates (see Section 14.8).

Fertilizers.   Small amounts of many kinds of salts, even those of the trace elements, stimulate nitrification. A reasonable balance of nitrogen, phosphorus, and potassium has been found to be helpful. Apparently, the stimulation of the organisms is much the same as is that of higher plants.

Applications of large quantities of ammonium nitrogen to strongly alkaline soils have been found to depress the second step in the nitrification reaction. Apparently the ammonia is toxic to the *Nitrobacter* under these conditions but does not affect adversely the *Nitrosomonas*. Consequently, nitrite accumulation may occur in soils very high in pH. Similarly, on such soils adverse effects may result from a compound such as urea, which supplies ammonium ions in the soil by hydrolysis.

Carbon/Nitrogen Ratio.   The significance of the carbon/nitrogen ratio has been rather fully considered (Section 8.14), so the explanation here may be brief. When microbes decompose plant and animal residues with high C/N ratios, they incorporate into their bodies all the inorganic nitrogen. It is thereby immobilized. Nitrification is thus more or less at a standstill because of a lack of ammoniacal nitrogen, this also having been swept up by the general purpose decay organisms. A serious competition with higher plants for nitrogen is thereby initiated.

After the carbonaceous matter has partially decomposed so that emerging material is no longer abundant and the C/N ratio of the remaining residue reduced, some of the immobilized nitrogen will be mineralized and ammonium compounds will again appear in the soil. Conditions are now favorable for nitrification and nitrates may again accumulate. Thus, the carbon/nitrogen ratio, through its selective influence on soil microorganisms, exerts a powerful control on nitrification and the presence of nitrate nitrogen in the soil (see Figure 8.5).

Pesticides. Nitrifying organisms are quite sensitive to some pesticides. If added at high rates many of these chemicals essentially exhibit nitrification whereas others slow the process down. Most studies suggest, however, that at ordinary field rates the majority of the pesticides have only minimal effects on the process. This subject is covered in greater detail in Chapter 20.

9.9 Fate of Nitrate Nitrogen

The nitrate nitrogen of the soil, whether added in fertilizers or formed by nitrification, may go in four directions (Figure 9.1). It may (a) be incorporated into microorganisms, (b) be assimilated into higher plants, (c) be lost in drainage, and (d) escape from the nitrogen cycle in a gaseous condition.

Use by Soil Organisms and Plants. Both plants and soil microorganisms readily assimilate nitrate nitrogen. However, if microbes have a ready food supply (for example, carbonaceous organic residues) they utilize the nitrates more rapidly than do higher plants. Thus, the higher plants have only what is left by the microorganisms and must await for the subsequent release of the nitrogen when microbial activity slows down (see Section 8.15). This is one of the reasons crops often are able during a growing season to recover only about half the fertilizer nitrogen added. Fortunately, some of the immobilized nitrogen is released during the following growing seasons.

Leaching and Gaseous Loss. Negatively charged nitrates are not absorbed by the negatively charged colloids that dominate most soils. Consequently, nitrates are subject to ready leaching from the soil, moving downward freely with the water. The amount of nitrate nitrogen lost in drainage water depends upon the climate and cultural conditions; it is low in unirrigated arid and semiarid regions and high in humid areas and where irrigation is practiced. Heavy nitrogen fertilization, especially for vegetables and other cash crops grown on coarse-textured soils, accentuates loss by this means. In recent years, widespread use of high levels of nitrogen fertilizers applied broadcast before the crop is planted has increased nitrate leaching losses. Coupled with concentrated losses from huge feedlots in some areas of the midwest and the great plains states, the nitrates in drainage water in a few areas may have increased to levels that could be harmful to humans and other animals. The need to take steps to minimize these losses is obvious. This subject is given further consideration in Section 20.11.

Gaseous loss of nitrogen from soils may occur in several forms. Perhaps the most important is through denitrification whereby nitrates are reduced to nitrogen oxide compounds and to elemental nitrogen. This process may be carried out by soil microorganisms or through purely chemical reactions. In alkaline soils ammonia gas may also be lost to the atmosphere in significant quantities. These loss mechanisms will each be discussed briefly.

9.10   Denitrification

Nitrates are subject to reduction in soils, especially in those that are poorly drained and low in aeration. The process is known as *denitrification*. Among the reduction products are nitrogen gases, which can be lost into the atmosphere. The nitrate reduction is due primarily to microbial action, although some chemical reduction occurs.

Reduction by Organisms.   The reduction of nitrate nitrogen to gaseous compounds is the most widespread type of volatilization. Biochemical reduction is most common. The microorganisms involved are common facultative anaerobic forms. They prefer elemental oxygen but under inadequate aeration can use the combined oxygen in nitrates and some of their reduced products. The exact mechanisms by which the reductions take place are not known. However, the general trend of the reactions may be represented as follows.

$$2\,NO_3^- \xrightarrow{-2\,[O]} 2\,NO_2^- \xrightarrow{-2\,[O]} 2\,NO \xrightarrow{-\,[O]} N_2O \xrightarrow{-[O]} N_2$$

Nitrates	Nitrites	Nitric oxide	Nitrous oxide	Elemental nitrogen

Each step in the reaction is triggered by a specific reductase enzyme. Thus, nitrate reductase, nitrite reductase, nitric oxide reductase, and nitrous oxide reductase all appear to be involved. It should be emphasized, however, that the reaction can stop at any stage and the gaseous products released (Figure 9.3).

Under field conditions, nitrous oxide and elemental nitrogen are lost in largest quantities, N_2O dominating if ample NO_2^- is present and if the soil is not too low in oxygen (Figure 9.4). Nitric oxide loss is generally not great and apparently occurs most readily under acid conditions.

It should be noted that N_2O can be formed in small quantities by certain *Nitrosomonas* bacteria during ammonia oxidation in aerobic soils. While this

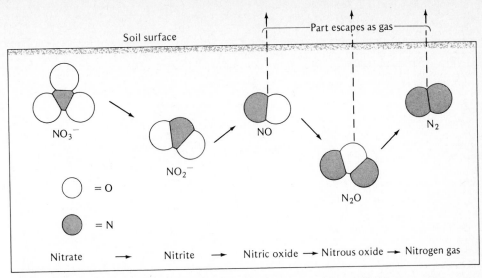

Soil surface

Part escapes as gas

NO_3^-

NO_2^-

NO

N_2O

N_2

○ = O

● = N

Nitrate → Nitrite → Nitric oxide → Nitrous oxide → Nitrogen gas

FIGURE 9.3 Illustration of the reductive nature of the denitrification process. Organisms take an oxygen from nitrate and leave the nitrite ion, which in turn loses an oxygen and produces nitric oxide, and so on. [Modified from Hughes (1980); used with permission of Deere & Company, Moline, IL.]

observation is of some scientific interest, it is not thought to be of much practical importance.

Chemical Reduction. There are nonmicrobial processes by which nitrogen may be reduced in soils to gaseous forms. For instance, nitrites in a slightly acid solution will evolve gaseous nitrogen when brought in contact with certain ammonium salts, with simple amino compounds such as urea, and even with lignins, phenols, and carbohydrates. The following reaction suggests what may happen to urea.

$$2\ HNO_2\ +\ CO(NH_2)_2\ \longrightarrow\ CO_2\ +\ 3\ H_2O\ +\ 2\ N_2 \uparrow$$

Nitrite   Urea

This type of gaseous loss is strictly chemical and does not require either the presence of microorganisms or adverse soil conditions. Its practical importance, however, is not too great.

Quantity of Nitrogen Lost Through Denitrification. Since gaseous nitrogen is unavailable to higher plants, any loss in this form is serious. As might be expected, the exact magnitude of the losses will depend upon the cultural and soil conditions. In well-drained humid-region soils that are not too heavily fertilized, the gaseous losses are probably less than those from leaching. Where drainage is restricted, and where large applications of ammonia and urea ferti-

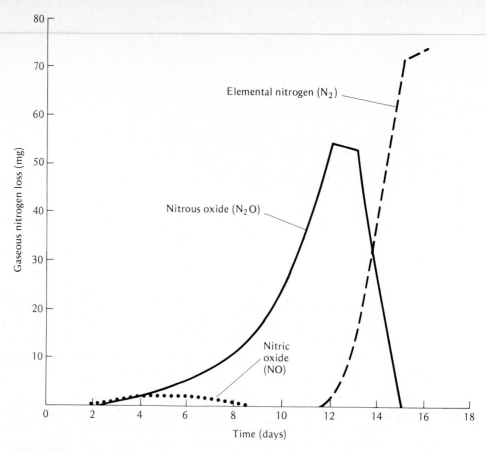

FIGURE 9.4 Denitrification loss of three nitrogen gases fron an anaerobic acid Norfolk sandy loam at 12.5% moisture. A closed system was used. Apparently N_2 was formed from the reduction of N_2O. Under field conditions much of the N_2O would probably have gone off as a gas and would not have remained to produce elemental N_2. [*From Cady and Bartholomew* (*1960*).]

lizer are made, substantial losses might be expected, 20–40% of nitrogen added not being too uncommon.

In flooded soils such as those used for rice culture, losses by denitrification may be very high. Often 60–70% of the applied fertilizer nitrogen is volatilized as oxides of nitrogen or elemental nitrogen. The reason for this high loss is simple. Ammonium-containing fertilizers added to or near the soil surface are oxidized to nitrates in the thin oxidized layer just below the soil–water interface. The nitrates then move down the profile into the reduced soil zone below, which is very low in oxygen (Figure 9.5). Denitrification then occurs and losses as N_2 and N_2O take place. Luckily this loss can be cut back by incorporating

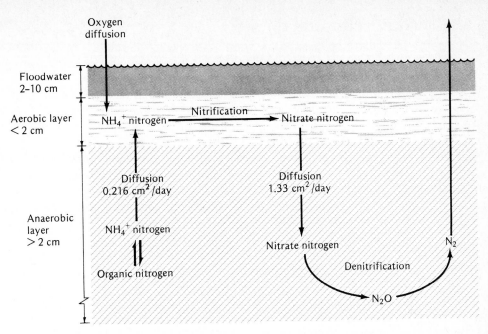

FIGURE 9.5 Nitrification–denitrification reaction and kinetics of the related processes controlling nitrogen loss from the aerobic–anaerobic layer of a flooded soil system. Nitrates, which form in the thin aerobic soil layer just below the soil–water interface, diffuse into the anaerobic (reduced) soil layer below and are denitrified to the N_2 and N_2O gaseous forms, which are lost to the atmosphere. Placing the urea or ammonium-containing fertilizers in the anaerobic layer greatly reduces the loss. [*From Patrick and Reddy (1977); used with permission of the Japanese Society of Soil Science and Plant Nutrition.*]

the fertilizer into the reduced zone, thereby largely preventing the formation of nitrates by nitrification. In some cases fertilizer use efficiency has been doubled by deep placement of the fertilizer.

Atmospheric pollution. The loss of nitrous oxide (N_2O) from the soil into the atmosphere is thought to have some deleterious effects on the environment. As the N_2O moves up into the stratosphere, it may participate in reactions that result in the destruction of ozone (O_3), a gas that helps shield the earth from harmful ultraviolet radiation from the sun. If this ozone shield were to be removed, the earth's surface would be warmed and serious ecological consequences then would result. While the validity of this hypothesis is still being evaluated, it suggests that N_2O loss from soils may have implications far beyond agriculture.

9.11 Nitrification Inhibitors[2]

Losses of nitrogen by volatilization as well as by leaching are generally higher when considerable nitrate nitrogen is present in the soil over long periods of time. Such losses are high when nitrogen fertilizers are applied in the fall for the succeeding crop or even in the spring on wet soils. To minimize them, it is desirable to keep much of the applied nitrogen in the ammonium rather than the nitrate form.

In recent years chemicals have been found that can inhibit the nitrification process. Several such chemicals have shown promise under controlled conditions (Table 9.2), but field performance has been spotty. One of the more effective of these chemicals is nitrapyrin [2-chloro-6-(trichloromethyl)pyridine] sold under the brand name of N-Serve. The compound inhibits oxidation of ammonia to nitrates. It is relatively inexpensive and can be applied at rates of 0.5 kg/ha or less. It is usually applied with anhydrous ammonia or incorporated into dry fertilizer such as urea.

TABLE 9.2 Examples of Inhibitors of Nitrification[a]

Common name or symbol	Chemical name
Dd, DNDN, Dicyan	Dicyandiamide
Thiourea, Tu	Thiourea
ASU	Guanylthiourea
Nitrapyrin, N-Serve	2-Chloro-6-(trichloromethyl)pyridine
ST	Sulfathiazole
MAST	2-Amino-4-methyl-6-trichloromethyl-1,3,5-triazine
ATC	4-Amino-1,2,4-triazole·HCl

[a] From Hauck (1980).

Response to nitrapyrin is variable and depends to a considerable extent on the soil texture and the time of year the nitrapyrin and the fertilizer are applied (Table 9.3). Poorest results are obtained with coarse-textured soils and best with those higher in clay. Reduction in nitrogen loss from the use of nitrapyrin is least when the fertilizer is placed on the soil surface alongside growing plants.

Sulfur-Coated Urea. Another means of reducing gaseous loss of nitrogen is by coating urea pellets with elemental sulfur. These sulfur-coated urea products have been found to be effective in increasing plant utilization of the nitrogen, especially on sandy soils and on some tropical soils, including rice paddies. The sulfur is strictly not a nitrification inhibitor but it slows down the rate of conversion of urea to ammonia and of ammonia to nitrates, thereby reducing

[2] For reviews of this subject, see ASA (1980).

TABLE 9.3 Probability of Crop Yield Increases[a] from Nitrification Inhibitors Applied with Ammonical Fertilizers at Different Times of the Year in the Eastern Part of the Midwest[b]

Soil texture	Time of application		
	Fall	Spring preplant	Spring sidedress
Sands	Poor	Fair	Fair
Loamy sands, sandy loams, and loams	Fair–good	Fair	Poor
Silt loams	Good	Fair	Poor
Clay loams and clays	Good	Good	Poor

[a] Chance of increase at any location any year, Poor = <20%; Fair = 20–60%; Good = >60%.
[b] Nelson and Huber (1980).

the availability of nitrates for the denitrification process. Unfortunately, however, the sulfur adds significantly to the cost of the urea fertilizer (25–50%). Also, it increases soil acidity, since oxidation of the sulfur in the soil produces sulfuric acid (see Section 9.27). These compounds will likely be economical only where high yield increases are obtained and/or where high crop values prevail.

9.12 Ammonia Volatilization

Application of ammonium-containing fertilizers or of urea, which quickly hydrolyzes to ammonia, can result in significant losses of ammonia gas, especially on sandy soils and on alkaline or calcareous soils. In upland soils this loss is most serious when the urea or ammonium-containing materials are applied to the soil surface. Not only are the opportunities for the ammonia to react with soil colloids minimized with surface applications, but surface soil temperatures are usually high, which enhances ammonia volatilization. Incorporating the fertilizers into the top few centimeters of soil can reduce ammonia loss by 25–75% over levels found when the materials are applied on the surface.

Gaseous ammonia loss from nitrogen fertilizers applied to the surface of paddy soils can be appreciable even on slightly acid soils. The applied fertilizer stimulates algae growing in the paddy water, which in turn can extract CO_2 from the water just as higher plants extract this gas from the atmosphere. The result is a marked increase in paddy water pH, levels above 9.0 not being uncommon. At these pH values ammonia is released from ammonium compounds and goes directly into the atmosphere. As with upland soils, this loss can be reduced significantly if the fertilizer is placed below the soil surface.

9.13 Biological Nitrogen Fixation

For centuries farmers have recognized that the yields of cereal crops such as wheat and corn (maize) are higher when these crops are grown after certain legumes such as clovers, and alfalfa (lucerne). We now know this beneficial effect is due not to the legume crop itself but to associated microorganisms that are able to fix atmospheric nitrogen into combined forms usable by the plants. This ability to fix atmospheric nitrogen is known to be shared by a number of organisms including several species of bacteria (not all of which are associated with legumes), a few actinomycetes and blue-green algae (cyanobacteria). The quantity of nitrogen fixed globally each year is enormous, having been estimated at about 175 million Mg (metric tons) (Table 9.4), which exceeds the amount applied in chemical fertilizers.

The Mechanism. Although biological nitrogen fixation is accomplished by a number of organisms, a common mechanism for the fixation appears to be involved. The overall effect of the process is to reduce nitrogen gas to amonia,

$$N_2 + 6\,H^+ + 6\,e^- \xrightarrow{\text{(nitrogenase)}} 2\,NH_3$$

Fe
Fe, Mo

TABLE 9.4 Biological Nitrogen Fixation from Different Sources[a]

Use	Area (10^6 ha)	Nitrogen fixed per year	
		Rate (kg/ha)	Total fixed (10^6 Mg)
Arable (cropped)			
Legumes	250	140	35
Nonlegumes	1,150	8	9
Permanent meadows and grassland	3,000	15	45
Forest and woodland	4,100	10	40
Unused	4,900	2	10
Ice covered	1,500	0	0
Total land	14,900	1	139
Sea	36,100	1	36
Grand total	51,000		175

[a] From Burns and Hardy (1975,).

which in turn is combined with organic acids to form proteins.

$$NH_3 + \text{organic acids} \longrightarrow \text{proteins}$$

The site of N_2 reduction is the enzyme *nitrogenase*, a two-protein complex consisting of a larger iron- and molybdenum-containing member and a smaller companion containing iron. Nitrogenase is indeed of great significance to humankind.

Fixation Systems. Biological nitrogen fixation occurs through a number of organism systems with or without direct association with higher plants (Table 9.5). Included are the following.

1. Symbiotic systems, nodule-forming:
   a. With legumes.
   b. With nonlegumes.
2. Symbiotic systems, nonnodule forming.
3. Nonsymbiotic systems.

Although the legume symbiotic systems have received most attention historically, recent findings suggest that the other systems are also very important world-wide and may even rival the legume-associated systems as suppliers of biological nitrogen to the soil. Each major system will be discussed briefly.

TABLE 9.5 Information on Different Systems of Biological Nitrogen Fixation[a]

N-fixing systems	Organisms involved	Plants involved	Site of fixation
Symbiotic			
Obligatory			
Legumes	Bacteria (*Rhizobium*)	Legumes	Root nodules
Nonlegumes (angiosperms)	Actinomycetes (*Frankia*)	Nonlegumes (angiosperms)	Root nodules
Associative			
Morphological involvement	Blue-green algae, bacteria	Various higher plants and microorganisms	Leaf and root nodules, lichens
Nonmorphological involvement	Blue-green algae, bacteria	Various higher plants and microorganisms	Rhizosphere (root environment) Phyllosphere (leaf environment)
Nonsymbiotic	Blue-green algae, bacteria	Not involved with plants	Soil, water independent of plants

[a] Modified from Burns and Hardy (1975).

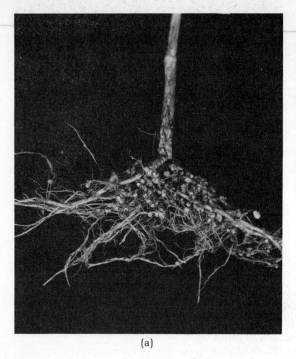

(a)

(b)

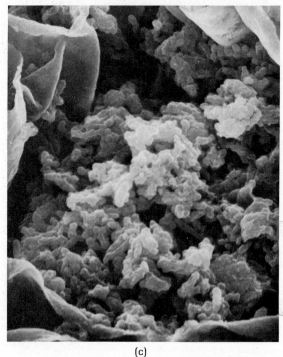

(c)

FIGURE 9.6 Photos illustrating soybean nodules.
In (a) the nodules are seen on the roots of the
soybean plant, and a closeup (b) shows a few of
the nodules associated with the roots. A scanning
electron micrograph shows a single plant cell within
the nodule stuffed with the bacterium *Rhizobium
japonicum.* [(c) *courtesy of W. J. Brill, University
of Wisconsin.*]

9.14 Symbiotic Fixation with Legumes

Bacteria of the genus *Rhizobium* in association with legumes provide the major biological source of fixed nitrogen in agricultural soils. These organisms invade the root hairs and the cortical cells, ultimately inducing the formation of nodules that serve as a home for the organisms (Figure 9.6). The host plant supplies the bacteria with carbohydrates for energy, and the bacteria reciprocate by supplying the plant with fixed nitrogen compounds. This living together or *symbiosis* is mutually beneficial.

Organisms Involved. There is considerable specificity between *Rhizobium* species and the host plants they will invade. A given *Rhizobium* species will inoculate some legumes but not others. This specificity of interaction is the basis for classifying rhizobia and their host plants into seven so-called cross-inoculation groups (Table 9.6). Those legumes that can be inoculated by a given *Rhizobium* species are included in the same cross-inoculation group. Thus *Rhizobium trifolii* inoculates *Trifolium* species (most clovers), *Rhizobium phaseoli* inoculates *Phaseolus vulgaris* (dry beans), and so on.

In areas where a given legume has been growing regularly, the appropriate species of *Rhizobium* may well be present in the soil. All too often, however, the natural *Rhizobium* population in the soil is too low, or the strain of the *Rhizobium* species present is not effective in stimulating nodule formation or in enhancing nitrogen fixation once the nodules have formed. It is necessary

TABLE 9.6 Cross-inoculation Groups of Legumes and Associated Rhizobia

Group	*Rhizobium* species	Legume
Alfalfa	*R. meliloti*	*Melilotus* (certain clovers), *Medicago* (alfalfa), *Trigonella* (fenugreek)
Clover	*R. trifolii*	*Trifolium* spp. (clovers)
Soybean	*R. japonicum*	*Glycine* max (soybeans)
Lupini	*R. lupini*	*Lupinus* (lupines), *Ornithopus* spp. (serradella)
Bean	*R. phaseoli*	*Phaseolus Vulgaris* (dry bean) *Phaseolus Coccineus* (runner bean)
Peas and vetch	*R. leguminosarum*	*Pisum* (peas), *vicia* (vetch *Lathyrus* (sweet pea), *Lens* spp. (lentil)
Cowpea miscellany	Various	*Vigna* (cowpea), *Lespedeza* (lespedeza), *Archis* (peanut), *Stylosanthes* (stylo), *Desmodium* (desmodium), *Cajanus* (pigeon pea), *Crotolaria* (crotalaria), *Pueraria* (kudzu)

TABLE 9.7 Average Yield of Soybeans as Affected by the Application of different Strains of Inoculant Containing *Rhizobium japonicum* (1975–76)[a]

The inoculant was applied in the row on a soil very low in organism numbers.

Inoculant strain	Inoculum population (10^7 MPN/g)[b]	Yield of soybeans (kg/ha)
No. 110 (single strain)	2.8	3565
No. 138 (single strain)	6.3	3468
Commercial[c]	2.3	3660
Control	—	1139

[a] From Bezdicek et al. (1978).
[b] MPN = most probable number.
[c] Specific strain not known.

in such circumstances to apply special cultures of rhizobia either by coating the legume seeds with the culture or by applying the inoculant directly to the soil. Effective and competitive strains of rhizobia are available commercially and are widely used. In situations where the natural rhizobia strains are low in numbers or are ineffective, significant yield increases are obtained from using inoculants (Table 9.7). In the United States most of the inoculant sold is used for soybeans, although some is used for forage legumes.

Quantity of Nitrogen Fixed. The rate of biological fixation of nitrogen is greatly dependent on soil and climatic conditions. The legume–rhizobia associations generally function best on soils that are not too acid and that are well supplied with phosphorus, potassium, and sulfur. Likewise, deficiencies of molybdenum, boron, cobalt, and iron will limit fixation rates. In contrast, high levels of available nitrogen, whether from the soil or added in fertilizers, tend to depress biological nitrogen fixation (Figure 9.7). Apparently the system to provide combined nitrogen tends to operate only when the nitrogen is needed by the plant.

From experiments around the world, it is possible to suggest typical levels of nitrogen fixation, not only by legume-associated organisms but by other mechanisms as well. Such levels are shown in Table 9.8. While these levels would not pertain to any specific ecological condition, they do give an idea as to the relative levels of fixation that occur. It is well to note that the legume complexes fix large quantities of nitrogen, a fact of extreme importance today and likely one of even greater importance in the future if energy costs for commercial fertilizer continue to rise.

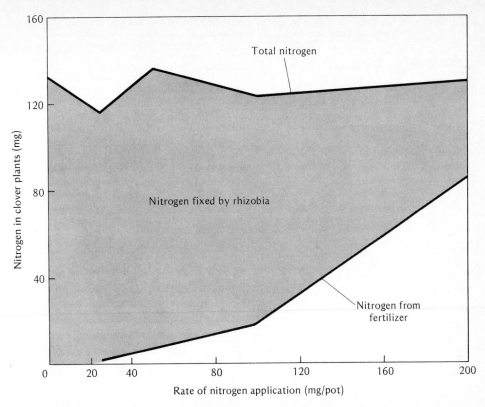

FIGURE 9.7 Influence of added inorganic nitrogen on the total nitrogen in clover plants, the proportion supplied by the fertilizer and that fixed by the rhizobium organisms associated with the clover roots. Increasing the rate of nitrogen application decreased the amount of nitrogen fixed by the organisms in this greenhouse experiment. [*From Walker et al.* (*1956*).]

9.15   Fate of Nitrogen Fixed by Legume Bacteria

The nitrogen fixed by the nodule organisms goes in four directions. First it is used directly by the host plant, which thereby benefits greatly from the symbiosis described in Section 9.14. Second, as the nitrogen passes into the soil itself, either by excretion or more probably by sloughing off of the roots and especially of the nodules, some of it is mineralized and becomes available almost immediately as ammonium or nitrate compounds. Because of the relatively low carbon/nitrogen ratios of such residues (about 20:1) mineralization takes place sufficiently rapid that some of the nitrogen is available to any crop being grown in association with the legume. The vigorous development of a grass in a legume–grass mixture is evidence of this rapid release (Figure

TABLE 9.8 Typical Levels of Nitrogen Fixation from Different Systems

Crop or plant	Associated organism	Typical levels of nitrogen fixation (kg/ha N per yr)
Symbiotic		
Legumes (nodulated)	Bacteria (*Rhizobium*)	
Alfalfa (*Medicago sativa*)		150–250
Clover (*Trifolium pratense* L.)		100–150
Soybean (*Glycine max* L.)		50–150
Cowpea (*Vigna unquiculata*)		50–100
Lupine (*Lupinus*)		50–100
Vetch (*Vicia vilbosa*)		50–125
Bean (*Phaseolus vulgaris*)		30– 50
Nonlegumes (nodulated)		
Alders (*Alnus*)	Actinomycetes (*Frankia*)	50–150
Species of *Gunnera*	Blue-green algae (*Nostoc*)	10– 20
Nonlegumes (nonnodulated)		
Pangola grass (*Degetaria decumbens*)	Bacteria (*Azospirillum*)	5– 30
Bahia grass (*Paspalum notatum*)	Bacteria (*Azotobacter*)	5– 30
Azolla	Blue-green algae (*Anabaena*)	150–300
Nonsymbiotic	Bacteria (*Azotobacter, Clostridium*)	5– 20
	Blue-green algae (various)	10– 50

9.8). Third, when a legume sod is turned under or when the legume is killed by a herbicide in a no-till cropping system, the entire root system is subject to breakdown. Some of the nitrogen is converted to ammonia and nitrate, and is assimilated by succeeding crops grown in the rotation.

The fourth pathway for the fixed nitrogen is to be incorporated into the bodies of the general-purpose decay organisms and finally into the soil organic matter. This process of immobilization was discussed in Section 9.4.

9.16 Do Legumes Always Increase Soil Nitrogen?

Because of their overall beneficial effect on crop production, one might assume that the legume/rhizobia symbiosis would result in an increase in soil nitrogen. It is true that the soil on which a legume has been grown will likely be higher in nitrogen than if a nonlegume had been grown during the same period. However, even the legume may not increase the soil nitrogen since the legume crop may utilize not only the nitrogen fixed by the rhizobia but additional nitrogen from the soil as well. This is true especially in soils with relatively high nitrogen contents. Crops such as beans and peas and even soybeans and peanuts may leave the land depleted of nitrogen, since they support low levels of fixation and in some cases their roots are removed in harvesting.

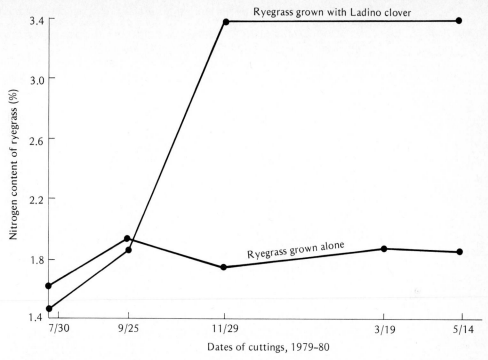

FIGURE 9.8 Nitrogen content of five field cuttings of ryegrass grown alone or with Ladino clover. For the first two harvests, nitrogen fixed by the clover was not available to the ryegrass and the nitrogen content of the ryegrass forage was low. In subsequent harvests, the fixed nitrogen apparently was available and was taken up by the ryegrass. This was probably due to the mineralization of dead Ladino clover root tissue. [*From Broadbent et al. (1982).*]

In general, however, the net draft of legumes on the soil nitrogen is less than that of nonlegumes, and in some situations soil nitrogen is actually increased through legume cropping. Consequently the use of legumes is encouraged in a rotation where the maintenance of nitrogen is important. Legumes are usually so economical in their use of soil nitrogen that a high-protein crop can be harvested with little or no draft on the soil nitrogen. Thus, they are nitrogen *savers*. This is an important fertility axiom.

9.17   Symbiotic Fixation with Nodule-Forming Nonlegumes

About 160 species from 13 genera of nonlegumes are known to develop nodules and to accommodate symbiotic nitrogen fixation (Bond, 1977). Included is an important group of angiosperms, listed in Table 9.9. The roots of these plants,

TABLE 9.9 Number and Distribution of Major Actinomycete-Nodulated Nonlegume Angiosperms[a]

In comparison there are about 13,000 legume species.

Genus	Family	Species[b] nodulated	Geographic distribution
Alnus	Betulaceae	33/35	Cool regions of the northern hemisphere
Ceanothus	Rhamnaceae	31/35	North America
Myrica	Myricaceae	26/35	Many tropical, subtropical and temperate regions
Casuarina	Casuarinaceae	24/25	Tropics and subtropics
Elaeagnus	Elaeagnaceae	16/45	Asia, Europe, N. America
Coriaria	Coriariaceae	13/15	Mediteranean to Japan, New Zealand, Chile to Mexico

[a] Selected from Torrey (1978).
[b] Number of species nodulated/total number of species in genus.

which are present in some forested areas and wet lands, are inoculated by soil actinomycetes of the genus *Frankia.* The actinomycetes invade the root hairs and distinctive nodules form. The rates of nitrogen fixation per hectare compare very favorably with those of the legume/rhizobia complexes (Table 9.8). On a worldwide basis the total nitrogen so fixed may even exceed that due to legumes (Table 9.4). While the rates of cycling of the fixed nitrogen may not be quite as high as those associated with agricultural crops, the total quantity fixed places the actinomycete-induced fixation on a par with rhizobia fixation.

Certain blue-green algae are known to develop nitrogen fixing symbiotic relations with green plants. One involves nodule formation of the stems of *Gunnera,* an angiosperm common in marshy areas of the southern hemisphere. Algae of the genus *Nostoc* are involved and the rate of fixation is typically 10–20 Kg/ha of nitrogen per year (Table 9.8).

9.18 Symbiotic Nitrogen Fixation Without Nodules

Recent studies have called attention to several significant nonlegume symbiotic nitrogen fixing systems that do not involve nodules. Among the most significant are those involving blue-green algae. One system of considerable practical importance is the *Azolla/Anabaena* complex, which flourishes in certain rice paddies of tropical and semitropical areas (Lumpkin and Plucknett, 1982). The *Anabaena* blue-green algae inhabit cavities in the leaves of the floating fern

Azolla and fix quantities of nitrogen comparable to these of the better rhizobia/ legume complexes (Table 9.8).

A more widespread but less intense nitrogen-fixing phenomena is that which occurs in the *rhizosphere* or root environment of nonlegumes and especially grasses (Dobereiner, 1978). The organisms responsible are bacteria, especially those of the *Spirillum* and *Azotabacter* genera (Table 9.8). The organisms use root exudates as sources of energy for their nitrogen-fixing activities. There are differences of opinion as to the rates of fixation that occur in the rhizosphere. Relatively high values have been observed in association with certain tropical grasses, but rates of 5–30 kg/ha per year are thought to be more typical. Even so, because of the vast areas of tropical grasslands, the total quantities of nitrogen fixed by rhizosphere organisms is likely very high (Table 9.4).

9.19   Nonsymbiotic Nitrogen Fixation

There exist in soils and water certain free-living microorganisms that are able to fix nitrogen. Since these organisms are not directly associated with higher plants the transformation is referred to as *nonsymbiotic* or *free-living*.

Fixation by Heterotrophs.   Several different groups of bacteria, and blue-green algae, are able to fix nitrogen nonsymbiotically. In upland mineral soils the major fixation is brought about by species of two genera of heterotrophic aerobic bacteria: *Azotobacter* (in temperate zone soils) and *Beijerinckia* (in tropical soils). Certain anaerobic bacteria of the genus *Clostridium* are also able to fix nitrogen. Because of pockets of low oxygen supply in most soils even when they are in the best of tilth, aerobic and anaerobic bacteria probably work side by side in many agricultural soils. The amount of nitrogen fixed by these heterotrophs varies greatly with the pH, soil nitrogen level, and sources of organic matter available to them for energy. *Azotobacter* species are sensitive to soil pH, their activity dropping off at pH values less than 6.0. *Clostridium* and *Beijerenckia* species are tolerant of wider pH levels. In any case, under normal agricultural conditions the rates of nitrogen fixation by these organisms are thought to be much lower than those associated with legumes, being in the range of 5–20 kg/ha per year (Table 9.8).

Among the autotrophs able to fix nitrogen are certain photosynthetic bacteria and blue-green algae. (Havelka et al., 1982). In the presence of light these organisms are able to fix their own CO_2 while simultaneously fixing nitrogen. The contribution of the photosynthetic bacteria is uncertain but those of blue-green algae are thought to be of some significance, especially in wetland areas and in rice paddies. In some cases, these algae have been found to fix sufficient nitrogen for moderate rice yields, but normal levels may be no more than

20–30 kg/ha per year. Nitrogen fixation by blue-green algae in upland soils also occurs but the level is much lower than is found under wetland conditions.

9.20 Addition of Nitrogen to Soil in Precipitation

The atmosphere contains ammonia and compounds released from the soil and plants as well as from the combustion of coal and petroleum products. Nitrates are there too in small quantities, probably resulting from electrical discharges (lightning) in the atmosphere. These atmosphere-borne nitrogen compounds are added to the soil through rain and snow. Although the rates of addition per hectare are small, the total quantity of nitrogen added is not insignificant.

The amount of ammonia and nitrates in precipitation varies markedly with location and with season. The additions are greater in the tropics than in humid temperate regions and larger in the latter than under semiarid climates. Rainfall additions of nitrogen are highest near cities and industrial areas and near huge animal feedlots. In Table 9.10 will be found some of the more important data regarding the amounts of nitrogen thus added to the soil in various parts of the world. The figures are for the most part from temperate regions.

The ammonium nitrogen added to the soil in precipitation is generally larger in amount than that in the nitrate form. The nitrate nitrogen is about the same for most locations, but the ammonium form shows wide variations. The range in total nitrogen added annually is 1–20 kg/ha. A figure of 5–8 kg/ha would be typical of that found in temperate regions. This annual acquisition of nitrogen in a readily available form to each hectare of land affords some aid in the maintenance of soil fertility.

TABLE 9.10 Amounts of Nitrogen Brought Down in Precipitation Annually[a]

Location	Years of record	Rainfall (in.)	Ammoniacal nitrogen		Nitrate nitrogen	
			kg/ha	lb/A	kg/ha	lb/A
Harpenden, England	28	28.8	2.96	2.64	1.49	1.33
Garford, England	3	26.9	7.20	6.43	2.16	1.93
Flahult, Sweden	1	32.5	3.72	3.32	1.46	1.30
Gröningen, Holland	—	27.6	5.08	4.54	1.64	1.46
Bloemfontein and Durban, South Africa	2	—	4.50	4.02	1.56	1.39
Ottawa, Canada	10	23.4	4.95	4.42	2.42	2.16
Ithaca, N.Y.	11	29.5	4.09	3.65	0.77	0.69

[a] From Lyon et al. (1952).

9.21 Reactions of Nitrogen Fertilizers

Nitrogen applied in fertilizers undergoes the same kinds of reactions as does nitrogen released by biochemical processes from plant residues. Most of the fertilizer nitrogen will be present in one or more of three forms: (a) nitrate, (b) ammonia, and (c) urea. The fate of each of these forms has already been discussed briefly. Thus, urea nitrogen is subject to ammonification, nitrification, and utilization by microbes and higher plants. Ammonium fertilizers can be oxidized to nitrates, fixed by the soil solids, or they can be utilized without change by higher plants or microorganisms. And nitrate salts can be lost by volatilization or leaching or they can be absorbed by plants or microorganisms.

High Concentrations.   One important fact should be remembered when dealing with the reaction of fertilizer nitrogen. The added fertilizer salts will almost invariably be localized and in higher concentrations than found in the bulk of the soil. For this reason, the usual reactions are sometimes subject to modification.

When anhydrous ammonia, ammonium-containing salts, or even urea is added to highly alkaline soils, some nitrogen loss in the form of free ammonia is likely. Also, under these conditions, the nitrification process is inhibited, only the first step proceeding normally. Nitrates may thus accumulate until much of the ammonium form has been oxidized. Only then will the second step, that of nitrate formation, take place at normal rate.

The addition of large amounts of nitrate-containing fertilizers may affect the processes of free fixation and gaseous nitrogen loss. In general, fixation by free-living organisms is depressed by high levels of mineral nitrogen. Gaseous losses, on the other hand, are often encouraged by abundant nitrates. Heavy nitrate fertilization would thus tend to increase losses of nitrogen from the soil.

In most soil situations the effects of higher localized concentrations of fertilizer materials on nitrogen transformations are not serious. Consequently, it can be assumed that fertilizer nitrogen will be changed in soils in a manner very similar to that of nitrogen released by biological transformations.

Soil Acidity.   Ammonium-containing fertilizers and those that form ammonia upon reacting in the soil have a tendency to increase soil acidity (see Section 18.11 for a more thorough discussion of this). The process of nitrification (see the equations in Section 9.7) releases hydrogen ions that become adsorbed on the soil colloids. For best crop growth in humid regions, continued and substantial use of acid-forming fertilizers must be accompanied by applications of lime.

The nitrate component of fertilizers does not increase soil acidity. In fact, nitrate fertilizers containing cations in the molecule (for example, sodium nitrate) have a slight alkalizing effect.

9.22   Practical Management of Soil Nitrogen

The problem of nitrogen control is twofold: (a) the maintenance of an adequate nitrogen supply in the soil and (b) the regulation of the nitrogen turnover to assure a ready availability to meet crop demands.

Nitrogen Balance Sheet.   Major gains and losses of available soil nitrogen are diagrammed in Figure 9.9. While the relative additions and losses by various mechanisms will vary greatly from soil to soil, the principles illustrated in the diagram are valid.

The major loss of nitrogen from most soils'is that removed in crop plants. A good crop of wheat or cotton may remove only 100 kg/ha of nitrogen, nearly half of which may be returned to the soil in the stalks or straw. A bumper silage corn crop, in contrast, may contain over 250 kg/ha and a good yield of alfalfa or of well-fertilized grass hay more than 300 kg/ha. It is obvious that modern yield levels require nutrient inputs far in excess of those of a generation ago.

Erosion, leaching, and volatilization losses are determined to a large degree by water management practices. Their magnitudes are so dependent upon specific situations that generalizations are difficult. However, soil and crop management practices which give optimum crop yields will likely hold these sources of nitrogen loss to a satisfactory minimum.

Meeting the Deficit.   In practice, nitrogen deficits are met from four sources—crop residues, farm manure, legumes, and commercial fertilizers. On dairy farms and beef ranches, much of the deficit will be met by the first three methods, fertilizers being used as a supplementary source. On most other types of farms, however, fertilizers will play a major role. Where vegetables and other cash

FIGURE 9.9   Major gains and losses of available soil nitrogen. The widths of the arrows indicate roughly the magnitude of the losses and the addition often encountered. It should be emphasized that the diagram represents average conditions only and that much variability is to be expected in the actual and relative quantities of nitrogen involved.

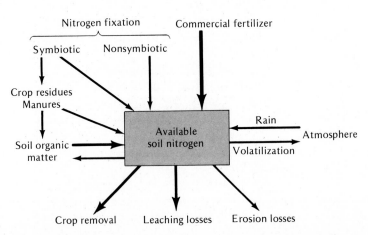

crops with high nutrient requirements are grown, the nitrogen deficit will be met almost entirely with commercial fertilizers. Even with the general field crops, modern yield levels can be maintained only through the extensive use of fertilizers.

Turnover Regulations.   By all odds, the more difficult of the two general problems of nitrogen control is the regulation of this element after it enters the soil. Availability at the proper time and in suitable amounts, with a minimum of loss, is the ideal. Even where commercial fertilizers are used to supply much of the nitrogen, maintaining an adequate but not excessive quantity of available nitrogen is not an easy task.

Soils under any given climate tend to assume what may be called a *normal* or *equilibrium content* of nitrogen. Thus, under ordinary methods of cropping and manuring, any attempt to raise the nitrogen content to a point materially higher than this normal will be attended by unnecessary waste due to drainage and other losses. At the same time, the nitrogen should be kept suitably active by the use of legumes and other organic materials with a low C/N ratio and by the applications of lime and commercial fertilizers.

This is essentially the recommendation already made for soil organic matter (Section 8.18), and it is known to be both economical and effective. In short, the practical problem is to supply adequate nitrogen to the soil, to keep it mobile, and to protect it from excessive losses caused by leaching, volatilization, and erosion.

9.23   Importance of Sulfur

Sulfur has long been recognized as essential for plant and animal growth. Although much is yet to be learned about the functions of this element, it is already known to be indispensable for many reactions in every living cell. Sulfur is a constituent of the amino acids methionine, cysteine, and cystine, deficiencies of which result in serious human malnutrition. The vitamins biotin, thiamine, and B_1 contain sulfur, and the structure of proteins is determined to a considerable extent by sulfur groups. The properties of certain protein enzymes, such as those concerned with photosynthesis and nitrogen fixation, are thought to be attributable to the type of sulfur linkages present. As with the other essential elements, sulfur plays a unique role in plant and animal metabolism.

Plants that are sulfur deficient are characteristically small and spindly. The younger leaves are light green to yellowish, and in the case of legumes nodulation of the roots is reduced. Protein quality, especially methionine content, suffers in sulfur-deficient plants. The maturity of fruits and seeds is delayed in the absence of adequate sulfur.

Deficiencies of Sulfur. It is only in recent years that deficiencies of sulfur have become common. Since it was first manufactured in 1840, sulfur-bearing superphosphate has helped meet the needs for this element. Likewise, ammonium sulfate, long a significant constituent of fertilizers, has been an important sulfur source. Atmospheric sulfur dioxide, a by-product of the combustion of sulfur-rich coals and residual fuel oils, has supplied large quantities of this element to both plants and soils. Thus, by seemingly incidental means the sulfur needs of crops in the past have been largely satisfied, especially in areas near industrial centers (Table 9.11).

In recent years, increased demand for high-analysis fertilizers has forced manufacturers to use alternatives to superphosphate and ammonium sulfate. Ammonium sulfate supplied 31% of the world's nitrogen in 1960, but only 8% in 1980; superphosphate as the world's source of phosphorus declined from 55% in 1960 to 20% in 1980 (IFDC, 1979). As a result, many sulfur-free fertilizers are on the market, and the average sulfur content of fertilizers has decreased. Even sulfur-containing pesticides, so commonly used a few years ago, have been largely replaced by organic materials free of sulfur.

The replacement of wood and coal for domestic heating by natural gas, electricity, and low-sulfur fuel oil has affected the amount and distribution of sulfur dioxide in the atmosphere. Intensified efforts in the United States to reduce air pollution in and around cities and industrial areas will likely further reduce the quantity of sulfur in the atmosphere. The recognition that clean air is a primary goal will necessitate finding alternative means of supplying sulfur for plant growth.

Coupled with these reductions in the supply of sulfur to soils and plants is the greater removal of this element in harvested crops. Yields have increased markedly during the past 20–25 years, and much of the sulfur removed in crops has not been returned. The quantity of sulfur thus removed is about the same as that of phosphorus. It is not surprising, therefore, that increased attention is being given to sulfur.

TABLE 9.11 Sulfur Collected in Precipitation at Rural, Urban and Industrial Locations in Southern and Midwestern United States[a]

State	Year	No. of locations	Sulfur collected (kg/ha per year)		
			Rural	Urban	Industrial
Iowa	1971–73	6	16.0	16.8	—
South Carolina	1973–75	15	8.2	11.3	11.2
Wisconsin	1969–71	20	16.0	42.0	168.0
Tennessee	1973–76	5	—	—	18.1

[a] Data from several sources cited by Jones and Suarez (1980).

Areas of Deficiency. Sulfur deficiencies are widespread ⌐. but are more prevalent in areas where soil parent materials a⌐ where extreme weathering and leaching has removed this elemen⌐ there is little replenishment of sulfur from the atmosphere. In many t⌐ countries one or more of these conditions prevail and sulfur-deficient area⌐ are common (IFDC, 1979). In the United States, deficiencies of sulfur are most common in the southeast, the northwest, California, and the Great Plains. In the northeast and in other areas with heavy industry and large cities, sulfur deficiencies do not seem to be widespread (Figure 9.10).

Crops vary in their sulfur requirements. Legume crops such as alfalfa, the clovers, and soybeans have high sulfur requirements as do cotton, sorghum, sugar beets, cabbage, turnips, and onions. Forests and grasses and cereals generally have lower sulfur requirements, although wheat in the northwestern states is often quite responsive to sulfur applications.

9.24   Natural Sources of Sulfur

There are three major natural sources from which plants can be supplied with available sulfur: (a) soil minerals, (b) sulfur gases in the atmosphere, and (c) organically bound sulfur. These will be considered in order.

Soil Minerals. There are several soil minerals in which sulfur is combined and from which it may be released for growing plants. For example, sulfides of iron, nickel, and copper are found in many soils, especially those with restricted drainage. They are quite abundant in soils of tidal marsh areas. Upon oxidation, the sulfides are changed to sulfates, which are quickly available for plant use.

In regions of low rainfall, sulfate minerals are common in soils. Accumulations of gypsum in the lower horizons of Mollisols and Aridisols (see Figure 13.10) are examples. Accumulations of soluble salts, including sulfates in the surface layers, are characteristic of saline soils (see Figure 6.14) of arid and semiarid regions. When the soils are dry, the salt accumulations are visible.

The accumulation of sulfates in subsoils is not limited to areas of low rainfall. For example, in the southeast, higher sulfate contents are often found in the subsoils than in the topsoils. Apparently, the sulfates are absorbed by the acid iron and aluminum oxide and kaolinitic clays that are usually present in larger quantities in the subsoil.

Atmospheric Sulfur.[3] The combustion of fuels, especially coal, releases sulfur dioxide and other sulfur compounds into the atmosphere. In fact, the content of sulfur in the air is generally directly related to the distance from industrial

[3] For a discussion of this topic see Terman (1978).

Canada

Alberta
British Columbia
Manitoba
Ontario
Saskatchewan

United States

Alabama	Missouri
Alaska	Montana
Arkansas	Nebraska
California	North Carolina
Colorado	North Dakota
Florida	Ohio
Georgia	Oklahoma
Hawaii	Oregon
Idaho	Pennsylvania
Illinois	South Carolina
Indiana	South Dakota
Iowa	Tennessee
Kansas	Texas
Louisiana	Virginia
Maryland	Washington
Michigan	Wisconsin
Minnesota	Wyoming
Mississippi	

Europe

Belgium	Ireland
Bulgaria	Italy
Czechoslovakia	Netherlands
Denmark	Norway
England	Poland
Finland	Spain
France	Sweden
Germany	U.S.S.R.
Iceland	Yugoslavia

Africa

Cameroon	Mozambique
Central African	Nigeria
Republic	Senegal
Chad	South Africa
Congo	Tanzania
Ghana	Togo
Guinea	Uganda
Ivory Coast	Upper Volta
Kenya	Zaire
Mali	Zambia
Malawi	Zimbabwe

South and Central America

Argentina
Brazil
Chile
Colombia
Costa Rica
El Salvador
Honduras
Puerto Rico
Venezuela
Windward Islands

Asia

Burma
India
Indonesia
Lebanon
Malaysia
Sri Lanka
Taiwan
Thailand

Australasia

Australia
Fiji
New Guinea
New Zealand
Solomon Islands

FIGURE 9.10 Locations throughout the world where sulfur deficiencies have been reported. Other sulfur-deficient areas undoubtedly exist, but have yet not been identified. [*Courtesy The Sulphur Institute.*]

centers (Figure 9.11). Except near seashores, where salt s̶,̶ ᵃ̶
significant quantities, and near marshes, which are sources of hy̶
the combustion of coal is the most significant source of atmosphe̶

Atmospheric sulfur becomes part of the soil-plant system in three w̶
Some of it is absorbed directly from the atmosphere by growing plants. A
considerable quantity is absorbed directly by the soil from the atmosphere,
and a similar amount is added with precipitation.

The quantity of sulfur absorbed directly by plants from the atmosphere
will vary with atmospheric and soil conditions. Experiments have shown that
even plants supplied with adequate soil sulfate can absorb 25–35% of their
sulfur from the atmosphere. If the soil sulfur is low and the atmospheric sulfur
high, about half of the plant sulfur can come from the atmosphere.

Data presented graphically in Figure 9.11 show a comparison of direct
soil absorption of sulfur and the addition of this element in precipitation. At
least in this instance, the quantity directly absorbed from the soil was somewhat greater than that from the precipitation.

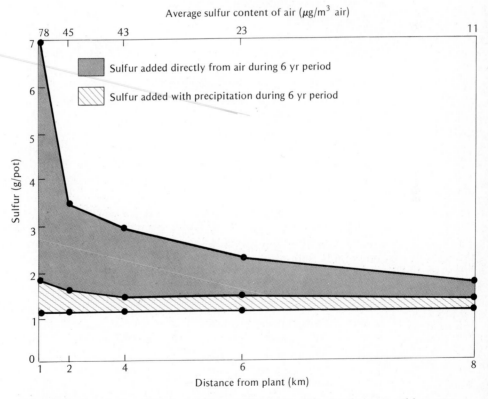

FIGURE 9.11 Sulfur added to soils as affected by distance from an oil-burning
industrial plant in Sweden. Note the rapid dropoff in sulfur added directly from the
air. [*From O. Johannson (1960).*]

The quantity of sulfur added in precipitation or absorbed directly varies according to the content of this element in the atmosphere. Samplings for sulfur in precipitation show variations in annual accretions from less than 1 to nearly 100 kg/ha (Coleman, 1966). Data in Table 9.11 indicate levels found in several locations in the United States.

It is not surprising that soils absorb atmospheric sulfur directly or that this element moves in through rain and snow. The sulfur dioxide forms sulfurous acid (H_2SO_3) when in contact with water or with water vapor. Any sulfur trioxide present would form sulfuric acid by the same process. These strong acids are readily absorbed by soils.

Recent environmental concerns about high sulfur levels in the atmosphere are of great practical significance to agriculture. On the one hand efforts to reduce atmospheric sulfur are welcomed in areas near industrial plants, where toxic effects of high sulfur levels on vegetation are noted. But in much wider areas drastic reductions in atmospheric sulfur would result in deficiencies of this element for optimum crop growth. In these cases the sulfur deficiency would need to be met by increasing the sulfur content of fertilizers, thereby increasing crop production costs. These matters are discussed further in Chapter 21.

Organic Bound Sulfur. In most humid region surface soils, about half the sulfur is in the organic form, although the proportion would be much lower in the subsoil. Just as is the case with nitrogen, however, all too little is known of the specific organic sulfur compounds present. The sulfur added in plant residues is mostly in the form of proteins, which are normally subject to rather ready microbial attack. In some manner, the sulfur (along with the nitrogen) is stabilized during humus formation. In this stable form or forms, it is protected from rapid release and thus cannot be lost or taken up by higher plants. The similarity between the general behavior of organic nitrogen and sulfur fractions is unique.

Although the specific sulfur compounds associated with soil organic matter are not known, the presence of some forms is strongly suspected. For example, sulfur-containing amino acids—cystine, cysteine, and methionine—and other compounds with direct C—S linkages are thought to be present, as are organic sulfates.

Forms of Sulfur. This discussion has identified three major forms of sulfur in soils and fertilizers: *sulfides, sulfates,* and *organic forms*. To these must be added the fourth important form, *elemental sulfur,* the starting point of most of the manmade chemical sulfur compounds. The following sections will show relationships among these forms.

9.25 The Sulfur Cycle

The major transformations that sulfur undergoes in soils are shown in 9.12. The inner circle shows the relationships among the four major forms of this element in soils and in fertilizers. The outer portions show the most important sources of sulfur and how this element is lost from the system.

Some similarity between the sulfur and nitrogen cycles is evident (compare Figure 9.1). In each case, the atmosphere is an important source of the element in question. Each is held largely in the organic fraction of the soil and each is dependent to a considerable extent upon microbial action for its various transformations.

Figure 9.12 should be referred to frequently as a more detailed examination

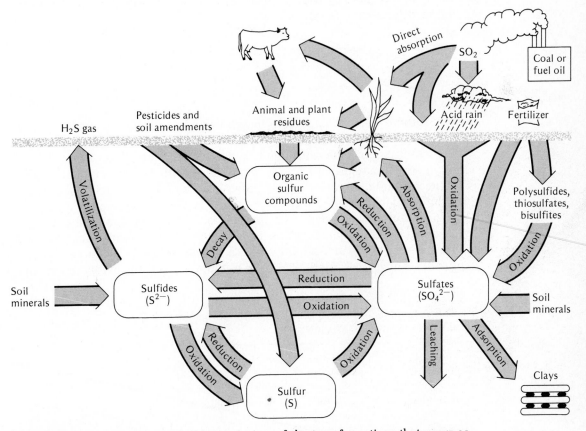

FIGURE 9.12   The sulfur cycle, showing some of the transformations that occur as this element changes form in soils, plants, and animals. It is well to keep in mind that, except for certain soils in arid areas, the great bulk of the sulfur is in the form of organic compounds.

is made of sulfur in plants and soils, beginning with the behavior of this element in soils.

9.26 Behavior of Sulfur Compounds in Soils

Mineralization. It is not surprising that sulfur behaves much like nitrogen as it is absorbed by plants and microorganisms and moves through the sulfur cycle. The organic forms of sulfur must be mineralized by soil organisms if the sulfur is to be used by plants. The rate at which this occurs is dependent upon the same environmental factors as affect nitrogen mineralization, including moisture, aeration, temperature, and pH. When conditions are proper for general microbial activity, sulfur mineralization occurs. The mineralization reaction might be expressed as follows.

$$\text{Organic sulfur} \rightarrow \text{Decay products} \rightarrow \text{Sulfates}$$

Proteins and other organic combinations $\quad\quad$ H$_2$S and other sulfides are simple examples

Immobilization. Immobilization of inorganic forms of sulfur occurs when low-sulfur, energy-rich organic materials are added to soils not plentifully supplied with inorganic sulfur. The mechanism is thought to be the same as for nitrogen. The energy-rich material stimulates microbial growth and the inorganic sulfur is synthesized into microbial tissue. Only when the microbial activity subsides does the inorganic sulfate form again appear in the soil solution.

These facts suggest that, like nitrogen, sulfur in soil organic matter may be associated with organic carbon in a reasonably constant ratio. The ratio among carbon, nitrogen, and sulfur for a number of soils on three different continents is given in Table 9.12. A ratio of 130:10:1.3 is reasonably representa-

TABLE 9.12 Mean Carbon/Nitrogen/Sulfur Ratios in a Variety of Soils Throughout the World[a]

Location	Description and number of soils	C/N/S ratio
North Scotland	Agricultural, noncalcareous (40)	147:10:1.4
Minnesota	Mollisols (6)	114:10:1.55
Minnesota	Spodosols (24)	132:10:1.22
Oregon	Agricultural, varied (16)	145:10:1.01
Eastern Australia	Acid soils (128)	152:10:1.21
Eastern Australia	Alkaline soils (27)	140:10:1.52

[a] From Whitehead (1964).

tive. Undoubtedly, there is a definite relationship among the co...
elements, a relationship that is in accord with their behavior in son...

9.27 Sulfur Oxidation and Reduction

During the microbial decomposition of organic sulfur compounds, sulfides are formed along with other incompletely oxidized substances such as elemental sulfur, thiosulfates, and polythionates. These reduced substances are subject to oxidation, just as are the ammonium compounds formed when nitrogenous materials are decomposed. The oxidation reactions may be illustrated with hydrogen sulfide and elemental sulfur.

$$H_2S + 2\,O_2 \rightarrow H_2SO_4 \rightarrow 2\,H^+ + SO_4^{2-}$$

$$2\,S + 3\,O_2 + 2\,H_2O \rightarrow 2\,H_2SO_4 \rightarrow 4\,H^+ + 2\,SO_4^{2-}$$

The oxidation of some sulfur compounds, such as sulfites (SO_3^{2-}) and sulfides (S^{2-}), can occur by strict chemical reactions. However, most of the sulfur oxidation occurring in soils is thought to be *biochemical* in nature. It is accomplished by a number of autotrophic bacteria including those of the genus *Thiobacillus*, five species of which have been characterized. Since the environmental requirement and tolerances of these five species vary considerably, the process of sulfur oxidation occurs over a wide range of soil conditions. For example, it occurs at pH values ranging from less than 2 to higher than 9. This is in contrast to the comparable nitrogen oxidation process, nitrification, where a rather narrow pH range near neutral is required.

Like nitrates, sulfates tend to be unstable in anaerobic environments. They are reduced to sulfides by a number of bacteria of two genera, *Desulfovibro* (five species) and *Desulfotomaculum* (three species). The organisms use the combined oxygen in sulfate to oxidize organic materials. A representative reaction is

$$\underset{\text{Organic alcohol}}{2\,R\text{---}CH_2OH} + \underset{\text{Sulfate}}{SO_4^{2-}} \rightarrow \underset{\text{Organic acid}}{2\,R\text{---}COOH} + 2\,H_2O + \underset{\text{Sulfide}}{S^{2-}}$$

In poorly drained soils, the sulfide ion would likely react immediately with iron, which in anaerobic conditions would be present in the ferrous form. This reaction might be expressed as follows.

$$Fe^{2+} + S^{2-} \rightarrow \underset{\substack{\text{Iron}\\\text{sulfide}}}{FeS}$$

Sulfur reduction takes place with compounds other than sulfates. For example, sulfites (SO_3^{2-}), thiosulfates ($S_2O_3^{2-}$), and elemental sulfur (S) are rather easily reduced to the sulfide form by bacteria and other organisms.

The oxidation and reduction of inorganic sulfur compounds are of great importance to growing plants. In the first place, these reactions determine to a considerable extent the quantity of sulfate present in soils at any one time. Since this is the form taken up by plants, the nutrient significance of sulfur oxidation and reduction is obvious. Second, the state of sulfur oxidation determines to a marked degree the acidity of a soil, a fact that will be considered further below.

Sulfur Oxidation and Acidity. Sulfur oxidation is an acidifying process. The preceding reactions of hydrogen sulfide and elemental sulfur illustrate this point. For every sulfur atom oxidized, two hydrogen ions result. This effect is utilized by adding sulfur to reduce the extreme alkalinity of certain alkali soils of arid regions and to reduce the pH of certain soils for the control of diseases such as potato scab (see Figure 7.11). It is also taken into consideration when making the choice of fertilizer, since some materials such as sulfur-coated urea are acid forming.

The acidifying effect of sulfur oxidation can bring about extremely acid soil conditions. For example, this has been known to occur when land is drained after it has been submerged for some time under brackish water or seawater. During the submerged period, sulfates in the water are reduced to sulfides, in which form they are stabilized generally as iron sulfides (Bloomfield and Coulter, 1973). Since there is a continuous supply of sulfates under these conditions, high levels of sulfides are built up. If there are periods of partial drying, elemental sulfur can form by partial oxidation of the sulfides. The sulfide and elemental sulfur content are hundreds of times higher than would be found in comparable upland soils.

When these areas are drained, the sulfides and/or elemental sulfur are quickly oxidized, forming sulfuric acid. The soil pH may drop to as low as 1.5, a level unknown in normal upland soils. Obviously plant growth cannot occur under these conditions. Furthermore, the quantity of limestone needed to neutralize the acidity is so high as to make this remedy completely uneconomical.

Sizeable areas of these kinds of soils, called *acid sulfate soils* or *cat-clays*, are found in southeast Asia and along the Atlantic coasts of South America and Africa. They also occur in tideland areas along the coasts of several other areas, including the southeastern part and the West Coast of the United States. So long as these soils are kept submerged, the soil reaction does not drop prohibitively. Consequently, production of paddy rice is sometimes possible under these conditions.

Another potentially harmful effect of acid production by sulfur is the development of acidity in fog or rain containing oxides of sulfur (Likens et al., [1979]). The sulfuric acid resulting from reactions of these oxides with water can lower the pH of the precipitation to 4 or below, creating problems for plants as well as people (Section 20.13). Fortunately, steps are being taken to reduce excess sulfur in the atmosphere.

9.28 Sulfate Retention and Exchange

Sulfate is the form in which plants absorb most of their sulfur from soils. Since this ion is quite soluble, it would be readily leached from the soil especially in humid regions, were it not for its adsorption by the soil colloids. As was pointed out in Chapter 4, most soils have some anion exchange capacity, which is associated with iron and aluminum oxide clays and to a limited extent with 1:1-type silicate clays. The sulfate is attracted by the positive charges that characterize acid soils containing these clays. It also reacts directly with hydroxyl groups associated with these clays. A general equation indicates how this may occur.

$$-Al \underset{OH}{\overset{OH}{\diamondsuit}} Al- + KHSO_4 \rightleftharpoons -Al \underset{\underset{K}{\overset{|}{SO_4}}}{\overset{OH}{\diamondsuit}} Al- + H_2O$$

This reaction takes place in acid soils. If the soil pH were increased and hydroxyl ions were added, the reaction would be driven to the left and sulfate would be released. Thus this is an ion exchange reaction, the hydroxyl being exchanged for an KSO_4^- anion.

Most soils will retain some sulfate, although the quantity held is generally small and its strength of retention is low compared to that of phosphate. Sulfate retention is generally higher in the subsoil than the topsoil, since iron and aluminum oxides are more prominent in subsoils. Soils of the southeast part of the United States tend to be higher in sulfate adsorption than elsewhere in the country because of their higher content of iron and aluminum oxides and 1:1-type silicate clays, especially in the subsoils.

Sulfur adsorption and exchange by soils is important since they can provide a ready supply of available sulfur for plants. Sulfur retention is also of some significance, since it helps the soil hold an otherwise mobile element that could easily be lost by leaching.

9.29 Sulfur and Soil Fertility Maintenance

In Figure 9.13 are depicted the major gains and losses of sulfur from soils. The problem of maintaining adequate quantities of sulfur for minerals is becoming increasingly important. Although this element is added to soils through adsorption from the atmosphere and as an incidental component of many fertilizers, each of these sources are apt to decline in the future. Even though chances for widespread sulfur deficiencies are generally less than for the three "fertilizer" elements, higher crop removal of sulfur makes it essential that farmers be on guard to prevent deficiencies of this element.

FIGURE 9.13 Major gains and losses of available soil sulfur. This represents average conditions from which considerable variation occurs in the field.

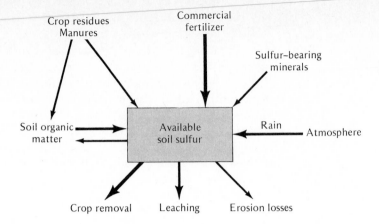

Situations where crops respond to sulfur applications are becoming more and more common. Less "incidental" fertilizer sulfur is being added in some areas, and additions of this element from the atmosphere will likely continue to decrease as air-pollution-abatement efforts are accelerated. Furthermore, steadily increasing crop yields are removing proportionately larger quantities of this and other elements. Although crop residues and farmyard manures can help replenish the sulfur removed, greater and greater dependence must be placed on fertilizer additions. Regular sulfur applications are necessary now for good crop yields in large areas far removed from industrial plants. There will certainly be an increased necessity for and use of sulfur in the future.

REFERENCES

ASA. 1980. *Nitrification Inhibitors—Potentials and Limitations*, ASA Special Publication No. 38 (Madison, WI: Amer. Soc. of Agron., Soil Sci. Soc. Amer.).

Bezdicek, D. F., D. W. Evans, A. Abede, and R. E. Witters. 1978. "Evaluation of peat and granular inoculum for soybean yield and N fixation under irrigation," *Agron. J.*, **70**:865–868.

Bloomfield, C., and J. K. Coulter. 1973. "Genesis and management of acid sulfate soils," *Adv. in Agron.*, **25**:265–326.

Bond, G. 1977. "Some reflections on *Alnus*-type root nodules," in W. Newton, J. R. Postgate, and C. Rodriguez-Barrueco (Eds.), *Recent Developments in Nitrogen Fixation* (New York: Academic Press).

Broadbent, F. E., T. Nakashima, and G. Y. Chang. 1982. "Estimation of nitrogen fixation by isotope dilution in field and greenhouse experiments," *Agron. J.*, **74**:625–628.

Burns, R. C., and R. W. F. Hardy. 1975. *Nitrogen Fixation in Bacteria and Higher Plants* (Berlin: Springer-Verlag).

Cady, F. B., and W. V. Bartholomew. 1960. "Sequential products of anaerobic denitrification in Norfolk soil material," *Soil Sci. Soc. Amer. Proc.*, 24:477– 82.

Coleman, R. 1966. "The importance of sulfur as a plant nutrient in world crop production," *Soil Sci.*, **101**:230–39.

Dalal, R. C. 1977. "Fixed ammonium and carbon-nitrogen ratios of some Trinidad soils," *Soil Sci.*, **124**:323–327.

Dobereiner, J. 1978. "Potential for nitrogen fixation in tropical legumes and grasses," in J. Dobereiner, R. H. Burris, and A. Hollaender (Eds.), *Limitations and Potentials for Biological Nitrogen Fixation in the Tropics* (New York: Plenum Press).

Hauck, R. D. 1980. "Mode of action of nitrification inhibitors," in *Nitrification Inhibitors—Potentials and Limitations,* ASA Publication No. 38 (Madison, WI: Amer. Soc. of Agron., Soil Sci. Soc. Amer.), pp. 19–32.

Havelka, U. D., M. G. Boyle, and R. W. F. Hardy. 1982. "Biological nitrogen fixation," in F. J. Stevenson (Ed.), *Nitrogen in Agricultural Soils,* Agronomy Series No. 22 (Madison, WI: Amer. Soc. of Agron., Crop Sci. Soc. Amer., Soil Sci. Soc. Amer.), pp. 365–422.

Hughes, H.A. 1980. *Conservation Farming* (Moline, IL: Deere & Company).

IFDC. 1979. *Sulfur in the Tropics* (Muscle Shoals, AL: International Fertilizer Development Institute).

Johannson, O. 1960. "On sulfur problems in Swedish agriculture," *Kgl. Lanabr. Ann.*, **25**:57–169.

Jones, U. S., and E. L. Suarez. 1980. "Impact of atmospheric sulfur deposition on agro-ecosystems," in D. S. Shriner et al. (Eds.) *Atmospheric Sulfur Deposition* (Ann Arbor, MI: Ann Arbor Science Publishers, Inc.).

Likens, G. E., R. F. Wright, J. N. Galloway, and T. J. Butler. 1979. "Acid rain," *Sci. Amer.*, **241**:43–51.

Lumpkin, T. A., and O. L. Plucknett. 1982. *Azolla as a Green Manure: Use and Management in Crop Production* (Boulder, CO: Westview Press).

Lyon, T. L., H. O. Buchman, and N. C. Brady. 1952. *The Nature and Properties of Soils,* 5th ed. (New York: Macmillan), p. 467.

Nelson, D. W., and D. M. Huber. 1980. "Performance of nitrification inhibitors in the midwest (east)," ch. 6 in *Nitrification Inhibitors—Potentials and Limitations,* ASA Special Publication No. 38 (Madison, WI: Amer. Soc. Agron., Soil Sci. Soc. Amer.).

Nommik, N., and K. Vahtras. 1982. "Retention and fixation of ammonium and ammonia in soils," ch. 4 in F. J. Stevenson (Ed.), *Nitrogen in Agricultural Soils,* Agronomy Series No. 22 (Madison, WI: Amer. Soc. of Agron., Crop Sci. Soc. Amer., Soil Sci. Soc. Amer.).

Patrick, W. H. Jr., and K. R. Reddy. 1977. "Fertilizer nitrogen reactions in flooded soils," in SEFMIA, Proceedings of the International Seminar on Soil Environment and Fertility Management in Intensive Agriculture (Tokyo: The Society of the Science of Soil and Manure), pp. 275–281.

Stevenson, F. J. (Ed.). 1982. *Nitrogen in Agricultural Soils*, Agronomy Series No. 22 (Madison, WI: Amer. Soc. Agron., Crop Sci. Soc. Amer., Soil Sci. Soc. Amer).

Terman, G. L. 1978. "Atmospheric sulphur—the agronomic aspects," *Tech. Bull.* 23 (Washington, DC: The Sulphur Institute).

Torrey, J. G. 1978. "Nitrogen fixation by actinomycete-induced angiosperms," *BioScience,* **28**:586–592.

Walker, T. W., et al. 1956. "Fate of labeled nitrate and ammonium nitrogen when applied to grass and clover grown separately and together," *Soil Sci.,* **81**:339–52.

Whitehead, D. C. 1964. "Soil and plant-nutrition aspects of the sulfur cycle," *Soils and Fertilizers,* **27**:1–8.

Supply and Availability of Phosphorus and Potassium

Next to nitrogen, the most critical elements in influencing plant growth and production throughout the world are phosphorus and potassium. Unlike nitrogen, however, these elements are not supplied through biochemical fixation, but must come from other sources to meet plant requirements. Included among these sources are (a) commercial fertilizer; (b) animal manures; (c) plant residues, including green manures; (d) human, industrial, and domestic wastes; and (e) native compounds of these elements, both organic and inorganic, already present in the soil. Since the first four sources are to be considered in later chapters, attention now will be focused on the ways and means of utilizing the body of soil as a source of these mineral elements.

10.1 Importance of Phosphorus[1]

With the possible exception of nitrogen, no other element is as critical in practical agriculture as is phosphorus. It is a component of the two compounds involved in the most significant energy transformations in plants, adenosine diphosphate (ADP) and adenosine triphosphate (ATP). ATP, synthesized from ADP through both respiration and photosynthesis, contains a high-energy phosphate group that drives most biochemical processes requiring energy. For example, the uptake of some nutrients (see Figure 1.15) and their transport within the plant, as well as the synthesis of different molecules, are energy-using processes that ATP helps to implement.

Phosphorus is an essential component of deoxyribonucleic acid (DNA), the seat of genetic inheritance in plants as well as animals, and of the various forms of ribonucleic acid (RNA) needed for protein synthesis. Obviously, phosphorus is essential for numerous metabolic synthesis processes.

It is difficult to state in detail all the beneficial effects of phosphorus on plants. Among the more significant ones are

1. Cell division and fat and albumin formation.
2. Flowering and fruiting, including seed formation.
3. Crop maturation, in which phosphorus counteracts the effects of excess nitrogen applications.
4. Root development, particularly of the lateral and fibrous rootlets.
5. Strength of straw in cereal crops, thus helping to prevent lodging.
6. Improvement of crop quality, especially of forages and of vegetables.

[1] For a review of the significance of this element see ISMA (1977).

10.2 The Phosphorus Cycle

The biological cycling of phosphorus from the soil to higher plants and return is illustrated in Figure 10.1. Note that the plant roots absorb inorganic and to a lesser extent organic phosphorus compounds and translocate them to above ground plant parts. The phosphorus in the plants is returned to the soil either in crop residues or in human and animal wastes. Microorganisms

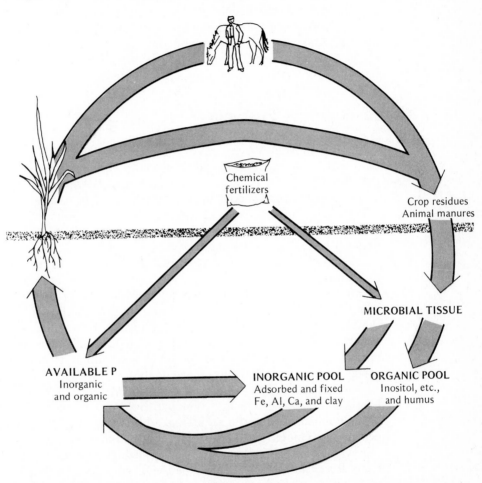

FIGURE 10.1 The phosphorus cycle in and above the soil surface. Crop residues, human wastes, animal manures, and chemical fertilizers are the primary sources of phosphorus. Two major pools of phosphorus are present in the soil: (a) an organic pool found in humus and plant and microbial residues and (b) an inorganic pool of insoluble Ca, Fe, and A compounds. In a representative mineral soil 40–60% of the phosphorus would be found in each of the organic and inorganic pools with about 1–2% in the microbial form and perhaps 0.01% in the available form.

decompose the residues and temporarily tie up at least part of the phosphorus in their body tissue. Ultimately it is released to become part of either the inorganic or the organic pool of this element in the soil. In either form the phosphorus is very slowly converted to the available forms that plant roots can absorb, thereby starting a repeat of the cycle.

In most soils, the amount of phosphorus in the available form at any one time is very low, seldom exceeding about 0.01% of the total phosphorus in the soil. It is for this reason that available phosphorus levels must be supplemented on most soils by adding chemical fertilizers. Unfortunately, much of the phosphorus thus supplied is converted to the less available inorganic pool, in which form it is released very slowly and is usable by plants only over a period of years. The problem of maintaining phosphorus in an available form will receive our attention next.

10.3   The Phosphorus Problem

While the phosphorus content of an average mineral soil compares favorably with that of nitrogen, it is much lower than potassium, calcium, or magnesium (see Table 1.3). Of even greater importance, however, is the fact that most of the phosphorus present in soils is not readily available to plants. Also, when soluble fertilizer salts of this element are supplied to soils their phosphorus is often fixed[2] or rendered insoluble or unavailable to higher plants, even under the most ideal field conditions (see Section 10.7).

Fertilizer practices in many areas exemplify the problem of phosphorus availability. Large quantities of phosphate fertilizers are applied to soils. The removal of phosphorus from soils by crops, however, is low compared to that of nitrogen and potassium, often being only one third or one fourth that of the latter elements. The necessity for high fertilizer dosage when relatively small quantities of phosphorus are being removed from soils indicates that much of the added phosphates becomes unavailable to growing plants. It also suggests some soil phosphorus accumulation.

In the United States, phosphorus added in fertilizers in the past 20 years has generally exceeded that removed by crops (Figure 10.2). In some areas, notably the eastern seaboard states, additions of phosphorus have greatly exceeded the removal of this element by crops. Since phosphorus is only sparingly removed by leaching, this utilization of phosphate fertilizers is obviously inefficient. Research has quantified this inefficiency by showing that less than 15% of fertilizer-applied phosphorus is normally taken up by the crop during the year the fertilizer was applied.

Briefly, then, the overall phosphorus problem is threefold: (a) a compara-

[2] Note that the term *fixed* has a different meaning for phosphorus and potassium than for nitrogen. *Nitrogen fixation* refers to the conversion of N_2 gas to combined forms that plants can utilize. *Phosphorus* and *potassium fixation* refers to the conversion of soluble forms of these elements to insoluble forms unavailable to plants.

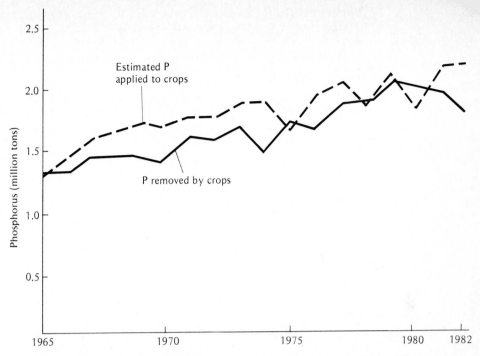

FIGURE 10.2 Estimates of fertilizer phosphorus applied in the United States from 1965 to 1982 compared to the removal of this element by crops. [*From Douglas (1982)*.]

tively low level of total phosphorus in soils, (b) low availability of this native phosphorus, and (c) a marked fixation of added soluble phosphates. Since crop removal of phosphorus is relatively low and world phosphate supplies are huge, the problem of supplying sufficient total phosphorus is a matter of economics and distribution. Increasing the availability of native soil phosphorus and retarding the fixation or reversion of added phosphates are, therefore, the problems of greatest scientific importance. These two phases will be discussed following a brief review of the phosphorus compounds present in soils.

10.4 Phosphorus Compounds in Soils[3]

Both inorganic and organic forms of phosphorus occur in soils and both are important to plants as sources of this element. The relative amounts in the two forms vary greatly from soil to soil but it is not at all uncommon for more than half the phosphorus to be in the organic form. Data in Table 10.1 give some idea of their relative proportions in mineral soils. The organic fraction

[3] For a review of soil phosphorus, see Larsen (1967).

TABLE 10.1 Total Phosphorus Content of Soils from Different Areas and the Percentage of Total Phosphorus in the Organic Form[a]

Soils	Number of samples	Total P (ppm)	Organic fraction (%)
Western Oregon soils			
Hill soils	4	357	66
Old valley-filling soils	4	1479	30
Recent valley soils	3	848	26
Iowa soils			
Mollisols	2	613	42
Alfisols	2	574	37
Afisols	2	495	53
Arizona soils			
Surface soils	19	703	36
Subsoils	5	125	34
Australia soils[b]			
Spodosol	1	398	65
Vertisol	1	362	86
Mollisol	1	505	75
Ghana soils[c]			
Antoakrom site	1	32	55
Bofoyedru site	1	47	63
Somiabra site	1	64	70
Obuasi site	1	51	67

[a] Source for Oregon, Iowa, and Arizona is quoted by Brady (1974).
[b] Source for Australia is Fares et al. (1974).
[c] Source for Ghana is Acquaye (1963).

generally constitutes 20–80% of the total. Despite the variation that occurs, it is evident that consideration of soil phosphorus would not be complete without some attention to both forms (Figure 10.3).

Inorganic Compounds. Most inorganic phosphorus compounds in soils fall into one of two groups: (a) those containing calcium, and (b) those containing *iron* and *aluminum*. The calcium compounds of most importance are listed in Table 10.2 The apatite minerals are the most insoluble and unavailable of the group. They may be found in even the more weathered soils, especially in their lower horizons. This fact is an indication of the extreme insolubility and consequent unavailability of the phosphorus contained therein. The simpler compounds of calcium, such as mono- and dicalcium phosphates, are readily available for plant growth. Except on recently fertilized soils, however, these compounds are present in extremely small quantities only because they easily revert to the more insoluble forms.

Much less is known of the exact constitution of the iron and aluminum phosphates contained in soils. The compounds involved are probably hydroxy phosphates such as strengite and variscite, although these two chemicals exist

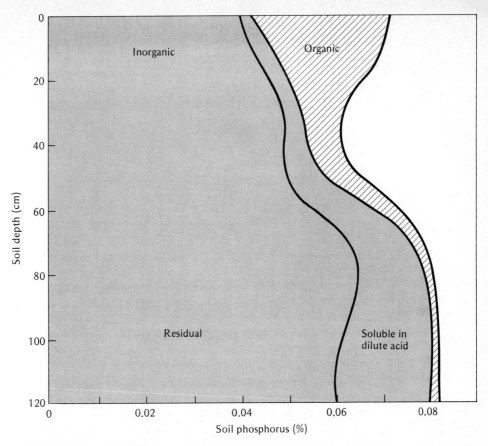

FIGURE 10.3 Distribution of phosphorus in organic and inorganic forms in an Iowa soil. The dilute-acid-soluble inorganic phosphorus is more readily available than the residual inorganic forms. In heavily fertilized soils the upper horizons would likely be much higher in inorganic phosphorus. [*From Black (1971).*]

only in very acid soils. Hydroxy phosphates are most stable in acid soils and are quite insoluble.

Many investigators have shown that phosphates react with certain iron or aluminum silicate minerals such as kaolinite. There is some uncertainty, however, as to the exact form in which this phosphorus is held in the soil. Most evidence indicates that it, too, is probably fixed as iron or aluminum phosphates, such as those described in the preceding paragraph.

Organic Phosphorus Compounds. There has been relatively less work done on the organic phosphorus compounds in soils, even though this fraction often comprises more than half of the total soil phosphorus. As a consequence, the nature of most of the organic-bound phosphorus in soils is not known. However, three main groups of organic phosphorus compounds found in plants are also

TABLE 10.2 Inorganic Calcium Compounds
of Phosphorus Often Found in Soils

Listed in order of increasing solubility.

Compound	Formula
Fluor apatite	$3\ Ca_3(PO_4)_2 \cdot CaF$
Carbonate apatite	$3\ Ca_3(PO_4)_2 \cdot CaCO_3$
Hydroxy apatite	$3\ Ca_3(PO_4)_2 \cdot Ca(OH)_2$
Oxy apatite	$3\ Ca_3(PO_4)_2 \cdot CaO$
Tricalcium phosphate	$Ca_3(PO_4)_2$
Octacalcium phosphate	$Ca_8H_2(PO_4)_6 \cdot 5H_2O$
Dicalcium phosphate	$CaHPO_4 \cdot 2H_2O$
Monocalcium phosphate	$Ca(H_2PO_4)_2$

present in soils. These are (a) phytin and phytin derivatives, (b) nucleic acids, and (c) phospholipids. While other organic phosphorus compounds are present in soils, the identity and amounts present are not known.

Phytin is a calcium-magnesium salt of phytic acid (inositol phosphoric acid), which has an empirical formula $(CH)_6(H_2PO_4)_6$. Phytin is the most abundant of the known organic phosphorus compounds in soils, although it generally accounts for no more than 30–40% of the total organic phosphorus. This compound is quite insoluble in soils.

Phosphate is a critical component of ribonucleic acid (RNA) and deoxyribonucleic acid (DNA), which occur in all plant residues. They are also thought to be present in soils, although at the relatively low levels of 1–2% of the soil organic phosphorus. Phospholipids are phosphorus-containing fatty compounds found in plants. They also make up only about 1–2% of the soil organic phosphorus. Obviously, nearly two-thirds of the organic phosphorus is in forms not as yet identified.

Phosphorus is present in phytin, nucleic acid and phospholipids in the phosphate form. Consequently, as microorganisms decompose these compounds, the phosphate that is released is available to react with inorganic compounds or to be taken up by plants.

10.5 Factors That Control the Availability of Inorganic Soil Phosphorus

The availability of inorganic phosphorus is largely determined by (a) soil pH; (b) soluble iron, aluminum, and manganese; (c) presence of iron-, aluminum-, and manganese-containing minerals; (d) available calcium and calcium minerals; (e) amount and decomposition of organic matter; and (f) activities of microorganisms. The first four factors are interrelated since soil pH drastically influences the reaction of phosphorus with the different ions and minerals.

10.6   pH and Phosphate Ions

As indicated in Section 6.10, the availability of phosphorus to plants is determined to no small degree by the ionic form of this element (Figure 10.4). The
ionic form in turn is determined by the pH of the solution in which the ion
is found. Thus, in highly acid solutions only $H_2PO_4^-$ ions are present. If the
pH is increased, first HPO_4^{2-} ions and finally PO_4^{3-} ions dominate. This situation
is represented by the equation

$$H_2PO_4^- \underset{H^+}{\overset{OH^-}{\rightleftharpoons}} H_2O + HPO_4^{2-} \underset{H^+}{\overset{OH^-}{\rightleftharpoons}} H_2O + PO_4^{3-}$$

(very acid solutions)   (very alkaline solutions)

At intermediate pH levels two of the phosphate ions may be present simultaneously. Thus, in solutions at pH 7.0, both $H_2PO_4^-$ and HPO_4^{2-} ions are found.
The $H_2PO_4^-$ ion is somewhat more available to plants than is the HPO_4^{2-}
ion. In soils, however, this relationship is complicated by the presence or ab-

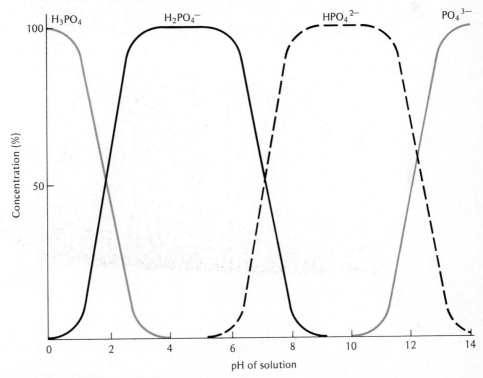

FIGURE 10.4   Relationship between solution pH and the relative concentrations of
three soluble forms of phosphate. In the pH range common for soils, the $H_2PO_4^-$ ions
predominate.

sence of other compounds or ions. For example, the presence of soluble iron and aluminum under very acid conditions, or calcium at high pH values, will markedly affect the availability of the phosphorus. Clearly, therefore, the effect of soil pH on phosphorus availability is determined in no small degree by the various cations present. The effect of these ions in acid soils will be discussed first.

10.7   Inorganic Phosphorus Availability in Acid Soils

Assume that there is either a nutrient solution or an organic soil very low in inorganic matter. Assume also that these media are acid in reaction but that they are low in iron, aluminum, and manganese. The H_2PO_4 ions, which would dominate under these conditions, would be readily available for plant growth. Normal phosphate absorption by plants would be expected so long as the pH was not too low.

Precipitation by Iron, Aluminum, and Manganese Ions.   If the same degree of acidity should exist in a normal mineral soil, however, quite different results would be expected (Figure 10.5). Some soluble iron, aluminum, and manganese

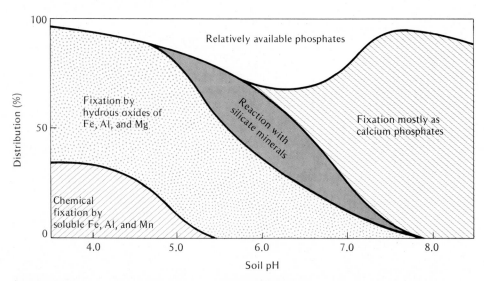

FIGURE 10.5   Inorganic fixation of added phosphates at various soil pH values. Average conditions are postulated, and it is not to be inferred that any particular soil would have exactly this distribution. The actual proportion remaining in an available form will depend upon contact with the soil, time for reaction, and other factors. It should be kept in mind that some of the added phosphorus may be changed to an organic form in which it would be temporarily unavailable.

are usually found in strongly acid mineral soils. Reaction with the $H_2PO_4^-$ ions would immediately occur, rendering the phosphorus insoluble and also unavailable for plant growth.

The chemical reactions occurring between the soluble iron and aluminum and the $H_2PO_4^-$ ions would result in the formation of hydroxy phosphates. This may be represented as follows, using the aluminum cation as an example.

$$Al^{3+} + H_2PO_4^- + 2H_2O \rightleftharpoons 2H^+ + Al(OH)_2H_2PO_4$$
$$\text{(soluble)} \qquad\qquad\qquad\qquad \text{(insoluble)}$$

In most strongly acid soils the concentration of the iron and aluminum ions greatly exceeds that of the $H_2PO_4^-$ ions. Consequently, the above reaction moves to the right, forming the insoluble phosphate. This leaves only minute quantities of the $H_2PO_4^-$ ion immediately available for plants under these conditions.

An interesting series of reactions occur when fertilizers containing $Ca(H_2PO_4)_2$ are added to soils, even those of relatively high pH (Lindsay and Stephenson, 1959) (Figure 10.6). The $Ca(H_2PO_4)_2$ in the fertilizer granules attracts water from the soil and the following reaction occurs.

$$Ca(H_2PO_4)_2 \cdot H_2O + H_2O \rightarrow CaHPO_4 \cdot 2H_2O + H_3PO_4$$

As more water is attracted, a H_3PO_4-laden solution with a pH of about 1.4 moves outward from the granule. This solution is sufficiently acid to dissolve and displace large quantities of iron, aluminum, and manganese. These ions react with the phosphate to form complex compounds, which later probably revert to the hydroxy phosphates of iron, aluminum, and manganese in acid soils and of calcium in neutral to alkaline soils. In any case, the immediate products of the addition to soils of a water-soluble compound [$Ca(H_2PO_4)_2 \cdot H_2O$] are a group of insoluble iron, aluminum, manganese, and calcium compounds. Even so, the phosphorus in these compounds is released quite readily for plant growth. It is only after these freshly precipitated compounds are allowed to "age" or to revert to more insoluble forms that availability to plants is greatly reduced.

Reaction with Hydrous Oxides. It should be emphasized that the $H_2PO_4^-$ ion reacts not only with the soluble iron, aluminum, and manganese but also with insoluble hydrous oxides of these elements, such as gibbsite ($Al_2O_3 \cdot 3H_2O$) and goethite ($Fe_2O_3 \cdot 3H_2O$). The actual quantity of phosphorus fixed by these minerals in acid soils quite likely exceeds that due to chemical precipitation by the soluble iron, aluminum, and manganese cations (Figure 10.5).

The compounds formed as a result of fixation by iron and aluminum oxides are likely to be hydroxy phosphates, just as in the case of chemical precipitation described above. Their formation can be illustrated by means of the following

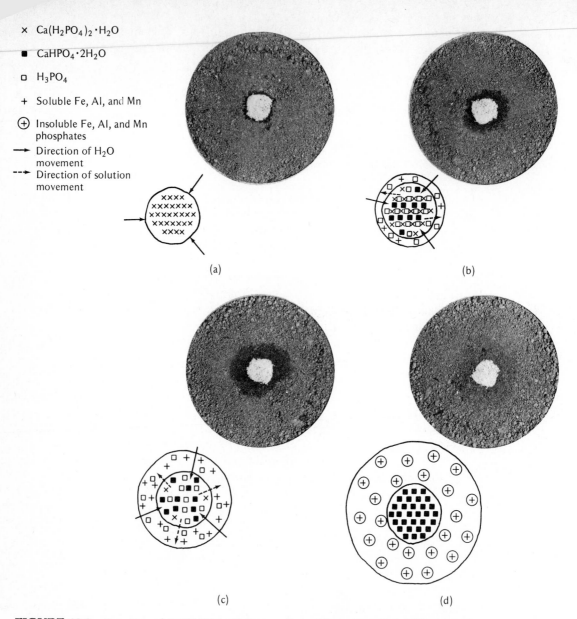

FIGURE 10.6 Reaction of $Ca(H_2PO_4)_2 \cdot H_2O$ granules with moist soil. (a) The granule has just been added to the soil and is beginning to absorb water from it. (b) In the moistened granule H_3PO_4 and $CaHPO_4 \cdot 2H_2O$ are being formed, and H_3PO_4 begins to move out into the soil as more soil water is being absorbed. (c) The H_3PO_4-laden solution moves into the soil, dissolving and displacing Fe, Al, and Mn and leaving insoluble $CaHPO_4 \cdot 2H_2O$ in the granule. (d) The Fe, Al, and Mn ions have reacted with the phosphate to form insoluble compounds, which, along with the residue of $CaHPO_4 \cdot 2H_2O$, are the primary reaction products. [*Photos courtesy G. L. Terman and National Plant Food Institute, Washington, DC*]

equation if the hydrous oxide of aluminum is represented as aluminum hydroxide.

$$\underset{\text{(soluble)}}{Al{\overset{\displaystyle OH}{\underset{\displaystyle OH}{\Big<}}}OH} + H_2PO_4^- \rightleftharpoons \underset{\text{(insoluble)}}{Al{\overset{\displaystyle OH}{\underset{\displaystyle H_2PO_4}{\Big<}}}OH} + OH^-$$

By means of this and similar reactions the formation of several basic phosphate minerals containing either iron or aluminum or both is thought to occur. Since several such compounds are possible, fixation of phosphorus by this mechanism probably takes place over a relatively wide pH range. Also the large quantities of hydrous iron and aluminum oxides present in most soils make possible the fixation of extremely large amounts of phosphorus by this means.

Thus, as both of the equations above show, the acid condition that would make possible the presence of the readily available $H_2PO_4^-$ ion in mineral soils at the same time results in conditions conducive to the vigorous *fixation* or precipitation of the phosphorus by iron, aluminum, and manganese compounds (Figure 10.6).

Fixation by Silicate Clays. A third means of fixation of phosphorus under moderately acid conditions involves silicate minerals such as kaolinite. Although there is some doubt about the actual mechanisms involved, the overall effect is essentially the same as when phosphorus is fixed by simpler iron and aluminum compounds. Some scientists visualize the fixation of phosphates by silicate minerals as a surface reaction between exposed —OH groups on the mineral crystal and the $H_2PO_4^-$ ions. Other investigators have evidence that aluminum and iron ions, removed from the edges of the silicate crystals, form hydroxy phosphates of the same general formula as those already discussed. This type of reaction might be expressed as follows.

$$\underset{\text{(in silicate crystal)}}{[Al]} + H_2PO_4^- + 2\,H_2O \rightleftharpoons 2\,H^+ + \underset{\text{(insoluble)}}{Al(OH)_2H_2PO_4}$$

Thus, even though phosphates react with different ions and compounds in acid soils, apparently similar insoluble iron and aluminum compounds are formed in each case. Major differences from soil to soil are probably due to differences in rate of phosphate precipitation and in the surface area of the phosphates once the reaction has occurred. This will be discussed again later.

Anion Exchange. The reactions described so far involved either the formation of insoluble chemical precipitates or the replacement of OH groups on the surface of either silicate clays or hydrous oxides. Phosphate can also participate in simple anion exchange. As was pointed out in Chapter 5, certain positively charged colloidal particles are the seat of exchange between two anions. This can be illustrated with $\diagdown_\diagup AlOH_2^+$ to represent the positively charged particle.

$$\left[\diagdown_\diagup AlOH_2^+\right]OH + H_2PO_4^- \rightleftharpoons \left[\diagdown_\diagup AlOH_2^+\right]H_2PO_4 + OH^-$$

positively
charged particle   positively
charged particle

Note that the reaction is reversible; that is, the OH^- ion can replace the $H_2PO_4^-$ ion and remove it from the colloidal surface to the soil solution. This reaction shows how anion exchange takes place and illustrates the importance of liming acid soils in helping to maintain a higher level of available phosphorus.

10.8 Inorganic Phosphorus Availability at High pH Values

The availability of phosphorus in alkaline soils is determined largely by the solubility of the calcium compounds in which the phosphorus is found. If an $H_2PO_4^-$-containing fertilizer such as concentrated superphosphate is added to an alkaline soil (say at pH = 8.0), the $H_2PO_4^-$ ion quickly reacts to form less soluble compounds. Three such compounds are dicalcium phosphate dihydrate (DCPD), $CaHPO_4 \cdot 2H_2O$; octacalcium phosphate (OCP), $Ca_8H_2(PO_4)_6 \cdot 5H_2O$; and tricalcium phosphate (TCP), $Ca_3(PO_4)_2$. While the specific mechanisms for the formation of these compounds are not well established, their relationship to each other can be shown rather simply.

$$Ca(H_2PO_4)_2 + CaCO_3 + H_2O \rightarrow \underset{\text{DCPD}}{2\,CaHPO_4 \cdot 2H_2O} + CO_2$$

$$\underset{\text{DCPD}}{6\,CaHPO_4 \cdot 2H_2O} + 2\,CaCO_3 + H_2O \rightarrow \underset{\text{OCP}}{Ca_8H_2(PO_4)_6 \cdot 5H_2O} + 2\,CO_2$$

$$\underset{\text{OCP}}{Ca_8H_2(PO_4)_6 \cdot 5H_2O} + CaCO_3 \rightarrow \underset{\text{TCP}}{3\,Ca_3(PO_4)_2} + CO_2 + 6\,H_2O$$

The solubility of the compounds and, in turn, the availability to plants of the phosphorus they contain decrease as the phosphorus changes from the $H_2PO_4^-$ ion to the DCPD and finally to the TCP.

Although the $Ca_3(PO_4)_2$ (TCP) thus formed is quite insoluble, it may be converted in the soil to even more insoluble compounds. Hydroxy, oxy, carbonate, and even fluor apatite compounds may be formed if conditions are favorable and if sufficient time is allowed (Table 10.2).

This type of reversion may occur in soils of the eastern United States which have been heavily limed. It is much more serious, however, in western soils, owing to the widespread presence of excess $CaCO_3$. The problem of utilizing phosphates in alkaline soils of the arid West is thus fully as serious as it is on highly acid soils in the East.

10.9 pH for Maximum Inorganic Phosphorus Availability

With insolubility of phosphorus occurring at both extremes of the soil pH range (Figure 10.5), the question arises as to the range in soil reaction in which minimum fixation occurs. The basic iron and aluminum phosphates have a minimum solubility around pH 3–4. At higher pH values some of the phosphorus is released and the fixing capacity somewhat reduced. Even at pH 6.5, however, much of the phosphorus is still probably chemically combined with iron and aluminum. As the pH approaches 6, precipitation as calcium compounds begins; at pH 6.5 the formation of insoluble calcium salts is a factor in rendering the phosphorus unavailable. Above pH 7.0, even more insoluble compounds, such as apatites, are formed.

These facts seem to indicate that maximum phosphate availability to plants is obtained when the soil pH is maintained in the 6.0–7.0 range (Figure 10.5). Even in this range, however, the fact should be emphasized that phosphate availability may still be very low and that added soluble phosphates are readily fixed by soils. The low recovery (perhaps 10–30%) by plants of added phosphates in a given season is partially due to this fixation.

10.10 Availability and Surface Area of Phosphates

When soluble phosphates are added to soils, two kinds of compounds form immediately: (a) fresh precipitates of calcium phosphate or iron and aluminum phosphates and (b) similar compounds formed on the surfaces of either calcium carbonate or iron and aluminum oxide particles. In each case, the total surface area of the phosphate-held particles is high, and consequently the ease of availability of the phosphorus is reasonably great. Thus, even though the water-soluble phosphorus in superphosphate may be precipitated in the soil in a matter of a few days, the freshly precipitated compounds will release much of their phosphorus to growing plants.

Effects of Aging. With time, changes take place in the reaction products of soluble phosphates and soils. These changes generally result in a reduction in surface area of the phosphates and a similar reduction in their availability. An increase in the crystal size of precipitated phosphates occurs in time. This decreases their surface area. Also, there is a penetration of the phosphorus held by calcium carbonate and iron or aluminum oxide particles into the particle

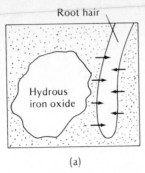
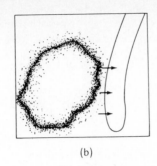
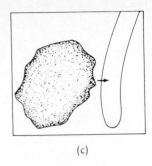

Root hair

Hydrous
iron oxide

(a) (b) (c)

FIGURE 10.7 How relatively soluble phosphates are rendered unavailable by compounds such as hydrous oxide. (a) The situation just after application of a soluble phosphate. The root hair and the hydrous iron oxide particle are surrounded by soluble phosphates. (b) Within a very short time most of the soluble phosphate has reacted with the surface of the iron oxide crystal. The phosphorus is still fairly readily available to the plant roots since most of it is located at the surface of the particle where exudates from the plant can encourage exchange. (c) In time the phosphorus penetrates the crystal and only a small portion is found near the surface. Under these conditions its availability is low.

itself (Figure 10.7). This leaves less of the phosphorus near the surface where it can be made available to growing plants. By these processes of aging, phosphate availability is reduced. Thus, the supply of available phosphorus to plants is determined not only by the kinds of compounds that form but also by their surface areas (Figure 10.8).

FIGURE 10.8 The effect of clay content on the percent recovery of fertilizer phosphorus by seven crops grown on four calcareous soils. Soils with higher clay contents tended to fix the phosphorus in forms not readily available to the crop plants. [*From Olsen et al. (1977); used with permission of the World Phosphate Industry Association, Paris, France.*]

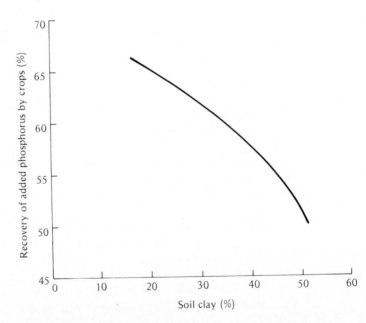

Relation to Soil Texture. Most of the compounds with which phosphorus reacts are in the finer soil fractions. As a consequence phosphorus fixation tends to be more pronounced in clay soils than in the coarser-textured ones. This fact is borne out by the data in Figure 10.8, which clearly illustrates the tendency of clays to reduce phosphate availability.

10.11 Phosphorus-Fixing Power of Soils

In light of the above discussion it is interesting to note the actual quantities of phosphorus that soils are capable of fixing. Data from three New Jersey soils presented in Table 10.3 emphasize the tremendous power of certain soils in this respect. For example, to satisfy the phosphorus-fixing power of the unlimed Collington soil, nearly 105 Mg of ordinary superphosphate containing 10% P would be required. Although liming definitely reduced the fixing capacity, the quantity of phosphorus fixed even on the limed soils is enormous. Thus, over 56 Mg of superphosphate would be required to satisfy completely the phosphorus-fixing power of a hectare–furrow slice of the limed Collington soil.

TABLE 10.3 Phosphorus-Adsorbing Power of Three New Jersey Soils Limed and Unlimed[a] (upper 15 cm)

Soil	Treatment	pH	Phosphorus fixing power (P)	
			kg/ha	lb/A
Sassafras	No lime	3.6	2780	2480
Sassafras	Lime	6.5	1361	1215
Collington	No lime	3.2	9173	8184
Collington	Lime	6.5	4920	4390
Dutchess	No lime	3.8	6727	6002
Dutchess	Lime	6.5	4309	3844

[a] From Toth and Bear (1947).

10.12 Influence of Soil Organisms and Organic Matter on the Availability of Inorganic Phosphorus

In addition to pH and related factors, organic matter and microorganisms strikingly affect inorganic phosphorus availability. Just as was the case with nitrogen, the rapid decomposition of organic matter and consequent high microbial population result in the *temporary* tying up of inorganic phosphates in microbial tissue.

Products of organic decay such as organic acids and humus are thought

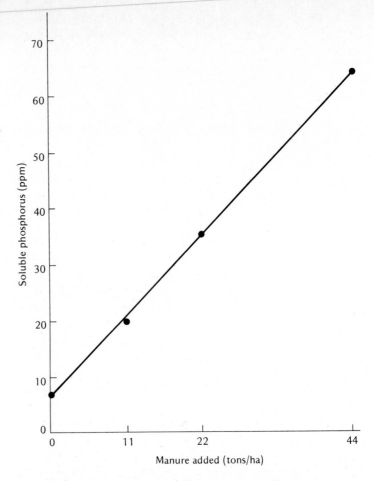

FIGURE 10.9 The effect of added organic matter (manure) on the soluble phosphorus level of soil at pH 7.2. As the manure decomposed, organic acids were released that formed stable complexes with iron and aluminum compounds and also affected the solubility of calcium phosphates. [*Data from El-Baruni and Olsen* (*1979*).]

to be effective in forming complexes with iron and aluminum compounds. This engagement of iron and aluminum reduces inorganic phosphate fixation to a remarkable degree. Figure 10.9 shows the effect of manure application on the soluble phosphorus level in soils. Apparently the manure was effective in releasing phosphorus after it had been fixed as Fe, Al, and Ca phosphates. Thus, organic decomposition products undoubtedly play an important role in inorganic phosphorus availability.

10.13   Availability of Organic Phosphorus

Only meager information has been obtained on the factors affecting the availability to higher plants of organic phosphorus compounds. Even though both *phytin* and *nucleic acids* can be utilized as sources of phosphorus, inorganic sources of this element are needed for normal production. Plants commonly suffer from a phosphorus deficiency even in the presence of considerable quantities of organic forms of this element. Just as with inorganic phosphates, the problem is one of availability.

Phytin behaves in the soil much as do the inorganic phosphates, forming iron, aluminum, and calcium phytates. In acid soils the phytin is rendered insoluble and thus unavailable because of reaction with iron and aluminum. Under alkaline conditions calcium phytate is precipitated and its phosphorus is rendered unavailable.

The fixation of nucleic acids involves an entirely different mechanism, but the end result—low phosphorus availability—is the same. Evidently, nucleic acids are strongly adsorbed by clays, especially montmorillonite.

This adsorption is particularly pronounced under acid conditions and results in a marked decrease in the rate of decomposition of the nucleic acids. Consequently, the available phosphorus supply from this source is low, especially in acid soils which contain appreciable amounts of montmorillonite.

The judicious application of lime to acid soils is thus fully as important in organic phosphorus nutrition as it is in rendering inorganic compounds available. Whether we are dealing with inorganic soil phosphates, added fertilizers, or organic materials, the importance of lime as a controlling factor in phosphate availability is clearly evident.

10.14 Intensity and Quantity Factors

It is important to be aware of the different forms in which phosphorus is found in soils. But it is even more important to know a soil's ability to maintain sufficiently high levels of phosphorus in the soil solution to assure satisfactory plant growth. The concentration of phosphorus in the soil solution is a measure of the *intensity* (I) *factor* of phosphorus nutrition. If this factor is maintained at about 0.2 ppm or above, maximum yield of most crop plants will occur (Table 10.4).

TABLE 10.4 Concentration of Phosphorus in Soil Solution That Provided 95% of Maximum Yield of Several Crops in Hawaii[a]

Crop	Soil	Approximate P in soil solution (ppm)
Cassava	Halii	0.005
Peanut	Halii	0.01
Corn	Halii	0.05
Soybean	Halii	0.20
Cabbage	Kula	0.04
Tomato	Kula	0.20
Head lettuce	Kula	0.30

[a] From Fox (1981).

As plant roots take up phosphorus from the soil solution, it is at least partially replenished by release of phosphorus from the labile forms, the soil solids that are in equilibrium with the soil solution. This source of soil solution replenishment is known as the *quantity (Q) factor* of phosphorus nutrition. While it involves all solid forms of phosphorus that releases this element to the solution, it is made up primarily of the slowly available forms shown in

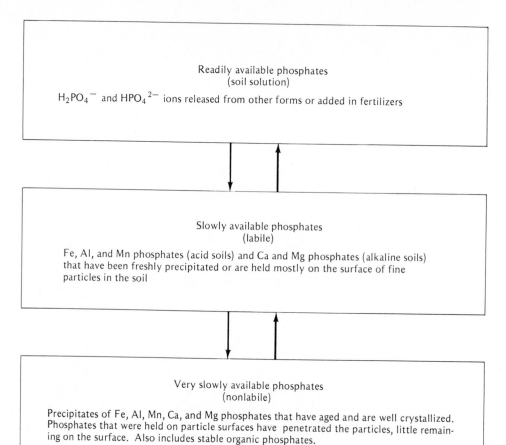

FIGURE 10.10 Classification of phosphorus compounds in soils in three major groups. Fertilizer phosphates are generally in the readily available phosphate (soil solution) group, but are quickly converted to the slowly available (labile) forms. These can be utilized by plants at first, but upon aging are rendered less available and are then classed as very slowly available (nonlabile). At any one time perhaps 80–90% of the soil phosphorus is in very slowly available forms. Most of the remainder is in the slowly available form since perhaps less than 1% would be expected to be readily available. Note that phosphorus moves from one form to another although the movement is slow.

Figure 10.10. They are freshly precipitated Fe, Al, Mn, and Ca compounds and surface-adsorbed phosphates that have not yet penetrated the particles on which they are held.

As shown in Figure 10.11, the Q level needed to provide a given intensity (I) or soil solution phosphorus level will vary from soil to soil. Tropical clays high in Fe and Al need a high Q level to assure an I level sufficient for normal growth. In contrast sandy soils with low clay and Fe and Al contents provide higher intensities with a given quantity of phosphorus.

When plant roots remove a given quantity of phosphorus from the soil solution (reduce the I factor), the level of I is drastically reduced in the sandy soil and may be reduced only slightly in the clay soil. The latter is thus said to be more highly buffered with respect to phosphorus than the sandy soil. The *potential buffering capacity* (PBC) of a soil is given by

$$PBC = \frac{\Delta Q}{\Delta I}$$

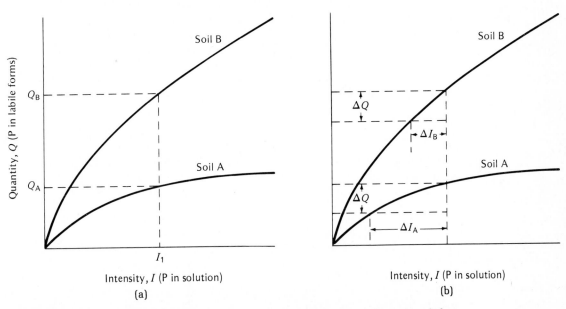

FIGURE 10.11 Relationship between the intensity factor (P in solution) and the quantity factor (P in solid labile forms that are in equilibrium with P in solution) on two soils. Soil A may be a sandy soil low in phosphorus-binding compounds, while soil B is a heavier-textured soil high in Fe, Al, Mn, or Ca that holds the phosphorus in relatively unavailable forms. (a) To provide a given level of P in the soil solution (intensity) a much higher quantity must be present in soil B than in soil A. (b) However, when a given quantity (ΔQ) of phosphorus is removed from the soil by a crop, the decrease in intensity is much greater for soil A (ΔI_A) than for soil B (ΔI_B). Soil B is more highly buffered than soil A.

ΔQ is the change in the quantity (Q factor), and ΔI is the change in the intensity (I factor).

The quantity/intensity relations of phosphorus in soils and the potential buffering capacity are important in determining the fertilizer phosphate requirements for high yields. We shall see in Section 10.22 that this same general relationship holds for potassium.

10.15 Practical Control of Phosphorus Availability

From a practical standpoint, the phosphorus-utilization picture is not too encouraging. The inefficient utilization of applied phosphates by plants has long been known. Most crops do not take up more than about 10–15% of the phosphorus added in fertilizers during the year the fertilizer is applied. This is due not only to the tendency of the soil to fix the added phosphorus but to the slow rate of movement of this element to plant roots in the soil.

Continuing application of phosphate fertilizers in time tends to increase the level of this nutrient in the soil and particularly its level in the labile forms which can release phosphorus to the soil solution. Thus, even though much of the phosphorus added in fertilizer is not used during the year of application, it may provide an important source in future years.

Liming and Placement of Fertilizers. The small amount of control that can be exerted over phosphate availability seems to be associated with *liming, fertilizer placement,* and *organic matter maintenance.* By holding the pH of soils between 6.0 and 7.0, the phosphate fixation can be kept at a minimum (Figure 10.5). To prevent rapid reaction of phosphate fertilizers with the soil, these materials are commonly placed in localized bands. In addition, phosphatic fertilizers are quite often granulated to retard still more their contact with the soil. The effective utilization of phosphorus in combination with animal

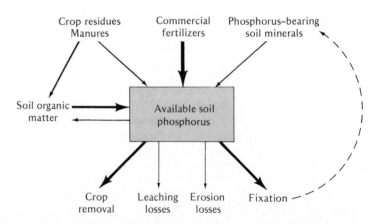

FIGURE 10.12 How the available phosphorus level in a soil is depleted and replenished. Note that the two main features are the addition of phosphate fertilizers and the fixation of this element in insoluble forms. It should be remembered that the amount of available phosphorus in the soil at any one time is relatively small, especially when compared to those of calcium, magnesium, and potassium.

manures is evidence of the importance of organic matter in increasing the availability of this element.

In spite of these precautions, a major portion of the added phosphates still reverts to less available forms (Figure 10.12). It should be remembered, however, that the reverted phosphorus is not lost from the soil and through the years is slowly made available to growing plants. This becomes an important factor, especially in soils that have been heavily phosphated for years.

In summary, maintaining sufficient available phosphorus in a soil largely narrows down to a twofold program: (a) the addition of phosphorus-containing fertilizers and (b) the regulation in some degree of the fixation in the soil of both the added and the native phosphates.

10.16   Potassium—The Third "Fertilizer" Element

The history of fertilizer usage in the United States shows that nitrogen and phosphorus received most of the attention when commercial fertilizers first appeared on the market. Although the role played by potassium in plant nutrition has long been known, the importance of potash fertilization has received full recognition only in comparatively recent years.

The reasons that a widespread deficiency of this element did not develop earlier are at least twofold. First, the supply of available potassium originally was so high in most soils that it took many years of cropping for a serious depletion to appear. Second, even though the potassium in certain soils may have been insufficient for optimum crop yields, production was much more drastically limited by a lack of nitrogen and phosphorus. With an increased usage of fertilizers carrying these latter elements, crop yields have been correspondingly increased. As a consequence, the drain on soil potassium has been greatly increased. This, coupled with considerable loss by leaching, has enhanced the demand for potassium in commercial fertilizers.

10.17   Effects of Potassium on Plant Growth[5]

Potassium plays many essential roles in plants. It is an activator of dozens of enzymes responsible for such plant processes as energy metabolism, starch synthesis, nitrate reduction, and sugar degradation. Because of its ease of transport across plant membranes, it is extremely mobile within the plant. Its relatively high concentration in plant cells helps it regulate the opening and closing of stomates in the leaves and the uptake of water by root cells.

Potassium is essential for photosynthesis and for starch formation and the translocation of sugars. It is necessary in the development of chlorophyll, although it does not, like magnesium, enter prominently into its molecular

[5] For further information on this topic see Mengel and Kirby (1980) and Sekhon (1978).

structure. This element is important to cereals in grain formation, as it aids in the development of plump, heavy kernels. Abundant available potassium also is absolutely necessary for tuber development. Therefore, the percentage of this element usually is comparatively high in mixed fertilizers recommended for potatoes. All root crops respond to liberal applications of potassium. As with phosphorus, it may be present in large quantities in the soil and yet exert no harmful effect on the crop.

By increasing crop resistance to certain diseases and by encouraging strong root and stem systems, potassium tends to prevent the undesirable "lodging" of plants and to counteract the damaging effects of excessive nitrogen. In delaying maturity, potassium works against undue ripening influences of phosphorus. In a general way, it exerts a balancing effect on both nitrogen and phosphorus, and consequently is especially important in a mixed fertilizer.

In considering potassium plant nutrition it should be noted that sodium has been found to partially take the place of potassium in the nutrition of certain plants. When there is a deficiency of potash, the native sodium of the soil, or that added in such fertilizers as nitrate of soda, may be very useful.

10.18   The Potassium Cycle

Figure 10.13 shows the major forms in which potassium is held in soils and the changes it undergoes as it is cycled through the soil and plant systems. The original sources of potassium are the primary minerals such as the micas and potash feldspar (microcline). As these minerals weather, their rigid lattice structures become more pliable. Potassium held between their 2:1-type layers is slowly made more available, first through nonexchangeable but slowly available forms, and finally through the readily exchangeable and the soil solution forms from which plant roots take up this element. At any one time most of the potassium is in the primary mineral and nonexchangeable forms. But the exchangeable and soil solution potassium are the forms of greatest immediate concern to plant growth. To maintain the levels of these last two forms high enough to encourage good crop production, it has become increasingly necessary to supplement soil-borne potassium with applications of chemical fertilizer. The sections that follow give greater details on the reactions involved in the potassium cycle.

10.19   The Potassium Problem

Availability of Potassium.   In contrast to the situation regarding phosphorus, most mineral soils, except those of a sandy nature, are comparatively high in *total* potassium. In fact, the total quantity of this element is generally greater than that of any other major nutrient element. Amounts as great as 35,000–55,000 kg potassium per hectare–furrow slice (31,225–44,600 lb per acre–furrow

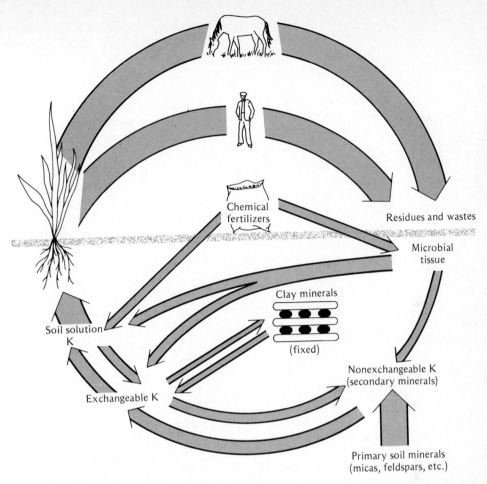

FIGURE 10.13 The major components of the potassium cycle. Primarily soil minerals such as the micas and feldspars are the original sources of potassium. They weather to fine micas (illite) and other silicate clays in which some of the potassium is held in a nonexchangeable, but slowly available form. The nonexchangeable potassium is slowly released to the exchangeable form and later to the soil solution, from which it is absorbed by plant roots, and is eventually recycled through plant residues and wastes to the soil. At any one time, most of the potassium is in primary minerals or in the nonexchangeable or fixed form. Chemical fertilizers are increasingly important sources of potassium.

slice) are not at all uncommon (see Table 1.3). Yet the quantity of potassium held in an easily exchangeable condition at any one time often is very small. Most of this element is held rigidly as part of the primary minerals or is fixed in forms that are at best only moderately available to plants. Also competition by microorganisms for this element contributes at least temporarily to its unavailability to higher plants. Thus, the situation in respect to potassium utilization parallels that of phosphorus and nitrogen in at least one way. A very large proportion of all three of these elements in the soil is insoluble and relatively unavailable to growing plants.

Leaching Losses. Unlike the situation with respect to phosphorus, however, much potassium is lost by leaching. An examination of the drainage water from mineral soils on which rather liberal fertilizer applications have been made will usually show considerable quantities of potassium. In extreme cases, the magnitude of this loss may approach that of potassium removal by the crop. For example, heavily fertilized sandy soils on which crops such as vegetables or tobacco are grown may suffer serious losses by leaching. Even on a representative humid region soil receiving only moderate rates of fertilizer, the annual loss of potassium by leaching is usually about 35 kg/ha (31 lb/A) (see Table 15.7).

Crop Removal. The third phase of the potassium problem concerns the quantity of this element in plants—in other words, its removal by growing crops. Under ordinary field conditions and with an adequate nutrient supply, potassium removal by crops is high, often being three to four times that of phosphorus and equaling that of nitrogen. The removal of 120–150 kg/ha (107–134 lb/A) of potassium by a 50 Mg/ha silage corn crop is not at all unusual.

Moreover, this situation is made even more critical by the fact that plants tend to take up soluble potassium far in excess of their needs if sufficiently large quantities are present. This tendency is termed *luxury consumption,* because the excess potassium absorbed apparently does not increase crop yields to any extent.

Examples of Luxury Consumption. The principles involved in luxury consumption are shown by the graph of Figure 10.14. For many crops there is a more or less direct relationship between the available potassium in the soil and the removal of this element by plants. The available potassium would, of course, include both that added and that already present. A certain amount of this element is needed for optimum yields and this is termed *required potassium.* All potassium above this critical level is considered a *luxury,* the removal of which is decidedly wasteful.

Under field conditions, luxury consumption becomes particularly wasteful. For example, to save labor, one may be tempted to supply potassium only during the first year of a three- or four-year perennial hay crop. Much of the potassium thus added is likely to be absorbed wastefully by the first crop of hay or even in the first cutting. Consequently, too little of the added potassium would remain for subsequent crops.

In summary, then, the problem of potassium economy in its most general terms is at least threefold: (a) a very large proportion of this element at a given time is relatively unavailable to higher plants; (b) because of the solubility

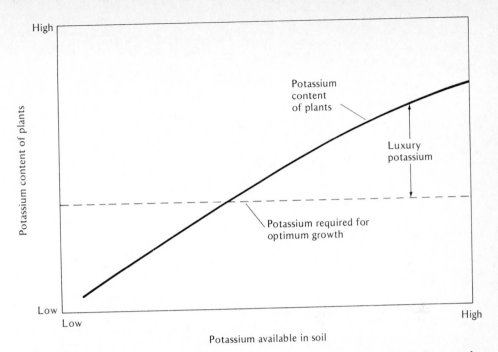

FIGURE 10.14   General relationship between the potassium content of plants and the available soil potassium. If excess quantities of potash fertilizers are applied to a soil, the plants will absorb potassium in excess of that required for optimum yields. This luxury consumption may be wasteful, especially if the crops are completely removed from the soil.

of its available forms, potassium is subject to wasteful leaching losses; and (c) the removal of potassium by crop plants is high, especially when luxury quantities of this element are supplied. With these ideas as a background, the various forms of potassium in soils and their availabilities will now be considered.

10.20   Forms and Availability of Potassium in Soils

For convenience, the various forms of potassium in soils can be classified on the basis of availability in three general groups: (a) *unavailable*, (b) *readily available*, and (c) *slowly available*. Although most of the soil potassium is in the first of these three forms, from an immediate practical standpoint the latter two are undoubtedly of greater significance.

The relationship among these three general categories is shown diagrammatically in Figure 10.15. Equilibrium tendencies presented therein are of vital practical importance, especially when dealing with the slowly and readily available forms. A slow change from one form to another can and does occur.

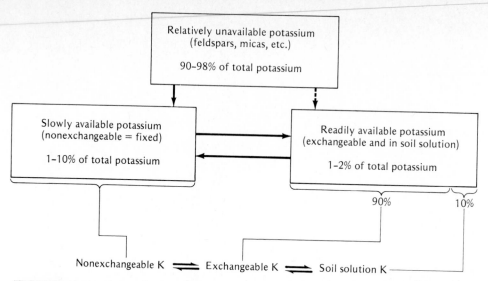

FIGURE 10.15 Relative proportions of the total soil potassium in unavailable, slowly available, and readily available forms. Only 1–2% is rated as readily available. Of this, approximately 90% is exchangeable and only 10% appears in the soil solution at any time. [*Modified from Attoe and Truog (1945).*]

This makes possible a fixation and conservation of added soluble potassium and a subsequent slow release of this element when the readily available supply is reduced.

Relatively Unavailable Forms. By far the greatest part (perhaps 90–98%) of all soil potassium in a mineral soil is in relatively unavailable forms (Figure 10.15). The compounds containing most of this form of potassium are the feldspars and the micas. These minerals are quite resistant to weathering and probably supply relatively insignificant quantities of potassium during a given growing season. However, their cumulative contribution over a period of years to the overall available potassium in the soil undoubtedly is of considerable importance. Potassium is gradually released to more available forms through the action of solvents such as carbonated water. Also, the presence of acid clay is of some significance in the breakdown of these primary minerals with the subsequent release of potassium and other bases (see Section 12.4).

Readily Available Forms. The readily available potassium constitutes only about 1–2% of the total amount of this element in an average mineral soil. It exists in soils in two forms: (a) potassium in the soil solution and (b) exchangeable potassium adsorbed on the soil colloidal surfaces. Although most of this available potassium is in the exchangeable form (approximately 90%), soil solution potassium is most readily absorbed by higher plants and is, of course, subject to considerable leaching loss.

As represented in Figure 10.15, these two forms of readily available potassium are in dynamic equilibrium. Such a situation is extremely important from a practical standpoint. In the first place, absorption of soil solution potassium by plants results in a temporary disruption of the equilibrium. Then to restore the balance, some of the exchangeable potassium immediately moves into the soil solution until the equilibrium is again established. On the other hand, when water-soluble fertilizers are added, the reverse of the above adjustment occurs. The exchangeable potassium can be seen as an important buffer mechanism for soil solution potassium.

Slowly Available Forms. In the presence of vermiculite, smectite, and other 2:1-type minerals the potassium of such fertilizers as muriate of potash not only becomes adsorbed but also may become definitely "fixed" by the soil colloids (Figure 10.16). The potassium as well as ammonium ions fit in between layers in the crystals of these normally expanding clays and become an integral part of the crystal. Potassium in this form cannot be replaced by ordinary exchange methods and consequently is referred to as *nonexchangeable potassium*. As such, this element is not readily available to higher plants. This form is in equilibrium, however, with the available forms and consequently acts as an extremely important reservoir of slowly available potassium. The entire equilibrium may be represented as follows.

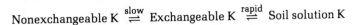

$$\text{Nonexchangeable K} \overset{\text{slow}}{\rightleftharpoons} \text{Exchangeable K} \overset{\text{rapid}}{\rightleftharpoons} \text{Soil solution K}$$

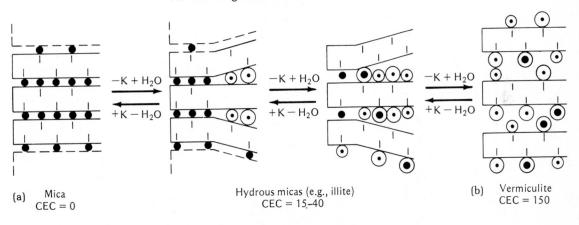

(a) Mica
CEC = 0

Hydrous micas (e.g., illite)
CEC = 15-40

(b) Vermiculite
CEC = 150

● Potassium ion, dehydrated   ◉•◉ Exchangeable cations, hydrated

FIGURE 10.16 Diagrams to illustrate the release of potassium from primary micas to fine-grained mica (illite) and then to vermiculite and the fixation of exchangeable potassium by reversing these release reactions. Note that the dehydrated K^+ ion is much smaller than the hydrated ions of Na^+, Ca^{2+}, Mg^{2+}, etc. Thus, when potassium is added to a soil containing 2:1-type minerals such as vermiculite, the reaction may go to the left and K^+ ions are held tightly (fixed) in between layers within the crystal, giving an illite-type structure. [*Modified from McLean (1978); used with permission of Potash Research Institute of India.*]

TABLE 10.5 Potassium Removal by Very Intensive Cropping and the Amount of This Element Coming from the Nonexchangeable Form

Soil	Total K used by crops		Percent derived from nonexchangeable form
	kg/ha	lb/A	
Wisconsin soils[a]			
Carrington silt loam	133	119	75
Spencer silt loam	66	59	80
Plainfield sand	99	88	25
Mississippi soils[b]			
Robinsonville fine silty loam	121	108	33
Houston clay	64	57	47
Ruston sandy loam	47	42	24

[a] Average of six consecutive cuttings of Ladino clover, from Evans and Attoe (1948).

[b] Average of eight consecutive crops of millet, from Gholston and Hoover (1948).

The importance of this adjustment to practical agriculture should not be overlooked. It is of special value in the conservation of added potassium. For example, when potassium-containing fertilizers are added to the soil, a large proportion of the soluble (soil solution) potassium becomes attached to the colloids. As a result, the above equilibrium shifts to the left, and some exchangeable ions are converted to the nonexchangeable form. Although this potassium is at least temporarily unavailable, it is not subject to leaching and thus a significant conservation is attained. Also, since the fixed potassium is slowly reconverted to the available forms later, it is by no means completely lost to growing plants.

Release of Fixed Potassium. The above reaction is perhaps of even greater practical importance in another way. The quantity of nonexchangeable or "fixed" potassium in some soils is quite large. The fixed potassium in such soils is continually released to the exchangeable form in amounts large enough to be of great practical importance. The data in Table 10.5 indicates the magnitude of the release of nonexchangeable potassium from certain soils. In several cases cited, the potassium removed by crops was supplied largely from nonexchangeable forms.

10.21   Factors Affecting Potassium Fixation in Soils

Several soil conditions markedly influence the amounts of potassium fixed. Among the factors are (a) the nature of the soil colloids, (b) wetting and drying, (c) freezing and thawing, and (d) the presence of excess lime.

Effects of Type of Clay, Moisture, and Temperature.   The ability of the various soil colloids to fix potassium varies widely. For example, 1:1-type clays such as kaolinite and soils in which these clay minerals are dominant fix little potassium. On the other hand, clays of the 2:1 type, such as vermiculite, smectite, and fine-grained mica (illite), fix potassium very readily and in large amounts. Even silt-sized fractions of some micaceous minerals fix and subsequently release potassium.

The potassium and ammonium ions are attracted between layers in the clay crystals by the same negative charges responsible for the internal adsorption of these and other cations. The tendency for fixation is greatest in minerals where the major source of negative charge is in the silica (tetrahedral) sheet. Thus vermiculite has a greater fixing capacity than montmorillonite (see Table 5.4 for formulas for these minerals).

The mechanism for potassium fixation may vary somewhat from one clay mineral to another. In the case of illite and some smectites and vermiculites, the potassium ions may penetrate the crystal lattice and be trapped when the layers contract during periods of drying. Clay minerals with a high negative charge caused by the Al^{3+}-for-Si^{4+} substitution (see Section 5.7) will strongly bind the K^+ ion and prevent its being exchanged. Some weathered micas may simply permit the K^+ ions to reenter the surface potassium sites that were vacated when the mineral weathered. It is easy to see why there would be such wide variability in the K-fixing capacity of clays.

The mechanism for potassium fixation is probably the same as that for fixation of the ammonium ion. These two ions do not have as high an affinity for water of hydration as do other cations such as Na^+ and Ca^{2+}. As a consequence, they can easily be dehydrated, and then their small size is such as to permit them to fit snugly between the silica sheets of adjoining layers in 2:1-type clay minerals (Figure 10.16). Once in place, these ions become trapped as a part of the rigid crystal structure, thereby preventing normal crystal lattice expansion and reducing the cation exchange capacity of the clay. The larger hydrated ions of cations such as Na^+, Ca^{2+} and Mg^{2+} are not able to fit between these layers and consequently escape fixation.

Alternate wetting and drying, and freezing and thawing has been shown to result in the fixation of potassium in nonexchangeable forms as well as its ultimate release to the soil solution. Although the practical importance of this is recognized, its mechanism is not well understood.

Influence of Lime.   Applications of lime sometimes result in an increase in potassium fixation of soils (Figure 10.17). Under normal liming conditions this may be more beneficial than detrimental because of the conservation of the potassium so affected. Thus, potassium in well-limed soils is not as likely to be leached out as drastically as is that in acid soils.

There are conditions, however, under which the effects of lime on the availability of potash are undesirable. For example, in soils where the negative charge is pH-dependent liming can greatly reduce the level of potassium in

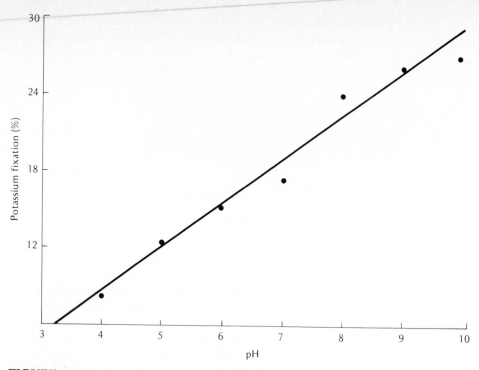

FIGURE 10.17 The effect of pH on the fixation of potassium in Indian soils. [*From Grewal and Kanwar (1976); used with permission of Indian Council of Agricultural Research.*]

the soil solution (Figure 10.18). Furthermore, high calcium levels in the soil solution may reduce potassium uptake by the plant. Finally, potassium deficiency has been noted in soils with excess calcium carbonate. Potassium fixation as well as cation ratios may be responsible for these adverse effects.

10.22 Intensity and Quantity Factors

As was the case for phosphorus, there is need to identify means by which the power of soils to supply potassium to plant roots can be described and measured. There are marked differences between the compounds in which these two elements are found in soils and in the rates at which they are made available to plants. But the concept of *intensity* and *quantity* discussed in Section 10.14 in relation to phosphorus is also useful in describing and measuring potassium-supplying power.

The intensity factor is a measure of the potassium that is immediately available to the root—the potassium in the soil solution. Since the absorption of the potassium ion by plant roots is affected by the activity in the soil solution

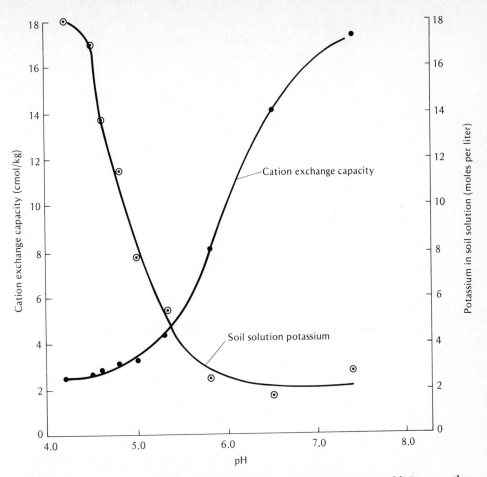

FIGURE 10.18 The influence of increased pH resulting from lime additions on the pH-dependent cation exchange capacity of a soil and the level of potassium in the soil solution. As the cation exchange capacity increases, some of the soil solution potassium is attracted to the adsorbing colloids. [*Data from Magdoff and Bartlett* (*1980*).]

of other cations and particularly of calcium and magnesium, some authorities prefer to use the ratio

$$\frac{[K^+]}{\sqrt{[Ca^{2+}] + [Mg^{2+}]}}$$

rather than the K^+ concentration, to indicate the intensity factor.

The quantity factor is a measure of the capacity of the soil to maintain the level of potassium in the soil solution over the period of time the crop is being produced. This capacity is due mainly to the exchangeable potassium, although some nonexchangeable forms release sufficient potassium during a growing season to provide a notable portion of the crop needs. The main point to remember, however, is that the quantity factor designates the potassium sources capable of helping to replenish the soil solution (intensity) level of potassium.

The concept of the buffering capacity of the soil, which indicates how the potassium level in the soil solution (*intensity*) varies with the amount of labile forms of this element (*quantity*), is useful in explaining differences in potassium supplying power of soils. The buffering capacity (B_{K^+}) is the ratio of changes in the quantity factor (ΔQ) to changes in the intensity factor (ΔI).

$$B_{K^+} = \frac{\Delta Q}{\Delta I}$$

This ratio is high for well-buffered soils and is low for those that are poorly buffered and are poor suppliers of potassium during a growth cycle. Clay-textured soils are well buffered, whereas that those are sandy are apt to be poorly buffered.

10.23   Practical Implications in Respect to Potassium

Frequency of Application.   One very important suggestion that is evident from the facts thus far considered is that frequent light applications of potassium are usually superior to heavier and less frequent ones. Such a conclusion is reasonable when considering the luxury consumption by crops, the ease with which this element is lost by leaching, and the fact that excess potassium is subject to fixation. Although the latter phenomenon has definite conserving features, these are in most cases entirely outweighed by the disadvantages of leaching and luxury consumption.

Potassium-Supplying Power of Soils.   A second very important suggestion is that full advantage should be taken of the potash-supplying power of soils. The idea that each pound of potassium removed by plants or through leaching must be returned in fertilizers may not always be correct. In some soils the large quantities of moderately available forms already present can be utilized. Where slowly available forms are not found in significant quantities, however, supplementary additions are necessary. Moreover, the importance of lime in reducing leaching losses of potassium should not be overlooked as a means of effectively utilizing the power of soils to furnish this element.

Potassium Losses and Gains.   The problem of maintaining soil potassium is outlined diagrammatically in Figure 10.19. Crop removal of potash generally exceeds that of the other essential elements, with the possible exception of

FIGURE 10.19 Gains and losses of *available* soil potassium under average field conditions. The approximate magnitude of the changes is represented by the width of the arrows. For any specific case the actual amounts of potassium added or lost undoubtedly may vary considerably from the above representation. As was the case with nitrogen and phosphorus, commercial fertilizers are important in meeting crop demands.

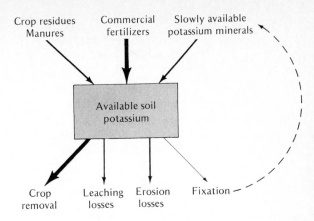

nitrogen. Annual losses from plant removal as great as 100 kg/ha of potassium are not uncommon, particularly if the crop is a legume and is cut several times for hay. As might be expected, therefore, the return of crop residues and manures is very important in maintaining soil potash. For example, 10 Mg of average animal manure supplies about 40 kg of potassium, fully equal to the amount of nitrogen thus supplied.

The annual losses of available potassium by leaching and erosion greatly exceed those of nitrogen and phosphorus. They are generally not as great, however, as the corresponding losses of available calcium and magnesium. The available potassium depletion by erosion assumes great importance considering that total potassium removal in this manner generally exceeds that of any other major nutrient element. The loss of potential sources of available potassium (soil minerals) cannot but eventually be serious.

Increasing Use of Potash Fertilizers. In the past, potash in fertilizers was added to supplement that returned in crop residues and that obtained from the slowly available forms of potassium carried by soil minerals. Fertilizer potash is now depended upon to supply much of the potassium needed for crop production. This is essentially true in cash-crop areas and in regions where sandy soils are prominent. Even in some heavier soils, the release of potassium from mineral form is much too slow to support maximum plant yields. Consequently, increased usage of commercial potash must be expected if yields are to be increased or even maintained.

REFERENCES

Acquaye, D. K. 1963. "Some significance of soil organic phosphorus mineralization in the phosphorus nutrition of cocoa in Ghana," *Plant and Soil,* **19**:65–80.

Attoe, O. J., and E. Truog. 1945. "Exchangeable and acid soluble potassium as regards availability and reciprocal relationships," *Soil Sci. Soc. Amer. Proc.,* **10**:81–86.

Black, C. A. 1971. *Behavior of Soil and Fertilizer Phosphorus in Relation to Water Pollution,* Jour. Paper J6373, Iowa Agr. and Home Econ. Exp. Sta.

Brady, N. C. 1974. *The Nature and Properties of Soils,* 8th ed. (New York: Macmillan).

Douglas, J. R. 1982. "U.S. fertilizer situation—1990," A paper at The Fertilizer Institute Seventh World Fertilizer Conference, Sept. 12–14, 1982, San Francisco, CA.

El-Baruni, B., and S. R. Olsen. 1979. "Effect of manure on solubility of phosphorus in calcareous soils," *Soil Sci.,* **112**:219–225.

Evans, C. E., and O. J. Attoe. 1948. "Potassium supplying power of virgin and cropped soils," *Soil Sci.,* **66**:323–34.

Fox, R. L. 1981. "External phosphorus requirements of crops," in *Chemistry in the Soil Environment,* ASA Special Publication No. 40 (Madison, WI: Amer. Soc. Agron. and Soil Sci. Soc. Amer.), pp. 223–239.

Gholston, L. E., and C. D. Hoover. 1948. "The release of exchangeable and nonexchangeable potassium from several Mississippi and Alabama soils upon continuous cropping," *Soil Sci. Soc. Amer. Proc.,* **13**:116–21.

Grewal, J. S., and J. S. Kanwar. 1976. *Potassium and Ammonium Fixation in Indian Soils* (Review), (New Delhi, India: Indian Council for Agricultural Research).

ISMA. 1977. "The importance of phosphorus in agriculture," 1977 issue of *Phosphorus in Agriculture* (ISMA Ltd., World Phosphate Industry Association, Paris, France).

Larsen, S. 1967. "Soil phosphorus," *Adv. in Agron.,* **19**:151–210.

Lindsay, W. L., and H. F. Stephenson. 1959. "Nature of the reactions of monocalcium phosphate monohydrate in soils: I. The solution that reacts with the soil; and II. Dissolution and precipitation reactions involving Fe, Al, Mn, and Ca," *Soil Sci. Soc. Amer. Proc.,* **23**:12–22.

Magdoff, F. R., and R. J. Bartlett. 1980. "Effect of liming acid soils on potassium availability," *Soil Sci.,* **129**:12–14.

McLean, E. O. 1978. "Influence of clay content and clay composition on potassium availability," in G. S. Sekhon (Ed.), *Potassium in Soils and Crops* (New Delhi, India: Potash Research Institute of India), pp. 1–19.

Mengel, K., and E. A. Kirkby. 1980. "Potassium in Crop Production," *Adv. in Agron.,* **33**:59–111.

Olsen, S. R., R. A. Bowman, and F. S. Watanabe. 1977. "Behavior of phosphorus in the soil and interactions with other nutrients," *Phosphorus in Agriculture,* **70**:31–46.

Sekhon, G. S. (ed.). 1978. *Potassium in Soils and Crops* (New Delhi, India: Potash Research Institute of India).

Toth, S. J., and F. E. Bear. 1947. "Phosphorus-adsorbing capacities of some New Jersey soils," *Soil Sci.,* **64**:199–211.

Micronutrient Elements

11

[*Preceding page*] Scientists have used highly sensitive analytical tools such as this atomic absorption spectrometer to identify essential micronutrients and to enhance and supplement the ability of soils to supply them. [*Courtesy Ray R. Weil, University of Maryland, College Park, MD.*]

Of the seventeen elements known to be essential for plant and microorganism growth, eight are required in such small quantities that they are called *micronutrients* or trace elements.[1] These are iron, manganese, zinc, copper, boron, molybdenum, cobalt, and chlorine. Other elements, such as silicon, vanadium, and sodium, appear to be helpful for the growth of certain species. Still others, such as iodine and fluorine, have been shown to be essential for animal growth but are apparently not required by plants. As better techniques of experimentation are developed and as purer salts are made available, it is likely that these and other elements may be added to the list of essential nutrients. Demands for efficiency of crop production will undoubtedly continue to encourage attention to these elements. Their field deficiencies have been noted in many countries. Table 11.1 lists crops for which micronutrient deficiencies have been noted in the United States along with the number of states reporting these deficiencies.

TABLE 11.1 Number of States Reporting Field Deficiencies of Six Micronutrients in Important Crops of the United States[a]

Crop group	Number of states reporting deficiency of					
	B	Cu	Zn	Mn	Fe	Mo
Forage legumes	36	1	1	2	2	15
Edible legumes	0	2	9	9	5	4
Soybeans	0	0	5	12	5	9
Corn and sorghum	3	5	26	8	8	0
Forage grasses	0	4	0	0	9	0
Small grains	0	4	3	11	4	0
Crucifers and various leafy vegetables	26	7	2	9	1	9
Solanaceous crops	6	3	1	6	1	1
Root and bulb crops	18	6	6	9	1	2
Cucurbits	1	0	0	4	1	2
Tree fruits	22	3	17	8	16	1
Small fruits	3	0	1	2	8	0
Nut crops	2	1	12	1	4	0
Ornamentals	0	0	2	0	18	1
Cotton	6	0	1	1	0	0
Tobacco	2	0	0	0	1	0
Other	0	3	4	1	5	0

[a] From a report of the Soil Test Commission of the Soil Science Society of America (1965) and Berger (1962).

[1] For review articles on this subject see Mortvedt et al. (1972) and Nicholas and Egan (1975).

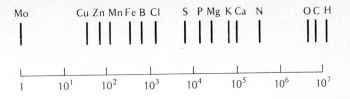

FIGURE 11.1 Relative numbers of atoms of the essential elements in alfalfa at bloom stage, expressed logarithmically. Note that there are more than 10 million hydrogen atoms for each molybdenum atom. Even so, normal plant growth would not occur without molybdenum. Cobalt is generally present in even smaller quantities in plants than is molybdenum. [*Modified from Viets* (1956).]

There are several reasons for the widespread concern for micronutrients.

1. Crop removal has in some cases lowered the level of these trace elements in the soil below that required for normal growth.
2. Improved crop varieties and macronutrient fertilizer practices have greatly increased crop production and thereby the micronutrient removal.
3. The trend toward high-analysis fertilizers has reduced the use of impure salts, which formerly contained some micronutrients.
4. Increased knowledge of plant nutrition has helped in the diagnosis of trace element deficiencies that formerly might have gone unnoticed.

11.1 Deficiency Versus Toxicity

One common characteristic of all the micronutrients is that they are required in very small amounts (Figure 11.1). Also, they are all harmful when the available forms are present in the soil in larger amounts than can be tolerated by plants or by animals consuming the plants. Thus the range of concentration of these elements in which plants will grow satisfactorily is not too great. Molybdenum, for example, may be beneficial if added at rates as little as 35–70 g/ha (0.5–1 oz/A), whereas applications of 3–4 kg/ha of available molybdenum may be toxic to most plants. Even those quantities present under natural soil conditions are in some cases excessive for normal crop growth. Although somewhat larger amounts of the other micronutrients are required and can be tolerated by plants, control of the quantities added, especially in maintaining nutrient balance, is absolutely essential. The concepts of deficiency, adequacy, and toxicity are illustrated in Figure 11.2 for several micronutrients.

11.2 Role of the Micronutrients

The specific roles of the various micronutrients in plant and microbial growth processes is not well understood. It is known, however, that several of the trace elements are effective through a number of enzyme systems (Table 11.2).

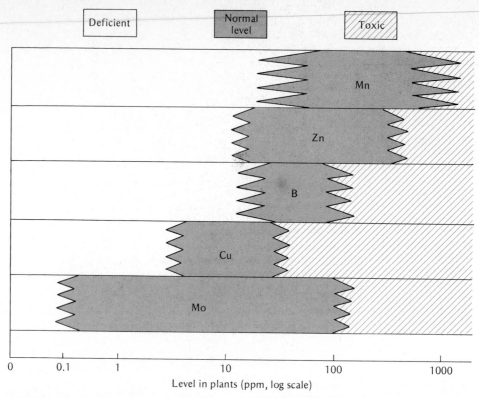

FIGURE 11.2 Deficient, normal, and toxic levels in plants for five micronutrients. [*Levels taken from Allaway (1968).*]

TABLE 11.2 Functions of Several Micronutrients in Higher Plants

Micronutrient	Functions in higher plants
Zinc	Present in several dehydrogenase, proteinase, and peptidase enzymes; promotes growth hormones and starch formation; promotes seed maturation and production.
Iron	Present in several peroxidase, catalase, cytochrome oxidase enzymes; found in ferredoxin, which participates in oxidation-reduction reactions (e.g., NO_3^- and SO_4^{2-} reduction, N fixation); important in chlorophyll formation.
Copper	Present in laccase and several other oxidase enzymes; important in photosynthesis, protein and carbohydrate metabolism, and probably nitrogen fixation.
Manganese	Activates decarboxylase, dehydrogenase and oxidase enzymes; important in photosynthesis, nitrogen metabolism, and nitrogen assimilation.
Boron	Activates certain dehydrogenase enzymes; facilitates sugar translocation and synthesis of nucleic acids and plant hormones; essential for cell division and development.
Molybdenum	Present in nitrogenase (nitrogen fixation) and nitrate reductase enzymes; essential for nitrogen fixation and nitrogen assimilation.
Cobalt	Essential for nitrogen fixation; found in vitamin B_{12}.

For example, copper, iron, and molybdenum are capable of acting as "electron carriers" in enzyme systems that bring about oxidation–reduction reactions in plants. Apparently such reactions, essential to plant development and reproduction, will not take place in the absence of these micronutrients. Zinc and manganese also function in enzyme systems that are necessary for important reactions in plant metabolism.

Molybdenum and manganese are essential for certain nitrogen transformations in microorganisms as well as in plants. Molybdenum is a component of

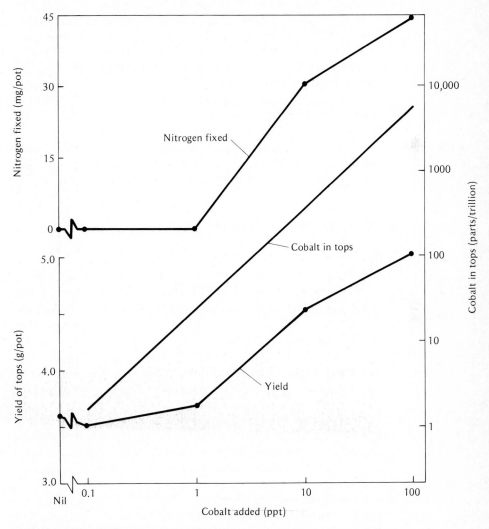

FIGURE 11.3 Effect of cobalt concentration in a nutrient solution (parts per thousand) on nitrogen fixation and on the yield and cobalt content of the tops of nodulated alfalfa plants. [*Redrawn from Wilson and Reisenauer* (*1963*).]

the enzyme nitrogenase, which is essential for the process of nitrogen fixation, both symbiotic and nonsymbiotic. It is also present in the enzyme *nitrate reductase,* which is responsible for the reduction of nitrates in soils and plants.

Zinc plays a role in protein synthesis, in the formation of some growth hormones, and in the reproduction process of certain plants. Copper is involved in both photosynthesis and respiration, and in the utilization of iron. A boron deficiency decreases the rate of water absorption, root growth, and translocation of sugars in plants. Iron is involved in chlorophyll formation and degradation and in the synthesis of proteins contained in the chloroplasts. Manganese seems to be essential for respiration and nitrogen metabolism.

The most recently recognized need is for chlorine and cobalt. The role of chlorine is still somewhat obscure; however, both root and top growth seem to suffer if it is absent. Cobalt is essential for the symbiotic fixation of nitrogen (Figure 11.3). This element is a component of vitamin B_{12}, believed necessary for the formation of a type of hemoglobin in nitrogen-fixing nodule tissue. In addition, legumes and other plants appear to have a cobalt requirement independent of nitrogen fixation, although the amount required is small compared to that for the nitrogen-fixation process.

It is obvious that the trace elements play a complex role in plant metabolism. As research uncovers new facts, the functions of these micronutrients will be more fully understood and undoubtedly more adequately respected.

11.3 Source of Micronutrients

Parent materials tend to influence in a practical way the micronutrient contents of soils, perhaps even more so than that of the macronutrients. Deficiencies of trace elements can frequently be related to low contents of the micronutrients in the parent rocks or transported parent material. Similarly, toxic quantities are commonly related to abnormally large amounts in the soil-forming rocks and minerals (see Figure 1.9).

Inorganic Forms.   The scant data available indicate that sources of the eight micronutrients vary markedly from area to area. Also, because of the extremely small quantities of some of these elements present in soils and in rocks, little is known about the specific minerals in which they are found. Except for iron and manganese, which are present in many soils in large total quantities, widespread, accurate analyses for micronutrients in soils and rocks have not been made. For these reasons only generalizations can be drawn concerning micronutrient sources (Table 11.3).

All of the micronutrients have been found in varying quantities in igneous rocks. Two of them, iron and manganese, have prominent structural positions in certain of the original silicate minerals. Others, such as cobalt and zinc, may also occupy structural positions as minor replacements for the major constituents of silicate minerals, including clays.

TABLE 11.3 Major Natural Sources of the Eight Micronutrients and Their Suggested Contents in a Representative Humid Region Surface Soil

Element	Major forms in nature	Analyses of soils Range (ppm)	Representative surface soil (ppm)
Iron	Oxides, sulfides, and silicates	10,000–100,000	25,000
Manganese	Oxides, silicates, and carbonates	20– 4,000	1,000
Zinc	Sulfides, carbonates, and silicates	10– 300	50
Copper	Sulfides, hydroxy carbonates, and oxides	2– 100	20
Boron	Borosilicates, borates	2– 100	10
Molybdenum	Sulfides, oxides, and molybdates	0.2– 5	2
Chlorine[a]	Chlorides	7– 50	10
Cobalt	Silicates	1– 40	8

[a] Much higher in saline and alkaline soils.

As mineral decomposition and soil formation occur, the mineral forms of micronutrients are changed just as macronutrients are. Oxides and in some cases sulfides of elements such as iron, manganese, and zinc are formed (Table 11.3). Secondary silicates, including the clay minerals, may contain considerable quantities of iron and manganese and smaller quantities of zinc and cobalt. The micronutrient cations released as weathering occurs are subject to colloidal adsorption, just as are the calcium or hydrogen ions. Anions such as the borate and molybdate may undergo adsorption or reaction in soils similar to that of the phosphates. Chlorine, by far the most soluble of the group, is added to soils in considerable quantities each year through rainwater. Its incidental addition in fertilizers and in other ways helps prevent the deficiency of chlorine under field conditions.

Organic Forms. Organic matter is an important secondary source of some of the trace elements. They seem to be held as complex combinations by the organic colloids. Copper is especially tightly held. In uncultivated profiles there is a somewhat greater concentration of micronutrients in the surface soil, much of it presumably in the organic fraction. Correlations between soil organic matter and copper, molybdenum, and zinc have been noted. Although the elements thus held are not always readily available to plants, their release through decomposition is undoubtedly an important fertility factor.

Forms in Soil Solution. Forms of micronutrients that tend to dominate the soil solution are found in Table 11.4. The specific forms present are determined largely by the pH and by soil aeration. Note that the cations are present in the form of either simple cations or hydroxy metal cations. The simple cations

TABLE 11.4 Forms of Micronutrients
Dominant in the Soil Solution[a]

Micronutrient	Dominant soil solution forms
Iron	Fe^{2+}, $Fe(OH)_2^+$, $Fe(OH)^{2+}$, Fe^{3+}
Manganese	Mn^{2+}
Zinc	Zn^{2+}, $Zn(OH)^+$
Copper	Cu^{2+}, $Cu(OH)^+$
Molybdenum	MoO_4^{2-}, $HMoO_4^-$
Boron	H_3BO_3, $H_2BO_3^-$
Cobalt	Co^{2+}

[a] From data in Lindsay (1972).

tend to be dominant under highly acid conditions. The more complex hydroxy metal cations form as the soil pH is increased.

Molybdenum is present in anionic forms similar generally to those of phosphorus. Although boron may also be present in anionic form, research suggests that undissociated boric acid (H_3BO_3) is the form that is dominant in the soil solution and is absorbed by higher plants.

11.4 General Conditions Conducive to Micronutrient Deficiency

Micronutrients are most apt to limit crop growth under the following conditions: (a) highly leached acid sandy soils, (b) muck soils, (c) soils very high in pH; and (d) soils that have been very intensively cropped and heavily fertilized with macronutrients only.

Strongly leached acid sandy soils are low in micronutrients for the same reasons are deficient in most of the macronutrients. Their parent materials were originally deficient in the elements and acid leaching has removed much of the small quantity of micronutrients originally present. In the case of molybdenum, acid soil conditions also have a markedly depressing effect on availability.

The micronutrient contents of organic soils are dependent upon the extent of the washing or leaching of these elements into the bog area as the deposits were formed. In most cases, this rate of movement was too slow to give deposits as high in micronutrients as are the surrounding mineral soils. Intensive cropping of muck soils and their ability to bind certain elements, notably copper, also accentuate trace-element deficiencies. Much of the harvested crops, especially vegetables, are removed from the land. Eventually the micro- as well as macronutrients must be supplied in the form of fertilizers if good crop yields are to be maintained. Intensive cropping of heavily fertilized mineral soils can also hasten the onset of micronutrient shortage, especially if the soils are coarse in texture.

The soil pH has a decided influence on the availability of all the micronutrients except chlorine, especially in well-aerated soils. Under very acid conditions, molybdenum is rendered unavailable; at high pH values all the cations are unfavorably affected. Overliming or a naturally high pH is associated with deficiencies of iron, manganese, zinc, copper, and even boron. Such conditions occur in nature in many of the calcareous soils of the west.

11.5   Factors Influencing the Availability of the Micronutrient Cations

Each of the five micronutrient cations (iron, manganese, zinc, copper, and cobalt) is influenced in a characteristic way by the soil environment. However, certain soil factors have the same general effects on the availability of all of them.

Soil pH.   The micronutrient cations are most soluble and available under acid conditions. In very acid soils there is a relative abundance of the ions of iron, manganese, zinc, and copper. In fact, under these conditions the concentrations of one or more of these elements is often sufficiently high to be toxic to common plants. As indicated in Chapter 6, one of the primary reasons for liming acid soils is to reduce the concentration of these ions.

As the pH is increased, the ionic forms of the micronutrient cations are changed first to the hydroxy ions of the elements and finally to the insoluble hydroxides or oxides. The following example uses the ferric ion as typical of the group.

$$Fe^{3+} \xrightarrow{OH^-} Fe(OH)^{2+} \xrightarrow{OH^-} Fe(OH)_2^+ \xrightarrow{OH^-} Fe(OH)_3$$

| Simple cation (soluble) | Hydroxy metal cations (soluble) | Hydroxide (insoluble) |

All of the hydroxides of the trace element cations are insoluble, some more so than others. The exact pH at which precipitation occurs varies from element to element and even between oxidation states of a given element. For example, the higher valent states of iron and manganese form hydroxides much more insoluble than their lower-valent counterparts. In any case, the principle is the same—at low pH values the solubility of micronutrient cations is at a maximum, and as the pH is raised, their solubility and availability to plants decrease. The desirability of maintaining an intermediate soil pH is obvious.

Zinc availability is reduced by magnesium-containing limestones. This may be due to interactions between the two cations in the soil or in the plant.

Oxidation State and pH. Three of the trace element cations are found in soils in more than one valent state. These are iron, manganese, and copper. The lower valent states are encouraged by conditions of low oxygen supply and relatively higher moisture level. They are responsible for the subdued subsoil colors, grays and blues in poorly drained soils in contrast to the bright reds, browns, and yellows of well-drained mineral soils.

The changes from one valent state to another are in most cases brought about by microorganisms and organic matter. In some cases the organisms may obtain their energy directly from the inorganic reaction. For example, the oxidation of manganese from the two-valent manganous form (Mn^{2+}) to MnO_2 can be carried on by certain bacteria and fungi. In other cases, organic compounds formed by the microbes may be responsible for the oxidation or reduction. In general, high pH favors oxidation, whereas acid conditions are more conducive to reduction.

At pH values common in soils the oxidized states of iron, manganese, and copper are generally much less soluble than are the reduced states. The hydroxides (or hydrous oxides) of these high-valent forms precipitate at lower pH values and are extremely insoluble. For example, the hydroxide of trivalent ferric iron precipitates at pH values near 3.0, whereas ferrous hydroxide does not precipitate until a pH of 6.0 or higher is reached.

The interaction of soil acidity and aeration in determining micronutrient availability is of great practical importance. The micronutrient cations are somewhat more available under conditions of restricted drainage. Flooded soils generally show higher availabilities than well-aerated soils. Very acid soils that are poorly drained often supply toxic quantities of iron and manganese. Such toxicity is less apt to occur under well-drained conditions. At the high end of the soil pH range, good drainage and aeration often have the opposite effect. Well-oxidized calcareous soils are sometimes deficient in available iron, zinc, or manganese even though there are adequate total quantities of these trace elements present. The hydroxides of the high-valent forms of these elements are too insoluble to supply the ions needed for plant growth. In contrast, at high soil pH values, molybdenum availability may be so high that the levels of this nutrient in the plants are toxic to the animals eating them.

There are marked differences in the sensitivity of different plant varieties to iron deficiency in soils with high pH (Olsen et al., 1981). This is apparently due to differences in their ability to solubilize iron immediately around the roots. Efficient varieties respond to iron stress by acidifying the immediate vicinity of the roots and by excreting compounds capable of reducing the iron to a more soluble form, thus increasing its availability (Figure 11.4).

Other Inorganic Reactions. Micronutrient cations interact with silicate clays in two ways. First, they may be involved in cation exchange reactions much like those of calcium or hydrogen. Second, they may be more tightly bound or fixed to certain silicate clays, especially of the 2:1 type. Zinc, manganese,

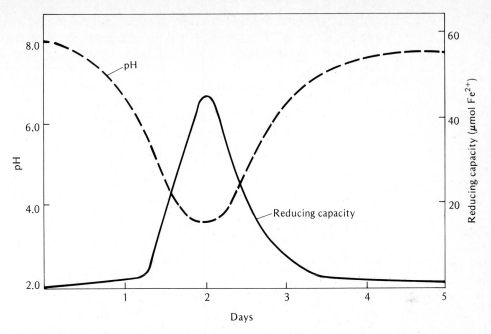

FIGURE 11.4 Response of one variety of sunflowers to iron deficiency. When the plant became stressed owing to iron deficiency, plant exudates lowered the pH and increased the reducing capacity immediately around the roots. Iron is solubilized and taken up by the plant, the stress is alleviated, and conditions return to normal. [*From Marschner et al.* (*1974*) *as reported by Olsen et al.* (*1981*); *used with permission of American Scientist.*]

cobalt, and iron ions are found as integral elements in these clays. Depending on the conditions, they may be released from the clays or fixed by them. The fixation may be serious in the case of cobalt and sometimes zinc since these two elements are present in soil in such small amounts (Table 11.3).

The application of large quantities of phosphate fertilizers can adversely affect the supply of some of the micronutrients. The uptake of both iron and zinc may be reduced in the presence of excess phosphates. From a practical standpoint, phosphate fertilizers should be used in only those quantities that are required for good plant growth.

Lime-induced chlorosis (iron deficiency) in fruit trees is encouraged by the presence of the bicarbonate ion. The chlorosis apparently results from iron deficiency in soils with high pH. In some way the bicarbonate ion interferes with iron metabolism.

Organic Combinations. Each of the four micronutrient cations may be held in organic combination. Microorganisms also assimilate them since they are apparently required for many microbial transformations. The organic compounds in which these trace elements are combined undoubtedly vary considerably, but they include proteins, amino acids, and constituents of humus, including the humic acids (see Section 8.10) and acids such as citric and tartaric. Among the most important are the *organic complexes,* combinations of the metallic cation and certain organic groups. These complexes may protect the micronutrients from certain harmful reactions, such as the precipitation of iron by phosphates and vice versa (see Section 10.12). On the other hand, complex formation may reduce micronutrient availability below that necessary for normal plant needs. These complexes, called *chelates,* are considered in the next section.

11.6 Chelates

Chelate, a term derived from a Greek word meaning "claw," is applied to compounds in which certain metallic cations are complexed or bound to an organic molecule. In complexed form the cations are protected from reactions with inorganic soil constituents that would make them unavailable for uptake by plants. Iron, zinc, copper, and manganese are among the cations that form chelate complexes. An example of an iron chelate ring structure is shown in Figure 11.5.

The effect of chelation can be illustrated with iron. In the absence of chelation, when an inorganic iron salt such as ferric sulfate is added to a calcareous soil, most of the iron is quickly rendered unavailable by reaction with hydroxide.

$$\underset{\text{(available)}}{Fe^{3+}} + 3\,OH^- \rightleftharpoons \underset{\text{(unavailable)}}{FeOOH} + H_2O$$

In contrast, if the iron is added in the form of an iron chelate, such as FeEDTA, the iron remains mostly in the chelate form, which is available for uptake by plants.

$$\underset{\text{(available)}}{FeEDTA} + 3\,OH^- \rightleftharpoons \underset{\text{(unavailable)}}{FeOOH} + EDTA^{3-} + H_2O$$

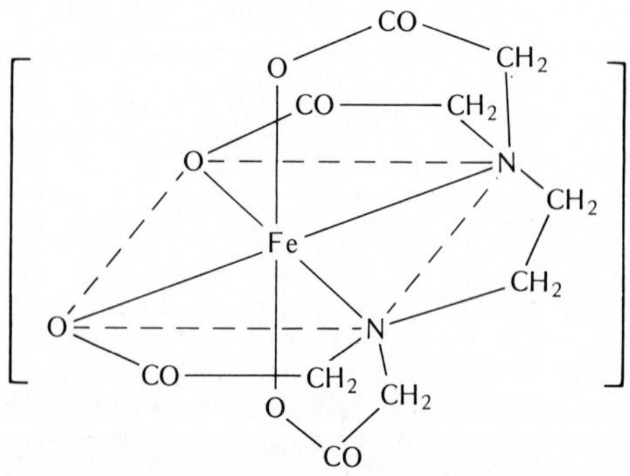

FIGURE 11.5 Structural formula for a common iron chelate, ferric ethylenediaminetetraacetate (FeEDTA). The iron is protected and yet can be utilized by plants.

The mechanism by which micronutrients from chelates are absorbed by plants is still obscure. The chelating agents in most cases are absorbed by growing plants, but the rate of their absorption is lower than that of the metals they carry. Thus, it would appear that the primary function of the chelate is to keep the metals available in the soil. At the same time, there is evidence that some of the benefit from the chelating agents is through increased translocation of the metals once they are absorbed by the plants.

Stability of Chelates. Some of the major chelating agents are listed in Table 11.5. They vary in their stability and in their suitability as sources of micronutrients. Except in soils of very high pH, iron chelates tend to be more stable than those of copper and zinc, which are in turn more stable than those of manganese. Thus, iron is more strongly attracted by the chelating agents than the other micronutrients. Consequently, if a zinc chelate is added to a soil with significant quantities of available iron, the following reaction may occur.

$$\text{Zn chelate} + \text{Fe}^{2+} \longrightarrow \rightleftharpoons \text{Fe chelate} + \text{Zn}^{2+}$$

Since the iron chelate is more stable than its zinc counterpart, the reaction goes to the right and the released zinc ion is subject to reaction with the soil. It is obvious that an added metal chelate must be stable within the soil if it is to have lasting advantage.

It should not be inferred that only iron chelates are effective. The chelates of other micronutrients, including zinc, manganese, and copper, have been used successfully to supply these nutrients. Apparently, replacement by iron in the soil is sufficiently slow to permit absorption by plants of the added trace element. Also, since spray and banded applications are often used to

TABLE 11.5 Common Chemical Names, Formulas, and Abbreviations for Major Chelating Agents[a]

Name	Formula	Abbreviation
Ethylenediaminetetraacetic acid	$C_{10}H_{16}O_8N_2$	EDTA
Diethylenetriaminepentaacetic acid	$C_{14}H_{23}O_{10}N_3$	DTPA
Cyclohexanediaminetetraacetic acid	$C_{14}H_{22}O_8N_2$	CDTA
Ethylenediaminedi(o-hydroxyphenylacetic acid)	$C_{18}H_{20}O_6N_2$	EDDHA
Hydroxyethylethylenediaminetriacetic acid	$C_{10}H_{18}O_7N_2$	HEDTA
Nitrilotriacetic acid	$C_6H_9O_6N$	NTA
Ethyleneglycol-bis(2-aminoethyl ether)tetraacetic acid	$C_{14}H_{24}O_{10}N_2$	EGTA
Citric acid	$C_6H_8O_7$	CIT
Oxalic acid	$C_2H_2O_4$	OX
Pyrophosphoric acid	$H_4P_2O_7$	P_2O_7
Triphosphoric acid	$H_5P_3O_{10}$	P_3O_{10}

[a] From Norvell (1972).

supply zinc and manganese, the possibility of reaction of these elements with iron in the soil can be reduced or eliminated.

The use of synthetic chelates in the United States is substantial in spite of the fact that they are quite expensive. Much of the use is to meet micronutrient deficiencies of citrus and other fruit trees. Although chelates may not replace the more conventional methods of supplying most micronutrients, they do offer some possibilities in special cases. Agricultural and chemical research will likely continue to increase the opportunities for their use.

Chelate stability also varies depending upon the specific chelating agent used. For the four micronutrients the following order of stability of some of the major chelates has been calculated by Norvell (1972).

Fe: EDDHA > DTPA, CDTA > EDTA > EGTA, HEDTA > NTA
Cu: DTPA > HEDTA, CDTA > EDTA, EDDHA > EGTA, NTA
Zn: DPTA > CDTA, HEDTA, EDTA > NTA > EGTA
Mn: DPTA, CDTA > EDTA, EGTA, HEDTA > NTA

11.7   Factors Influencing the Availability of the Micronutrient Anions

Unlike the cations needed in trace quantities by plants, the anions seem to have relatively little in common. Chlorine, molybdenum, and boron are quite different chemically, so little similarity would be expected in their reaction in soils.

Chlorine. Chlorine has only recently been found to be essential for plant growth, in spite of the fact that it is used in larger quantities by most crop plants than any of the micronutrients except iron. Its essentialness was not noted earlier since chlorine occurs widely as an impurity in fertilizer salts used for field research, and significant quantities of chlorine are added to soil annually through precipitation.

Most of the chlorine in soils is in the form of simple, soluble chloride salts such as potassium chloride. The chloride ions are not tightly adsorbed by negatively charged clays and as a result are subject to movement both upward and downward with the water in the profile. In humid regions, one would expect little chlorine to remain in the soil since it would be leached out. In semiarid and arid regions, a higher concentration might be expected, the amount reaching the point of salt toxicity in some of the poorly drained saline soils. In most well-drained areas, however, one would not expect a high chlorine content in the surface of arid-region soils.

Except under conditions where toxic quantities of chlorine are found in soils, there are apparently no common situations under field conditions that reduce the availability and utilization of this element. Accretions of chlorine from the atmosphere are believed to be sufficient to meet crop needs. Salt spray alongside ocean beaches evaporates, leaving sodium chloride dust, which moves into the atmosphere, to be returned later dissolved in snow and rain. The amount added to the soil in this way varies according to the distance from the salty body of water and other factors. It is likely that a figure of about 10 kg/ha per year is the minimum that can be expected in most situations, and that the average would be about 20 kg. In any case, this form of accretion, plus that commonly added as an incidental component of commercial fertilizers, should largely prevent a field deficiency of chlorine.

Boron. The availability and utilization of boron are determined to a considerable extent by pH. Boron is most soluble under acid conditions. It apparently occurs in acid soils in part as boric acid (H_3BO_3), which is readily available to plants. In quite acid sandy soils, soluble boron fertilizers may be leached downward with comparative ease. Evidently, the element is not fixed under these conditions. In heavier soils, especially if they are not too acid, this rapid leaching does not occur.

At higher pH values, boron is less easily utilized by plants. This may be due to lime-induced fixation of this element by clay and other minerals, since the calcium and sodium borates are reasonably soluble. In any case, overliming can and often does result in a deficiency of boron.

Boron is held in organic combinations from which it may be released for crop use. The content of this nutrient in the topsoil is generally higher than that in the subsoil. This may in part account for the noticeably greater boron deficiency in periods of dry weather. Apparently, during drought periods plant roots are forced to exploit only the lower soil horizons, where the boron content is quite low. When the rains come, plant roots again can absorb boron from the topsoil, when its concentration is highest.

Molybdenum. Soil pH is the most important factor influencing the availability and plant uptake of molybdenum. The following equation shows the forms of this element present at low and high soil pH:

$$H_2MoO_4 \underset{+H^+}{\overset{+OH^-}{\rightleftharpoons}} HMoO_4^- + H_2O \underset{+H^+}{\overset{+OH^-}{\rightleftharpoons}} MoO_4^{2-} + H_2O$$

At low pH values the relatively unavailable H_2MoO_4 and $HMoO_4^-$ forms are prevalent, whereas the more readily available MoO_4^{2-} anion is dominant at pH values above 5 or 6. The MoO_4^{2-} ion is subject to adsorption by oxides of iron and possibly aluminum just as is phosphate, but calcium molybdate is much more soluble than its phosphorus counterpart.

The liming of acid soils will usually increase the availability of molybdenum (Figure 11.6). The effect is so striking that some researchers, especially those in Australia and New Zealand, argue that the primary reason for liming very acid soils is to supply molybdenum. Furthermore, in some instances an ounce or so of molybdenum added to acid soils has given about the same

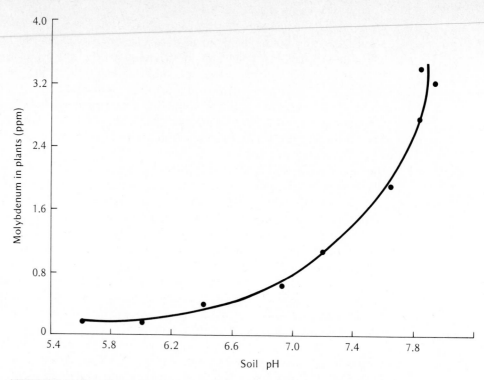

FIGURE 11.6 Effect of soil pH on the uptake of molybdenum by barley plants. [*Drawn from unpublished data of U. C. Gupta and J. A. MacLeod, Research Station, Charlottetown, Prince Edward Island, Canada. Quoted by Gupta and Lipsett* (*1981*).]

increase in the yield of legumes as has the application of several tons of lime.

The utilization of phosphate by plants seems to favor that of molybdenum and vice versa. For this reason, molybdate salts are often applied along with superphosphate to molybdenum-deficient soils. This practice apparently encourages the uptake of both elements and is a convenient way to add the extremely small quantities of molybdenum required.

A second common anion, the sulfate, seems to have the opposite effect on plant uptake of molybdenum. Sulfate reduces molybdenum uptake, although the specific mechanisms for this antagonism is not yet known.

11.8 Need for Nutrient Balance

Nutrient balance among the trace elements is as essential as, but even more difficult to maintain than, macronutrient balance. Some of the plant enzyme systems that depend on micronutrients require more than one element. For example, both manganese and molybdenum are needed for the assimilation of nitrates by plants. The beneficial effects of combinations of phosphates and molybdenum have already been discussed. Apparently, some plants need zinc and phosphorus for optimum utilization of manganese. The utilization of boron and calcium is dependent upon the proper balance between these two nutrients. A similar relationship exists between potassium and copper and between potassium and iron in the production of good-quality potatoes. Copper utilization is favored by adequate manganese, which in some plants is assimilated only if zinc is present in sufficient amounts. Of course, the effects of these and other nutrients will depend upon the specific plant being grown, but the complexity of the situation can be seen from the examples cited.

Antagonism. Some enzymatic and other biochemical reactions requiring a given micronutrient may be "poisoned" by the presence of a second trace element in toxic quantities. For example,

1. Excess copper or sulfate may adversely affect the utilization of molybdenum.
2. Iron deficiency is encouraged by an excess of zinc, manganese, copper, or molybdenum.
3. Excess phosphate may encourage a deficiency of zinc, iron, or copper, but enhance the absorption of molybdenum.
4. Heavy nitrogen fertilization intensifies copper and zinc deficiencies.
5. Excess sodium or potassium may adversely affect manganese uptake.
6. Excess lime reduces boron uptake.
7. Excess iron, copper, or zinc may reduce the absorption of manganese.

Some of these antagonistic effects may be utilized effectively in reducing toxicities of certain of the micronutrients. For example, copper toxicity of citrus groves caused by residual copper from insecticide sprays may be reduced by adding iron and phosphate fertilizers. Sulfur additions to calcareous soils containing toxic quantities of soluble molybdenum may reduce the availability and hence the toxicity of molybdenum.

These examples of nutrient interactions, both beneficial and detrimental, emphasize the highly complicated nature of the biological transformations in which micronutrients are involved. The total acreage where unfavorable nutrient balances require special micronutrient treatment is increasing with more intensive cropping of soils.

11.9 Soil Management and Micronutrient Needs

Although the characteristics of each micronutrient are quite specific, some generalizations with respect to management practices are possible.

In seeking the cause of plant abnormalities one should keep in mind the conditions under which micronutrient deficiencies or toxicities are likely to occur. Sandy soils, mucks, and soils having very high or very low pH values are suspect. Areas of intensive cropping and heavy macronutrient fertilization may be deficient in the micronutrients.

Changes in Soil Acidity. In very acid soils, one might expect toxicities of iron and manganese and deficiencies of phosphorus and molybdenum. These can be corrected by liming and by appropriate fertilizer additions. Calcareous soils may have deficiencies of iron, manganese, zinc, and copper and, in a few cases, a toxicity of molybdenum.

No specific statement can be made concerning the pH value most suitable for all the elements. However, medium-textured soils generally supply adequate quantities of micronutrients when the soil pH is held between 6 and 7. In sandy soils, a somewhat more acid reaction may be justified since the total quantity of micronutrients is low and even at pH 6.0 some cation deficiencies may occur.

Soil Moisture. Drainage and moisture control can influence micronutrient solubility in soils. Improving the drainage of acid soils will encourage the formation of the oxidized forms of iron and manganese. These are less soluble and, under acid conditions, less toxic than the reduced forms.

Moisture control at high pH values can have the opposite effect. High moisture levels maintained by irrigation may result in the chemical reduction of high-valent compounds, the oxides of which are extremely insoluble. Flooding a soil will favor the reduced forms, which are more available to growing plants. Poor drainage also increases the availability of molybdenum, in some soils to the point of producing plants with toxic levels of this element.

Fertilizer Applications. The most common management practice to overcome micronutrient deficiencies (and some toxicities) is the application of commercial fertilizers. Examples of fertilizer materials applied for each of the micronutrients are shown in Table 11.6. The materials are most commonly applied to the soil, although in recent years foliar sprays and even seed treatments have been used. Foliar sprays of dilute inorganic salts or organic chelates are more effective than soil treatments where high soil pH and other factors render the soil-applied nutrients unavailable. Treating seeds with small dosages (20–40 g/ha) of molybdenum has brought about quite satisfactory results on molybdenum-deficient acid soils.

The micronutrients can be applied to the soil either as separate materials or incorporated in standard macronutrient carriers. Unfortunately the solubility of Cu, Fe, Mn, and Zn can be reduced by such incorporation, but B and Mo remain in reasonably soluble condition. Liquid macronutrient fertilizers containing polyphosphates (see Section 18.5) encourage the formation of complexes that protect added micronutrients from adverse chemical reactions.

Economic responses to micronutrients are becoming more widespread as intensity of cropping increases. For example, responses of fruits, vegetables, and field crops to zinc and iron applications have been noted in areas with

TABLE 11.6 A Few Commonly Used Fertilizer Materials That Supply Micronutrients[a]

Micronutrient	Commonly used fertilizers		Nutrient content (%)
Boron	Borax	$Na_2B_4O_7 \cdot 10H_2O$	11
	Sodium pentaborate	$Na_2B_{10}O_{16} \cdot H_2O$	18
Copper	Copper sulfate	$CuSO_4 \cdot 5H_2O$	25
Iron	Ferrous sulfate	$FeSO_4 \cdot 7H_2O$	19
	Iron chelates	NaFeEDDHA	6
Manganese	Manganese sulfate	$MnSO_4 \cdot 3H_2O$	26–28
	Manganese oxide	MnO	41–68
Molybdenum	Sodium molybdate	$Na_2MoO_4 \cdot 2H_2O$	39
	Ammonium molybdate	$(NH_4)_6Mo_7O_{24} \cdot 4H_2O$	54
Zinc	Zinc sulfate	$ZnSO_4 \cdot H_2O$	35
	Zinc oxide	ZnO	78
	Zinc chelate	$Na_2ZnEDTA$	14

[a] Selected from Murphy and Walsh (1972).

TABLE 11.7 Areas of Micronutrient Deficiency, Range in Recommended Rates of Application in the Deficient Areas, and Some Crops Having a High Requirement for Micronutrients

Micronutrient	Area deficient in United States[a] (10^6 ha)	Common Range in recommended application rates (kg/ha)	Crops having a high requirement
Iron	1.54	0.5–10	Blueberries, cranberries, rhododendron, peaches, grapes, nut trees
Manganese	5.26	5–30	Dates, beans, soybeans, onions, potatoes, citrus
Zinc	2.63	0.5–20	Citrus and fruit trees, soybeans, corn, beans
Copper	0.24	1–20	Citrus and fruit trees, onions, small grains
Boron	4.85	0.5–5	Alfalfa, clovers, sugar beets, cauliflower, celery, apples, other fruits
Molybdenum	0.61	0.05–1	Alfalfa, sweet clover, cauliflower, broccoli, celery

[a] Calculated from Burgess (1966).

neutral to alkaline soils. Even on acid soils, deficiencies of the elements have been demonstrated. Molybdenum, which has been used for some time for forage crops and for cauliflower and other vegetables, has received attention in recent years in soybean-growing areas, especially those with acid soils. Seed treatments of about 18 g/ha have given good responses. These examples, along with those from muck areas and sandy soils where micronutrients have been used for years, illustrate the need for these elements if optimum yields are to be maintained.

Marked differences in crop needs for micronutrients make fertilization a problem where rotations are being followed. On general-crop farms, vegetables are sometimes grown in rotation with small grains and forages. If the boron fertilization is adequate for a vegetable crop such as red beets or even for alfalfa, the small-grain crop grown in the rotation is apt to show toxicity damage. These facts emphasize the need for specificity in determining crop nutrient requirements and for care in meeting these needs (Table 11.7).

Micronutrient Availability.   Major sources of micronutrients and the general reactions that make them available to higher plants and microorganisms are summarized in Figure 11.7. Original and secondary minerals are the primary sources of these elements, while the breakdown of organic forms releases ions to the soil solution. The micronutrients are utilized by higher plants and microorganisms in important life-supporting processes. Removal of nutrients in harvested crops reduces the soluble ion pool, and if it is not replenished with chemical fertilizers, nutrient deficiencies will result. These fertilizers, although intended primarily for supplying the macronutrients, are increasingly important sources of micronutrients as well.

FIGURE 11.7   Diagram of soil sources of soluble forms of micronutrients and their utilization by plants and microorganisms.

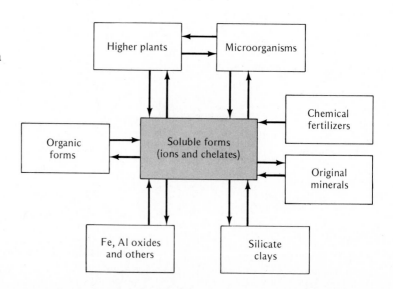

Worldwide Management Problems. Micronutrient deficiencies have been diagnosed in most crop production areas of the United States and Europe. However, in some developing countries, particularly in the tropics, the extent of these deficiencies is much less well known. Limited research suggests that there may be large areas with deficiencies of one or more of these elements. The management principles established for the economically developed countries should be helpful in alleviating these deficiencies.

REFERENCES

Allaway, W. H. 1968. "Agronomic controls over the environmental cycling of trace elements," *Adv. in Agron.*, **20**:235–274.

Anonymous. 1965. "A survey of micronutrient deficiencies in the U.S.A. and means of correcting them," *Report of the Soil Test Commission of Soil Sci. Soc. Amer.*, Madison, WI.

Berger, K. C. 1962. "Micronutrient deficiencies in the United States," *J. Agr. and Food Chem.*, **10**:178–181.

Burgess, W. D. 1966. *The Market for Secondary and Micronutrients* (New York: Allied Chemical Corp.).

Gupta, U. C., and J. Lipsett. 1981. "Molybdenum in soils, plants, and animals," *Adv. in Agron.*, **34**:73–116.

Lindsay, W. L. 1972. "Inorganic phase equilibria of micronutrients in soils," in J. J. Mortvedt, P. M. Giordano, and W. L. Lindsay (Eds.), *Micronutrients in Agriculture* (Madison, WI: Soil Sci. Soc. Amer.).

Marschner, H., A. Kalisch, and V. Romheld. 1974. "Mechanism of iron uptake in different plants species," *Proc. 7th Int. Colloquium on Plant Analysis and Fertilizer Problems,* Hannover, West Germany.

Mortvedt, J. J., P. M. Giordano, and W. L. Lindsay (Eds.) 1975. *Micronutrients in Agriculture* (Madison, WI: Soil Sci. Soc. Amer.).

Murphy, L. S., and L. M. Walsh. 1972. "Correction of micronutrient deficiencies with fertilizers," in J. J. Mortvedt, P. M. Giordano, and W. L. Lindsay (Eds.), *Micronutrients in Agriculture* (Madison, WI: Soil Sci. Soc. Amer.).

Nicholas, D. J. D., and A. R. Egan (Eds.). 1975. *Trace Elements in Soil-Plant-Animal Systems* (New York: Academic Press).

Norvell, W. A. 1972. "Equilibria of metal chelates in soil solution," in J. J. Mortvedt, P. M. Giordana, and W. L. Lindsay (Eds.), *Micronutrients in Agriculture* (Madison, WI: Soil Sci. Soc. Amer.).

Olsen, R. A., R. B. Clark, and J. H. Bennett. 1981. "The enhancement of soil fertility by plant roots," *Amer. Scientist*, **69**:378–384.

Viets, F. J. Jr. 1965. "The plants' needs for and use of nitrogen," in *Soil Nitrogen* (*Agronomy,* 10) (Madison, WI: Amer. Soc. Agron.).

Wilson, D. O., and H. M. Reisenauer. 1963. "Cobalt requirements of symbiotically grown alfalfa," *Plant and Soil,* **19**:364–373.

Origin, Nature, and Classification of Parent Materials

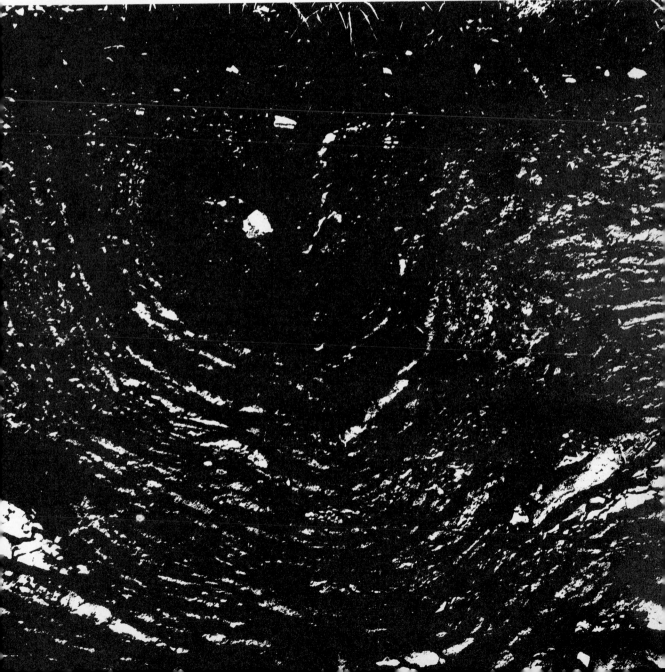

[*Preceding page*] Parent materials such as the highly weathered rock (largely diorite) under this Alabama soil have great influence on soil properties. [*Courtesy USDA Soil Conservation Service.*]

The influence of weathering is evident on all sides. Nothing escapes it. It breaks up the country rocks, modifies or destroys their physical and chemical characteristics, and carries away the soluble products and even some of the solids as well. The unconsolidated residues, the *regolith,* are left behind. But weathering goes further. It synthesizes a soil from the upper layers of this heterogeneous mass. Hence, the upper part of the regolith may be designated as *parent material* of soils. These parent materials are not always allowed to remain undisturbed on the site of their development. Climatic agencies may shift them from place to place until they are allowed to rest long enough for a soil profile to develop. A study of weathering and of the parent materials that result provides a basis for understanding the soil components and is a necessary introduction to soil formation and classification.

The consideration of parent materials will begin with a brief review of the kinds of rocks and minerals from which the regolith and in turn the soils have formed.

12.1 Classification and Properties of Rocks

The rocks found in the earth's outer surface are commonly classified as *igneous, sedimentary,* or *metamorphic.* Those of igneous origin are formed from molten magma and include such common rocks as *granite* and *diorite* (Figure 12.1). They are composed of primary minerals such as (a) quartz, (b) the feldspars,

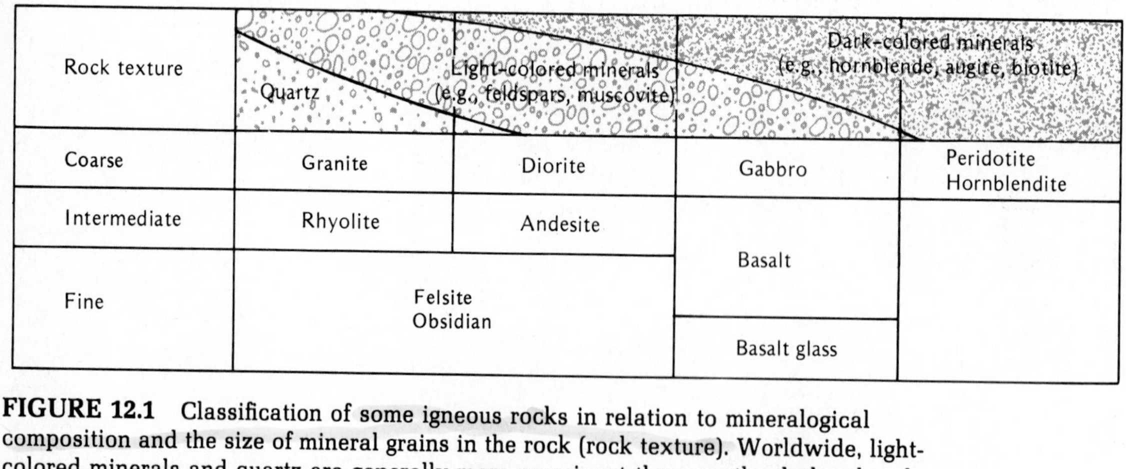

Rock texture	Quartz	Light-colored minerals (e.g. feldspars, muscovite)		Dark-colored minerals (e.g. hornblende, augite, biotite)
Coarse	Granite	Diorite	Gabbro	Peridotite Hornblendite
Intermediate	Rhyolite	Andesite	Basalt	
Fine		Felsite Obsidian	Basalt glass	

FIGURE 12.1 Classification of some igneous rocks in relation to mineralogical composition and the size of mineral grains in the rock (rock texture). Worldwide, light-colored minerals and quartz are generally more prominent than are the dark-colored minerals.

TABLE 12.1 Some of the More Important Sedimentary and Metamorphic Rocks and the Minerals Commonly Dominant in Them

Sedimentary rocks	Dominant mineral	Metamorphic rocks	Dominant mineral
Limestone	Calcite ($CaCO_3$)	Gneiss	(varies)[a]
Dolomite	Dolomite [$CaMg(CO_3)_2$]	Schist	(varies)[a]
Sandstone	Quartz (SiO_2)	Quartzite	Quartz (SiO_2)
Shale	Clays	Slate	Clays
Conglomerate	(varies)[b]	Marble	Calcite ($CaCO_3$)

[a] The minerals present are determined by the original rock, which has been changed by metamorphism. Primary minerals present in the igneous rocks commonly dominate these rocks, although some secondary minerals are also present.

[b] Small stones of various mineralogical makeup will be cemented into conglomerate.

and (c) the dark-colored minerals, including biotite, augite, and hornblende. In general, gabbro and basalt (Figure 12.2), which are high in the dark-colored iron- and magnesium-containing minerals, are more easily weathered than are the granites and other lighter-colored rocks.

Sedimentary rocks have resulted from the deposition and recementation of weathering products of other rocks. For example, quartz sand weathered from a granite and deposited on the shore or beach of a prehistoric sea may through geological changes have become cemented into a solid mass called a *sandstone*. Similarly, recemented clays are termed *shale*. Other important sedimentary rocks are listed in Table 12.1 with their dominant minerals. The resistance of a given sedimentary rock to weathering is determined by its particular dominant minerals and by the cementing agent.

Metamorphic rocks are those that have formed by the metamorphism or change in form of other rocks. Igneous and sedimentary masses subjected to tremendous pressures and high temperature have succumbed to metamorphism. Igneous rocks are commonly modified to form *gneisses* and *schists;* those of sedimentary origin such as sandstone and shale may be changed to *quartzite* and *slate*. Some of the common metamorphic rocks are shown in Table 12.1. As with those of igneous and sedimentary origin, the particular mineral or minerals that dominate a given metamorphic rock will influence its resistance to weathering (see Table 12.2 for a listing of the more common minerals).

12.2 Weathering—A General Case

Before considering the various kinds of parent materials that develop from rocks, attention will be given to the general trends of weathering responsible for the formation of the regolith and the soils. Some of the changes which take place during weathering are indicated in Figure 12.2. Study it carefully,

TABLE 12.2 The More Important Primary and Secondary Minerals Found in Soils Listed in Order of Decreasing Resistance to Weathering Under Conditions Common in Humid Temperate regions

The original minerals are also found abundantly in igneous and metamorphic rocks. Secondary minerals are commonly found in sedimentary rocks.

Primary Minerals		Secondary Minerals		
		Geothite	$FeOOH$	Most resistant
		Hematite	Fe_2O_3	
		Gibsite	$Al_2O_3 \cdot 3H_2O$	
Quartz	SiO_2			
		Clay minerals	Al silicates	
Muscovite	$KAl_3Si_3O_{10}(OH)_2$			
Orthoclase	$KAlSi_3O_8$			
Biotite	$KAl(Mg,Fe)_3Si_3O_{10}(OH)_2$			
Albite	$NaAlSi_3O_8$			
Hornblende[a]	$Ca_2Al_2Mg_2Fe_3Si_6O_{22}(OH)_2$			
Augite[a]	$Ca_2(Al,Fe)_4(Mg,Fe)_4Si_6O_{24}$			
Anorthite	$CaAl_2Si_2O_8$			
Olivine	$(Mg,Fe)_2SiO_4$			
		Dolomite	$CaMg(CO_3)_2$	
		Calcite	$CaCO_3$	
		Gypsum	$CaSO_4 \cdot 2H_2O$	Least resistant

[a] The given formula is only approximate since the mineral is so variable in composition.

keeping in mind that it deals primarily with weathering as the process occurs in temperate regions.

Weathering is basically a combination of destruction and synthesis. Rocks, the original starting point in the weathering process, are first broken down physically into smaller rocks and eventually into the individual minerals of which they are composed. Simultaneously, rock fragments and the minerals therein are attacked by chemical forces and are changed to new minerals either by minor modifications (alterations) or by complete chemical changes. These changes are accompanied by a continued decrease in particle size and by the release of soluble constituents, which are subject to loss in drainage waters or recombination into new (secondary) minerals.

The new minerals that are formed are shown (Figure 12.2) in two groups: (a) the silicate clays and (b) the very resistant end products, including iron and aluminum oxides. Along with the very resistant primary minerals such as quartz, these two groups dominate temperate region soils.

The Processes. Two basic processes are involved in the changes indicated in Figure 12.2. These are mechanical and chemical. The former is often designated as *disintegration*, the latter as *decomposition*. Both are operative, moving

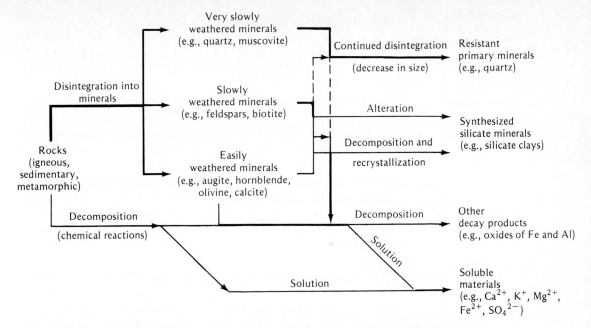

FIGURE 12.2 Pathways of weathering that takes under moderately acid conditions common in humid temperate regions. Major weathering pathways are indicated by the solid lines, minor pathways by broken lines. As one would expect, climate modifies the exact relationships. In arid regions physical breakdown (disintegration) would dominate, and soluble ions would not be lost in large quantities. In humid regions decomposition becomes more important, especially under tropical conditions.

from left to right in the weathering diagram. Disintegration results in a decrease in size of rocks and minerals without appreciably affecting their composition. By decomposition, however, definite chemical changes take place, soluble materials are released, and new minerals are synthesized or are left as resistant end products. Mechanical and chemical processes may be outlined as follows:

1. Mechanical (disintegration):
   a. Temperature—differential expansion of minerals, frost action, and exfoliation.
   b. Erosion and deposition—by water, ice, and wind.
   c. Plant and animal influences.
2. Chemical (decomposition):
   a. Hydrolysis.
   b. Hydration.
   c. Acidification.
   d. Oxidation.
   e. Dissolution.

12.3   Mechanical Forces of Weathering

Temperature Effects.   Variations of temperature, especially if sudden or wide, greatly influence the disintegration of rocks. During the day rocks become heated and at night often cool much below the temperature of the air. This warming and cooling is particularly effective as a disintegrating agent. Rocks are aggregates of minerals that differ in their coefficients of expansion upon being heated. With every temperature change, therefore, differential stresses are set up that eventually must produce cracks and rifts, thus encouraging mechanical breakdown.

Because heat is conducted slowly, the outer surface of a rock is often at a markedly different temperature from the inner and more protected portions. This differential heating and cooling sets up lateral stresses that, in time, may cause the surface layers to peel away from the parent mass. This phenomenon is referred to as *exfoliation* and at times is sharply accelerated by the freezing of included water.

The presence of water, if freezing occurs, greatly increases these mechanical effects. The force developed by the freezing of water is equivalent to about 1465 Mg/m² (150 tons/ft²), an almost irresistible pressure. It widens cracks in huge boulders and dislodges mineral grains from smaller fragments. The influence of temperature is not ended when rocks are reduced to fragments.

Influence of Water, Ice, and Wind.   Rain water beats down upon the land and then travels oceanward, continually shifting, sorting, and reworking unconsolidated materials of all kinds. When loaded with such sediments, water has a tremendous cutting power, as is amply demonstrated by the gorges, ravines, and valleys the country over. The rounding of sand grains on an ocean beach is further evidence of the abrasion that accompanies water movement.

Ice is an erosive transporting agency of tremendous capacity, and next to water is perhaps the most important and spectacular physical agent of weathering. A look at Greenland or Alaska will illustrate its power. The abrasive action of glaciers as they move under their own weight disintegrates rocks and minerals alike. Glaciers affect the underlying solid rock and also grind and mix unconsolidated materials that they pick up as they move (see Figure 12.10). Even though they are not so extensive at the present time, glaciers in ages past have been responsible for the transportation and deposition of parent materials over millions of acres. The importance of ice as a mechanical agent is not to be underestimated (see Section 12.12).

Wind has always been an important carrying agent and, when armed with fine debris, exerts an abrasive action also. Dust storms of almost continental extent have occurred in the past, with the result that tons of material have been filched from one section and transferred to another. As dust is transferred and deposited, abrasion of one particle against another occurs. The rounded rock remnants in some arid areas of the west are caused largely by wind action.

Plants. Simple plants, such as mosses and lichens, grow upon exposed rock, there to catch dust until a thin film of highly organic material accumulates. Higher plants sometimes exert a prying effect on rock, which results in some disintegration. This is most obvious in the case of tree roots in rocky areas. Such influences, as well as those exerted by animals, are of little import in producing parent material when compared to the drastic physical effects of water, ice, wind, and temperature changes.

12.4 Chemical Processes of Weathering

Scarcely has the disintegration of rock material begun when its decomposition usually is also apparent. This is especially noticeable in warm humid regions, where chemical and physical processes are particularly active and markedly accelerate each other.

Chemical weathering is affected by a series of reactions involving numerous weathering agents. Among the most important of these agents is water (with its omnipresent solutes), oxygen, and the organic chemicals released by higher plants and microorganisms. These agents commonly act in concert as primary minerals such as the feldspars and micas are converted to secondary minerals, including silicate clays, and to soluble forms carrying sufficient essential elements to support plant growth. Although examples will be given of specific types of reaction taking place as weathering occurs, it is well to keep in mind that in nature the processes tend to occur simultaneously.

Water and Its Solutions. Water and its dissolved salts and acids are perhaps the most pervasive factors in the weathering of minerals. Through *hydrolysis*, *hydration*, and *dissolution*, water is able to enhance the degradation, alteration, and resynthesis of minerals. A simple example is the hydrolysis of microcline, a potassium-containing feldspar.

$$\underset{\text{(solid)}}{KAlSi_3O_8} + \underset{\text{(liquid)}}{H_2O} \rightarrow \underset{\text{(solid)}}{HAlSi_3O_8} + \underset{\text{(solution)}}{K^+ + OH^-}$$

$$\underset{\text{(solid)}}{2\,HAlSi_3O_8} + \underset{\text{(liquid)}}{14\,H_2O} \rightarrow \underset{\text{(solid)}}{Al_2O_3 \cdot 3H_2O} + \underset{\text{(solution)}}{6\,H_4SiO_4}$$

Note that these reactions illustrate dissolution and hydration as well as hydrolysis. The potassium released is soluble and is subject to adsorption by soil colloids, to uptake by plants and to removal in the drainage water. Likewise, the silicic acid (H_4SiO_4) is soluble and can be slowly removed in drainage water or is subject to recombination with other compounds to form secondary minerals such as the silicate clays. Hydration is evident through the formation of the hydrated aluminum oxide ($Al_2O_3 \cdot 3H_2O$). Iron oxides are also commonly hydrated.

Acid Solution Weathering. Weathering is accelerated by the presence of the hydrogen ion in percolating waters, such as that contained in carbonic acid. For example, this acid results in the chemical solution of calcite in limestone; the following reaction illustrates what takes place.

$$CaCO_3 + H_2CO_3 \rightarrow Ca^{2+} + 2\ HCO_3^-$$

Calcite   (solution)   (bicarbonate solution)
(solid)

Other acids much stronger than carbonic are also present in most humid region soils. They include very dilute inorganic acids such as HNO_3 and H_2SO_4 as well as some organic acids. Also present are the hydrogen ions associated with soil clays, which are available for reaction with other soil minerals.

An example of the reaction of hydrogen ions with soil minerals is that of an acid clay with a feldspar such as anorthite.

$$CaAl_2Si_2O_8 + \begin{array}{c} H \\ H \end{array} \boxed{Micelle} \rightarrow H_2Al_2Si_2O_8 + Ca\ \boxed{Micelle}$$

Anorthite   (solid)   Acid silicate   (solid)
(solid)   (solid)

The mechanism is simple; the hydrogen ions of the acid clays replace the bases of the fresh minerals that they may closely contact. The unstable acid silicate that results is subject to dissolution and subsequent reaction to form silicate clays. Thus, "clay begets clay" by a mechanism that depends upon the removal of bases from the weathering mass of minerals by insoluble inorganic acids.

Oxidation. Concomitant with the action of water in weathering is that of oxidation. It is particularly manifest in rocks carrying iron, an element that is easily oxidized. In some minerals iron is present in reduced ferrous (Fe^{2+}) form. If oxidation to the ferric ion takes place while iron is still part of the crystal lattice, other ionic adjustments must be made since a three-valent ion is replacing a two-valent one. These adjustments result in a less stable crystal, which is then subject to both disintegration and decomposition.

In other cases, ferrous iron may be released from the crystal and is almost simultaneously oxidized to the ferric form. An example of this is the hydration of olivine and the release of ferrous oxide, which may be immediately oxidized to ferric oxide (geothite).

$$3\ MgFeSiO_4 + 2\ H_2O \rightarrow H_4Mg_3Si_2O_9 + SiO_2 + 3\ FeO$$

Olivine   Serpentine   Ferrous oxide

$$4\ FeO + O_2 + 2\ H_2O \rightarrow FeOOH$$

Ferrous oxide   Geothite

Dissolution. As the processes of chemical weathering progress, significant changes take place within the crystals of the minerals being weathered. When ions are removed or are oxidized within the crystal, the rigidity of the mineral lattice structure is weakened and the mechanical breakdown is made easier. This provides a favorable environment for further chemical reactions. Ions or compounds solubilized in one soil layer are carried downward in the leaching water and can thereby influence the weathering of lower layers. Table 12.3 suggests the comparative losses through solution of constituents in a granite and a limestone. The metallic cations are most readily removed, but silicon, iron, and aluminum are also subject to slow dissolution.

Integrated Weathering Processes. As previously stressed, the chemical weathering processes just described occur simultaneously and are interdependent. For example, hydrolysis of a given primary mineral may release iron that is quickly oxidized to the ferric form and in turn is hydrated to give a hydrous oxide of iron. Hydrolysis may also release silicic acid and aluminum or iron compounds, which can be recombined to form secondary silicate minerals such as the silicate clays. These reactions are illustrated in a general way in Figure 12.3, which shows how a hypothetical primary silicate mineral is changed by chemical weathering. In studying the figure, keep in mind that each of the products of chemical weathering is important in determining soil characteristics, including those important to growing plants. The silicate clays and iron and aluminum oxides play major roles in determining soil physical properties as well as the availability and stable supply of essential nutrient elements. The soluble products of chemical weathering can be lost from the soil by leaching, can participate in the genesis of new minerals, and can be taken

TABLE 12.3 Comparative Losses of Mineral Constituents as Weathering Takes Place in a Granite and in a Limestone

The losses are compared to that of aluminum, which was considered in these cases to have remained constant during the weathering process.

Granite to clay[a]		Limestone to clay[b]	
Constituent	Comparative loss (%)	Constituent	Comparative loss (%)
CaO	100.0	CaO	99.8
Na_2O	95.0	MgO	99.4
K_2O	83.5	Na_2O	76.0
MgO	74.7	K_2O	57.5
SiO_2	52.5	SiO_2	27.3
Fe_2O_3	14.4	Fe_2O_3	24.9
Al_2O_3	0.0	Al_2O_3	0.0

[a] From Merrill (1879).
[b] From Diller (1898).

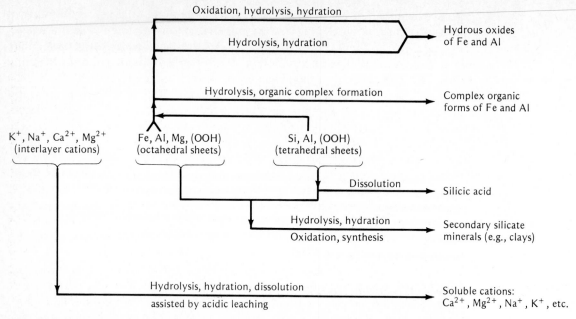

FIGURE 12.3 Illustration of chemical weathering of a hypothetical silicate mineral—(K^+, Na^+, Ca^{2+}, Mg^{2+}, Fe, Al, Mg, (OOH), Si, Al, (OOH)—showing the various chemical processes responsible for the weathering. Si, with some Al, and associated O^{2-} and OH^- ions are assumed to be in the tetrahedral sheets; Fe, Al, and Mg and associated O^{2-} and OH^- in the octahedral sheets; and the K^+, Na^+, Ca^{2+}, and Mg^{2+} ions between the mineral layers. Silicon is removed by dissolution, forming silicic acid, or is combined with Al, Fe, and Mg in secondary silicate minerals (clays). Iron may be oxidized and with aluminum hydrated to form hydrous oxides or complex organic compounds. Note the dominance of water-dependent reactions. Silicate clays and iron and aluminum oxides are secondary minerals formed as chemical weathering proceeds. Soluble metal cations, complex organic ions of Fe and Al, and even a soluble silicon compound are products of these reactions and are subject to removal from the soil by leaching.

up by plants and microorganisms. Figures 12.2 and 12.3 should be useful in visualizing how weathering helps produce the specific soil components that are studied in this text.

12.5 Factors Affecting Weathering of Minerals

The following general factors have a significant effect on the weathering process: (a) climatic conditions, (b) physical properties, and (c) chemical characteristics of the rocks and minerals. Each will be discussed briefly.

Climatic Conditions. Climatic conditions, more than any other factor, will control the kind and rate of weathering that take place if sufficient time is allowed. Under conditions of low rainfall, for example, mechanical processes of weathering dominate, resulting in decreased particle size with little change in composition. The presence of more moisture encourages chemical as well as mechanical changes, resulting in new minerals and soluble products. In humid temperate regions, silicate clays are among the minerals synthesized. They account for much of the strong agricultural character of soils in these areas.

Weathering rates are generally more rapid in humid tropical areas, where year-round high temperatures accelerate the weathering processes. As a result, the primary silicate minerals have succumbed to the intense weathering. Even quartz, the most resistant of the common primary minerals, disappears in time under these conditions. The more resistant products of chemical weathering such as the hydrous oxides of iron and aluminum are prominent in these soils.

Climate also largely controls the dominant types of vegetation present. In this way it indirectly influences the biochemical reactions in soils and in turn their effect upon mineral weathering. For example, soils developing under conifer trees whose needles are low in metallic cations are commonly more acid than soils developed under grassland or most deciduous trees. More will be said about this later (see Section 13.1).

Physical Characteristics.   Particle size, hardness, and degree of cementation are three physical characteristics that influence weathering. In rocks, large crystals of different minerals encourage disintegration, since there is some variation in the amount of expansion and contraction occurring with each mineral as temperatures change. The resulting stress helps to develop cracks and break these rocks into their mineral components. Finer-grained materials are apparently more resistant to mechanical breakdown.

Particle size of the rocks also influences the chemical breakdown of minerals. In general, a given mineral is more susceptible to decomposition when present in fine particles than in large grains. The much larger surface area of finely divided material presents greater opportunity for chemical attack. For example, quartz particles of sand size are extremely resistant to chemical weathering. In contrast, clay-sized quartz, although not subject to ready decomposition, is not as resistant as many of the other minerals in similar-sized particles.

Hardness and cementation influence weathering. Thus, a dense quartzite or a sandstone cemented firmly by a slowly weathered mineral will resist mechanical breakdown and present a small amount of total surface area for chemical activity. Porous rocks such as volcanic ash or coarse limestone, on the other hand, are readily broken down into smaller particles. These have in total a larger surface area for chemical attack and are obviously more easily decomposed.

Chemical and Structural Characteristics.   Chemical and crystalline characteristics also determine the ease of decomposition. Minerals such as gypsum $(CaSO_4 \cdot 2H_2O)$, which are slightly soluble in water, are quickly removed if there is adequate rainfall. Water charged with carbonic acid likewise dissolves less soluble minerals such as calcite and dolomite (Section 12.4). Consequently, these minerals are seldom found in the surface soils in areas with even moderate rainfall.

The dark-colored primary minerals, sometimes called ferromagnesian because of their iron and magnesium contents, are more susceptible to chemical weathering than are the feldspars and quartz. The presence of ferrous iron (and perhaps less tightness of crystal packing) helps account for the more rapid breakdown of the ferromagnesian minerals. That ferrous iron is subject to oxidation is evidenced by the staining of minerals as they weather.

Tightness of packing of the ions in the crystal units of minerals influences weathering rates. For example, olivine and biotite, which are relatively easily weathered, have crystal units less tightly packed than zircon and muscovite, comparable minerals quite resistant to weathering.

Climatic and biotic conditions will likely determine the relative stability of the various soil-forming minerals. Consequently, one cannot present a listing of minerals based on their resistance to weathering under all climatic conditions. However, studies (Barshad, 1955) of minerals remaining in the soil and the regolith under various environmental conditions have led to the following general order of weathering resistance of the sand- and silt-sized particles of some common minerals.

Quartz (most resistant) > muscovite and potassium feldspars > sodium and calcium feldspars > biotite, hornblende, and augite > olivine > dolomite and calcite > gypsum

This order would be changed slightly depending on the climate and other environmental conditions. This listing accounts for the absence of gypsum, calcite, and dolomote in soils of humid regions and for the predominance of quartz in the coarser fraction of most soils.

12.6 Weathering in Action—Genesis of Parent Materials

A physical weakening, caused usually by temperature changes, initiates the weathering process, but it is accompanied and supplemented by certain chemical transformations. Such minerals as the feldspars, mica, hornblende, and the like suffer hydrolysis and hydration, while part of the combined iron is oxidized and hydrated. The minerals soften, lose their lustre, and become more porous. If hematite or geothite is formed, the decomposing mass becomes definitely red or yellow. Otherwise, the colors are subdued.

Coincident with these changes, such cations as calcium, magnesium, so-

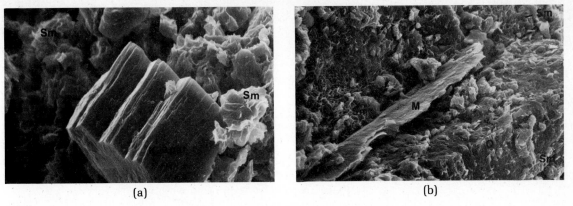

FIGURE 12.4 Scanning electron micrographs illustrating silicate clay formation from weathering of a granite rock in southern California. (a) A potassium feldspar (K-spar) is shown surrounded by smectite (Sm) and vermiculate. (b) Mica (M) and quartz (Q) associated with smectite (Sm). *[Courtesy J. R. Glasmann, Union Oil Company, Brea, CA.]*

dium, and potassium are solubilized and are removed by leaching, leaving a residue deprived of its easily soluble bases. As the process goes on, all but the most resistant of the original minerals disappear and their places are occupied by (a) secondary silicates that recrystallize into silicate clay (Figure 12.4) and (b) more resistant products, such as iron and aluminum oxides (Figure 12.2).

The intensity of the various forces will fluctuate with climate. Under arid conditions, the physical forces will dominate and the resultant soil material will contain the less-weathered minerals. Temperature changes, wind action, and water erosion will be accompanied by a minimum of chemical action.

In a humid region, however, these forces are more varied, and practically all of them will be active. Vigorous chemical changes will accompany disintegration, and clayey materials will be more evident. Moreover, the processes will be accelerated and intensified by decaying organic matter.

The forces of both physical and chemical weathering lose their intensity from the surface downward in soils. The lower organic matter content is a factor, as is the decrease in porosity and aeration of the underlayers. This differentiation with depth is the forerunner of a profile development as soil genesis takes place.

12.7 Geological Classification of Parent Materials

Two groups of inorganic parent materials are recognized: (a) *sedentary* (formed in place) and (b) *transported,* which may be subdivided according to the agencies of transportation and deposition as indicated below and diagrammed in Figure 12.5.

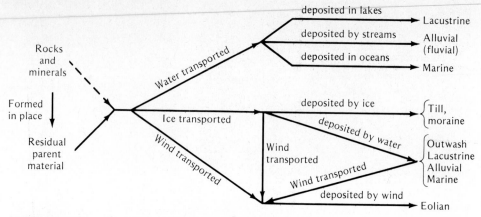

FIGURE 12.5 How various kinds of parent material are formed, transported, and deposited.

1. Sedentary
   Still at original site   Residual
2. Transported
   By gravity   Colluvial

   By water   Alluvial
                 Marine
                 Lacustrine

   By ice   Glacial (till, moraine)

   By wind   Eolian

These terms properly relate only to the placement of the parent materials, but they are also applied to the soils that have been developed by the weathering of these deposits—for example, glacial soils, alluvial soils, and residual soils. Such groupings are very general, however, since a wide diversity occurs within each soil group.

12.8 Residual Parent Material

Residual parent material develops in place from the underlying rock below and is rarely transported to another site. Typically, it has experienced long and often intense weathering. In a warm, humid climate, it is likely to be thoroughly oxidized and well leached. It is generally comparatively low in calcium because this constituent has been leached out. Red and yellow colors are characteristic when weathering has been intense as in hot humid areas.

In cooler and especially drier climates, residual weathering is much less drastic and the oxidation and hydration of the iron may be hardly noticeable. Also, the calcium content is higher and the colors of the debris subdued. Tremendous areas of this type of debris are found on the Great Plains and in other regions of the western United States.

Residual materials are of wide distribution on all continents. In the United States, the physiographic map (Figure 12.6) shows six great eastern and central provinces where residual materials are prominent: (a) Piedmont Plateau, (b) Appalachian Mountains and plateaus, (c) the limestone valleys and ridges near the Appalachians, (d) the limestone uplands, (e) the sandstone uplands, and (f) the Great Plains region. The first three groups alone encompass about 10% of the area of the United States. In addition, great expanses of these sedentary accumulations are found west of the Rocky Mountains.

A great variety of soils occupy these regions covered by residual debris since climate and vegetation, two of the determining factors in soil characterization, vary so radically over this great area. It has already been emphasized that the profile of a mature soil is largely a reflection of climate and its accompanying vegetation.

12.9 Colluvial Debris

Colluvial debris is made up of the fragments of rock detached from the heights above and carried down the slopes mostly by gravity. Frost action has much to do with the development of such deposits. Talus slopes, cliff detritus, and similar heterogeneous materials are good examples. Avalanches are made up largely of such accumulations.

Parent material developed from colluvial accumulation is usually coarse and stony, since physical rather than chemical weathering has been dominant. At the base of slopes in regions of medium- to fine-textured material such as loess, some superior soils develop. However, colluvial materials are generally not of great importance in the production of agricultural soils because of their small area, their inaccessibility, and their unfavorable physical and chemical characteristics.

12.10 Alluvial Stream Deposits

There are three general classes of alluvial deposits: (a) floodplains (fluvial deposits), (b) alluvial fans, and (c) deltas. They will be considered in order.

Floodplains (Fluvial Deposits). A stream on a gently inclined bed usually begins to swing from side to side in variable curves, depositing alluvial material

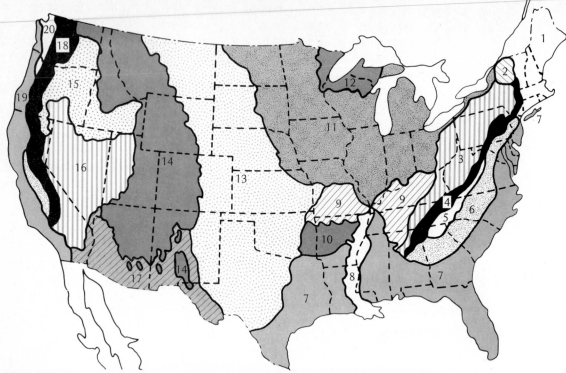

FIGURE 12.6 Generalized physiographic and regolith map of the United States. The regions are as follows (major residual areas italicized).

1. New England: mostly glaciated metamorphic rocks.
2. Adirondacks: glaciated metamorphic and sedimentary rocks.
3. *Appalachian Mountains and plateaus:* shales and sandstones.
4. *Limestone valleys and ridges:* mostly limestone.
5. Blue Ridge mountains: sandstones and shales.
6. *Piedmont Plateau:* metamorphic rocks.
7. Atlantic and Gulf coastal plain: sedimentary rocks with sands, clays, and limestones.
8. Mississippi floodplain and delta: alluvium.
9. *Limestone uplands:* mostly limestone and shale.
10. *Sandstone uplands:* mostly sandstone and shale.
11. Central lowlands: mostly glaciated sedimentary rocks with loess, a wind deposit of great agricultural importance (see Figure 12.14).
12. Superior uplands: glaciated metamorphic and sedimentary rocks.
13. *Great Plains region:* sedimentary rocks.
14. *Rocky Mountain region:* sedimentary, metamorphic, and igneous rocks.
15. Northwest intermountain: mostly igneous rocks; loess in Columbia and Snake river basins (see Figure 12.11).
16. Great Basin: gravels, sands, alluvial fans from various rocks; igneous and sedimentary rocks.
17. Southwest arid region: gravel, sand, and other debris of desert and mountain.
18. *Sierra Nevada and Cascade mountains:* igneous and volcanic rocks.
19. *Pacific Coast province:* mostly sedimentary rocks.
20. Puget Sound lowlands: glaciated sedimentary.
21. California central valley: alluvium and outwash.

on the inside of the curves and cutting on the opposite banks. This results in *oxbows* and *lagoons*, which are ideal for the further deposition of alluvial matter and development of swamps. This state of meander naturally increases the probability of overflow at high water, a time when the stream is carrying much suspended matter. Part of this sediment is deposited over the flooded areas, the coarser near the channel, building up natural levees, and the finer farther away in the lagoons and slack water. Thus, there are two distinct types of first bottom deposits—*meander* and *flood*. As might be expected, floodplain deposits are variable, ranging texturally from gravel and sands to silt and clay.

If there is a change in grade, a stream may cut down through its already well-formed alluvial deposits, leaving *terraces* above the floodplain on one or both sides. Often two or more terraces of different heights may be detected along some valleys, marking a time when the stream was at these elevations.

Floodplain deposits are found to a certain extent beside every stream, the greatest development in the United States occurring along the Mississippi River (Figure 12.7). This area varies from 20 to 75 miles in width, and from Cairo, Illinois, to the Gulf is over 500 miles long. The soils derived from such sediments usually are very rich, but they may require drainage and protection from overflow.

Equally productive soils are found in floodplains of other countries. The Nile River valley in Egypt, the Euphrates, Ganges, Indus, and Brahmaputra river valleys of Asia, and the Amazon Basin in Brazil are good examples. These floodplain deposits are ideal for the production of wetland rice, the food crop for most of the low income people of these areas.

Alluvial Fans. Where streams descend from uplands, a sudden change in gradient sometimes occurs as the stream emerges at the lower level. A deposition of sediment is thereby forced, giving rise to *alluvial fans*. Fan material generally is gravelly and stony, somewhat porous, and well drained.

Alluvial fan debris is found over wide areas in arid and semiarid regions. The soils therefrom, when irrigated and properly handled, often prove very productive. In humid regions, especially in certain glaciated sections, such deposits also occur in large enough areas to be of considerable agricultural importance.

Delta Deposits. Much of the finer sediment carried by streams is not deposited in the floodplain but is discharged into the body of water to which the stream is tributary. Unless there is sufficient current and wave action, some of the suspended material accumulates, forming a delta. Such delta deposits are by no means universal, being found at the mouths of only a small proportion of the rivers of the world. A delta often is a continuation of a floodplain, its front so to speak, and is not only clayey in nature but is likely to be swampy as well. The deltas of the Nile and Po rivers are good examples of these conditions.

FIGURE 12.7 Floodplain and delta of the Mississippi River. This is the largest continuous area of alluvial soil in the United States.

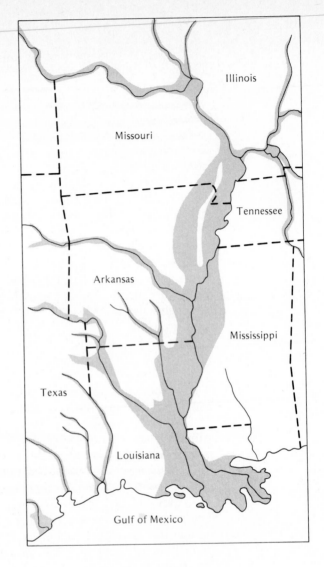

Delta sediments, where they occur in any considerable acreage and are subject to flood control and drainage, become rather important agriculturally. The combination deltas and floodplains of the Mississippi, Ganges, Po, Tigris, and Euphrates rivers are striking examples. The Nile Valley of Egypt, for centuries the granary of Rome, bespeaks the fertility and productivity of soils originating from such parent material.

12.11   Marine Sediments

Much of the sediment carried away by stream action is eventually deposited in the oceans, seas, and gulfs, the coarser fragments near the shore, the finer particles at a distance. Also, considerable debris is torn from the shoreline by the pounding of the waves and the undertow of the tides. If there have been changes in shoreline, the alternation of beds will show no regular sequence and considerable variation in topography, depth, and texture. These deposits have been extensively raised above sea level along the Atlantic and Gulf coasts of the United States and elsewhere, and have given origin to large areas of valuable soils (Figure 12.8).

Marine deposits have been worn and weathered by a number of agencies. First, the weathering and erosion necessary to throw them into stream suspension were sustained. These were followed by the sorting and solvent action of the stream itself. Next, the sediment was swept into the ocean to be deposited and stratified after being pounded and eroded by the waves for years. At last came the emergence of the deposits above the sea and the final action of weathering. The latter effects are of great significance since they determine the topography and, to a considerable extent, the chemical and physical nature of the resultant parent material.

Marine sediments, although having been subjected to weathering a shorter time than some of the residual debris, are generally more worn and usually carry less of the mineral nutrient elements. They may be sandy, as are the deposits along the Atlantic seaboard of the United States. But in the Atlantic and Gulf coastal flatwoods and the interior pinelands of Alabama and Mississippi, clayey deposits are not uncommon.

In the continental United States, the marine deposits of the Atlantic and Gulf coastal provinces (Figure 12.6) occupy approximately 11% of the country and are very diversified, owing to source of material, age, and the climatic conditions under which they now exist. Severe leaching occurs in times of heavy rainfall. In spite of this, however, their soils support a great variety of

FIGURE 12.8   Location of marine sediments and their relation to the uplands. The emerged coastal plain has already suffered some disection from stream action.

crops when adequately supplied with organic matter, properly cultivated, and
carefully fertilized.

12.12   The Pleistocene Ice Age

During the Pleistocene, northern North America, northern and central Europe,
and parts of northern Asia were invaded by a succession of great ice sheets.
Certain parts of South America and areas in New Zealand and Australia were

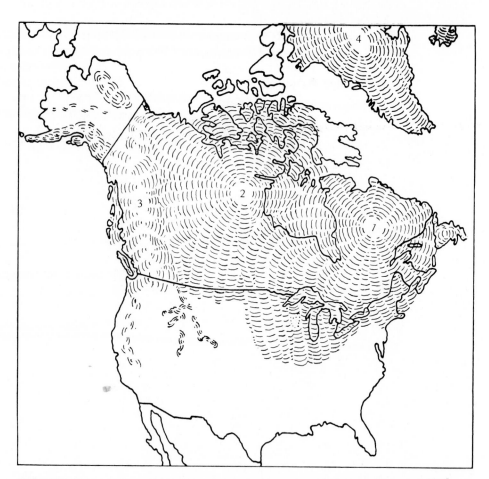

FIGURE 12.9   Maximum development of continental glaciation in North America.
The four centers of ice accumulation are numbered. Apparently the eastern and central
United States were invaded from the Labradorian (1) and Hudson Bay (2) centers.
Note the markedly southerly advance of the ice in the upper Mississippi Valley, where
topography offered little resistance. To the east the Appalachian highlands more or
less blocked the ice invasion.

similarly affected. Antarctica undoubtedly was capped with ice much as it is today. The sea level was about 130 m (425 ft) lower than it is today.

It is estimated that the Pleistocene ice at its maximum extension covered perhaps 20% of the land area of the world. It is surprising to learn that present-day glaciers, which we consider as mere remnants of the Great Ice Age, occupy about one third as much land surface. The present volume of ice, however, is much less, since our living glaciers are comparatively thin and are definitely on the wane. Even so, if present-day glaciers were to melt, the world sea level would increase by about 65 m (210 ft).

In North America, the major centers of ice accumulation were in central Labrador and the western Hudson Bay region, with a minor concentration in the Canadian Rockies. From the major centers, great continental glaciers pushed outward in all directions, but especially southward, covering from time to time most of what is now Canada, southern Alaska, and the northern part of the contiguous United States. The southernmost extension was down the Mississippi Valley, where the least resistance was met because of the lower and smoother topography (Figure 12.9).

Europe and central North America apparently sustained four distinct ice invasions. In the United States they are identified as Nebraskan, Kansan, Illinoian, and Wisconsin. The glacial debris of the eastern United States and Canada is practically all of Wisconsin age. These invasions were separated by long interglacial ice-free intervals that are estimated to have covered a total time period considerably longer than the periods of glaciation. Some of the interglacial intervals evidently were times of warm or semitropical climate in regions that are now definitely temperate. The total length of the Pleistocene ice age is estimated at 1–1.5 million years (Table 12.4). According to recent studies using radioactive carbon (^{14}C), the glacial ice disappeared from northern Iowa and central New York possibly only about 12,000 years ago. We may be now enjoying the mildness of another interglacial period.

TABLE 12.4 Nomenclature Used to Identify Periods of Glaciation and Interglaciation in North America During the Pleistocene Period

Period Name	Type of period	Approximate date starting
Holocene	Interglacial	10,000 B.C.
Wisconsin	Glacial	
Sangamonian	Interglacial	
Illinoian	Glacial	
Yarmouthian	Interglacial	
Kansan	Glacial	
Aftonian	Interglacial	
Nebraskan	Glacial	1–1.5 million B.C.

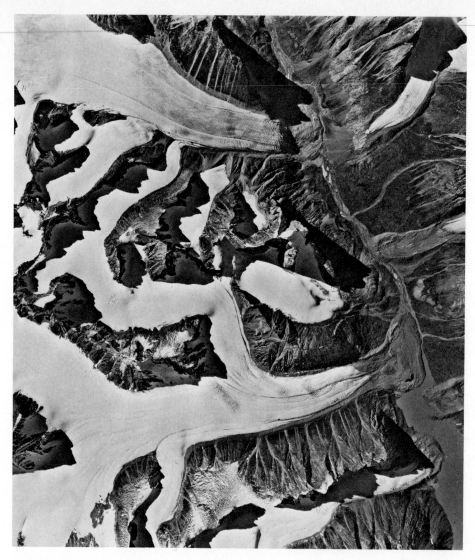

FIGURE 12.10 Tongues of a modern-day glacier in Canada. Note the evidence of transport of materials by the ice and the "glowing" appearance of the two major ice lobes. [*Photo* A–16817–102 *courtesy National Air Photo Library, Surveys and Mapping Branch, Canadian Department of Energy, Mines and Resources.*]

Of the major interglacial periods, the first (Aftonian) persisted perhaps 200,000 years, the second (Yarmouthian) may have lasted for 300,000 years, and the third (Sangamonian), which preceded the Wisconsin ice advance, possibly endured for 125,000 years. Assuming the Pleistocene age to have continued for not more than 1 million years, it is obvious that glacial ice occupied what is now the north central and northeastern United States less than half of the period designated as the Great Ice Age.

As the glacial ice was pushed forward, it conformed to the unevenness of the areas invaded. It rose over hills and mountains with surprising ease. Not only was the existing regolith with its mantle of soil swept away but hills were rounded, valleys filled, and, in some cases, the underlying rocks were severely ground and gouged. Thus, the glacier became filled with rock wreckage, carrying much of its surface and pushing great masses ahead (Figure 12.10). Finally, when the ice melted and the region again was free, a mantle of glacial drift remained—a new regolith and fresh parent material for soil formation.

The area covered by glaciers in North America is estimated at 10.4 million km² (4 million mi²) and about 20% of the United States is influenced by the deposits. An examination of Figures 12.9 and 12.11 will indicate the magnitude of the ice invasion at maximum glaciation in this country.

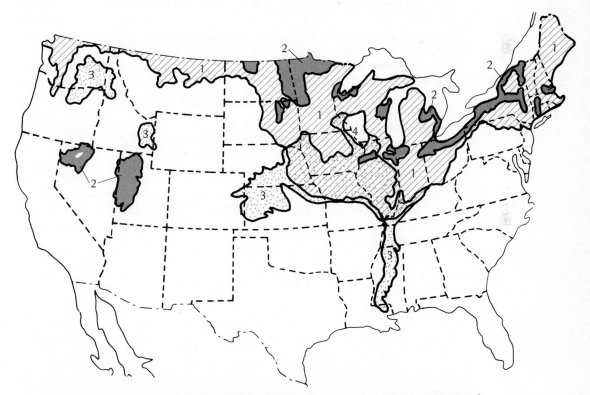

FIGURE 12.11 Areas in the United States covered by the continental ice sheet and the deposits either directly from, or associated with, the glacial ice. (1) Till deposits of various kinds. (2) Glacial-lacustrine deposits. (3) The loessial blanket—note that the loess overlies great areas of till in the Midwest. (4) An area, mostly in Wisconsin, that escaped glaciation; it is partially loess covered.

12.13 Glacial Till and Associated Deposits

The materials deposited directly by the ice are commonly called *glacial till.* Till is a mixture of rock debris of great diversity, especially as to size of particles. Boulder clay, so common in glaciated regions, is typical of the physical heterogeneity expected.

Glacial till is found mostly as irregular deposits called *moraines,* of which there are various kinds. *Terminal* moraines characterize the southernmost extension of the various glacial lobes when the ice margin was stationary long enough to permit an accumulation of debris (Figure 12.12). Many other moraines of a *recessional* nature are found to the north, marking points where the ice front became stationary for a time as it receded by melting. Although moraines of this type are generally outstanding topographic features, they give rise to soils that are rather unimportant because of their small area and unfavorable physiography.

The *ground moraine,* a thinner and more level deposition laid down as the ice front retreated rapidly, is of much more importance. It has the widest extent of all glacial deposits and usually possesses a favorable agricultural topography. Associated with the moraine in certain places are such special features as *kames* (conical hills or short ridges of sand and gravel deposited by ice), *eskers* (long narrow ridges of coarse gravel deposited by ice-walled streams coming from the glacier), and *drumlins* (cigar-shaped hills composed of till and oriented parallel to the direction of ice movement).

An outstanding feature of glacial till materials is their variability. This is because of the diverse ways by which the debris was laid down, of differences in the chemical composition of the original materials, and of fluctuations in the grinding action of the ice. As might be expected, the soils derived from such soil material are most heterogeneous. Such variations indicate that the term "glacial soil" is of value only in suggesting the mode of deposition of the parent materials. It indicates practically nothing about the characteristics of a soil so designated.

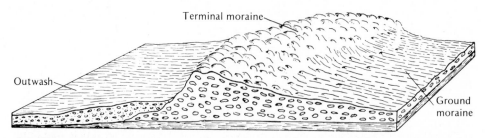

FIGURE 12.12 Relationship between a terminal moraine, its outwash, and its associated ground moraine. Note the differences in topography and in the general nature of the three deposits. The ground moraine is the most widely distributed. The direction of ice retreat is to the right.

12.14 Glacial Outwash and Lacustrine Sediments

Torrents of water were constantly gushing from the great Pleistocene ice sheet, especially during the summer. The vast loads of sediment carried by such streams were either dumped immediately or carried to other areas for deposition. So long as the water had ready egress, it flowed rapidly away to deposit its load downstream.

Outwash Plains.   A type of deposit of great importance is the *outwash* plain, formed by streams heavily laden with glacial sediment (Figure 12.12). This sediment is usually assorted and therefore variable in texture. Such deposits are particularly important in valleys and on plains, where the glacial waters were able to flow away freely. These valley fills are common in the United States, both north and south of the terminal moraine.

Glacial Lake (Lacustrine) Deposits.   In many cases the ice front came to a standstill where there was no such ready escape for the water, and ponding occurred as a result of damming action of the ice (Figure 12.13). Often, very large lakes were formed that existed for many years. Particularly prominent were those south of the Great Lakes in New York, Ohio, Indiana, Michigan,

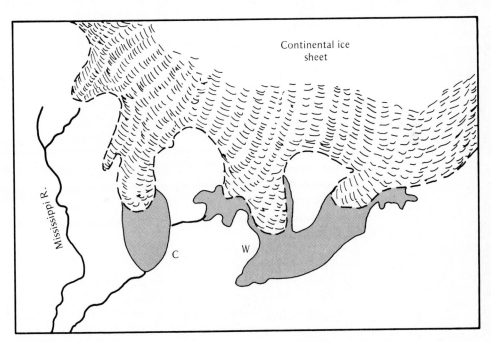

FIGURE 12.13   When conditions of topography were favorable, the Wisconsin ice sheet acted as a great waning dam. A stage in the development of glacial lakes in Chicago (C) and Warren (W) is represented. [*After Daly* (*1934*).]

and in the Red River Valley (Figure 12.10). The latter lake, called Glacial Lake Agassiz, was about 1200 km (745 mi) long and 400 km (248 mi) wide at its maximum extension. The others, individually much smaller, covered great stretches of country. Large lakes also occurred in the intermountain regions west of the Rockies as well as in the Connecticut River Valley in New England and elsewhere.

The deposits that were made in these glacial lakes range from coarse delta materials near the shore to fine silts and clay in the deeper and stiller waters. As a consequence, the soils developed from these lake sediments are most heterogeneous. Even so, large areas of fertile fine-textured soils have developed from these materials. Because of climatic differences, weathering has been variable, and profile contrasts are great. Extending westward from New England along the Great Lakes to the broad expanse of the Red River Valley, these deposits have produced some of the most important soils of the northern states. In the intermountain regions of the United States, they have given rise to agriculturally important soils, especially when irrigated.

12.15 Glacial Eolian Deposits

During the glaciation, much fine material was carried miles below the front of the ice sheets by streams that found their source within the glaciers. This sediment was deposited over wide areas by the overloaded rivers. When added to the great stretches of unconsolidated till in the glaciated regions and the residual material devoid of vegetation on the Great Plains, these sediments presented unusually favorable conditions for wind erosion in dry weather.

Origin and Location of Loess.   Wind erosion was responsible for the deposition of a very important soil material in the midwest and in parts of the northwest. In the Mississippi Valley, extensive deposits of glacial outwash left on river floodplains during the summer were subject to wind movement during the winter when melting had stopped and the floodplains were dry. This wind-blown material, called *loess*, was deposited in the uplands, the thickest deposits being found where the valleys were widest. The fine material high in silt-sized particles covered existing soils and parent materials, both original and glacial in origin.

Loess is found over wide areas in the United States, concealing the unconsolidated material below. It covers eastern Nebraska and Kansas, southern and central Iowa and Illinois, northern Missouri, and parts of southern Ohio and Indiana, besides a wide band extending southward along the eastern border of the Mississippi River (Figures 12.11 and 12.14). Extensive loessial deposits also occur in the Palouse region in Washington and Idaho and other areas of the northwestern United States. Along the Missouri and Mississippi rivers, its accumulation has formed great bluffs which impart a characteristic topogra-

FIGURE 12.14 Approximate distribution of loess in central United States. The soil that has developed therefrom is generally a silt loam, often somewhat sandy. Note especially the extension down the eastern side of the Mississippi River and the irregularities of the northern extensions. Smaller areas of loess occur in Washington, Oregon, and Idaho (see Figure 12.11).

phy to the region. Farther from the rivers, the deposits, an upland type, are shallower and smoother in topography.

Nature of Loess. Loess is usually silty in character and has a yellowish buff color unless it is very markedly weathered or carries a large amount of humus.

The larger particles are usually unweathered and angular. Quartz seems to predominate, but large quantities of feldspar, mica, hornblende, and augite are found. The vertical walls and ridges formed when this deposit is deeply eroded constitute one of its most striking physical characteristics.

Loess has given rise to soils of considerable diversity. A comparison of the loessial area of the midwest (Figure 12.14) with the map of soil orders (Figure 13.8) will show the presence of loess in six distinct soil regions. This situation, while indicating the probability of great differences in the fertility and productivity of loess soils, especially emphasizes the influence of climate as a strong determinant of soil characteristics.

Other Eolian Deposits. Eolian deposits other than loess occur, such as volcanic ash and sand dunes. Soils from volcanic ash occur in Montana, Oregon, Idaho, Nebraska, and Kansas. They are light and porous and are generally of less agricultural value than soils developed from loess.

Sand dunes are of little value under any conditions and become a menace to agriculture if they are moving. Examples of such deposits are found in the great Sahara and Arabian deserts. Smaller areas occur in Nebraska, Colorado, New Mexico, and along the eastern seaboard.

12.16 Agricultural Significance of Glaciation

The Pleistocene glaciation in the United States has been generally beneficial, and especially so for agriculture. The leveling and filling actions when drift was abundant have given a smoother topography more suited to farming operations. The same can be said for the glacial lake sediments and loess deposits. Also, the parent materials thus supplied are geologically fresh, and the soils derived from them usually are not drastically leached. Young soils generally are higher in available nutrients and, under comparable conditions, superior in crop-producing power.

Glaciation was somewhat less beneficial to areas near the centers of ice accumulation. For example, in eastern Canada the moving ice picked up unconsolidated materials (including soils) and transported them to areas in the northern United States. While these materials generally contributed to the productivity of soils in the United States, the glaciation left large areas of shallow soils in eastern Canada.

In glaciated areas there is great variability in the nature of the soils that develop. On glacial outwash plains the soils may be coarse textured, even high in gravel, resulting in low moisture retention and droughtiness. Glacial lacustrine clays are often not well drained because of their high clay content. Some glacial tills are also quite compact, and the soils developing on them tend to have imperfect drainage (Table 12.5). Soils developed on loess tend to be erosion prone even though they may be quite productive. These differences

TABLE 12.5 Relationship Between Certain Parent Materials in Pennsylvania andSoil Drainage Class[a]

Parent material class	Number of profiles	Percent of profiles in drainage classes	
		Well and moderately well drained	Somewhat poorly, poorly, and very poorly drained
Alluvium	23	81	19
Limestone	23	100	0
Till	52	65	35
Shale	77	74	26
Sandstone	21	100	0

[a] From Petersen et al. (1971).

emphasize the definite influence of parent material on the kinds of soil that develop.

12.17 Conclusion

Together with climate, the nature and properties of parent materials are the most significant factors affecting the kind and quality of the world's soils. Knowledge of these parent materials, their source or origin, mechanisms for weathering, and means of transport is essential if we are to gain an understanding of soils—and especially if we are to classify them properly. These facts should be kept in mind as we turn to a study of soil formation and classification in the next chapter.

REFERENCES

Barshad, I. 1955. "Soil development," in F. E. Bear (Ed.), *Chemistry of the Soil* (New York: Reinhold).

Daly, R. A. 1934. *The Changing World of the Ice Age* (New Haven: Yale University Press).

Diller, J. S. 1898. *Educational Series of Rock Specimens,* Bull. 150, U.S. Geol. Surv., p. 385.

Merrill, G. P. 1879. *Weathering of Micaceous Gneiss,* Bull. 8, Geol. Soc. Amer., p. 160.

Petersen, G. W., R. L. Cummingham, and R. P. Matelski. 1971. "Moisture characteristics of Pennsylvania soils: III. Parent material and drainage relationships," *Soil Sci. Soc. Amer. Proc.,* **35:**115–119.

Soil Formation, Classification, and Survey

[*Preceding page*] An examination of a soil profile such as this one in Nebraska can be used to help classify soils as well as to judge their potential for crop production. [*Courtesy USDA Soil Conservation Service.*]

To study soils satisfactorily and to use them for the benefit of humankind, some sort of classification is necessary. The value of field experimental work is restricted and may even be misleading unless the relation of one soil to another is known. The crop requirements in any region depend to a marked degree on the soils in question and on their profile similarities and differences.

In arriving at such an understanding, three phases must be considered: (a) soil genesis, or the evolution of a soil from its parent material; (b) soil classification, in this case especially as it applies to the United States; and (c) soil survey, its interpretation and utilization. Soil genesis will receive attention first, starting with the five major factors influencing soil formation.

13.1 Factors Influencing Soil Formation

Studies of soils throughout the world have shown that the kinds of soil that develop are largely controlled by five major factors:

1. Climate (particularly temperature and precipitation).
2. Living organisms (especially native vegetation and human beings).
3. Nature of parent material:
    a. Texture and structure.
    b. Chemical and mineralogical composition.
4. Topography of area.
5. Time that parent materials are subjected to soil formation.

In fact, soils are often defined in terms of these factors as "dynamic natural bodies having properties derived from the combined effect of *climate* and *biotic activities,* as modified by *topography,* acting on *parent materials* over periods of *time.*"

Climate. Climate is perhaps the most influential of these factors. It determines the nature of the weathering that occurs. For example, temperature and precipitation affects the rates of chemical and physical processes responsible for profile development. Consequently, if allowed ample opportunity, climatic influences eventually tend to dominate the soil formation picture.

For every 10°C rise in temperature, the rate of chemical reactions doubles. Furthermore, biochemical changes by soil organisms are sensitive to temperature as well as moisture. The influence of temperature and effective moisture on the organic matter contents of soil (see Figure 8.8) in the Great Plains of the United States is evidence of the strong influence of climate on soil characteristics. The very modest profile development characteristic of arid areas contrasted with the deep-weathered profiles of the humid tropics is further evidence of climatic control.

Much of the influence of climate is due to the measure of control that it exercises over natural vegetation. In humid regions, plentiful rainfall provides an environment favorable for the growth of trees (Figure 13.1). In contrast, grasslands are the dominant native vegetation in semiarid regions. Thus, climate exerts much of its influence through a second soil-forming factor, the living organisms (compare Figures 13.1 and 13.10).

Living Organisms. The major role of biotic activity in profile differentiation cannot be overemphasized. Organic matter accumulation, profile mixing, nutrient cycling, and structural stability are all enhanced by the activities of

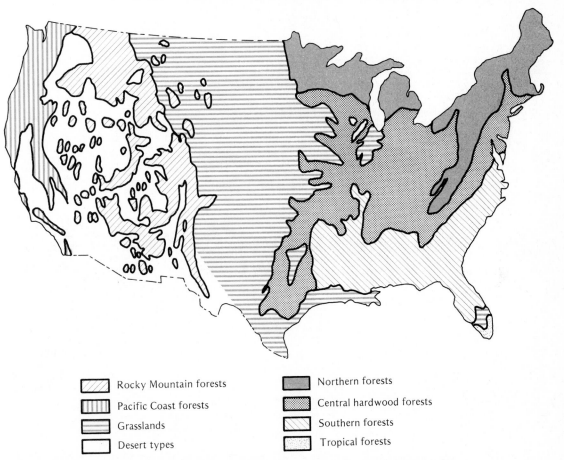

Rocky Mountain forests
Pacific Coast forests
Grasslands
Desert types

Northern forests
Central hardwood forests
Southern forests
Tropical forests

FIGURE 13.1 General types of natural vegetation in the United States. Note the relationship between the dominant vegetation and the kinds of soil that develop (see Figures 13.8 and 13.10). [*Redrawn from a more detailed map by the U.S. Geological Survey.*]

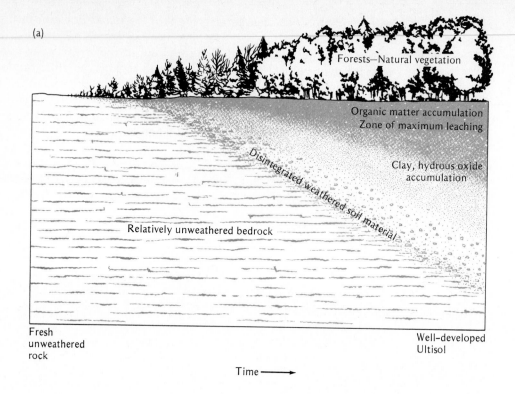

(a)

Forests—Natural vegetation

Organic matter accumulation
Zone of maximum leaching

Clay, hydrous oxide
accumulation

Disintegrated weathered soil material

Relatively unweathered bedrock

Fresh
unweathered
rock

Well-developed
Ultisol

Time ⟶

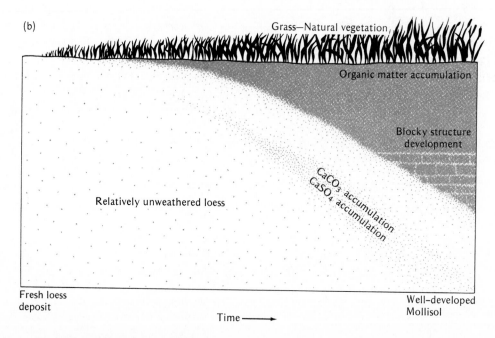

(b)

Grass—Natural vegetation

Organic matter accumulation

Blocky structure
development

$CaCO_3$ accumulation
$CaSO_4$ accumulation

Relatively unweathered loess

Fresh loess
deposit

Well-developed
Mollisol

Time ⟶

FIGURE 13.2 [*opposite*] How two soil profiles may have developed (a) from the weathering in place of solid rock and (b) from the weathering of wind-deposited loess. Organic matter accumulation in the upper horizons occurs in time, the amount and distribution depending on the type of natural vegetation present. Clay and iron oxide accumulate and characteristic structures develop in the lower horizons. The end products differ markedly from the soil materials from which they form.

organisms in the soil. Also, nitrogen is added to the soil system by microorganisms alone or in association with plants. And vegetative cover reduces natural erosion rates, thereby slowing down the rate of mineral surface removal.

The most striking evidence of the effect of vegetation on soil formation is seen in comparing properties of soils formed under grassland and forest vegetation (Figure 13.2). For example, the organic matter content of the "grassland" soils is generally much higher, especially in the subsurface horizons. This gives the soil darker color and higher moisture- and cation-holding capacity as compared to the "forest" soil. Also, structural stability tends to be encouraged by the grassland vegetation.

The mineral element content in the leaves, limbs, and stems of the natural vegetation strongly influences the characteristics of the soils that develop. Coniferous trees, for example, tend to be low in metallic cations such as calcium, magnesium, and potassium. The cycling of nutrients from the litter falling from these trees (Figure 13.3) will be low compared to that of some deciduous trees, the litter of which is much higher in bases. Consequently, soil acidity is much more likely to develop under pine vegetation with its low base status than under ash or elm trees. Also the removal of bases by leaching is encouraged by coniferous vegetation.

As might be expected, there is an interaction among the natural vegetation, other soil organisms, and the characteristics of the soil that develops. Thus, as soils develop under native grasslands, nonsymbiotic nitrogen fixers (*Azotobacter*) tend to flourish. This provides an opportunity for an increase in both the nitrogen and organic matter contents of the soil. Vegetation is thus seen to be a critical factor in determining soil character.

Human activities should not be overlooked as factors influencing soil formation. The removal of natural vegetation by cutting trees or by cultivating the soil abruptly modifies the soil-forming factors. Likewise, irrigation of a soil in an arid area drastically influences the soil-forming factors, as does adding fertilizer and lime to soil of low fertility. Although human activities have taken place only in recent geological times, they have had significant influences on soil-forming processes in some areas.

Parent Material. The nature of the *parent material* profoundly influences the characteristics of even highly weathered soils. For example, the texture of sandy soils is determined largely by parent materials. In turn, the downward movement of water is controlled largely by the texture of the parent material. The chemical and mineralogical compositions of parent material often not only determine the effectiveness of the weathering forces but in some instances partially control the natural vegetation. The presence of limestone in a humid region soil will delay the development of acidity, a process that the climate encourages. In addition, the leaves of species of trees found on limestone materials are relatively high in metallic cations or bases. As these high-base leaves fall to the soil surface, are decomposed, and become incorporated into the soil, they further delay the process of acidification or, in this particular case, the progress of soil development.

Parent material has a marked influence on the type of clay minerals in the soil profile. In the first place, the parent material itself may contain clay minerals, perhaps from a previous weathering cycle. Second, as our study of clay minerals indicated, the nature of parent material definitely influences the nature of the clay that develops. Illite, for example, tends to form from the mild weathering of potash-containing micas, whereas smectite is favored by base-rich minerals high in calcium and magnesium.

Certainly the kind of soil that develops would be markedly affected by the nature of the clay minerals present. These in turn are greatly influenced by parent material.

Topography. The *topography* of the land may be such as to hasten or delay the work of climatic forces. In smooth flat country excess water is removed less rapidly than if the landscape were rolling. Rolling topography encourages some natural erosion of the surface layers, which, if extensive enough, may eliminate the possibility of a deep soil. On the other hand, if water stands for part or all of the year on a given area, the climatic influences become relatively ineffective in regulating soil development.

There is a definite interaction among topography, vegetation, and soil formation. In the grassland–forest transition zones, trees commonly occupy the slight depressions in an otherwise prairie vegetation. This is apparently a moisture effect. As would be expected, however, the nature of the soil in the depressions is quite different from that in the uplands. Topography, therefore, is significant not only as a modifier of climate and vegetative effects but often as a major control in local areas.

Time. The length of *time* that materials have been subjected to weathering plays a significant role in soil formation. A comparison of the soils of a glaciated region with those in a comparable area untouched by the ice sheet illustrates the significance of time. The influence of parent material is much more apparent in the soils of glaciated regions, where insufficient time has elapsed since the ice retreated to permit the full development of soils.

Soils located on alluvial or lacustrine materials generally have not had as much time to develop as the surrounding upland soils. Some coastal-plain parent materials have been uplifted only in recent geological time, and consequently the soils thereon have had relatively little time for weathering.

The interaction of time with other factors affecting soil formation must be emphasized. The time required for the development of a horizon will definitely be related to the parent material, the climate, and the vegetation. It is easy to visualize the interdependence of these factors in determining the kind of soil that develops.

13.2 Weathering and Soil Development Processes

Mineral soils have originated from the unconsolidated materials (the regolith) that cover the country rock. The weathering processes of disintegration and decomposition that have given rise to this regolith are in general destructive processes. Thus, rocks and minerals are destroyed or altered. And soluble nutrients are subject to loss by leaching. It seems incongruous that these same seemingly destructive processes can promote the genesis of natural bodies called soils, and yet this is the case.

There are no distinct stages in the development of soils. Even if stages are identified, they seem to overlap and blend together to give a continuum of genetic processes. Only a few of the obvious happenings can be identified as movement occurs from solid rock or recently deposited soil material to a well-developed soil profile.

Disintegration and Synthesis. The process of disintegration of solid rock makes possible a foothold for living organisms. Decomposing minerals release nutrients that nourish simple plant and animal forms. Alteration of primary minerals to silicate and other clays destroys one mineral while giving birth to another. These clays hold reserves of nutrients and water, permitting plants to gain a foothold. Residues from the plants return to the weathering mass and are altered to humus, which exceeds even clay in its nutrient- and water-holding capacities.

Thus, *silicate clays, humus*, and *living organisms*, together with life-giving *water*, become prime constituents as soil characteristics develop. They affect markedly the kind and extent of layering or horizon differentiation that occurs. A general description of how soil layers form will illustrate this point.

Organisms and Organic Matter. As soon as plants gain a foothold in a weathering rock or in recently deposited soil material, the development of a *soil profile* has begun. Residues of plants and animals decay and are mixed with mineral matter by living organisms. The upper part of the soil mass becomes slightly darker in color than the deeper layers. The first evidence of layering occurs and it assumes structural stability due to the presence of the organic matter. The surface horizon begins to appear in the young soil.

Decaying organic matter accelerates soil layering in other ways. Acids released from organic decomposition enhance the breakdown of base-containing minerals, yielding soluble nutrients, and secondary minerals such as the silicate clays and the oxides of iron and aluminum. These products may simply enrich the upper layers in which they are formed, or they may be moved downward by percolating waters, eventually to accumulate as layers at a lower depth in the developing soil. This downward movement and accumulation again dictates layer or horizon formation. Upper horizons may be depleted of nutrients and clay-sized minerals, whereas deeper layers tend to be enriched in these same constituents (Figure 13.2).

As these chemical and physical changes occur, living organisms, including higher plants and animals, continue to play a vital role. They physically manipulate, move, and bind the soil particles and structural units, helping to provide stability of the horizons. At the same time, by moving through the soil they help mix materials from adjacent horizons, thereby reversing the process of distinct horizon differentiation.

Nutrient Recycling. Living organisms together with soil water provide an important mechanism for stabilizing the acid-base ratio of the soil solution as weathering occurs. This is done by *recycling nutrients* through the soil (Figure 13.3). Soluble elements are absorbed by plants from the soil, translocated to the upper plant parts, released again upon the death of the plant, and moved downward into the soil by percolating water, ready to be recycled again. The nutrient content of the plant parts and the amount of rainfall thereby establish the ionic environment of the weathering soil. Plants high in mineral elements help to maintain a high metallic cation concentration in the soil solution. Plants of low nutrient content cannot maintain such high concentrations, especially if they grow, as they often do, in areas of high rainfall and high leaching potential. Nutrient recycling thus helps control the acid-base balance of the weathering solutions and the ultimate horizons which develop.

FIGURE 13.3 Nutrient recycling is an important factor in determining the relationship between vegetation and the soil that develops. If residues from the vegetation are low in bases, acid weathering conditions are favored. High-base-containing plant residues tend to neutralize acids, thereby favoring only slightly acid to neutral weathering conditions.

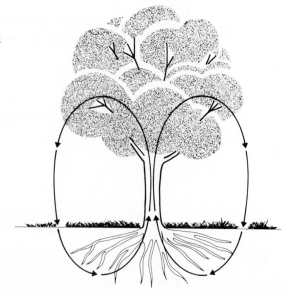

Water's Role.   Soil water is active from the very beginning in helping to enhance the development of soils. Its presence is essential for plant growth and for most of the chemical reactions whereby mineral breakdown occurs. Water movement in the regolith is of no less significance to soil development. This is shown by the nutrient recycling just described and by the downward movement of silicate clays, oxides of iron and aluminum, and salts of various kinds. In regions of low rainfall, upward movement and evaporation of water result in the accumulation of salts at the soil surface or at some point below the surface. Water is seen to be the principal transport within the soil body.

Water affects the nature of soil horizons in other important ways. If water can freely drain from the weathering area, aerobic conditions exist. The resulting soil is generally weathered to a considerable depth and contains oxidized minerals and nutrient elements, and root penetration is uninhibited by excess water. In wet areas, oxidative weathering is minimized and undecomposed organic matter tends to accumulate at or near the mineral surface. Reduced conditions characterize the soil horizons, which are exploited to only a limited extent by growing plants except for rice and other water-loving species. Horizon differentiation is indeed affected directly by excess water.

Acquired Versus Inherited Characteristics.   Each of these examples illustrates how, over a period of time, significant changes take place in the parent material. Layers or horizons develop that may be leached of certain minerals or may be enhanced by the deposition or the formation in place of others. Thus, the weathered material has some characteristics that are *acquired* in contrast to those *inherited* from the parent material.

The presence of quartz is perhaps the most common example of an *inherited* characteristic, although the presence of any mineral, such as feldspar, mica, or silicate clays carried over unchanged into a soil, is a good illustration. Color also is often inherited, as in the case of a red parent sandstone or shale. Examples of *acquired* characteristics include organic matter, clay formed as the soil developed, and other products of weathering such as red and yellow iron oxides. In the early stages of soil formation inherited soil properties dominate, but as the soil develops acquired characteristics become more prominent and eventually may become dominant.

13.3   The Soil Profile

The layering or horizon development described in the previous section eventually gives rise to natural bodies called *soils*. Each soil is characterized by a given sequence of these horizons. A verticle exposure of this sequence is termed a *soil profile*. Attention will now be given to the major horizons making up soil profiles and the terminology used to describe them.

For convenience in study and description, the layers resulting from soil-forming processes are grouped under five heads: O, A, E, B, and C. Singly and in combination these capital letters are used to designate the *master horizons* in soils. Subordinate distinctions of these master horizons (to be described later) are designated by lowercase letters. A common sequence of occurrence of horizons within the profile is shown in Figure 13.4.

Organic (O) Horizons. The O group are organic horizons that form above the mineral soil. They result from litter derived from dead plants and animals. Occurring commonly in forested areas, they are generally absent in grassland regions. The specific horizons are

Oi: Organic horizon wherein the original forms of the plant and animal residues are only slightly decomposed.
Oe: Organic horizon, residues intermediately decomposed.
Oa: Organic horizon, residues highly decomposed.

Eluvial Horizons. This group are mineral horizons that lie at or near the surface and are characterized as zones of maximum leaching or *eluviation* (from the Latin *ex* or *e*, out, and *lavere*, to wash) and of maximum organic matter accumulation. Beginning with the surface, these master horizons are designated as A, E, and so on (Figure 13.4).

A: Topmost mineral horizon, containing a strong admixture of humified organic matter, which tends to impart a darker color than that of the lower horizons.

FIGURE 13.4 Theoretical mineral soil profile, showing the major horizons that may be present in a well-drained soil in the temperate humid region. Any particular profile may exhibit only some of these horizons and the relative depths vary. In addition, however, it may exhibit more detailed subhorizons than indicated here. The solum includes the A, E, and B plus some fragipans and duripans of the C horizon.

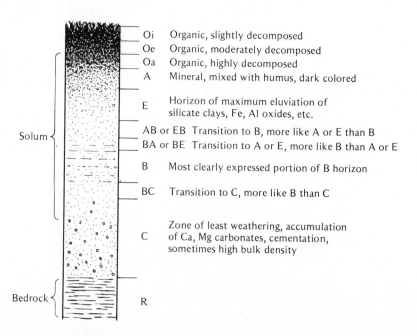

Oi   Organic, slightly decomposed
Oe   Organic, moderately decomposed
Oa   Organic, highly decomposed
A   Mineral, mixed with humus, dark colored

E   Horizon of maximum eluviation of silicate clays, Fe, Al oxides, etc.

AB or EB   Transition to B, more like A or E than B
BA or BE   Transition to A or E, more like B than A or E

B   Most clearly expressed portion of B horizon

BC   Transition to C, more like B than C

C   Zone of least weathering, accumulation of Ca, Mg carbonates, cementation, sometimes high bulk density

R

Solum

Bedrock

E: Horizon of maximum eluviation of clay, iron, and aluminum oxides and a corresponding concentration of resistant minerals, such as quartz, in the sand and quartz sizes. It is generally lighter in color than the A horizon. (Formerly called A_2 horizon.)

AB (or EB): Transition layer between A (or E) and B with properties more nearly like those of A (or E) than of the underlying B. Sometimes absent. (Formerly called A_3 horizon.)

E/B: (*not shown*) Horizon that would qualify as E except for including parts constituting less than 50% of the volume that qualify as B.

A/C: (*not shown*) Transition layer between A and C, having subordinate properties of both A and C but not dominated by properties of either.

B (Illuvial) Horizons. The B horizons follow and include layers in which *illuviation* (from the Latin *il,* in, and *lavere*) has taken place from above or even below. This is the region of maximum accumulation of materials such as iron and aluminum oxides and silicate clays. In arid regions, calcium carbonate, calcium sulfate, and other salts may accumulate in the lower B.

The B horizons are sometimes referred to as the *subsoil*. This is incorrect, however, since the B horizons may be incorporated at least in part in the plow layer or they may be considered below the plow layer in the case of soils with deep A horizons.

The specific B horizons are as follows.

BA (or BE): Transition layer between A (or E) and B with properties more nearly like B than A (or E). Sometimes absent. (Formerly called B_1 horizon.)

B/E: (*not shown*) Qualifies as B in more than 50% of its volume including parts that qualify as E.

B: Zone of most clearly expressed properties of B horizon. Commonly a zone of maximum accumulation of clays and hydrous oxides, which may have moved down from upper horizons or may have formed in place. Organic matter content is somewhat higher than that of E, but is definitely lower than that of A. Maximum development of blocky or prismatic structure or both. Bw is a color or structural B horizon.

BC: Transition layer between B and C with properties more like those of B than those of C below.

CB: (*not shown*) Transition layer between C and B with properties more like those of C than those of B.

C Horizon. The C horizon is the unconsolidated material underlying the solum (A and B). It may or may not be the same as the parent material from which the solum formed. It is outside the zones of major biological activities and is little affected by solum-forming processes. Its upper layers may in time become a part of the solum as weathering and erosion continue.

R: Underlying consolidated rock.

Subordinate Distinctions. The characteristics of the master horizons are further specified by lowercase letters.

a Highly decomposed organic matter.
b Buried soil horizon.
c Concretions or nodules.
e Intermediately decomposed organic matter.
f Frozen soil.
g Strong gleying (mottling).
h Illuvial accumulation of organic matter.
i Slightly decomposed organic matter.
k Accumulation of carbonates.
m Strong cementation.
n Accumulation of sodium.
o Residual accumulation of sesquioxides.
p Plowing or other disturbance.
q Accumulation of silica.
r Weathered or soft bedrock.
s Illuvial accumulation of sesquioxides.
t Accumulation of clay.
v Plinthite.
w Color or structural B.
x Fragipan character.
y Accumulation of gypsum.
z Accumulation of salts.

Horizons in a Given Profile. It is at once evident that the profile of any one soil probably will not show all of the horizons that collectively are cited in Figure 13.4. The profile may be immature or may be overly influenced by some local condition such as poor drainage, texture, or topography. Again, some of the horizons are merely transitional and may at best be very indistinct. With so many factors involved in soil genesis, a slight change in their coordination may not encourage the development of some layers. As a result, only certain horizons are consistently found in well-drained and uneroded mature soils. The ones most commonly found are Oe or Oa, if the land is forested; A or E, or both, depending on circumstances; B or Bw and C. Conditions of soil genesis will determine which others are present and their clarity of definition.

When a virgin soil is put under cultivation, the upper horizons become the furrow slice. The cultivation, of course, destroys the original layered condition of this portion of the profile and the furrow slice becomes more or less

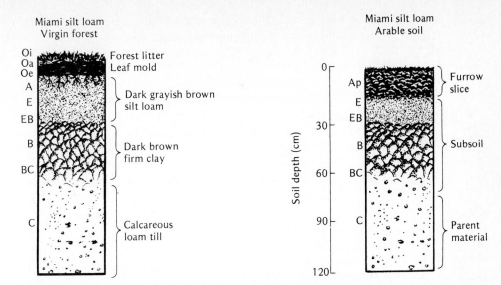

FIGURE 13.5 Generalized profile of the Miami silt loam, one of the Alfisols or Gray-Brown Podzolic soils of the eastern United States, before and after land is plowed and cultivated. The surface layers are mixed by tillage and are termed the Ap (plowed) horizon. If erosion occurs, they may disappear, at least in part, and some of the B horizon will be included in the furrow slice.

homogeneous. In some soils the A and E horizons are deeper than the furrow slice (Figure 13.5). In other cases where the A is quite thin, the plowline is just at the top of or even down in the B.

Many times, especially on cultivated land, serious erosion has produced a *truncated* profile. As the surface soil was swept away, the plowline was gradually lowered in order to maintain a sufficiently thick furrow slice. Hence, the furrow slice, in many cases, is almost entirely within the B zone and the C horizon is correspondingly near the surface. Many farmers, especially in the southern states, are today cultivating the B horizons without realizing the ravages of erosion. In profile study and description, such a situation requires careful analysis.

13.4 Concept of Individual Soils

Soils that have developed by the various processes discussed in this and the preceding chapter differ greatly from place to place. They vary in many profile characteristics, such as degree of horizon differentiation, depth, clay and organic matter contents, and wetness. These differences are noted not only from continent to continent or region to region but from one part of a given field to another. In fact, notable differences sometimes occur within a matter of a few feet. Furthermore, the changes in soil properties occur gradually as one moves from one location to another.

To study intelligently the soil at different places on the earth's surface and to communicate information about it in an orderly manner, systems of soil classification have been developed. These permit classification of *the soil* into a number of individual units or natural bodies which are called *soils*. This involves two basic steps. First, the kind and range of soil properties that are to characterize each soil unit is determined. Second, each unit is given a name, such as Cecil clay, Barnes loam, or Miami silt loam. In this way, when a given soil unit is discussed by name, its identity and characteristics are known by the soil scientist just as botanists do when plants are designated by their species names.

To develop a useful classification system and to establish the kinds and

FIGURE 13.6 A schematic diagram to illustrate the concept of pedon and of the soil profile that characterizes it. Note that several contiguous pedons with similar characteristics are grouped together in a larger area (outlined by broken lines) called a polypedon or individual soil. Several individual soils are present in this landscape.

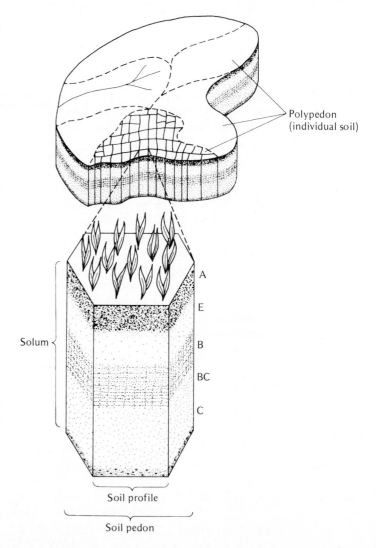

ranges of properties that are to characterize given soil units, the soil must be studied in the field. Even after tentative classes have been established, they must be tested in the field to be certain of their utility in a classification scheme.

Pedon. It is obvious that the whole soil or even large areas of it cannot be studied at one time. Therefore, small three-dimensional bodies of soil are used. Such a unit is large enough so that the nature of its horizons can be studied and the range of its properties identified. It varies in size from about 1 to 10 square meters and is called a *pedon* (rhymes with "head on"; from the Greek *pedon*, ground). It is a three-dimensional body of soil whose lateral dimensions are large enough to permit study of horizons and of its physical and chemical composition (Figure 13.6).

Because of its very small size, a pedon obviously cannot be used as the basic unit for a workable field soil classification system. However, a group of pedons, termed a *polypedon,* closely associated in the field and similar in their properties are of sufficient size to serve as a basic classification unit, or a *soil individual.* Such a grouping approximates what in the United States has been called a *soil series.* More than 13,000 soil series have been characterized in this country. They are the basic units used in the field classification of the nation's soils.

Two extremes in the concept of soils have now been identified. One extreme is that of a natural body called *a* soil, characterized by a three-dimensional sampling unit (pedon), related groups of which are termed a soil series. At the other extreme is *the* soil, a collection of all these natural bodies that is distinct from water, solid rock, and other natural parts of the earth's crust. These two extremes represent opposite ends of elaborate soil classification schemes that are used to organize knowledge of soils.

13.5   Soil Classification in the United States

The history of the classification of soils suggests three kinds of concepts: (a) soils as a habitat for crops, (b) soils as a superficial mantle of weathered rock, and (c) soils as natural bodies. The first of these, because of its relation to practical agriculture, is the oldest. From the time man first began cultivating his own crops he noticed differences in the productive capacity of soils and classified them, if only in terms of "good" and "bad" for his purposes. The early Chinese, Egyptian, Greek, and Roman civilizations acknowledged differences in soils as media for plant growth. Such recognition is common even today as soils are described as being good "cotton," "soybean," "alfalfa," "wheat," or "rice" soils.

The concept of soils as the uppermost mantle of weathered rock may too have been in the minds of our ancient ancestors. However, it was not until geology became a science in the late eighteenth and early nineteenth centuries that this concept became prominent. While in no way replacing the "plant media" concept, it brought into being such terms of classification as "sandy" and "clayey" soils as well as "limestone" soils and "lake-laid" soils. These terms have a geological connotation.

The Russian soil scientist V. V. Dokuchaev and his associates were the first to develop the concept of soils as natural bodies that were subject to classification. Further, they noted a definite relationship among climate, vegetation, and soil characteristics. Dokuchaev's monograph in 1883 developed this thesis. Some of the terms used until recent times in the United States to describe soils were of Russian origin, examples being *Chernozem* (black earth) and *Podzol* (under ash).

The first organized attempts in the United States to classify soils and to map them in the field began in 1898 and were dominated by the crop habitat and geological concepts. The practical agricultural significance of soils and their classification seemed to have had highest priority, but the study of soils in the field forced a partial geological bias as well. Although the natural body concept as developed by the Russians was known to a few American scientists late in the nineteenth century, it was not until the early part of the twentieth century that it received attention. C. F. Marbut of the U.S. Department of Agriculture promoted this concept along with the importance of the soil profile in identifying the natural bodies. His soil classification scheme, presented first in 1927 and in more detail in 1935, was based on the concept of soils as natural bodies.

With the background laid by Marbut, scientists of the U.S. Department of Agriculture developed in 1938 (and improved in 1949) a system of soil classification based on the natural body concept. This system also had a strong genetic bias, that is, the classification of a given soil was dependent to a considerable degree on the supposed processes by which it had formed. The difficulty of ascertaining for certain which genetic processes were responsible for a given soil was one of the weaknesses of this system, although it served very well for a period of about 25 years.

The Soil Survey Staff of the U.S. Department of Agriculture, working for years with cooperating soil scientists in other countries, developed a comprehensive system of soil classification that was adopted in the United States in 1965 and is used to at least some degree by scientists in 31 other countries (Cline, 1980). This system will be used in this text.

13.6   New Comprehensive Classification System— Soil Taxonomy

The latest comprehensive soil classification system, called *Soil Taxonomy*, (Soil Survey Staff, 1975) maintains the natural body concept and has two other major features that make it most useful. First, the primary bases for identifying different classes in the system are the properties of soils, properties that can

PLATE I. Photographs of profiles of each soil order in *Soil Taxonomy*. [*Used with the permission of the Soil Science Society of America.*]

1. Alfisols—a Typic Hapludalf from southern Michigan.
2. Aridisols—a Typic Haplargid from central Arizona.

   [*Overleaf*]
3. Entisols—a Typic Udifluvent from southwest Wisconsin.
4. Histosols—a Limnic Medisaprist from southern Michigan.
5. Inceptisols—a Typic Dystrochrept from West Virginia.
6. Mollisols—a Typic Argiudoll from central Iowa.
7. Oxosols—a Tropeptic Haplorthox from central Puerto Rico.
8. Spodosols—a Typic Haplorthod from northern New York.
9. Ultisols—a Typic Hapludult from western Arkansas.
10. Vertisols—a Udic Chromustert from Lajas Valley, Puerto Rico.

7

9

8

10

PLATE II. Landsat satellites have provided valuable remote sensing information on the earth's soils and other natural resources. In the *upper* photo of an area in Tippecanoe County, Indiana, a false color composite of three spectral bands (two visible, one infrared) is used to show wooded areas (red), cultivated soils developed under prairie vegetation (dark), and cultivated soils developed under forest vegetation (light areas). The *lower* photo shows an area along the Chester River in Queen Annes County, Maryland. The modern thematic mapper (TM) scanner used provided reasonably good detail, and a false color composite of three spectral bands showing wooded areas in brown and red and areas with little soil cover in green and blue. Soil differences are obvious in some fields. [*Upper photo courtesy M. F. Baumgardner, Purdue University; lower photo courtesy Goddard Space Flight Center, NASA.*]

be measured quantitatively either in the field or in the laboratory. Furthermore, the measurements so obtained can be verified by others. This lessens the likelihood of controversy over the place of a given soil in the classification system—which is common when scientists deal with systems where genesis or presumed genesis is the basis for the classification.

The second significant feature of *Soil Taxonomy* is the nomenclature employed, especially for the broader classification categories. The names give a definite connotation of the major characteristics of the soils in question—a connotation easily understood in many languages since Latin or Greek root words are the bases for the names. Consideration will be given to the nomenclature used after brief reference is made to the major criteria for the system—soil properties.

Bases of Soil Classification. *Soil Taxonomy* is based on the properties of soils as they are found today. While one of the objectives of the system is to group soils similar in genesis, the specific criteria used to place soils in these groups are those of soil properties. The advantage of this system over those based primarily on soil genesis or presumed soil genesis (Smith, 1963) are as follows.

1. It permits classification of soils rather than soil-forming processes.
2. It focuses on the soil rather than related sciences such as geology and climatology.
3. It permits the classification of soils of unknown genesis—only the knowledge of their soil properties is needed.
4. It permits greater uniformity of classification as applied by a large number of soil scientists. Differences in interpretation of how a soil was formed do not influence its classification under this scheme.

It should not be assumed that *Soil Taxonomy* ignores soil genesis. Since soil properties, the basis for the new system, are often related directly to soil genesis, it is difficult to emphasize soil properties without at least indirectly emphasizing soil genesis as well.

It is not possible to generalize with respect to the kinds of soil properties used as criteria for *Soil Taxonomy*. All of the chemical, physical, and biological properties presented in this text are used in this classification scheme. A few examples of the criteria used include moisture, temperature, color, texture, and structure of the soil. Chemical and mineral properties such as the contents of organic matter, clay, iron and aluminum oxides, and salts, the pH, the percentage base saturation, and soil depth are other important criteria for classification.

Among the most significant of the properties used as a basis for classification is the presence or absence of certain diagnostic soil horizons. Because of their prominence in helping to determine the place of a soil in the classification system, they will be given somewhat more detailed attention.

Diagnostic Horizons. To illustrate that soil properties are the primary criteria for classifying soil under *Soil Taxonomy*, brief mention will be made of certain *diagnostic* surface and subsurface horizons. The diagnostic surface horizons are called *epipedons* (from the Greek *epi*, over, and *pedon*, soil). The epipedon includes the upper part of the soil darkened by organic matter, the upper eluvial horizons, or both. It may include part of the B horizon (see Section 13.3) if the latter is significantly darkened by organic matter. Six epipedons are recognized but only four are of any importance in the soils of the United States. The other two, called *anthropic* and *plaggen,* are the result of man's intensive use of soils. They are found in parts of Europe and probably Asia. The major features of the six surface horizons are given in Table 13.1.

Many subsurface horizons characterize different soils in the system. Those that are considered diagnostic horizons are shown along with their major features in Table 13.1. Each of these layers can be used as a distinctive property to help place a soil in its proper class.

TABLE 13.1 Major Features of Diagnostic Horizons in Mineral Soils Used to Differentiate at the Higher Levels of *Soil Taxonomy*

Diagnostic horizon (and designation)	Major feature
Surface horizons = epipedons	
Mollic (A)	Thick, dark colored, high base saturation, strong structure
Umbric (A)	Same as Mollic except low base saturation
Ochric (A)	Light colored, low organic content, may be hard and massive when dry
Histic (O)	Very high in organic content, wet during some part of year
Anthropic (A)	Man-modified Mollic-like horizon, high in available P
Plaggen (A)	Man-made sod-like horizon created by years of manuring
Subsurface horizons	
Argillic (Bt)	Silicate clay accumulation
Natric (Btn)	Argillic, high in sodium, columnar or prismatic structure
Spodic (Bhs)	Organic matter, Fe and Al oxide accumulation
Cambic (B)	Changed or altered by physical movement or by chemical reactions
Agric (A or B)	Organic and clay accumulation just below plow layer resulting from cultivation
Oxic (Bo)	Highly weathered, primarily mixture of Fe, Al oxides, and 1:1-type minerals
Duripan (m)	Hard pan, strongly cemented by silica
Fragipan (x)	Brittle pan, usually loamy textured, weakly cemented
Albic (E)	Light colored, clay and Fe, Al oxides mostly removed
Calcic (k)	Accumulation of $CaCO_3$ or $CaCO_3 \cdot MgCO_3$
Gypsic (y)	Accumulation of gypsum
Salic (z)	Accumulation of salts

TABLE 13.2 Comparison of the Classification of a Common Cultivated Plant, White Clover (*Trifolium repens*), and a Soil, Miami Series

Plant classification			Soil classification	
Phylum	Pterophyta		Order	Alfisols
Class	Angiospermae		Suborder	Udalfs
Subclass	Dicotyledoneae	Increasing specificity	Great Group	Hapludalfs
Order	Rosales		Subgroup	Typic Hapludalfs
Family	Leguminosae		Family	Fine loamy, mixed, mesic
Genus	*Trifolium*		Series	Miami
Species	*repens*		Phase[a]	Miami, eroded phase

[a] Technically not a class in *Soil Taxonomy* but used in field surveying.

Categories of the System. There are six categories of classification in *Soil Taxonomy:* (a) order (the broadest category), (b) suborder, (c) great group, (d) subgroup, (e) family, and (f) series (the most specific category). These categories may be compared with those used for the classification of plants. The comparison would be as shown in Table 13.2, where white clover (*Trifolium repens*) and Miami silt loam are the examples of plants and soils, respectively.

Just as *Trifolium repens* identifies a specific kind of plant, the Miami series identifies a specific kind of soil. The similarity continues up the classification scale to the highest categories—that of phylum for plants and of order for soils. With this general background a brief description of the six soil categories and the nomenclature used in identifying them is presented.

The *order* category is based largely on soil-forming processes as indicated by the presence or absence of major diagnostic horizons. A given order includes soils whose properties suggest that they are not too dissimilar in their genesis. As an example, soils developed under grassland vegetation have the same general sequence of horizons and are characterized by a thick, dark, epipedon (surface horizon) high in bases. They are thought to have been formed by the same general genetic processes and are thereby usually included in the same order, Mollisol. There are 10 soil orders in *Soil Taxonomy* (Table 13.3).

Suborders are subdivisions of orders that emphasize genetic homogeneity. Thus, wetness, climatic environment, and vegetation are characteristics that help determine the suborder in which a given soil is found. Some 47 soil suborders are recognized in the United States.

Diagnostic horizons (Section 13.6) are used to differentiate the *great groups* in a given suborder. Soils in a given great group are thought to have the same kind and arrangement of these horizons. More than 230 great groups are recognized in the United States.

Subgroups are subdivisions of the great groups. The central concept of a great group makes up one subgroup (Typic). Other subgroups may have characteristics that are intergrades between those of the central concept and

TABLE 13.3 Names of Soil Orders in *Soil Taxonomy* with Their Derivation and Major Characteristics

Formative element

Name	Derivation	Pronunciation	Major characteristics
Entisols	Nonsense symbol	rec*ent*	Little profile development, ochric epipedon common
Inceptisols	L. *inceptum,* beginning	inc*ep*tion	Embryonic soils of humid regions, Cambic horizon, few diagnostic features
Mollisols	L. *mollis,* soft	*moll*ify	Mollic epipedon, high base saturation, dark soils of grasslands
Alfisols	Nonsense symbol	Ped*alf*er	Argillic or natric horizon; high to medium base saturation, forest soil
Ultisols	L. *ultimus,* last	*ult*imate	Argillic horizon, low base saturation, forest soil
Oxisols	Fr. *oxide,* oxide	*ox*ide	Oxic horizon, no argillic horizon, highly weathered
Vertisols	L. *verto,* turn	in*vert*	High in dark swelling clays; deep cracks when soil dry
Aridisols	L. *aridus,* dry	Ar*id*	Dry soil, Ochric epipedon, sometimes argillic or natric horizon
Spodosols	Gk. *Spodos,* wood ash	*Pod*zol; odd	Spodic horizon with Fe, Al, and humus accumulation
Histosols	Gk. *Histos,* tissue	*Hist*ology	Peat or bog; >20% organic matter

those of other orders, suborders, or great groups. More than 1200 subgroups are recognized, about 1000 of which are found in the United States.

In the *family* category are found soils with a subgroup having similar physical and chemical properties affecting their response to management and especially to the penetration of plant roots. Differences in texture, mineralogy, temperature, and soil depth are bases for family differentiation.

The *series* category is essentially the same as has been used for years in the United States. Ideally, it includes only one polypedon, but in the field it may include parts of other contiguous and closely related polypedons. The series is a subdivision of the family. Its differentiating characteristics are based primarily on the kind and arrangement of horizons. There are about 13,000 soils series recognized in the United States (Figure 13.7).

Nomenclature.   An important feature of the system is the nomenclature used to identify different soil classes. The names of the classification units are combi-

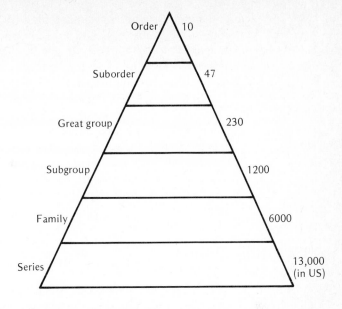

FIGURE 13.7 The categories of *Soil Taxonomy* and approximate number of units in each category. [*Data from USDA, personal correspondence.*]

Order — 10
Suborder — 47
Great group — 230
Subgroup — 1200
Family — 6000
Series — 13,000 (in US)

nations of syllables, most of which are derived from Latin or Greek and are root words in several modern languages. Since each part of a soil name conveys a concept of soil character or genesis, the name automatically describes the general kind of soil being classified. For example, soils of the order *Aridisol* (Latin *aridus*, dry, and *solum*, soil) are characteristic of arid or dry places. Those of the order *Inceptisols* (Latin *inceptum*, beginning, and *solum*, soil) are soils with only the beginnings of profile development. Thus, the names of orders are combinations of (a) formative elements, which generally define the characteristics of the soils, and (b) the ending *sol*.

The names of suborders automatically identify the order of which they are a part. For example, soils of the suborder *Aquolls* are the wetter soils (Latin *aqua*, water) of the Mollisol order. Likewise, the name of the great group identifies the suborder and order of which it is a part. *Argiaquolls* are Aquolls with clay or argillic (Latin *argilla*, white clay) horizons.

The nomenclature as it relates to the different categories in the classification system might be illustrated as follows:

Mollisol———order
Aquoll————suborder
Argiaquoll————great group
Typic Argiaquoll————subgroup

Note that the three letters *oll* identify each of the lower categories as being in the Mollisol order. Likewise, the suborder name *Aquoll* is included as part of the great group and subgroup name. Given only the subgroup name, the great group, suborder, and order to which the soil belongs are automatically known.

Family names in general identify groups of soil series similar in texture, mineral composition, and in soil temperature at a depth of 50 cm. Thus the name *Typic Argiaquoll, fine, mixed, mesic,* applies to a family in the *Typic Argiaquoll* subgroup with a fine texture, mixed clay mineral content, and mesic (8–15°C) soil temperature.

Soil series names have local significance since they normally identify the particular locale in which the soil is found. Thus, names such as Fort Collins, Cecil, Miami, Norfolk, and Ontario are used to identify the soil series. When the textural class name of the surface horizon is added to that of the series, a soil phase name has been identified. Fort Collins loam and Cecil clay are examples. Likewise, Cecil, eroded phase, is a Cecil series that has been eroded.

With this brief explanation of the nomenclature of the new system, the order category of the system will now be considered.

13.7 Soil Orders

Ten orders are recognized. With the exception of one order (Entisols), which roughly corresponds to the azonal soils of the 1949 system, the new orders bear little resemblance to those formerly used. The names of these orders and their major characteristics are shown in Table 13.3. Note that all order names have a common end, *sol* (from the Latin *solum,* soil).

Figure 13.8 is a general soil map showing the distribution of soil orders of the United States. Similarly, Figure 13.9 is a general soil map of the world. Profiles of each soil order are shown in color on Plate I (following page 430).

Entisols (little if any profile development). These are mineral soils without natural genetic horizons or with only the beginnings of such horizons. Some have an ochric epipedon and a few man-made anthropic or agric epipedons. The extremes of highly productive soils on recent alluvium and infertile soils on barren sands, as well as shallow soils on bedrock, are included. The common characteristic of all Entisols is lack of significant profile development.

Soils of this order are found under a wide variety of climatic conditions in the United States (Figure 13.8). For example, in the Rocky Mountain region and in southwest Texas, shallow, medium-textured Entisols (Orthents) over hard rock are common. They are used mostly as rangeland. Sandy Entisols (Psamments) are found in Florida, Alabama, and Georgia and typify the sand hill section of Nebraska. Psamments are used mostly for grazing in the drier climates. They may be forested or used for cropland in humid areas. Some of the citrus-, vegetable-, and peanut-producing areas of the South are typified by Psamments.

Entisols are probably found under even more widely varied climatic conditions outside the United States (Figure 13.9). Psamments are typical of the shifting sands of the Sahara Desert and Saudi Arabia. Large areas of Psamments dominate parts of southern Africa and of central and north central Australia.

BOROLLS

M2a Udic subgroups of Borolls with Aquolls and Ustorthents; gently sloping.

M2b Typic subgroups of Borolls with Ustipsamments, Ustorthents, and Boralfs; gently sloping.

M2c Aridic subgroups of Borolls with Borollic subgroups of Argids and Orthids, and Torriorthents; gently sloping.

M2S Borolls with Boralfs, Argids, Torriorthents, and Ustolls; moderately sloping or steep.

UDOLLS

M3a Udolls, with Aquolls, Udalfs, Aqualfs, Fluvents, Psamments, Ustorthents, Aquepts, and Albolls; gently or moderately sloping.

USTOLLS

M4a Udic subgroups of Ustolls with Orthents, Ustochrepts, Usterts, Aquents, Fluvents, and Udolls; gently or moderately sloping.

M4b Typic subgroups of Ustolls with Ustalfs, Ustipsamments, Ustorthents, Ustochrepts, Aquolls, and Usterts; gently or moderately sloping.

M4c Aridic subgroups of Ustolls with Ustalfs, Orthids, Ustipsamments, Ustorthents, Ustochrepts, Torriorthents, Borolls, Ustolls, and Usterts, gently or moderately sloping.

M4S Ustolls with Argids and Torriorthents; moderately sloping or steep.

XEROLLS

M5a Xerolls with Argids, Orthids, Fluvents, Cryoboralfs, Cryoborolls, and Xerorthents; gently or moderately sloping.

M5S Xerolls with Cryoboralfs, Xeralfs, Xerorthents, and Xererts; moderately sloping or steep.

SPODOSOLS

AQUODS

S1a Aquods with Psammaquents, Aquolls, Humods, and Aquults; gently sloping.

ORTHODS

S2a Orthods with Boralfs, Aquents, Orthents, Psamments, Histosols, Aquepts, Fragiochrepts, and Dystrochrepts; gently or moderately sloping.

S2S1 Orthods with Histosols, Aquents, and Aquepts; moderately sloping or steep.

S2S2 Cryorthods with Histosols; moderately sloping or steep.

S2S3 Cryorthods with Histosols, Andepts and Aquepts; gently sloping to steep.

ULTISOLS

AQUULTS

U1a Aquults with Aquents, Histosols, Quartzipsamments, and Udults; gently sloping.

HUMULTS

U2S Humults with Andepts, Tropepts, Xerolls, Ustolls, Orthox, Torrox, and Rock land; gently sloping to steep.

UDULTS

U3a Udults with Udalfs, Fluvents, Aquents, Quartzipsamments, Aquepts, Dystrochrepts, and Aquults; gently or moderately sloping.

U3S Udults with Dystrochrepts; moderately sloping or steep.

VERTISOLS

UDERTS

V1a Uderts with Aqualfs, Eutrochrepts, Aquolls, and Ustolls, gently sloping.

USTERTS

V2a Usterts with Aqualfs, Orthids, Udifluvents, Aquolls, Ustolls, and Torrerts; gently sloping.

Areas with Little Soil

X1 Salt flats.

X2 Rock land (plus permanent snow fields and glaciers).

Slope Classes

Gently sloping—Slopes mainly less than 10%, including nearly level.

Moderately sloping—Slopes mainly between 10 and 25%.

Steep—Slopes mainly steeper than 25%.

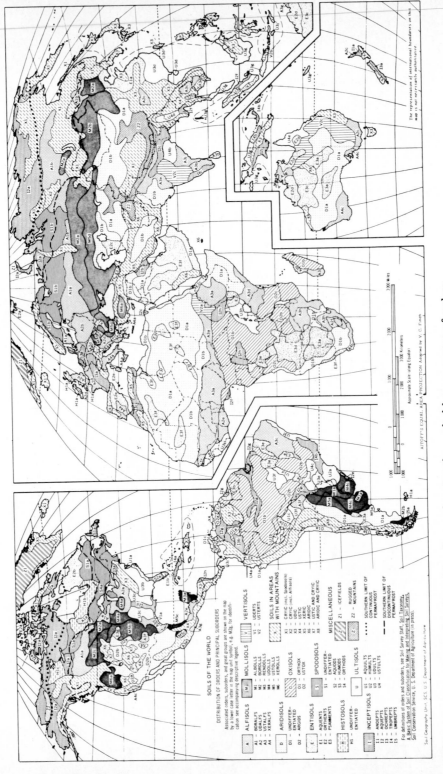

FIGURE 13.9 Generalized world soil map showing the probable occurrence of orders and suborders according to *Soil Taxonomy*. [*Courtesy USDA Soil Conservation Service, Soil Survey Staff.*]

A **ALFISOLS**—soils with argillic or natric horizon, medium-high bases; forest areas.

A1 BORALFS—cold.
  A1a—with Histosols, cryic temperature regimes common.
  A1b—with Spodosols, cryic temperature regimes.

A2 UDALFS—temperate to hot, usually moist.
  A2a—with Aqualfs.
  A2b—with Aquolls.
  A2c—with Hapludults.
  A2d—with Ochrepts.
  A2e—with Troporthents.
  A2f—with Udorthents.

A3 USTALFS—temperate to hot, dry more than 90 cumulative days during periods when temperature is suitable for plant growth.
  A3a—with Tropepts.
  A3b—with Troporthents.
  A3c—with Tropustults.
  A3d—with Usterts.
  A3e—with Ustochrepts.
  A3f—with Ustolls.
  A3g—with Ustorthents.
  A3h—with Ustox.
  A3j—Plinthustalfs with Ustorthents.

A4 XERALFS—temperate or warm, moist in winter and dry more than 45 consecutive days in summer.
  A4a—with Xerochrepts.
  A4b—with Xerorthents.
  A4c—with Xerults.

D **ARIDISOLS**—soils of dry areas.

D1 ARIDISOLS—undifferentiated.
  D1a—with Orthents.
  D1b—with Psamments.
  D1c—with Ustalfs.

D2 ARGIDS—with horizons of clay accumulation.
  D2a—with Fluvents
  D2b—with Torriorthents.

E **ENTISOLS**—soils without pedogenic horizons.

E1 AQUENTS—seasonally or perennially wet.
  E1a—Haplaquents with Udifluvents.
  E1b—Psammaquents with Haplaquents.
  E1c—Tropaquents with Hydraquents.

E2 ORTHENTS—loamy or clayey textures, many shallow to rock.
  E2a—Cryorthents.
  E2b—Cryorthents with Orthods.
  E2c—Torriorthents with Aridisols.
  E2d—Torriorthents with Ustalfs.
  E2e—Xerorthents with Xeralfs.

E3 PSAMMENTS—sand or loamy sand textures.
  E3a—with Aridisols.
  E3b—with Orthox.
  E3c—with Torriorthents
  E3d—with Ustalfs.
  E3e—with Ustox.
  E3f—shifting sands.
  E3g—Ustipsamments with Ustolls.

H **HISTOSOLS**—organic soils.

H1 HISTOSOLS—undifferentiated.
  H1a—with Aquods.
  H1b—with Boralfs.
  H1c—with Cryaquepts.

I **INCEPTISOLS**—soils with few diagnostic features.

I1 ANDEPTS—amorphous clay or vitric volcanic ash or pumice.
  I1a—Dystrandepts with Ochrepts.

I2 AQUEPTS—seasonally wet.
  I2a—Cryaquepts with Orthents.
  I2b—Halaquepts with Salorthids.
  I2c—Halaquepts with Humaquepts.
  I2d—Haplaquepts with Ochraqualfs.
  I2e—Humaquepts with Psamments.
  I2f—Tropaquepts with Hydraquents.
  I2g—Tropaquepts with Plinthaquults.
  I2h—Tropaquepts with Tropaquents.
  I2j—Tropaquepts with Tropudults.

I3 OCHREPTS—thin, light-colored surface horizons and little organic matter.
  I3a—Dystrochrepts with Gragiochrepts.
  I3b—Dystrochrepts with Orthox.
  I3c—Xerochrepts with Xerolls.

I4 TROPEPTS—continuously warm or hot.
  I4a—with Ustalfs.
  I4b—with Tropudults.
  I4c—with Ustox.

I5 UMBREPTS—dark-colored surface horizons with medium to low base supply.
  I5a—with Aqualfs.

M **MOLLISOLS**—soils with nearly black, organic-rich surface horizons and high base supply.

M1 ALBOLLS—light gray subsurface horizon over slowly permeable horizon; seasonally wet.
  M1a—with Aquepts.

M2 BOROLLS—cold.
  M2a—with Aquolls.
  M2b—with Orthids.
  M2c—with Torriorthents.

M3 RENDOLLS—subsurface horizons with much calcium carbonate but no accumulation of clay.

Figure 13.9 (Cont.)

M4 UDOLLS—temperate or warm, usually moist.
   M4a—with Aquolls.
   M4b—with Eutrochrepts.
   M4c—with Humaquepts.

M5 USTOLLS—temperate to hot, dry more than 90 cumulative days in year.
   M5a—with Argialbolls.
   M5b—with Ustalfs.
   M5c—with Usterts.
   M5d—with Ustochrepts.

M6 XEROLLS—cool to warm, moist in winter and dry more than 45 consecutive days in summer.
   M6a—with Xerorthents.

O **OXISOLS**—highly weathered soils with oxic horizon.

O1 ORTHOX—hot, nearly always moist.
   O1a—with Plinthaquults.
   O1b—with Tropudults.

O2 USTOX—warm or hot, dry for long periods but moist more than 90 consecutive days in the year.
   O2a—with Plinthaquults.
   O2b—with Tropustults.
   O2c—with Ustalfs.

S **SPODOSOLS**—soils of forest areas with spodic horizon, low bases.

S1 SPODOSOLS—undifferentiated.
   S1a—cryic temperature regimes; with Boralfs.
   S1b—cryic temperature regimes; with Histosols.

S2 AQUODS—seasonally wet.
   S2a—Haplaquods with Quartzipsamments.

S3 HUMODS—with accumulations of organic matter in subsurface horizons.
   S3a—with Hapludalfs.

S4 ORTHODS—with accumulations of organic matter, iron, and aluminum in subsurface horizons.
   S4a—Haplorthods with Boralfs.

U **ULTISOLS**—soils in forest areas with subsurface horizons of clay accumulation and low base supply.

U1 AQUULTS—seasonally wet.
   U1a—Ochraquults with Udults.
   U1b—Plinthaquults with Orthox.
   U1c—Plinthaquults with Plinthaquox.
   U1d—Plinthaquults with Tropaquepts.

U2 HUMULTS—temperate or warm and moist all of year; high content of organic matter.
   U2a—with Umbrepts.

U3 UDULTS—temperate to hot; never dry more than 90 cumulative days in the year.
   U3a—with Andepts.
   U3b—with Dystrochrepts.
   U3c—with Udalfs.
   U3d—Hapludults with Dystrochrepts.
   U3e—Rhodudults with Udalfs.
   U3f—Tropudults with Aquults.
   U3g—Tropudults with Hydraquents.
   U3h—Tropudults with Orthox.
   U3j—Tropudults with Tropepts.
   U3k—Tropudults with Tropudalfs.

U4 USTULTS—warm or hot; dry more than 90 cumulative days in the year.
   U4a—with Ustochrepts.
   U4b—Plinthustults with Ustorthents.
   U4c—Rhodustults with Ustalfs.
   U4d—Tropustults with Tropaquepts.
   U4e—Tropustults with Ustalfs.

V **VERTISOLS**—soils with high content of swelling clays; deep, wide cracks develop during dry periods.

V1 UDERTS—usually moist in some part of most years; cracks open less than 90 cumulative days in the year.
   V1a—with Usterts.

V2 USTERTS—cracks open more than 90 cumulative days in the year.
   V2a—with Tropaquepts.
   V2b—with Tropofluvents.
   V2c—with Ustalfs.

X **Soils in areas with mountains**—Soils with various moisture and temperature regimes; many steep slopes; relief and total elevation vary greatly within short distances and with changes in altitude; vertical zonation common.

X1 Cryic great groups of Entisols, Inceptisols, and Spodosols.
X2 Boralfs and cryic great groups of Entisols and Inceptisols.
X3 Udic great groups of Alfisols, Entisols, Inceptisols, and Ultisols.
X4 Ustic great groups of Alfisols, Inceptisols, Mollisols, and Ultisols.
X5 Xeric great groups of Alfisols, Entisols, Inceptisols, Mollisols, and Ultisols.
X6 Torric great groups of Aridisols and Entisols.
X7 Ustic and cryic great groups of Alfisols, Entisols, Inceptisols, and Mollisols; ustic great groups of Ultisols; cryic great groups of Spodosols.
X8 Aridisols, torric and cryic great groups of Entisols, and cryic great groups of Spodosols and Inceptisols.

Z **MISCELLANEOUS**
Z1 Icefields.
Z2 Rugged mountains—mostly devoid of soil (includes glaciers, permanent snowfields, and, in some places, areas of soil).

Entisols having medium to fine textures (Orthents) are found in northern Quebec and parts of Alaska, Siberia, and Tibet. Orthents are typical of some mountain areas such as the Andes in South America and some of the uplands of an area extending from Turkey eastward to Pakistan.

As might be expected, the agricultural productivity of the Entisols varies, greatly depending on their location and properties. When adequately fertilized and when their water supply is controlled, some of these soils are quite productive. However, restrictions on their depth, clay content, or water balance limit the intensive use of large areas of these soils.

Inceptisols (few diagnostic features). The profile development of soils in this order is more advanced than that of the Entisol order but less so than that of the other orders. Inceptisol profiles contain horizons that are thought to form rather quickly and result mostly from alteration of parent materials. The horizons do not represent extreme weathering. Horizons of marked accumulation of clay, and iron and aluminum oxides are absent in this order. Many agriculturally useful soils are included along with others whose productivity is limited by factors such as imperfect drainage.

Inceptisols are found in the United States and on each of the continents (Figures 13.8 and 13.9). For example, some Inceptisols called Andepts (productive soils developed from volcanic ash) are found in a sizable area of Oregon, Washington, Idaho, and in Ecuador and Colombia in South America. Some called Ochrepts (Greek *ochros*, pale), with thin, light-colored surface horizons, extend from southern New York through central and western Pennsylvania, West Virginia, and eastern Ohio. Ochrepts dominate an area extending from southern Spain through central France to central Germany. They are also present in Chile, North Africa, eastern China, and western Siberia.

Tropepts (Inceptisols of tropical regions) are found in northwestern Australia, central Africa, southwestern India, and in southwestern Brazil. Areas of wet Inceptisols or Aquepts (Latin *aqua*, water) are found along the Amazon and Ganges rivers.

As might be expected, there is considerable variability in the natural productivity of Inceptisols. For instance, those found in the Pacific Northwest are quite fertile and provide us with some of our best wheat lands. In contrast, some of the low-organic-containing Ochrepts in southern New York and northern Pennsylvania are not naturally productive. They have been allowed to reforest following earlier periods of crop production.

Mollisols (dark soils of grasslands, mollic epipedon, base rich). This order includes some of the world's most important agricultural soils. They are characterized by a *mollic epipedon,* or surface horizon, which is thick, dark, and dominated by divalent cations (Table 13.3). They may have an argillic (clay), natric, albic, or cambic horizon but not an oxic or spodic one. The surface horizons generally have granular or crumb structures and are not hard when the soils are dry. This justifies the use of a name which implies softness.

Most of the Mollisols have developed under prairie vegetation (Figure 13.10). Grassland soils of the central part of the United States, lying between the Aridisols on the West and Alfisols or Spodosols on the east make up the central core of this order. However, some soils developed under forest vegetation have a mollic epidedon and are included among the Mollisols. They, along

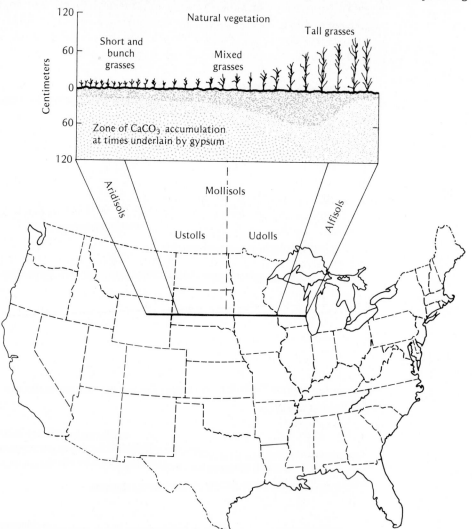

FIGURE 13.10 Correlation between natural grassland vegetation and certain soil orders is graphically shown for a strip of territory in north central United States. The control, of course, is climate. Note the deeper organic matter layer and deeper zone of calcium accumulation as one proceeds from the drier areas in the west toward the more humid region where prairie soils are found. Alfisols develop under grassland vegetation, but more commonly occur under forests and have lighter-colored surface horizons.

with the grassland soils that dominate this order, are among the more productive cultivated soils of the world. The temperature and climate conditions in areas where Mollisols are found are shown in Figure 13.11.

Mollisols are dominant in the Great Plains states (Figure 13.8). Those in the eastern and more humid part of this region are called Udolls (Latin *udus*, humid). A region extending from North Dakota to southern Texas is characterized by Ustolls (Latin *ustus*, burnt), which are intermittently dry during the summer. Farther west in parts of Idaho, Utah, Washington, and Oregon are found sizable areas of Xerolls (Greek *xeros*, dry), which are the driest of the

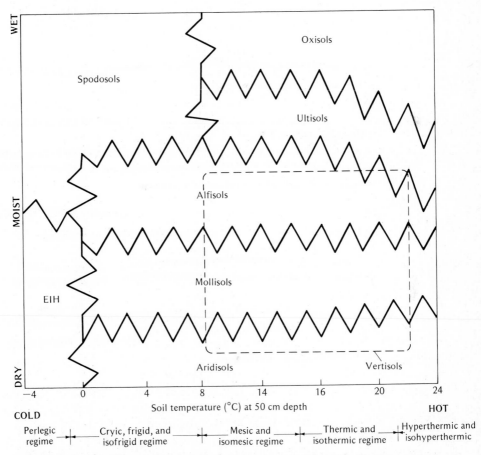

FIGURE 13.11 Diagram showing the climatic (rainfall and soil temperature) conditions under which seven of the soil orders are found. The other three (Entisols, Inceptisols, and Histosols) may be found under any of the climate and temperature conditions shown (including the area labeled EIH). Vertisols are found only where clayey materials are in abundance and where the climate and temperature conditions inside the box with broken lines prevail.

Mollisols. This soil order characterizes a larger land area in the United States than any other soil order (see Table 13.4).

The largest area of Mollisols outside the United States is that stretching from east to west across the heartland of the Soviet Union (Figure 13.9). Other sizable areas are found in Mongolia and northern China and in northern Argentina, Paraguay, and Uruguay.

The native fertility of Mollisols dictates that they be rated among the world's best soil. When first cleared for cultivation, their high native organic matter released sufficient nitrogen and other nutrients to produce high crop yields even without fertilization. Yields on these soils were unsurpassed by other unirrigated areas. Even today, when moderate to heavy fertilization increases the productivity of naturally infertile soils in more humid areas, Mollisols still rate among the best. Photographs of profiles of two Mollisols are shown in Figure 13.12.

Alfisols (argillic or natric horizon, medium–high bases, forest soils. Alfisols are moist mineral soils having no mollic epipedon or oxic or spodic horizons. They have gray to brown surface horizons (commonly an Ochric epipedon), medium to high base status, and contain an illuvial horizon in which silicate clays have accumulated. The clay horizon is generally more than 35% base saturated. This horizon is termed *argillic* if only silicate clays are present and *natric* if, in addition to the clay, it is more than 15% saturated with sodium and has prismatic or columnar structure.

The Alfisols appear to be more strongly weathered than the Inceptisols but less so than the Spodosols. They are formed mostly in cool to hot humid region areas (Figure 13.11) under native deciduous forests, although grass is the native vegetation in some cases.

Some of our best agricultural soils are in the Alfisol order. It is typified by Udalfs (Latin *udus,* humid) in Ohio, Indiana, Michigan, Wisconsin, Minnesota, Pennsylvania, and New York. It includes sizable areas of Xeralfs (Greek *xeros,* dry) in central California; some cold-climate Boralfs (Greek *boreas,* northern) in the Rockies; Ustalfs (Latin *ustus,* burnt) in areas of hot summers, including Texas and New Mexico; and wet Alfisols, Aqualfs (Latin *aqua,* water), in parts of the midwest.

Alfisols occur in other countries having climatic environments similar to those of areas just mentioned (Figure 13.9). A large area dominated by Boralfs is found in northern Europe stretching from the Baltic States through western Russia. A second large area is found in Siberia. Ustalfs are prominent in the southern half of Africa, eastern Brazil, eastern India, and in southeastern Asia. Large areas of Udalfs are found in central China, in England, France and Central Europe, and southeastern Australia. Xeralfs are prominent in southwestern Australia, Italy, and central Spain.

In general, Alfisols are quite productive soils. Their medium to high base status, generally favorable texture, and location (except for some Xeralfs) in humid and subhumid regions all favor good crop yields. In the United States,

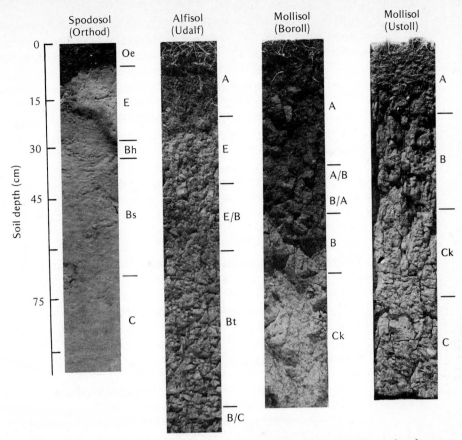

FIGURE 13.12 Monoliths of profiles representing four soil orders. The suborder names are also shown (in parentheses). Note the spodic horizons in the Spodosol characterized by humus (Bh) and iron (Bs) accumulation. In the Alfisol is found the illuvial clay horizon (Bt). The thick dark surface horizon (mollic epipedon) characterizes both Mollisols. Note that the zone of calcium accumulation (Ck) is higher in the Ustoll, which has developed in a dry climate. The E/B horizon in the Alfisol has characteristics of both A and B horizons.

these soils rank favorably with the Mollisols and Ultisols in their productive capacity.

Ultisols (argillic horizon, low bases, forest soils). Ultisols are usually moist soils that develop under warm to tropical climates (Figure 13.11). They are more highly weathered and acidic than the Alfisols but generally are not so acid as the Spodosols. They have argillic (clay) horizons with base saturations lower than 35% and their mean annual temperatures at 50 cm depth are above 8°C. Except for the wetter members of the order, their subsurface horizons are commonly red or yellow in color, evidence of accumulation of free oxides of iron. They still have some weatherable minerals, however, in contrast to the Oxisols. Ultisols are formed on old land surfaces, normally under forest vegetation, although savannah or even swamp vegetation is common.

Most of the soils of the southeastern part of the United States fall in this order. Udults, the moist but not wet Ultisols, extend from the east coast (Maryland to Florida) to and beyond the Mississippi River Valley and are the most extensive of soils in the humid southeast. Humults (high in organic matter) are found in Hawaii, eastern California, Oregon, and Washington. Xerults (Ultisols in Mediterranean-type climates) are found locally in southern Oregon and northern and western California.

Ultisols are prominent on the east and northeast coasts of Australia. Large areas of Udults are located in southeastern Asia and in southern China. Important areas are also found in southern Brazil and Paraguay.

Although Ultisols are not naturally as fertile as Alfisols or Mollisols, they respond well to good management. They are located mostly in regions of long-growing seasons and of ample moisture for good crop production. Their clays are usually of the 1:1 type along with oxides of iron and aluminum, which assures ready workability. Where adequate chemical fertilizers are applied, these soils are quite productive. In the United States the better Ultisols compete well with Mollisols and Alfisols as first-class agricultural soils.

Oxisols (oxic horizon, highly weathered). These are the most highly weathered soils in the classification system (Figure 13.11). Their most important diagnostic feature is the presence of a deep *oxic* subsurface horizon—a horizon generally very high in clay-size particles dominated by hydrous oxides of iron and aluminum. Weathering and intense leaching have removed a large part of the silica from silicate minerals in this horizon, leaving a high proportion of the oxides of iron and aluminum. Some quartz and 1:1-type silicate clay minerals remain, but the hydrous oxides are dominant. The clay content of these soils is very high, but the clays are of the nonsticky type. The depth of weathering in Oxisols is much greater than for most of the other soils—16 m or more having been observed.

Those soils that in recent years have been termed Latosols and some of those called Ground Water Laterites are included among the Oxisols. They occupy old land surfaces and occur mostly in the tropics. Relatively less is known of the Oxisols than of most of the other soil orders. They occur in large geographic areas, however, and millions of people in the tropics depend upon them for their food and fiber production.

The largest known areas of Oxisols occur in South America and Africa (Figure 13.9). Orthox (oxisols having a short dry season or none) occur in northern Brazil and neighboring countries. An area of Ustox (hot, dry summers) nearly as large occurs in Brazil to the south of the Orthox. Oxisols are found in the southern two thirds of Africa, being located on old land surfaces of this area.

The management of Oxisols presents both problems and opportunities.

Most of them have not been cleared of their native vegetation or used for modern cultivation. They are mostly either still covered with native vegetation or have been tilled by primitive methods. The few instances where modern farming techniques have been used have met with mixed success. Heavy fertilization, especially with phosphorus-rich materials, is required. Deficiencies of micronutrients have also been observed frequently. In some areas torrential rainfall makes cultural practices that leave the soil bare extremely harmful.

Although extensive research will be needed to improve utilization of these soils, experience up to now indicates that their potential for food and fiber production is far in excess of that currently being realized. In both Brazil and central Africa selected areas of these soils have been demonstrated to be high in productivity when they are properly managed.

Vertisols (dark swelling clays). This order of mineral soils is characterized by a high content (>30%) of swelling-type clays to a depth of 1 m, which in dry seasons causes the soils to develop deep, wide cracks. A significant amount of material from the upper part of the profile may slough off into the cracks, giving rise to a partial "inversion" of the soil. This accounts for the term *invert*, which is used to characterize this order in a general way.

In the past, most of these soils have been called *Grumusols*. Because of their excessive shrinking, cracking, and shearing they are generally unstable and present problems when used for building foundations, highway bases, and even for agricultural purposes.

Vertisols are found mostly in subhumid to semiarid climates with moderate to high temperatures (Figure 13.11). There are several small but significant areas of Vertisols in the United States (Figure 13.8). Two are located in humid areas, one in eastern Mississippi and western Alabama and the other along the southeast coast of Texas. These soils are of the Udert (Latin *udus*, humid) suborder because, owing to the moist climate, cracks do not persist for more than three months of the year.

Two Vertisol areas are found in east-central and southern Texas, where the climate is drier. Since cracks persist for more than three months of the year, the soils belong to the Ustert (Latin *ustus*, burnt) suborder, characteristic of areas with hot, dry summers.

In India, Ethiopia, the Sudan, and northern and eastern Australia large areas of Vertisols are found (see Figure 13.9). Smaller areas are found in sub-Saharan Africa and in Mexico, Venezuela, Bolivia, and Paraguay. These latter soils probably are of the Ustert suborder, since dry weather persists long enough for the wide cracks to stay open for periods of three months or longer.

In each of the major Vertisol areas listed above some of the soils are used for crop production. Even so, their very fine texture and their marked shrinking and swelling characteristics make them less suitable for crop production than soils in the surrounding areas. They are sticky and plastic when wet and hard when dry. As they dry out following a rain, the period of time when they can be plowed or otherwise tilled is very short. This is a limiting factor even in the United States, where powerful tractors and other mechanical equipment make it possible to plow, prepare seedbeds, plant, and cultivate very quickly. The limitation becomes even more severe in India and the Sudan, where slow-moving animals or human power are used to till the soil. Not only can the cultivators not perform tillage operations on time, but they are limited to the use of small, near-primitive tillage implements because their animals cannot pull larger equipment through the "heavy" soil.

In spite of their limitations, Vertisols are widely tilled, especially in India and in the Sudan. Sorghum, corn, millet, and cotton are crops commonly grown. Unfortunately, the yields are generally low. Further research and timely soil management are essential if these large soil areas are to help produce the food crops these countries so badly need.

Aridisols (soils of dry areas). These mineral soils are found mostly in dry climates (Figure 13.11). Except where there is ground water or irrigation, the soils layers are dry throughout most of the year. Consequently, they have not been subjected to intensive leaching. They have an ochric epipedon generally light in color and low in organic matter. They may have a horizon of accumulation of calcium carbonate (calcic), gypsum (gypsic), or even more soluble salts (salic). If groundwater is present, conductivity measurements show the presence of soluble salts.

Aridisols include most of the soils of the arid regions of the world, such as those formerly designated as Desert, Reddish Desert, Sierozem, Reddish Brown, and Solonchak (see Table 13.10).

A large area of Aridisols called Argids (Latin *argilla,* white clay), which have a horizon of clay accumulation, occupies much of the southern parts of California, Nevada, Arizona, and central New Mexico. The Argids also extend down into northern Mexico. Smaller areas of Orthids (Aridisols without argillic or natric horizons) are found in several western states.

Vast areas of Aridisols are present in the Sahara desert in Africa, the Gobi and Taklamakan deserts in China, and the Turkestan desert of the Soviet Union. Most of the soils of southern and central Australia are Aridisols, as are those of southern Argentina, southwestern Africa, West Pakistan, and the Middle East countries.

Without irrigation, Aridisols are not suitable for growing cultivated crops. Some areas are used for sheep or goat grazing, but the production per unit area is low. Where irrigation water is available Aridisols can be made most productive. Irrigated valleys in arid areas are among the most productive in the world.

Spodosols (spodic horizon, forests, low bases). Spodosols are mineral soils that have a spodic horizon, a subsurface horizon with an accumulation of organic matter, and oxides of aluminum with or without iron oxides. This illuvial horizon usually occurs under an eluvial horizon, normally an albic horizon, which is light in color and has therefore been described by the term "wood ash."

These soils form mostly on coarse-textured, acid parent materials subject to ready leaching. They occur only in humid climates and are most common where it is cold and temperate (Figure 13.11). Forests are the natural vegetation under which most of these soils have developed.

Species low in metallic ion contents, such as pine trees, seem to encourage the development of Spodosols. As the litter from these low-base species decomposes, strong acidity develops. Percolating water leaches acids down the profile, and the upper horizons succumb to this intense acid leaching. Most minerals except quartz are removed. In the lower horizons, oxides of aluminum and iron as well as organic matter precipitate, thus yielding the interesting Spodosol profiles.

Many of the soils in the northeastern United States, as well as those of northern Michigan and Wisconsin, belong to this order (Figure 13.8). Most of them are Orthods, the "common" Spodosols described above. Some, however, are Aquods since they are seasonally saturated with water and possess characteristics associated with this wetness, such as very high surface organic accumulation, mottling in the albic horizon, and the development of a hard pan (duripan) in the albic horizon. Important areas of Aquods occur in Florida.

The area of Spodosols in the northeast extends up into Canada. Other large areas of this soil order are found in northern Europe and Siberia. Smaller but important areas are found in the southern part of South America and in cool mountainous areas of temperate regions.

Spodosols are not naturally fertile. When properly fertilized, however, these soils can become quite productive. For example, the productive "potato" soils of northern Maine are Spodosols, as are some of the vegetable-producing soils of Florida, Michigan, and Wisconsin. Even so, the low native fertility of most Spodosols makes them uncompetitive for tilled crops. They are covered mostly with forests, the vegetation under which they originally developed.

Histosols (peat or bog). The last of the soil orders includes the organic soils (bog soils) as well as some of the half-bog soils. These soils have developed in a water-saturated environment. Histosols can form in any climate from the equator to permafrost areas so long as water is available. They contain a minimum of 12% organic carbon (by weight) if the clay content is low and 18% if the clay content is more than 50%. In virgin areas, the organic matter retains much of the original plant tissue from. Upon drainage and cultivation of the area, the original plant tissue form tends to disappear.

Less progress has been made on the classification of these soils than of the other orders. However, four suborders have been established and are being tested in the field (see Table 13.5). These soils are of great practical importance in local areas, being among the most productive, especially for vegetable crops. Their characteristics are given more detailed consideration in Chapter 14.

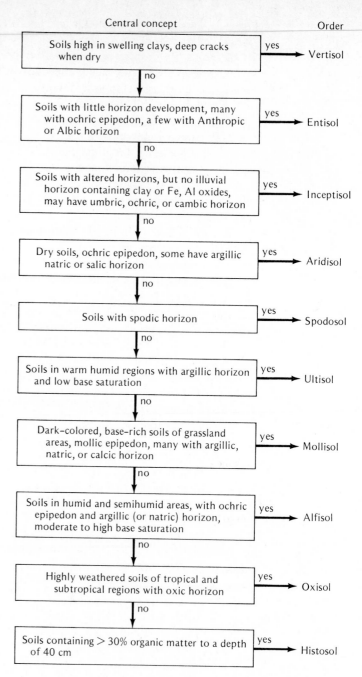

FIGURE 13.13 Simplified key showing the central concept of the ten soil orders in *Soil Taxonomy*.

Central concept

Order

Soils high in swelling clays, deep cracks when dry — yes → Vertisol

no

Soils with little horizon development, many with ochric epipedon, a few with Anthropic or Albic horizon — yes → Entisol

no

Soils with altered horizons, but no illuvial horizon containing clay or Fe, Al oxides, may have umbric, ochric, or cambic horizon — yes → Inceptisol

no

Dry soils, ochric epipedon, some have argillic natric or salic horizon — yes → Aridisol

no

Soils with spodic horizon — yes → Spodosol

no

Soils in warm humid regions with argillic horizon and low base saturation — yes → Ultisol

no

Dark-colored, base-rich soils of grassland areas, mollic epipedon, many with argillic, natric, or calcic horizon — yes → Mollisol

no

Soils in humid and semihumid areas, with ochric epipedon and argillic (or natric) horizon, moderate to high base saturation — yes → Alfisol

no

Highly weathered soils of tropical and subtropical regions with oxic horizon — yes → Oxisol

no

Soils containing > 30% organic matter to a depth of 40 cm — yes → Histosol

Summary. The soil map of the United States based on *Soil Taxonomy* (Figure 13.8) identifies the regions in which the various soil orders and suborders are found. Likewise, the generalized world soil map (Figure 13.9) shows the probable locations of the different soil orders throughout the world. Plate I shows typical profiles of soils in each of the soil orders.

A simple key to soil orders in *Soil Taxonomy* is shown in Figure 13.13. This key helps illustrate the relationships among the soil orders.

The approximate land area for each of the ten soil orders in the United States and in the world is shown in Table 13.4. Note that the percentages of Mollisols and Inceptisols are higher than the others in the United States

TABLE 13.4 Approximate Land Areas of Different Soil Orders and Suborders of the United States and of the World.[a]

Order/Suborder	% of ice-free land area		Order/Suborder	% of ice-free land area	
	U.S.	World		U.S.	World
Alfisols	13.5	14.0	Mollisols	25.1	9.0
Aqualfs	(1.0)		Aquolls	(1.3)	
Boralfs	(3.0)		Borolls	(5.0)	
Udalfs	(5.9)		Udolls	(4.8)	
Ustalfs	(2.6)		Ustolls	(8.9)	
Xeralfs	(1.10)		Xerolls	(5.2)	
Aridisols	11.6	18.0	Oxisols	0.012	9.0
Argids	(8.7)		Orthox	(0.005)	
Orthids	(2.9)		Ustox	(0.007)	
Entisols	8.0	12.0	Spodosols	4.8)	4.0
Aquents	(0.2)		Aquods	(0.7)	
Fluvents	(0.3)		Orthods	(4.1)	
Orthents	(5.3)				
Psamments	(2.2)		Ultisols	12.8	10.0
			Aquults	(1.2)	
Histosols	0.5	1.0	Humults	(0.8)	
			Udults	(9.8)	
Inceptisols	18.2	20.0	Xerults	(1.0)	
Andepts	18.2				
Aquepts	(11.4)		Vertisols	1.0	3.0
Ochrepts	(4.3)		Uderts	(0.4)	
Tropepts	(0.01)		Usterts	(0.6)	
Umbrepts	(0.7)		Xererts	(0.008)	
			Misc. land	4.5	—

[a] Source: USDA Soil Conservation Service, Soil Survey Staff, Washington, DC (personal communication).

whereas the percent of land area on which Aridisols and Oxisols predominate is higher worldwide.

13.8   Soil Suborders, Great Groups, and Subgroups

Suborders.   The ten orders just described are subdivided into about 47 suborders as shown in Table 13.5. The characteristics used as a basis for subdividing into the suborders are those that give the class the greatest genetic homogeneity. Thus, soils formed under wet conditions are generally identified under separate suborders (e.g., Aquents, Aquerts, Aquepts), as are the drier soils (e.g., Ustalfs,

TABLE 13.5   Soil Orders and Suborders in *Soil Taxonomy*

Note that the ending of the suborder names identifies the order in which the soils are found.

Order	Suborder	Order	Suborder
Entisols	Aquents *water*	Ultisols	Aquults
	Arents *plowed*		Humults *organic*
	Fluvents *river plain*		Udults
	Orthents *common*		Ustults
	Psamments *sand*		Xerults
Inceptisols	Andepts *dark volcanic*	Oxisols	Aquox
	Aquepts *organic*		Humox
	Ochrepts *pale*		Orthox
	Plaggepts *man-made (age)*		Torrox *hot, dry*
	Tropepts *continually warm*		Ustox " "
	Umbrepts *dark*		
		Vertisols	Torrerts
Mollisols	Albolls *bleached*		Uderts *humid*
	Aquolls		Usterts
	Borolls *cool*		Xererts
	Rendolls *?*		
	Udolls *humid*	Aridisols	Argids *mix horizon*
	Ustolls *dry, hot*		Orthids
	Xerolls *xeric*		
		Spodosols	Aquods
Alfisols	Aqualfs		Ferrods *iron*
	Boralfs		Humods
	Udalfs		Orthods
	Ustalfs		
	Xeralfs	Histosols	Fibrists *least decompose*
			Hemists *intermediate*
			Saprists *most decomp*
			Folists *mass of leaves*

Ustults). This arrangement also provides a convenient device for grouping soils outside the classification system (e.g., the *wet* and *dry* soils).

To determine the relationship between suborder names and soil characteristics, reference should be made to Table 13.6. Here the formative elements for suborder names are identified, as is their connotation. Thus, the *Boroll* suborder (Greek *boreas*, northern) (Table 13.6) includes *Mollisols* found in cool climates. Likewise, soils in the *Udult* suborder (Latin *udus*, humid) are Ultisols of the more humid climates. Identification of the primary characteristics of each of the other suborders can be made by cross reference to Tables 13.5 and 13.6.

Great Groups.   The *great groups* are subdivisions of suborders. They are defined largely on the presence or absence of diagnostic horizons and the arrange-

TABLE 13.6   Formative Elements in Names of Suborders in *Soil Taxonomy*

Formative element	Derivation	Connotation of formative element
alb	L. *albus*, white	Presence of albic horizon (a bleached eluvial horizon)
and	Modified from Ando	Ando-like
aqu	L. *aqua*, water	Characteristics associated with wetness
ar	L. *arare*, to plow	Mixed horizons
arg	L. *argilla*, white clay	Presence of argillic horizon (a horizon with illuvial clay)
bor	Gk. *boreas*, northern	Cool
ferr	L. *ferrum*, iron	Presence of iron
fibr	L. *fibra*, fiber	Least decomposed stage
fluv	L. *fluvius*, river	Floodplains
fol	L. *folia*, leaf	Mass of leaves
hem	Gk. *hemi*, half	Intermediate stage of decomposition
hum	L. *humus*, earth	Presence of organic matter
ochr	Gk. base of *ochros*, pale	Presence of ochric epipedon (a light surface)
orth	Gk. *orthos*, true	The common ones
plagg	Modified from Ger. *plaggen*, sod	Presence of plaggen epipedon
psamm	Gk. *psammos*, sand	Sand textures
rend	Modified from Rendzina	Rendzina-like
sapr	Gk. *sapros*, rotten	Most decomposed stage
torr	L. *torridus*, hot and dry	Usually dry
ud	L. *udus*, humid	Of humid climates
umbr	L. *umbra*, shade	Presence of umbric epipedon (a dark surface)
ust	L. *ustus*, burnt	Of dry climates, usually hot in summer
xer	Gk. *xeros*, dry	Annual dry season

TABLE 13.7 Formative Elements for Names of Great Groups and Their Connotation

These formative elements combined with the appropriate suborder names give the great group names.

Formative element	Connotation	Formative element	Connotation	Formative element	Connotation
acr	Extreme weathering	gibbs	Gibbsite	psamm	Sand texture
agr	Agric horizon	gyps	Gypsic horizon	quartz	High quartz
alb	Albic horizon	gloss	Tongued	rhod	Dark red colors
and	Ando-like	hal	Salty	sal	Salic horizon
arg	Argillic horizon	hapl	Minimum horizon	sider	Free iron oxides
bor	Cool	hum	Humus	sombr	Dark horizon
calc	Calcic horizon	hydr	Water	sphagn	Sphagnum moss
camb	Cambic horizon	luv, lu	Illuvial	sulf	Sulfur
chrom	High chroma	med	Temperate climates	torr	Usually dry
cry	Cold	nadur	See *natr* and *dur*	trop	Continually warm
dur	Duripan	natr	Natric horizon	ud	Humid climates
dystr, dys	Low base saturation	ochr	Ochric epipedon	umbr	Umbric epipedon
		pale	Old development	ust	Dry climate, usually hot in summer
eutr, eu	High base saturation	pell	Low chroma		
		plac	Thin pan	verm	Wormy, or mixed by animals
ferr	Iron	plagg	Plaggen horizon		
fluv	Floodplain	plinth	Plinthite	vitr	Glass
frag	Fragipan			xer	Annual dry season
fragloss	See *frag* and *gloss*				

ments of those horizons. These horizon designations are included in the list of formative elements for the names of great groups shown in Table 13.7. Note that these formative elements refer to epipedons such as mollic and ochric (see Table 13.1), to subsurface horizons such as argillic and natric, and to pans such as duripan and fragipan. Remember that the great group names are made up of these formative elements attached as prefixes to the names of suborders in which the great groups occur. Thus, a *Ustoll* with a *natric* horizon (high in sodium) belongs to the *Natrustoll* great group. As might be expected, the number of great groups is high, more than 230 having been identified.

The names of selected great groups from three orders are given in Table 13.8. This list illustrates again the utility of the *Soil Taxonomy*, especially the nomenclature it employs. The names identify the suborder and order in which the great groups are found. Thus, Argiudolls are Mollisols of the Udoll suborder characterized by an argillic horizon. Cross reference to Table 13.7 identifies the specific characteristics separating the great group classes from each other. Careful study of these two tables will show the utility of this classification system.

Subgroups. Subgroups are subdivisions of great groups. There are more than 1200 subgroups recognized, some 1000 of which are in the United States. They

TABLE 13.8   Examples of Names of Great Groups of Selected Suborders of the Mollisol, Alfisol, and Ultisol Orders

The suborder name is identified as the italicized portion of the great group name.

	Great group characterized by		
	Argillic horizon	Minimum horizon	Old development
Mollisols			
1. Aquolls (wet)	Argi*aquolls*	Hapl*aquolls*	—
2. Udolls (moist)	Argi*udolls*	Hapl*udolls*	Pale*udolls*
3. Ustolls (dry)	Argi*ustolls*	Hapl*ustolls*	Pale*ustolls*
4. Xerolls (Med.)[a]	Argi*xerolls*	Hapl*oxerolls*	Pale*xerolls*
Alfisols			
1. Aqualfs (wet)	—		
2. Udalfs (moist)	Arg*udalfs*	Hapl*udalfs*	Pale*udalfs*
3. Ustalfs (dry)	—	Hapl*ustalfs*	Pale*ustalfs*
4. Xeralf (Med.)[a]	—	Hapl*oxeralfs*	Pale*xeralfs*
Ultisols			
1. Aqults (wet)	—	—	Pale*aquults*
2. Udults (moist)	—	Hapl*udults*	Pale*udults*
3. Ustults (dry)	—	Hapl*ustults*	Pale*ustults*
4. Xerults (Med.)[a]	—	Hapl*oxerults*	Pale*xerults*

[a] Med. = Mediterranean climate; distinct dry period in summer.

permit characterization of the core concept of a given great group and of gradations from that central concept to other units of the classification system. The subgroup most nearly representing the central concept of a great group is termed *Typic.* Thus, the *Typic Hapludolls* subgroup typifies the Hapludolls great group. A Hapludoll with restricted drainage would be classed as an *Aquic Hapludoll.* One with evidence of intense animal (earthworm) activity would fall in the *Vermic Hapludolls* subgroup.

Some intergrades may have properties in common with other orders or with other great groups. Thus, the *Entic Hapludolls* subgroup intergrades toward the Entisols order. The subgroup concept illustrates very well the flexibility of this classification system.

13.9   Soil Families and Series

Families.   The family category of classification is based on properties important to the growth of plant roots. The criteria used may vary from one subgroup to another but include broad textural classes, mineralogical classes, temperature classes, and thickness of the soil penetrable by roots. Terms such as fine, fine loamy, sandy, and clayey are used to identify the broad textural

TABLE 13.9 Characteristics of Currently Recognized Families in the United States of a Selected Subgroup, Typic Paleudults[a]

Particle-size class	Mineralogy	Soil temperature
Loamy skeletal	Siliceous	Mesic
Loamy skeletal	Siliceous	Thermic
Loamy skeletal over clayey	Siliceous	Mesic
Clayey skeletal	Mixed	Mesic
Coarse loamy	Siliceous	Hyperthermic
Coarse loamy	Siliceous	Mesic
Coarse loamy	Siliceous	Thermic
Fine loamy	Siliceous	Mesic
Fine loamy	Siliceous	Thermic
Fine silty	Siliceous	Thermic
Clayey	Kaolinitic	Mesic
Clayey	Kaolinitic	Thermic
Clayey	Mixed	Isohyperthermic
Clayey	Mixed	Mesic
Clayey	Mixed	Thermic
Clayey	Oxidic	Isohyperthermic

[a] From Beinroth et al. (1980).

classes. Terms used to describe the mineralogical classes include montmorillonitic, kaolinitic, siliceous, gypsic, and mixed. For temperature classes, terms such as frigid, mesic, and thermic are used. Thus, a Typic Argiudoll from Iowa, fine in texture, having a mixture of clay minerals, and being located in a temperate region, might be classed in the *Typic Argiudolls fine, mixed, mesic* family. In contrast, a sandy-textured Typic Cryorthod, high in quartz and located in eastern Canada, might be considered in the *Typic Cryorthods sandy, siliceous, frigid* family. About 6000 families have been identified in the United States. In Table 13.9 are shown families of the Typic Paleudults subgroup.

Series. The families are subdivided into the lowest category of the system called *series.* One series may be differentiated from another in a family through differences in one or more but not necessarily all of the soil properties. Series are, of course, established on the basis of profile characteristics. This requires a careful study of the various horizons as to number, order, thickness, texture, structure, color, organic content, and reaction (acid, neutral, or alkaline). Such features as hardpan at a certain distance below the surface, a distinct zone of calcium carbonate accumulation at a certain depth, or striking color characteristics greatly aid in series identification.

In the United States each series is given a name, usually from some city, village, river, or country, such as Fargo, Muscatine, Cecil, Mohave, Ontario, and the like. There are more than 13,000 soil series in the United States.

The complete classification of a Mollisol, the Brookston series, is given in Figure 13.14. This illustrates how *Soil Taxonomy* can be used to show the

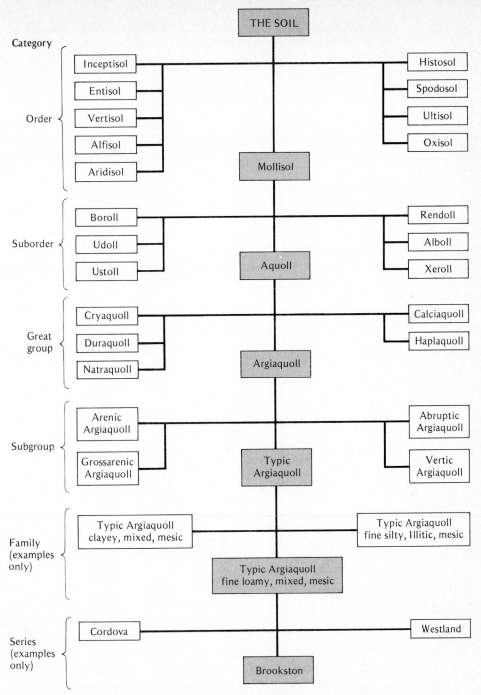

FIGURE 13.14 Diagram illustrating the classification of a Mollisol, the Brookston series found in Ohio. In each category other taxa (classification units) are shown to illustrate how this soil fits in the overall classification scheme.

relationship between *the* soil, a comprehensive term covering all soils, and *a* specific soil series. This figure is deserving of study.

13.10   Soil Phases, Associations, and Catenas

Soil Phase.   Although technically not included as a class in *Soil Taxonomy*, a *phase* is a subdivision on the basis of some important deviation that influences the use of the soil, such as surface texture, erosion, slope, stoniness, or soluble salt content. It really marks a departure from the normal or key series already established. Thus, a Cecil sandy loam, 3–5% slope, and a Hagerstown silt loam, stony phase, are examples of soils where distinctions are made in respect to the phase.

Soil Associations and Catenas.   Two more means of classifying associated soils should be mentioned. In the field, soils of different kinds commonly occur

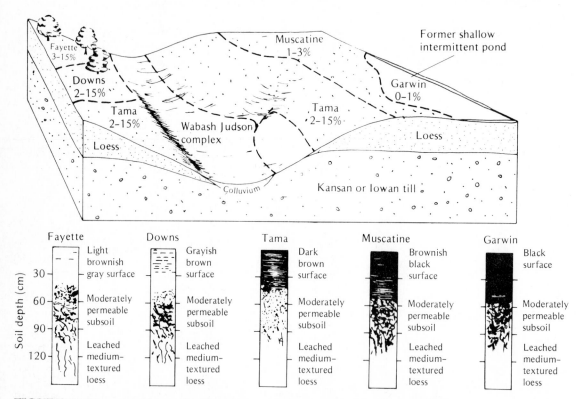

FIGURE 13.15   Association of soils in Iowa. Note the relationship of soil type to (1) parent material, (2) vegetation, (3) topography, and (4) drainage. Two Alfisols (Fayette and Downs) and three Mollisols (Tama, Muscatine, and Garwin) are shown. [*From Riecken and Smith (1949).*]

460   13   Soil Formation, Classification, and Survey

together. Such an *association* of soils may consist of a combination from different soil orders (Figure 13.15). Thus, a shallow Entisol on steep uplands may be found alongside a well-drained Alfisol or a soil developed from recent alluvium. The only requirement is that the soils occur together in the same area. Soil associations are important in a practical way, since they help deter-

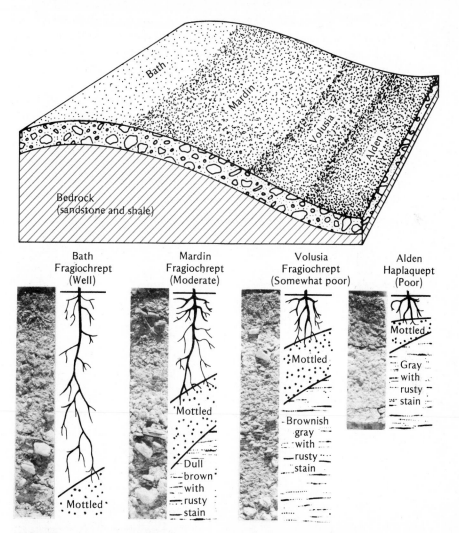

FIGURE 13.16 Monoliths showing four soils of a drainage catena (below) and a diagram showing their topographic association in the field (above). Note the decrease in the depth of the well-aerated zone (above the mottled layers) from the Bath soil (left) to the Alden (right), which remains poorly aerated throughout the growing season. These soils are all developed from the same parent material and differ only in drainage and topography. The Volusia soil as pictured was cultivated, while the others were located on virgin sites. Note that all four soils belong to the Inceptisol order.

mine combinations of land-use patterns, which must be utilized to support a profitable agriculture.

Well-drained, imperfectly drained, and poorly drained soils, all of which have developed from the same parent materials under the same climatic conditions, often are closely associated under field conditions. This association on the basis of drainage or of differences in relief is known as a *catena* and is very helpful in practical classification of soils in a given region. The relationship can be seen by referring to Figure 13.16, where the Bath-Mardin-Volusia-Alden catena is shown. Although all four or five members of a catena are seldom found in a given area, the diagram illustrates the relationship of drainage to topography.

The practical significance of the soil catena can be seen in Figure 13.17, where the tolerance of several crops to differences in soil drainage is shown. Recommendations for crop selection and management are commonly made on the basis of soil drainage class. Such recommendations, along with careful study of farm soil survey maps, are invaluable to farmers in making the best choices of soil and crop compatibility.

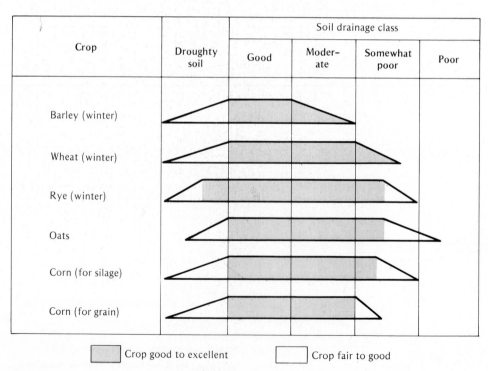

FIGURE 13.17 Relationship between soil drainage class and the growth of certain grain crops. [*From "Cornell Recommends" (Anon., 1962).*]

13.11 Other Classification Systems

Soil Taxonomy has been developed only in recent years and is used wholly or in part by scientists from 31 countries (Cline, 1980). Other soil classification systems, some new and others having been used for years, are employed extensively in countries other than the United States. Such systems have been developed and are in use in the Soviet Union, France, Canada, Australia, and elsewhere. Also, scientists at the Food and Agriculture Organization (FAO) of the United Nations have prepared a legend for a World Soils Map. It is far beyond the scope of this text to make direct comparison among these classification systems. However, a general idea of equivalent classes of *Soil Taxonomy*,

TABLE 13.10 Soil Orders in *Soil Taxonomy* and Some Approximate Equivalents in the 1949 System (Used in the United States) and in the Legend for the FAO World Soil Map

Soil taxonomy orders	Comparable U.S. 1949 system[a]	Units in FAO world soil map
Entisols	Azonal	Regosols, Arenosols, Fluvisols, Lithosols
Inceptisols	Ando, Sol Brun acide, some Brown Forest	Cambisols, Andosols, Fluvisols
Mollisols	Chestnut, Chernozem, Brunizem (Prairie), Rendzinas, some Brown Forest and associated Solnetz	Chernozems, Kastanozems, some Phaeozems
Alfisols	Gray-Brown Podzolic, Gray Wooded, Non-Calcic Brown and Degraded Chernozems, Planosols	Luvisols, Nitosols, Acrisols with high base saturation
Ultisols	Red-Yellow Podzolic, Reddish-Yellow Lateritic soils of U.S.	Acrisols and part of Nitosols and Luvisols all with low base saturation
Oxisols	Lateritic Soils, Latosols	Ferralsols
Vertisols	Grumusol	Vertisols
Aridisols	Desert, Reddish Desert, Sierozem, Solonchak, some Brown and Reddish Brown	Yermosols, Xerosols Solonchaks
Spodosols	Podzol, Brown Podzolic and Ground Water Podzol	Podzols
Histosols	Bog Soils	Histosols

[a] In addition to those listed, some orders included soils with restricted drainage classed as Low-Humic Gley and Humic Gley. Refer to the Glossary for brief descriptions of orders in the 1949 system.

the old 1949 U.S. system, and the World Soils Map legend used by the FAO is shown in Table 13.10. These should be useful to relate these three attempts to classify soils.

FIGURE 13.18 Soil survey maps are made using aerial photographs such as this one from Winneshiek County, Iowa. The field soil scientist is able to visualize quickly the topographic features of the section in which he is working. [*Courtesy USDA Soil Conservation Service, Soil Survey Staff.*]

13.12 Soil Survey and Its Utilization

The classifications as outlined in the preceding pages are used in soil survey, the function of which is to classify, locate on a base map, and describe the nature of soils as they occur in the field. The soils in the United States are classified into *series* and *phases* on the basis of their profile characteristics. The soil surveyor, since his work is localized, concerns himself mostly with series and phase separations, giving the broader regional distinctions only general consideration.

Field Mapping. As the survey of any area progresses, the location of each soil unit is shown on a suitable base map. The maps used in the United States are aerial base maps such as the one shown in Figure 13.18. They have considerable advantage over the ordinary contour maps in that land cover and field boundaries show very clearly. Thus, the surveyor can quickly locate his position on the map and can indicate readily the soil boundaries (Figure 13.19).

Using the procedure outlined above, a field map is thus obtained which gives not only the soil series but also the phase, which furnishes information as to slope and severity of erosion. Thus, the map becomes of even greater practical value.

The field map, after the separations have been carefully checked and correlated, is now ready for reproduction. In older maps this reproduction was made in color. For more recent maps a reproduction is made of the aerial photograph with the appropriate soil boundaries drawn in. When published, the map accompanies a bulletin containing a description of the topography, climate, agriculture, and soils of the area under consideration. Each soil series is minutely described as to profile and suggestions as to practical management for agricultural and nonagricultural uses are usually made. Such a field map is shown in Figure 13.20, where the relationship of topography to soil associations can be seen readily.

Use of Soil Surveys. Soil survey maps and bulletins are useful in many ways. Scientists use them to facilitate research on crop production, land evaluation and zoning, and human settlements. Extension specialists and country agricultural agents often base their recommendations for land use and crop management on soil surveys and on interpretive bulletins based on them. Engineers and hydrologists can select superior roadbeds and building sites and can make estimates of water runoff and infiltration using soil survey maps and reports. Homeowners can use the reports in selecting trees and shrubs for landscaping. The reports contain interpretive tables indicating limitations for soil use as cropland, pasture, woodland wildlife, community development, sewage disposal, recreation, and engineering uses. Table 13.11 illustrates how soil surveys are used to determine land suitability for agricultural and other purposes.

The soil survey is perhaps of greatest practical value in land classification

(a)   (b)

FIGURE 13.19 (a) Soil survey maps are prepared by soil scientists examining soils in the field using a soil auger and other diagnostic tools. (b) Mapping units are outlined on a topographic map, first in the field and later more permanently in the map room. [*Courtesy USDA Soil Conservation Service, Soil Survey Staff.*]

for agricultural and other uses. The classification of greatest importance to agriculturists is that in use by the U.S. Soil Conservation Service because of its widespread application.

13.13   Land Capability Classification

At this point it may be well to note the difference between *soil* and *land*. Soil is the more restrictive term, referring to a collection of natural bodies with depth as well as breadth whose characteristics may be only indirectly related to their current vegetation and use. Land is a broader term that includes

FIGURE 13.20 Field examination makes possible the delineating of soil boundaries, which are then indicated on the aerial photo (Figure 13.18). The soil legend identifies the soil name (first two letters), the slope (second capital letter), and the degree of erosion (the number). Thus, OsC2 is an Orwood silt loam (Os) that has a 5–9% slope (C) and is moderately eroded (2). [*Courtesy USDA Soil Conservation Service, Soil Survey Staff.*]

among its characteristics not only the soil but other physical attributes, such as water supply, existing plant cover, and location with respect to cities, means of transportation, and so on. Thus, we have *forestland, bottomland,* and *grasslands,* which may include a variety of soils.

Soil survey maps and reports have become two of the bases for systems of land capability or suitability classification. These systems are based on the assumption that every hectare of land should be used in accordance with its *capability* and *limitations*. Land is classified according to the most suitable sustained use that can be made of it while providing for adequate protection from erosion or other means of deterioration. Thus, an area where the soils are deep and well drained, with a stable surface structure, and where the slope is only 1–2% may be intensively cropped almost indefinitely with little danger of erosion or loss in productivity. It may also be used with little limitation as a site for a building, a roadbed, or a septic tank field. Such an area has great capabilities and few limitations in the use to which it can be put. In contrast, an area on which shallow or poorly drained soils are found or where steep slopes are prevalent has limited capabilities and many limitations as to its use. Only crops that can tolerate wetness can be grown. The area makes a poor building site unless artificial drainage is installed, but it may be quite suitable for a wildlife refuge. Clearly, the characteristics of soils are one of the criteria for identifying the best land use.

Several land-use classification systems have been developed around the world, some relating to agricultural use and others to forestry, recreation, wildlife use, building construction and other engineering purposes, and sewage disposal. We shall consider the land use classification system developed by the U.S. Soil Conservation Service. Although its primary utilization is for agriculture, it is also useful for other purposes.

Capability Classes.   Under the system set up by the U.S. Soil Conservation Service, eight land capability classes are recognized. These classes are numbered from I to VIII. Soils having greatest capabilities for response to management and least limitations in the ways they can be used are in Class I. Those with least capabilities and greatest limitations are found in Class VIII (Figures 13.21 and 13.22). A brief description of the characteristics and safe use of soils in each class follows.

Class I.   Soils in this land class have few limitations upon their use. They can be cropped intensively, used for pasture, range, woodlands, or even wildlife preserves. The soils are deep, well drained, and the land is nearly level. They are either naturally fertile or have characteristics that encourage good response of crops to applications of fertilizer.

The water-holding capacity of soils in Class I is high. In arid and semiarid areas, soils having all the other favorable characteristics mentioned above, may be in Class I if they are irrigated by a permanent irrigation system.

The soils in Class I need only ordinary crop management practices to maintain their productivity. These include the use of fertilizer and lime and the return of manure and crop residues, including green manures. Crop rotation are also followed.

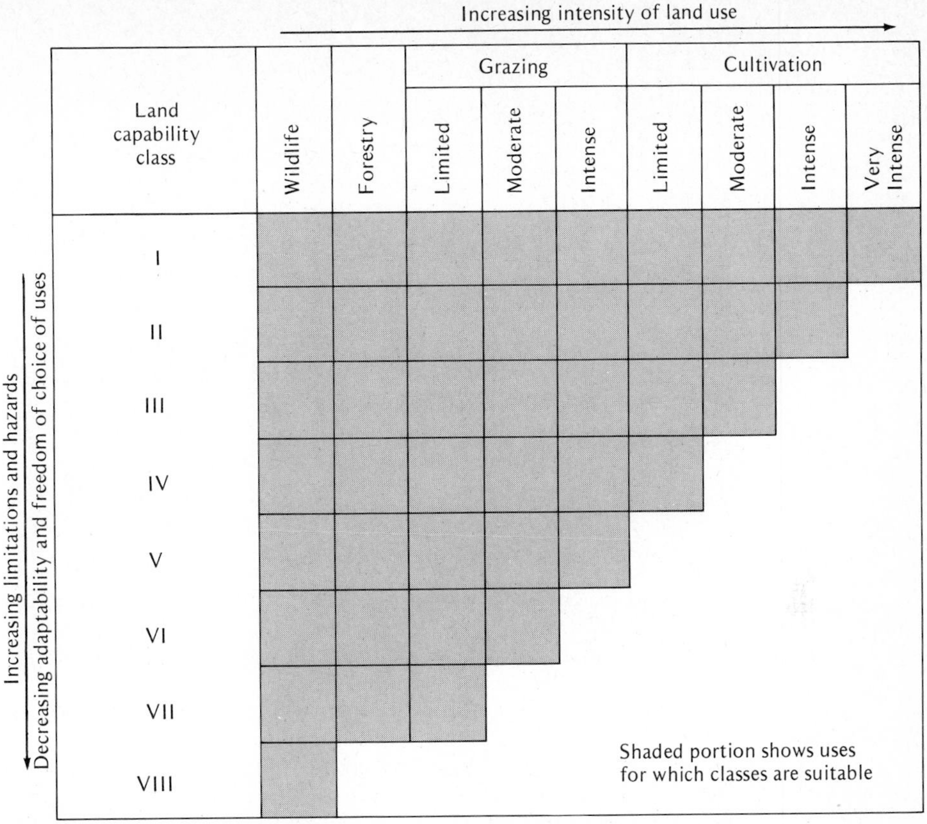

Land capability class	Wildlife	Forestry	Grazing			Cultivation			
			Limited	Moderate	Intense	Limited	Moderate	Intense	Very Intense
I	▓	▓	▓	▓	▓	▓	▓	▓	▓
II	▓	▓	▓	▓	▓	▓	▓	▓	
III	▓	▓	▓	▓	▓	▓	▓		
IV	▓	▓	▓	▓	▓	▓			
V	▓	▓	▓	▓	▓				
VI	▓	▓	▓	▓					
VII	▓	▓	▓						
VIII	▓								

(Left vertical axis: Increasing limitations and hazards / Decreasing adaptability and freedom of choice of uses)

Shaded portion shows uses for which classes are suitable

FIGURE 13.21 Intensity with which each land capability class can be used with safety. Note the increasing limitations on the safe uses of the land as one moves from class I to class VIII. [*Modified from Hockensmith and Steele (1949).*]

Class II. Soils in this class have some limitations that reduce the choice of plants or require moderate conservation practices. These soils may be used for the same crops as Class I. However, they are capable of sustaining less intensive cropping systems, or, with the same cropping systems, they require some conservation practices.

The use of soils in Class II may be limited by one or more factors, such as (a) gentle slopes, (b) moderate erosion hazards, (c) inadequate soil depth, (d) less than ideal soil structure and workability, (e) slight to moderate alkaline or saline conditions, and (f) somewhat restricted drainage.

The management practices that may be required for soils in Class II include terracing, strip cropping, contour tillage, rotations involving grasses and legumes, and grassed waterways. In addition, those conservation practices that are used on Class I land are also generally required for soils in Class II.

FIGURE 13.22 Several land capability classes in San Mateo County, California. A range is shown from the nearly level land in the foreground (class I), which can be cropped intensively, to the badly eroded hillsides (classes VII and VIII). Although topography and erosion hazards are emphasized here, it should be remembered that other factors—drainage, stoniness, droughtiness—also limit soil usage and help determine the land capability class. [*Courtesy USDA Soil Conservation Service.*]

Class III. Soils in Class III have severe limitations that reduce the choice of plants or require special conservation practices or both. The same crops can be grown on Class III land as on Classes I and II. The amount of clean, cultivated land is restricted, however, as is the choice of the particular crop to be used. Crops that provide soil cover such as grasses and legumes must be more prominent in the rotations used.

Limitations in the use of soils in Class III result from factors such as (a) moderately steep slopes, (b) high erosion hazards, (c) very slow water permeability, (d) shallow depth and restricted root zone, (e) low water-holding capacity, (f) low fertility, (g) moderate alkalinity or salinity, and (h) unstable soil structure.

Soils in Class III often require special conservation practices. Those mentioned for Class II land must be employed, frequently in combination with restrictions in kinds of crops. Tile or other drainage systems may also be needed in poorly drained areas.

Class IV. Soils in this class can be used for cultivation, but there are very severe limitations on the choice of crops. Also, very careful management may be required. The alternative uses of these soils are more limited than for Class III. Close-growing crops must be used extensively, and row crops cannot be grown safely in most cases. The choice of crops may be limited by excess moisture as well as by erosion hazards.

The most limiting factors on these soils may be one or more of the following: (a) steep slopes, (b) severe erosion susceptibility, (c) severe past erosion, (d) shallow soils, (e) low water-holding capacity, (f) poor drainage, and (g) severe alkalinity or salinity. Soil conservation practices must be applied more frequently than on soils in Class III. Also, they are usually combined with severe limitations in choice of crop.

Class V. Soils in Classes V–VIII are generally not suited to cultivation. Those in Class V are limited in their safe use by factors other than erosion hazards. Such limitations include (a) frequent stream overflow, (b) growing season too short for crop plants, (c) stony or rocky soils, and (d) ponded areas where drainage is not feasible. Often, pastures can be improved on this class of land.

Class VI. Soils in this class have extreme limitations that restrict their use largely to pasture or range, woodland, or wildlife. The limitations are the same as those for Class IV land, but they are more rigid.

Class VII. Soils in Class VII have very severe limitations that restrict their use to grazing, woodland, or wildlife. The physical limitations are the same as for Class VI except they are so strict that pasture improvement is impractical.

Class VIII. In this land class are soils that should not be used for any kind of commercial plant production. Their use is restricted to recreation, wildlife, water supply, or aesthetic purposes. Examples of kinds of soils or landforms included in Class VIII are sandy beaches, river wash, and rock outcrop.

Subclasses. In each of the land capability classes are *subclasses* that have the same kind of dominant limitations for agricultural uses. The four kinds of limitations recognized in these subclasses are risks of erosion (e); wetness, drainage, or overflow (w): root-zone limitations (s); and climatic limitations (c). Thus, a soil may be found in Class III (e), indicating that it is in Class III because of risks of erosion.

Land Use Capability in the United States. In 1977, the U.S. Department of Agriculture made a national inventory of soil and water conservation needs for the United States (USDA, 1981). This inventory included information on land capability classification. In Figure 13.23 summary information from this inventory is presented.

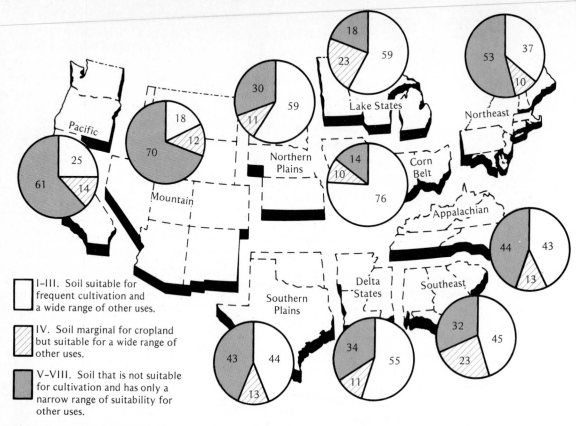

FIGURE 13.23 Percentages of land in capability Classes I–VIII by crop production region. [*From USDA* (*1981*).]

About 44% of the land in the 50 United States (248 million hectares) is suitable for regular cultivation (Classes I, II, and III). Another 13% (76 million ha) is marginal for growing cultivated crops. The remainder is suited primarily for grasslands and forests and is used mostly for this purpose.

This land classification scheme illustrates the use that can be made of soil surveys in a practical way. The many soils delineated on a map by the soil surveyor are viewed in the light of their safest and best longtime use. The eight land capability classes have become the starting point in the development of farm plans so useful to thousands of American farmers.

13.14 Soil Interpretations for Nonfarm Uses

Soil surveys can also be used to identify the potential for nonfarm uses of land. Examples of such soil interpretation are suitability for building sites, waste disposal, highway beds, woodlands and recreational sites, and wildlife.

TABLE 13.11 Suitability of Different Soil Areas for Production of Three Crops and for Nonfarm Uses[a]

Based on number of 2-ha parcels per square mile with only moderate problems:
>15 = good; 5–15 = fair; <5 = poor.

Soil area	Corn	Vegetables	Alfalfa	Buildings with basements	Septic tank filter fields	Summer camp sites
Honeoye	Good	Good	Good	Good	Good	Good
Madrid	Good	Good	Good	Good	Good	Good
Bath	Fair	Poor	Fair	Good	Good	Good
Volusia	Poor	Poor	Poor	Fair	Poor	Good
Chenango	Fair	Poor	Good	Good	Good	Good
Erie	Poor	Poor	Poor	Poor	Poor	Good
Ovid	Fair	Poor	Fair	Fair	Poor	Fair

[a] Modified from Cline and Marshall (1977).

It is obvious that soil surveys are important tools in regional and local land-use planning. Table 13.11 illustrates the suitability of selected soil areas for both farm and nonfarm uses. Note that for some soils (Honeoye and Madrid), high quality for crop production and for nonfarm uses are closely correlated. Other soils, however (e.g., Bath and Chenango), rate higher for nonfarm than for farm uses. This is not surprising since chemical and physical properties needed for good crop production may be quite different from those needed for waste disposal, for example. The high permeability required for waste disposal is the same characteristic that may be associated with droughtiness for crop production. In any case, these soil interpretations illustrate the usefulness of soil surveys to farmers and nonfarmers alike.

REFERENCES

Anonymous. 1962. "Cornell Recommends," an annual publication of the New York State College of Agricuture, Cornell University.

Beinroth, F. H., G. Uehara, J. A. Silva, R. W. Arnold, and F. B. Cady. 1980. "Agrotechnology transfer in the tropics based on Soil Taxonomy," *Adv. in Agron.,* **33:**303–339.

Cline, M. G. 1980. "Experience with soil taxonomy of the United States," *Adv. in Agron.,* **33:**193–226.

Cline, M. G., and R. L. Marshall. 1977. "Soils of New York landscapes," Inf. Bull. 119, Physical Sciences, Agronomy 6, New York State College of Agriculture and Life Sciences, Cornell University.

Hockensmith, R. D., and J. G. Steel. 1949. "Recent trends in the use of the land-capability classification," *Soil Sci. Soc. Amer. Proc.,* **14**:383–388.

Riecken, F. F., and G. D. Smith. 1949. "Principal upland soils of Iowa, their occurrence and important properties," *Agron.* 49 *Revised,* Iowa Agr. Exp. Sta.

Simonson, R. W. 1968. "Concept of soil," *Adv. in Agron.,* **20**:1–47.

Smith, G. D. 1963. "Objectives and basic assumptions of the new soil classification system," *Soil Sci.,* **96**:6–16.

Soil Survey Staff. 1975. *Soil Taxonomy: A Basic System of Soil Classification for Making and Interpreting Soil Surveys* (Washington, DC: U.S. Department of Agriculture, Soil Conservation Service).

USDA. 1981. "Soil, water and related resources in the United States," 1980 Appraisal Part II, Soil and Water Resources Conservation Act (Washington, DC: U.S. Department of Agriculture).

Organic Soils (Histosols): Their Nature, Properties, and Utilization

14

On the basis of organic content, two general groups of soils are commonly recognized—*mineral* and *organic*. The mineral soils vary in organic matter content from a mere trace to as high as 20% by weight (12% C) if their clay content is low, or even 30% (18% C) if the soil is high in clay. Those soils with higher organic matter are termed organic soils and are classed as *Histosols* in the comprehensive soil classification system.[1] Those that are cultivated probably average about 80% organic matter.

Organic soils are by no means so extensive as are mineral soils, yet their total area is quite large, more than 300 million ha worldwide. Their use in favored localities for the intensive production of crops, particularly high-value crops such as vegetables, is very important. Although less is known about the physical, chemical, and biological characteristics of organic soils than of mineral soils, in general, the same edaphological principles hold for both.

14.1 Genesis of Organic Deposits

Marshes, bogs, and swamps provide conditions suitable for the accumulation of organic deposits. The favorable water environment in and adjacent to such areas encourages the growth of many plants, such as pondweed, cattails, sedges, reeds and other grasses, mosses, shrubs, and also trees. These plants in numberless generations thrive, die, and sink down to be covered by the water in which they have grown. The water shuts out the air, preventing rapid oxidation, and thus acts as a partial preservative.

Layering of Peat Beds. As one generation of plants follows another, layer after layer of organic residue is deposited in the swamp or marsh (Figure 14.1). The constitution of these successive layers changes as time goes on since a sequence of different plant life occurs. Thus, deep-water plants may be supplanted by reeds and sedges, these by various mosses, and these in turn by shrubs, until finally forest trees, either hardwoods or conifers, with their characteristic undergrowth, may become dominant. The succession is by no means regular or definite, as a slight change in climate or water level may alter the sequence entirely.

The profile of an organic deposit is therefore characterized by layers, differing not only as to their degree of decomposition but also as to the nature of the original plant tissue. In fact, these layers later may have become soil hori-

---

[1] For a review of the characteristics and classification of these soils see Soil Science Society of America (1974).

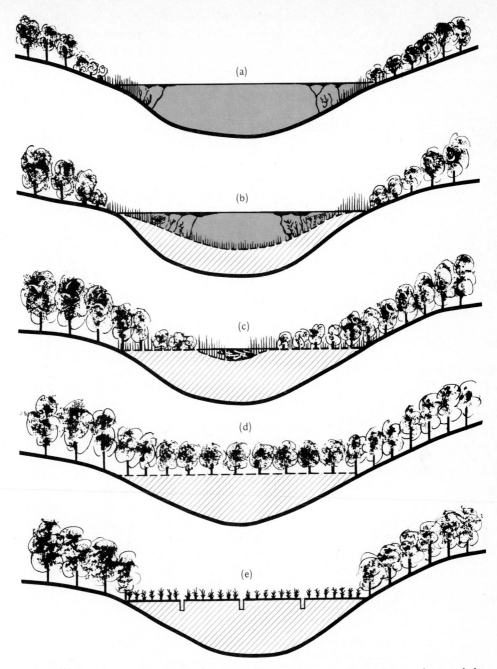

FIGURE 14.1 Four stages in the development of a typical woody peat bog and the area after clearing and draining. (a) Nutrient runoff from the surrounding uplands encourages aquatic plant growth, especially around the pond edges. (b, c) Organic debris fills in the bottom of the pond. (d) Eventually trees cover the entire area. (e) When the land is cleared and a drainage system installed, the area becomes a most productive muck soil.

zons. Their final character is determined in part by the nature of the original materials and in part by the type and degree of decomposition. Thus, the profile characteristics of organic soils, as with those dominantly mineral in nature, are in part inherited and in part acquired.

14.2 Area and Distribution of Peat Accumulations

Peat deposits are found all over the world where conditions are favorable for their formation, but they are most extensive in areas with cold climates. About 80% of the world's more than 300 million hectares of organic deposits are found in the Soviet Union and Canada, although most of these deposits are unexploited (Moore and Bellamy, 1974). Organic deposits occupy perhaps 10 million ha in Finland, 7.5 million in the United States, 3 million in Norway, and about 1.5 million each in the United Kingdom, Sweden, Poland, and Germany. In recent years attention has been called to the organic deposits found in the tropics, especially in coastal swamps. In Indonesia alone, 12 million ha are found.

The total acreage of peat deposits in the "lower 48" United States is between 10 and 12 million ha. In Alaska and northern Canada, and in northern Asia and Europe as well, there occur great areas of peat derived from sedges and mosses called *muskeg*. In many cases the subhorizons of such accumulations are perpetually frozen. They are an important feature of the tundra areas that characterize these regions.

Peat in Nonglaciated United States. Organic deposits of the nature described occur in many parts of the United States. In Florida the Everglades, spread out over an extensive plain, contain considerably over 800,000 ha (2 million A) of saw grass (sedge) accumulations. Along the Atlantic coastal plain great marsh deposits, as well as forested areas, are found, especially in North Carolina. In California there are the tube-reed beds of the great central valley, approximately 120,000 ha (300,000 A) in extent. Louisiana alone possesses about 1.2 million ha (3 million A) of organic soils. All of these are southward and outside the glaciated areas of the United States. They are related only indirectly to the glaciation through a change in climate and a rise in ocean level due to the melting of the ice.

Peat in Glaciated United States. About 75% of the peat deposits of the contiguous United States occur in the glaciated areas. Minnesota, Wisconsin, and Michigan are especially favored in this respect, their combined acreage running well above 4.8 million ha (12 million A). Washington, with about 800,000 ha (2 million A), ranks highest of all the western states. Indiana, Massachusetts, New York, and New Jersey fall into the 120,000–200,000 ha class. Other states, especially Iowa, Illinois, and Maine, contain smaller but often very important areas.

The Pleistocene glaciation, by impeding drainage, led to the formation of swamps and bogs. As the climate became milder and gradually attained its present status, conditions became ideal for swamp vegetation to flourish. As a result, certain parts of the glaciated region are organic accumulations ranging from a few centimeters to 15 m (50 ft) in depth. Some large areas, tens of square kilometers in extent, are found in Minnesota, while in other localities the peat lies in isolated patches or in long ribbons, as in southeastern Wisconsin and in central New York.

14.3   Peat Parent Materials

Peat, regardless of its stage of decomposition, may conveniently be classified according to its parent materials under three general heads.[2]

1. *Sedimentary peat.* Mixtures of water lilies, pondweed, hornwort, pollen plankton, etc.
2. *Fibrous peat.* Sedges of various kinds; mosses (sphagnum, hypnum, and others); reeds and other grasses; cattails, both latifolia and angustifolia; and their mixtures.
3. *Woody peat.* Deciduous and coniferous trees and their undergrowth.

Sedimentary Peat.   Sedimentary peat usually accumulates in comparatively deep water and therefore generally is found well down in the profile. Sometimes it is intermixed with the other types of peat nearer the surface. Sedimentary peat seems to be derived from plant materials that humify rather freely. Owing to the nature of the original tissue and perhaps also to the type of decay, a highly colloidal and characteristically compact and *rubbery* substance develops; its moisture capacity is high, perhaps four or five times its dry weight. Water thus imbibed is held tenaciously and therefore this peat dries out very slowly. The colloidal materials of sedimentary peat dry irreversibly—that is, when once dry, this peat absorbs water very slowly and persistently remains in a *hard* and *lumpy* condition.

Sedimentary peat is thus very undesirable as a soil because its unfavorable physical condition renders it unsatisfactory for use in the growing of plants. Even small amounts in the furrow slice lower the desirability of the peat for agricultural purposes. Fortunately, in most cases, it occurs well down in the

---

[2] In Germany peats are classified in a general way into *high-moor* and *low-moor* in reference to the shape of the deposits. The high-moors are convex, that is, raised in the center; the low-moors are concave. The former are usually quite acid and low in calcium, the latter less acid and quite high in exchangeable calcium. In England the corresponding terms are *moor* and *fen*. The term *heathland* refers to a shallow acid peat.

profile and its presence usually is unnoticed or ignored unless it obstructs drainage or otherwise interferes with the agricultural utilization of the peat deposit.

Fibrous Peat. A number of fibrous peats occur, often in the same swamp deposit. They are all high in water-holding capacity and may exhibit varying degrees of decomposition. They differ among themselves, especially as to their filamentous or fibrous physical nature. Undecomposed moss and sedge are fine enough to be used in greenhouses and nurseries and as a source of organic matter for gardens and flower beds. Reed and cattails, however, are somewhat coarse, especially the latter.

Fibrous materials as they decay may make satisfactory field soils, although their productivity will vary. Moss peats are almost invariably quite acid and relatively low in ash and nitrogen. The sedges are intermediate in these respects, while cattail peats are not so acid and have a better nutrient balance. Fibrous peats may occur at the surface of the organic accumulation of which they are a part or well down in the profile. They usually lie above the sedimentary deposit when this type of peat is present. Nevertheless, if just the right fluctuation of conditions has occurred, a stratum of sedimentary peat may lie embedded within the fibrous peat horizons and rather near the surface.

Woody Peat. Since trees are the vegetation present in many swamp deposits, woody peat is usually at the surface of the organic accumulation. This is not an invariable rule, however, since a rise in water level might kill the trees and so favor reed, sedge, or cattail as to give a layer of fibrous material over the woody accumulation. It is not particulary surprising, therefore, to find subhorizon layers of woody peat.

Woody peat is brown or black in color when wet, according to the degree of humification. It is loose and open when dry or merely moist and decidedly nonfibrous in character. Virgin deposits often are notably granular. It is thus easily distinguished from the other two types of peat unless the samples are unusually well disintegrated and decomposed.

Woody peat develops from the residues not only of deciduous and coniferous trees but also from the shrubs and other plants that occupy the forest floor of the swamp. Maple, elm, tamarack, hemlock, spruce, cedar, pine, and other trees occur as the climax vegetation in swamps of temperate regions. In spite of the great number of plants that contribute to its accumulation, woody peat is rather homogeneous unless it contains admixtures of fibrous materials.

The water capacity of woody peat is somewhat lower than that of sedge peat, which in turn is much less than that of moss peat. For that reason, woody peat is less desirable than the others for use in greenhouses and nurseries, where such materials are used as a means of moisture control and a compost conditioner. Woody residues produce a field soil, however, that is quite superior and much prized for the growing of vegetables and other crops. Unfortunately, such woody peat in the United States is confined mostly to Wisconsin, Michigan, and New York.

14.4 Uses of Peat

Nurseries, Greenhouses, and Lawns. In countries other than the Soviet Union and Ireland, horticultural and field agricultural uses of peat predominate. Peat products are used directly or are mixed with soil for potted plants and for home flower and vegetable gardens. They are also used extensively for commercial greenhouse production and as a mulch for home gardens and lawns. The high moisture and nutrient holding capacity, and the good physical condition of the peat is ideal to enhance the growth of greenhouse and home garden plants. Fiber pots made of compressed peats are used to start plants, which are later transplanted to soil without being removed from the pot. Peat exploitation is a profitable business in Europe and North America.

Peat as Fuel. By far the largest consumer of peat for fuel is the Soviet Union, where energy for some power stations is derived from peat. Small quantities are also used for this purpose in Ireland. The use of peat as domestic fuel is declining, but is still prevalent in remote areas of the Soviet Union, Ireland, Finland, Scotland, and West Germany.

Field Soils. In the United States, the most extensive use of peat is as a field soil, especially for vegetable production. Thousands of acres are now under cultivation, often producing two crops a year. Peats for such use should be well decomposed, the decay often having gone so far as to make it almost impossible to identify with certainty the various plants that have contributed to the accumulation. As the following pages indicate, the edaphological interest in peat relates mainly to its use as a natural soil. At the same time, its use as a soil supplement for greenhouse, nursery, and home gardening has definite edaphological implications.

14.5 Classification of Organic Soils

Two types of classification are employed, one using traditional field terminology and the other the terminology of the *Soil Taxonomy* classification system. In each case, considerable emphasis is placed on the stage of breakdown of original plant materials.

Peat Versus Muck. As a practical matter, organic soils are commonly differentiated on the basis of their state of decomposition. Those deposits that are slightly decayed or nondecayed are termed *peat;* those that are markedly decomposed are called *muck.* In peat soils, one can identify the kinds of plants

FIGURE 14.2 Graph showing the line separating organic from nonorganic soil materials. Soils with clay and organic matter contents above the line are organic soils, those below are mineral soils. [*From McKenzie (1974); used with permission of the Soil Science Society of America.*]

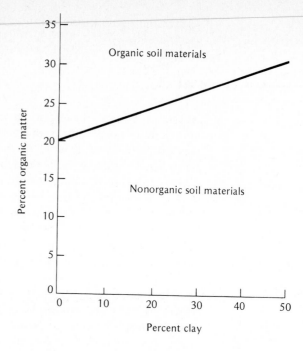

in the original deposit, especially in the upper horizons. By contrast, mucks are generally decomposed to the point where the original plant parts cannot be identified.

Peat soils are coarse or fine-textured depending on the nature of the deposited plant residues. Well-decomposed mucks, on the other hand, are usually quite fine since the original plant material has broken down. When dry, they may be quite powdery and subject to wind erosion.

Soil Taxonomy—Orders. In this system organic soils are identified as the order *Histosols* (see Section 13.7). This order contains a minimum of 20% organic matter (12% carbon) if the clay content is low, and a minimum of 30% (18% carbon) if as much as 50% clay is present (Figure 14.2).

Four suborders have been established. The *Fibrist* suborder includes soils in which the undecomposed fibrous organic materials are easily identified even after a sample of the moist material is rubbed between the thumb and finger. Their bulk densities are low, commonly less than 0.1 Mg/m³. They have high water-holding capacities and are brown and yellow in color. They usually develop under cold climates and other environmental conditions that discourage decomposition.

The most highly decomposed organic materials are found in soils classed in the *Saprist* suborder. The original plant fibers have mostly disappeared. The bulk densities of these soils are commonly 0.2 Mg/m³ or more. Water-holding capacities are the lowest for organic soils. Their color is commonly very dark gray to black and they are quite stable in their physical properties.

The *Hemist* suborder includes soils intermediate in their properties between those of the Fibrists and Saprists. They are more decomposed than the Fibrists but less so than the Saprists. They have intermediate values for bulk density (between 0.1 and 0.2) and water-holding capacity. Their colors are also intermediate between those of the other two suborders.

The *Folist* suborder consists of well-drained Histosols consisting primarily of a surface organic horizon developed from leaf litter, twigs and branches that rest on rock or unconsolidated materials such as gravel, stones and boulders. The mineral pore space is filled or partly filled with the organic materials. Folists are found mostly in the humid tropics.

In addition to visual appearance, a chemical test is used to ascertain the degree of decomposition of the original plant material in organic soils. An alkaline solution of sodium pyrophosphate dissolves certain organic acids which are the products of decomposition. The amount of the acids dissolved by the agent is related to tissue decomposition. This method provides some quantification of an otherwise subjective judgment of decomposition.

Drawings of organic soil profiles representative of the Saprist and Fibrist suborders are shown in Figure 14.3. The Fibrist soil has a high percentage of undecomposed fibrous materials and is very acid. The cultivated Saprist soil has a nearly neutral pH, low undecomposed fiber percentage, and is darker in color.

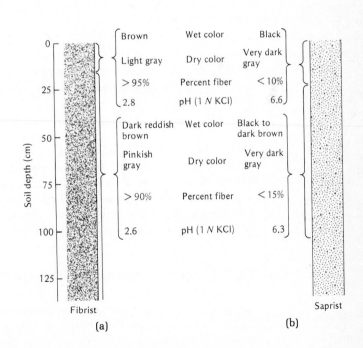

FIGURE 14.3 Two organic soil profiles showing distinguishing characteristics of each. (a) An uncultivated soil of the Fibrist suborder with a high percentage of undecomposed fiber and high acidity. (b) In contrast, a cultivated soil of the Saprist suborder with only 10–20% undecomposed fibers and reasonably low acidity. Both soils are found in Minnesota. [*Modified from Farnham and Finney (1966)*.]

TABLE 14.1 Names of Great Groups in the Four Suborders of Histosols

Note that soil temperature as the primary criterion for differentiating among these great groups. High sphagnum content in cool climate fibrist areas and the presence of acid-forming sulfur compounds in hemists are also differentiating characteristics.

	Fibrists	Saprists	Hemists	Folists
Mean annual temperature <8°C Frost in profile (frigid and isofrigid)	Cryofibrists Sphagnofibrists[a]	Cryosaprists	Cryohemists	Cryofolists
Mean annual temperature <8°C No frost (frigid)	Borogibrists Sphagnofibrists[c]	Borosaprists	Borohemists	Borofolists
Mean annual temperature >8°C (mesic, thermic, hyperthermic)	Medifibrists	Medisaprists	Medihemists	
Mean annual temperature >8°C and <5°C between mean summer and winter soil temperatures (isomesic, isothermic and isohyperthermic)	Tropofibrists	Troposaprists	Tropohemists	Tropofolists
Presence of sulficid materials[b] No climatic constraints			Sulfohemists Sulfihemists	

[a] Fibrists derived primarily from sphagnum.
[b] Sulfur compounds that result in development of acid sulfate soils when the area is drained; sulfohemists are acid sulfate soils, sulfihemists are potential sulfate soils.
[c] See Section 13.9 for soil temperature classes.

Great Groups and Other Categories. The classification of Histosols at the great group level and below is still being field tested, but the great groups currently identified are shown in Table 14.1. Note that soil temperature is the primary criteria for differentiating among the great groups, although high sphagnum peat content and the presence of acid forming sulfur compounds are also considered. Subgroups are determined by a number of properties such as the presence of more than one kind of organic material, of distinct mineral layers in the profile, of underlying rock, of permafrost or of open water. Families are differentiated on the bases of properties such as particle size, type of intermixed minerals, soil pH, soil depth, and temperature. The applicability of this classification scheme is being tested.

14.6 Physical Characteristics of Field Peat Soils

Color. The color of a typical cultivated organic soil, dark brown or intensely black, is perhaps the first physical characteristic that attracts attention. Although the original materials may be gray, brown, or reddish brown, dark humic compounds appear as decomposition takes place. In general, the changes that the organic matter undergoes seem to be somewhat similar to those occurring in the organic residues of mineral soils in spite of the restricted aeration of the peat deposit.

Bulk Density. The second outstanding characteristic is the light weight of the representative organic soil when dry. The bulk density compared with mineral surface soils is low, 0.20–0.30 Mg/m³ being common for a well-decomposed organic soil. A cultivated surface mineral soil will usually fall within 1.25–1.45 Mg/m³. A hectare–furrow slice of the depth common in mineral soils, though variable, may be considered to weigh from 450,000 to perhaps 550,000 kg when dry. Compared with the 2.25 or 2.5 million kg ordinarily considered as the dry weight of a hectare–furrow slice of a representative mineral soil, such figures seem small indeed.

Water-Holding Capacity. A third important property of organic soil is its high water-holding capacity on a weight basis. While a dry mineral soil will adsorb and hold from one fifth to two fifths its weight of water, an organic soil will retain two to four times its dry weight of moisture. Undecayed or only slightly decomposed moss or sedge peat, in contrast with soils developed from these materials, has a much greater water-holding capacity. Not uncommonly, such materials can hold water to the extent of 12, 15, or even 20 times their dry weights. This explains in part their value in greenhouse and nursery operations.

Unfortunately organic soils in the field do not greatly surpass mineral soils in their capacity to supply plants with water for two reasons. First, their amounts of unavailable water are much higher proportionately than that of mineral soils. Second, the light weight of organic soils means that a given *volume* of organic soil may not adsorb much more water than the same volume of a mineral soil. Thus a given volume of peat soil at optimum moisture will supply only slightly more water to plants than a comparable mineral soil.

Structure. A fourth outstanding characteristic of a typical woody or fibrous organic soil is its almost invariably loose physical condition. While decayed organic matter is largely colloidal and possesses high adsorptive powers, its cohesion and plasticity are rather low. Most organic soils are therefore porous, open, and easy to cultivate. These characteristics make it especially desirable for vegetable production. However, during dry periods, light, loose peat, whose structure has been destroyed by cultivation, may drift badly in a high wind and extensive crop damage may result. Peat may also become ignited when dry. Such a fire often is difficult to extinguish and may continue for several years.

In suggesting that the surface layer of an arable peat soil is likely to be granular, it is not to be inferred that the whole profile is in this structural condition also. Far from it. All sorts of physical arrangements are encountered in the subhorizons, such as laminated, vertical, fragmental, fibrous, and rubbery.

The variability from horizon to horizon is determined not only by the character of the original materials but also by the nature and degree of decomposition.

14.7 The Colloidal Nature of Organic Soil

The colloidal nature of organic soil material has already been emphasized (Section 5.3). The same graphic formula employed for mineral soils can be used to represent the colloidal complex of peat.

$$
\begin{matrix}
\text{Ca} \\
\text{H} \\
\text{Mg}
\end{matrix}
\boxed{\text{Micelle}}
$$

The micelle is humus, and the metallic cations are adsorbed in the same order of magnitude as in mineral soils: Ca > Mg > K or Na (see Section 5.4).

Colloidal properties.   Organic colloids have specific surface areas two to ten times those of mineral soil colloids. On a weight basis maximum cation exchange capacities are correspondingly greater for the organic colloids. But on a volume basis, as is seen in Table 14.2, most of this difference disappears. In fact, a cubic meter of vermiculite has about twice the cation exchange capacity of an equal volume of organic colloid. But the fine-grained micas and hydrous oxide clays have lower capacities than the humus, even on a volume basis.

TABLE 14.2   Representative Maximum Cation Exchange Capacities of an Organic Colloid and Several Inorganic Colloids

On a weight basis, the CEC of humus greatly exceeds that of the minerals, but on a volume basis both vermiculite and smectites have higher capacities.

	Cation exchange capacity	
Colloid	Weight basis (cmol/kg)	Volume basis (cmol/L)
Humus (organic)	300	75
Vermiculite	120	150
Smectite	90	113
Fine-grained micas	25	31
Kaolinite	5	6
Hydrous oxides	3	4

Since plants, whether in the field or in the greenhouse, absorb their nutrients from a given volume rather than weight of soil, the volume basis for evaluating cation exchange capacity appears to be more appropriate. In some deep, artificially drained peat soils, however, these loose open soils encourage deep root penetration and a resulting large volume of soil from which nutrients are absorbed.

Under acid conditions, the cation exchange capacities of the 2:1-type silicate clay colloids exceeds that of organic colloids even more than is shown in Table 14.2. These mineral colloids are characterized by permanent negative charges, thereby assuring cation adsorption and exchange at all pH levels. In contrast, the negative charges on the organic colloids are pH-dependent, and are very low under acid soil conditions.

Forms of Nutrient Elements.   In mineral soils, most of the essential nutrients are held in relatively unavailable forms, only small amounts being in the exchangeable and soil solution forms. As the data in Table 14.3 illustrate, the opposite is true for organic soils. High percentages of calcium, magnesium, and potassium are held in exchangeable form by the organic colloids. Thus, most of these soil-held cations are available for plant growth and for soil microorganisms.

Strength of Acids.   In general, the colloidal complex of organic colloids when saturated with hydrogen will develop a lower soil pH than will acid mineral clays similarly charged. In other words, the organic complex is the stronger acid. This means that at the same percentage base saturation, peat has a lower pH than mineral soils.

Buffering and pH.   On a weight basis, organic soils have a higher buffering capacity against pH change than do mineral soils. Consequently, considerably more lime or sulfur is needed to change the pH of a given weight of peat than that of a mineral soil. Again, however, on a volume basis, the differences

TABLE 14.3   Percent of Three Cations Commonly Found in Exchangeable Forms in Organic and Mineral Soils

The remainder is held in the solid framework of the soils.

Cation	Percent in exchangeable form	
	Mineral soil	Organic soil
Calcium	25	80
Magnesium	7	70
Potassium	0.5	35

TABLE 14.4 Exchange Data for Representative Organic and Inorganic Soils

Exchange characteristics	Weight basis (cmol/kg)		Volume basis (cmol/L)	
	Organic soil	Mineral soil	Organic soil	Mineral soil
Exchangeable Ca	150	8	38	10
Other exchangeable bases, M	40	3	10	4
Exchangeable H and Al	60	5	15	6
Cation exchange capacity	250	16	63	20
Percentage base saturation	76	69	76	69
pH	5.0–5.2	5.6–5.8	5.0–5.2	5.6–5.8

are much less. The plow layer of a peat soil may be even more responsive to pH change with liming than that of mineral soils.

It is well to note that the pH of a peat soil at a given percent base saturation is generally lower than that of a representative mineral soil. Data in Table 14.4 illustrate this point. The representative organic soil has a higher percentage base saturation than the mineral soil, yet the pH of the organic soil is lower. Note also that the difference in cation exchange capacity between the organic and mineral soils is much less when they are compared on a volume basis.

14.8 Chemical Composition of Organic Soils

Organic soils are quite variable chemically, but they have some characteristics in common, as is shown by data from fifteen organic soils (Table 14.5). Also, a representative analysis of the surface layer of an arable organic soil is given in Table 14.6, but with the understanding that it is merely an example. The chemical analysis of the upper layer of a representative humid region mineral soil is cited for comparison and contrast.

Nitrogen and Organic Matter. The high nitrogen and organic contents of organic soils are self-evident and need no further emphasis. But two interrelated features justify consideration. First, organic soils have a high carbon/nitrogen ratio, the minimum being in the neighborhood of 20. In contrast, this equilibrium ratio for a cropped mineral soil is usually around 10–12.

Second, organic soils show exceedingly vigorous nitrification in spite of their high carbon/nitrogen ratio. In fact, the nitrate accumulation, even in a low-lime peat, is usually greater than that of a representative mineral soil. Apparently, this is due to the presence of adequate calcium, and the inactivity of part of the carbon. Thus, the *effective* carbon/nitrogen ratio of organic soils may be as narrow as that of mineral soils. As a result, competitive general-

TABLE 14.5 Chemical Analyses of Certain Representative Peat Soils[a]

Values in percent based on dry matter.

Source	Nature	Organic matter	N	P_2O_5	K_2O	CaO
Minnesota	Low lime	93.0	2.22	0.18	0.07	0.40
Minnesota	High lime	79.8	2.78	0.24	0.10	3.35
Michigan	Low lime	77.5	2.10	0.26	—	0.16
Michigan	High lime	85.1	2.08	0.25	—	6.80
German	Low lime	97.0	1.20	0.10	0.05	0.35
German	High lime	90.0	2.50	0.25	0.10	4.00
Austria	Low lime	93.3	1.40	0.10	0.06	0.45
Austria	High lime	83.8	2.10	0.18	0.13	2.38
New York	Low lime	94.2	1.26	0.15	0.10	0.60
New York	High lime	81.5	2.56	0.19	0.28	6.51
Minnesota	Low organic	59.7	2.35	0.36	0.17	2.52
Minnesota	High organic	94.0	1.70	0.16	0.04	0.31
Florida	Sawgrass peat	87.1	2.79	0.41	0.04	5.20
Canada	Peat soil	74.3	2.19	0.20	0.16	—
Washington	Woody sedge peat	89.2	3.52	0.43	0.09	1.29

[a] From Lyon et al. (1952).

purpose heterotrophic organisms are not excessively encouraged, and the nitrifiers are given ample opportunity to oxidize the ammoniacal nitrogen.

Phosphorus and Potassium. The phosphorus and potassium of peat are both low, the latter exceedingly so in comparison with a mineral soil. Even the phosphorus is actually less in kilograms per hectare–furrow slice (Table 14.6). On this basis, the representative mineral soil is cited as containing 1100 kg of phosphorus. An equivalent layer of organic soil (say 550,000 kg) would furnish only 825 kg of phosphorus. This explains why, in growing crops on an organic soil, phosphorus as well as potash must be applied in large amounts.

TABLE 14.6 Analyses for a Representative Peat and Mineral Surface Soils

Constituent	Peat		Mineral	
	g/100 g soil	kg/HFS	g/100 g soil	kg/HFS
Organic matter	80.00	440,000	4.0	88,000
Nitrogen (N)	2.50	13,750	0.15	3,300
Phosphorus (P)	0.15	825	0.05	1,100
Potassium (K)	0.10	550	1.70	37,400
Calcium (Ca)	2.00	11,000	0.40	8,800
Magnesium (Mg)	0.30	1,650	0.30	6,600
Sulfur (S)	0.60	3,300	0.04	880

Calcium and pH. The high calcium content of organic soils is apparently due to the high content of this element in the seepage water that enters the swamp from the surrounding uplands. The calcium is not subject to leaching, as with mineral soils, and is adsorbed by the organic materials. Exchangeable calcium is an outstanding characteristic of many organic soils developed in areas when there is substantial calcium in the surrounding mineral uplands.

Even when reasonable levels of calcium are available, the majority of organic soils are distinctly acid, often very markedly so. It is not all uncommon to find organic soils with pH values of 5.5 or less even though the calcium content is as high as 2.5–3.0%. With high cation adsorption capacities, organic soils at a low percentage base saturation can be carrying exceptionally large amounts of exchangeable calcium even though they are decidedly acid.

Magnesium and Sulfur. The percentage of magnesium in organic soil is usually no greater than that of a mineral soil. The amount per hectare–furrow slice, however, is much less because of the low dry weight of peat by volume (Table 14.6). Organic soils, long intensively cropped, may develop a magnesium deficiency unless fertilizers carrying this constituent have been used. Since most organic soils are seldom limed, there is little chance of adding magnesium in this particular way.

The abundance of sulfur in organic soils is expected since plant tissue always contains considerable sulfur. Consequently, organic deposits such as peat should be comparatively high in this constituent. When sulfur oxidation is vigorous, as is usually the case with arable organic soils, sulfates may accumulate. Sulfur is abundant enough in most organic soils to reduce the possibility of this element being a limiting factor in plant growth.

Anomalous Features. Organic soils, in comparison with mineral soils, exhibit three somewhat anomalous features. First, organic soils possess a high carbon/nitrogen ratio and yet support vigorous nitrification. Second, organic soils that are high in calcium still may be definitely acid. And third, in the presence of a high hydrogen ion concentration, nitrate accumulation takes place far beyond that common in mineral soils with the same low pH. This last feature indicates that the hydrogen ion concentration does not in itself impede nitrification. Apparently, the high calcium content and the low content of soluble iron, aluminum, and manganese in organic soils account for this anomaly.

14.9 Bog Lime—Its Importance

Some organic soils are underlain at varying depths by a soft, impure calcium carbonate called *bog lime* or *marl*. Bog lime, when it contains numerous shells, is often termed *shell marl*. Such a deposit may come from the shells of certain of the Mollusca or from aquatic plants such as mosses, algae, and species of *Chara*. These organisms are able to precipitate the calcium as insoluble calcium carbonate. It seems that these plants and animals occupied the basin before or in some cases during the formation of the peat, the marl resulting from the accumulation of their residues on the bottom of the basin or farther up in the profile.

When marl is present, it not only may supply calcium to the circulating waters but, if high in the profile, may actually become mixed with the surface peat. As a result, such organic soils are likely to be low in acidity or even alkaline. In general, an alkaline peat is not considered as highly desirable for intensive culture, especially the growing of vegetables, as are moderately acid peats.

14.10   Factors That Determine the Value of Peat and Muck Soils

Drainage and Water Table.   The value of peat agriculturally will depend on a number of factors. Of first consideration is the possibility of drainage and control of the water table sufficient to allow an adequate aeration of the root zone during the growing season. It may be advantageous to raise or lower the water table at various times during the season. For instance, celery at setting is benefited by plenty of moisture. As the crop matures, the water table should be gradually lowered to accommodate the root development.

Depth and Quality.   Peat settles considerably during the first few years after drainage and clearing for cultivation and continues to subside appreciably thereafter. Subsidence resulting from oxidation of the organic matter in the peat continues to be a problem if the peat deposit is not deep. Rates of subsidence of up to 5 cm a year have been noted (Parent et al., 1982). A depth of 1–1.5 m of organic material is desirable, especially if calcareous clay or marl underlies the deposit.

The quality of the peat as indicated by the degree of decay and the nature of the original plant materials is important. The presence of the rubbery sedimentary type in the furrow slice is especially undesirable (Section 14.3). Woody peat, on the other hand, is generally considered more desirable than that coming from cattails, reeds, and other plants, and it usually is highly prized wherever it occurs. Much of the acreage of this type of peat soil in the United States lies in Wisconsin, Michigan, and New York.

14.11   Preparation of Peat for Cropping

Since organic soils are often forested or covered with shrubs and other plants, the first step is to drain and clear the land. If it is then used as pasture for a few years, the roots and stumps are given time to decay, and their removal is relatively easy when the land is finally fitted for cultivation.

A peat area may be burned over to remove the brush and other debris, but this should be done while the land is wet and not likely to catch fire. Such a fire is often difficult to extinguish and may ruin the deposit by destroying the more fertile surface layer.

After breaking the peat soil, preferably with a heavy plow drawn by a tractor powerful enough to mash down second-growth brush or sapling trees, it is advisable to grow such crops as corn, oats, or rye for a year or two, as they do well on raw and uneven peat lands. Once the peat is adequately weathered, freed of roots and stumps and all hummocks eliminated, it is ready for vegetable production. More thorough drainage is now required and is usually obtained by a system of ditches. Sometimes tile drains or even mole drains are used (see Section 16.6).

14.12   Management of Peat Soils

All sorts of vegetable crops may be grown on organic soils—celery, lettuce, spinach, onions, potatoes, beets, carrots, asparagus, and cabbage perhaps being the most important, along with such specialized crops as peppermint. In some cases, peat is used for field crops but the high-valued vegetable and nursery crops are more common. Certain nursery stocks do well on peat. In Europe, organic soils have been used for pasture and meadows. In fact, almost any crop will grow on organic soil if properly managed. For a view of peat under cropping see Figure 14.4.

Structural Management.   Plowing is not ordinarily necessary every year since peat is porous and open unless it contains considerable silt and clay. In fact, a cultivated organic soil generally needs packing rather than loosening. The longer a peat has been cropped the more important compaction is likely to be. Cultivation tends to destroy the original granular structure, leaving the soil in a powdery condition when dry. It is then susceptible to wind erosion, a very serious problem in some sections (Figure 14.4).

For this reason, a roller or packer is an important implement in the management of such land. The compacting of organic soils allows the roots to come into closer contact with the soil and facilitates the rise of water from below. It also tends to reduce the blowing of the soil during dry weather, although a windbreak of some kind is much more effective. The cultivation of peat, while easier than for mineral soils, is carried on in much the same way and should be shallow, especially after the root development of the crop has begun.

Use of Lime.   Lime, which so often must be used on mineral soils, ordinarily is less necessary on organic soils unless they have developed in regions low in calcium in the surrounding uplands. On acid mucks containing appreciable quantities of inorganic matter, however, the situation is quite different. The highly acid conditions result in the dissolution of iron, aluminum, and manga-

FIGURE 14.4 Windbreaks, such as these in Michigan, help protect valuable muck land from blowing. The unprotected field in the lower left has been wetted by sprinkler to prevent its blowing. [*Courtesy USDA Soil Conservation Service.*]

nese in toxic quantities. Under these conditions, large amounts of lime may be necessary to obtain normal plant growth.

Commercial Fertilizers. Of much greater importance than lime are commercial fertilizers. In fact, these materials are needed for the production of most crops, especially vegetables. As organic soils are very low in phosphorus and potassium, these elements must be added. Since vegetables usually are rapid-growing plants, succulence often being an essential quality, large amounts of readily available nitrogen are necessary. The nitrogen of newly cleared peat is often available rapidly enough to supply this need, especially for oats, rye, corn, wheat, and similar crops. Such land requires at the beginning only a small amount of nitrogen with the phosphoric acid and potash. After organic soils have been cropped for a few years, decay and nitrification are frequently too slow to meet the crop demand for nitrogen. Under such conditions this element is needed in larger amounts and a fertilizer containing more nitrogen as well as phosphoric acid and potash is usually recommended.

Micronutrients. Peat soils are in need of not only potassium, phosphorus, and nitrogen, but often some of the trace elements as well. Copper sulfate and salts of manganese and zinc are used to meet plant needs on both peat and muck soils. Boron deficiencies are also becoming evident. Common salt in addition seems to give good results for beets, celery, and cabbage. Perhaps both the sodium and the chlorine play important rôles in crop nutrition on these soils.

14.13   Organic Versus Mineral Soils

In assigning organic soils to a separate chapter, one is encouraged to think of them as distinctly and even radically different from most mineral soils. In many respects this certainly is true. Yet, fundamentally, the same types of change occur in the two groups; nutrients become available in much the same way and their management is based upon the same principles of fertility and water management.

REFERENCES

Buckman, H. O., and N. C. Brady. 1952. *The Nature and Properties of Soils*, 5th ed. (New York: Macmillan).

Farnham, R. S., and H. R. Finney. 1966. "Classification of organic soils," *Adv. in Agron.*, **17**:115–62.

McKenzie, W. E. 1974. "Criteria used in soil taxonomy to classify organic soils," in *Histosols: Their Characteristics, Classification and Use*, SSSA Special Publication No. 6 (Madison, WI: Soil Sci. Soc. Amer.), pp. 1–10.

Moore, P. D., and D. J. Bellamy. 1974. *Peatlands* (London: Unwin).

Parent, L. E., J. A. Millette, and G. R. Mehuys. 1982. "Subsidence and Erosion of a histosol," *Soil Sci. Soc. Amer. J.*, **46**:404–408.

Soil Science Society of America. 1974. *Histosols: Their Characteristics, Classification and Use*, SSSA Special Publication No. 6 (Madison, WI: Soil Sci. Soc. Amer.).

Losses of Soil Moisture and Their Regulation

[*Preceding page*] Drip irrigation of this young grape vineyard illustrates the need to conserve soil moisture by making the water available to the plant without exposing it to losses through evaporation from the soil. [*Courtesy USDA Soil Conservation Service.*]

Wise use and management of water are perhaps the most critical factor in schemes to increase food supplies, especially in the population-heavy developing nations. At the same time, industrial and domestic requirements for water have expanded dramatically, creating strong competition for traditional agricultural uses. This competition forces increased emphasis on efforts to manage and conserve water and to prevent its misuse.

The central role of soils and plants in the hydrological cycle[1] is illustrated in Figure 15.1. This figure also shows the interrelationship among the various uses of water. Further, it emphasizes the need to minimize the loss of water from plant and soil surfaces in both the vapor and liquid forms.

15.1 Interception of Rainwater by Plants

Rainfall data are often dealt with as though all precipitation reaches the soil. Such is far from the case, expecially where the vegetative cover is dense. Part of the rain or snow is intercepted by the plant leaves and stems and is evaporated directly into the atmosphere without reaching the soil.

The extent of plant interception of precipitation is determined by a number of factors. In areas with perennial standing cover, such as range and natural grasslands and especially forests, the annual interception is much more than where annual field crops are grown. Forested areas, chiefly those with year-round foliage (coniferous) are particularly effective in intercepting precipitation. One third to one half of the annual precipitation in such forests commonly does not reach the soil, being intercepted by the plant parts and evaporated directly into the atmosphere.

Field and vegetable crops that occupy the land for only part of the year intercept much less of the annual precipitation. Even so, it is not at all uncommon to find that 10–20% of the seasonal rainfall falling on these crops does not reach the soil (Table 15.1). During periods of light rainfall a high proportion of the rain may be intercepted in a dense stand of a crop like alfalfa. While such losses are unavoidable, they must be taken into account when considering soil–water–plant relations.

15.2 The Soil–Plant–Atmosphere Continuum

The emergence of the concept of water potential brought with it the realization that the same basic principles applied for the retension and movement of water whether it be in soil, in plants, or in the atmosphere. Further came the recogni-

[1] Excellent reviews of soil–water–plant relations are given in Kozlowski (1968–72).

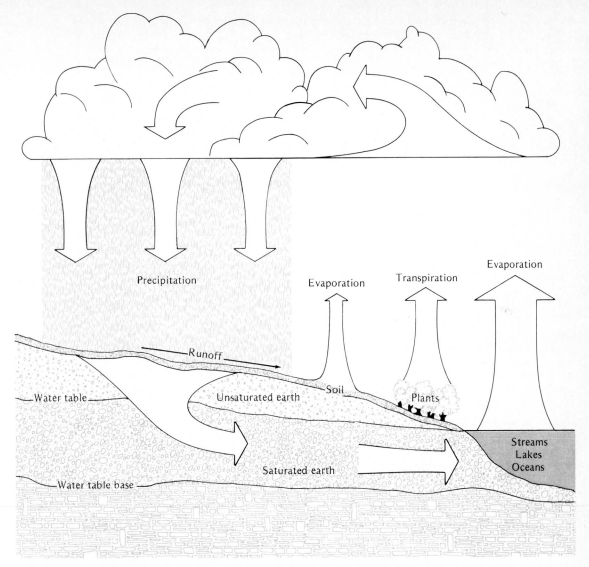

FIGURE 15.1   Diagram of the hydrologic cycle showing the interrelationships among soil, plants, and the atmosphere. Only water is lost from soils in evaporation and transpiration from plant surfaces. Solic particles are often included in surface runoff water, and nutrients in percolation water. *[From Science of Food and Agriculture (1983); used with permission of the Council for Agricultural Science and Technology, Ames, IA.]*

tion of an unified and dynamic soil–plant–atmosphere system in which water flows from areas of high to those of low free energy. The system has been termed the *soil–plant–atmosphere continuum* (SPAC) (Phillip, 1966).

Figure 15.2 illustrates in a general way the function of the SPAC. Water moves from the soil to the plant to the atmosphere and back to the soil. In

TABLE 15.1 Seasonal Interception of Rainfall by Crops at Bethany, Missouri, and Sussex, New Jersey[a]

Average of 3 years' records for alfalfa and corn and 1 year for soybeans. Values in percent of total seasonal rainfall for each crop.

Fate of rainfall	Alfalfa	Corn	Soybean
Direct to soil	64.7	70.3	65.0
Ran down stem	13.7	22.8	20.4
Total to soil	78.4	93.1	85.4
Remainder to atmosphere	21.6	6.9	14.6

[a] From Haynes (1954).

FIGURE 15.2 Soil–plant–atmosphere continuum. In recent years scientists have come to realize that a common set of basic principles governs relationships among water, soil, plants, and the atmosphere. This continuum is best illustrated by following water as it is added to soils through precipitation or irrigation, as it behaves in soils, is lost directly to the atmosphere from soils, or is absorbed by plants, transported upward, and subsequently evaporated into the atmosphere. Water behavior in all cases is subject to the same basic physical and chemical laws.

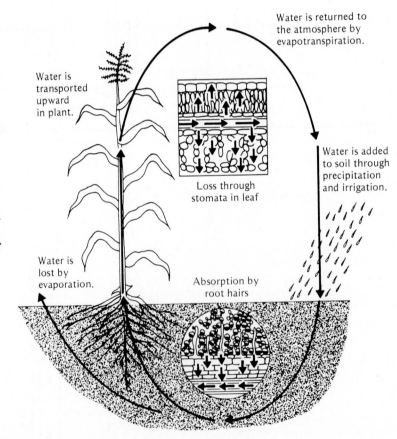

Water is returned to the atmosphere by evapotranspiration.

Water is transported upward in plant.

Loss through stomata in leaf

Water is added to soil through precipitation and irrigation.

Water is lost by evaporation.

Absorption by root hairs

this cycle or continuum the movement of water is determined by the moisture potential and by the resistance to flow set up by features such as the size and configuration of pores and the permeability of membranes.

SPAC Energy Relations.   In Chapter 3, the free energy level of water was seen to be a major controlling factor in determining soil–water behavior. The same can be said for soil–plant and plant–atmosphere relations. As water moves through the soil to plant roots, into the roots, across cells into stems, up the plant xylem to the leaves, and is evaporated from the leaf surfaces, its tendency to move is determined by differences in free energy levels of the water, or by the moisture tension.

The moisture tension must be lower in the soil than in the plant roots if water is to be absorbed from the soil. Likewise, movement to the xylem, up the stem through the xylem, and to the leaf cells is related to differences in moisture tension (Figure 15.3). The negative pressures found in stem xylem of plants suggest this relationship. Water tensions in the xylem may reach 20 bars at the wilting point for field crops and are commonly 20 to 25 bars in the upper limbs of tall trees growing in soils amply supplied with water. Desert plants will exist with tensions of 20 to 80 bars, although growth appears to stop at tensions below about 20 bars.

Two Points of Resistance.   Changes in water tension illustrated in Figure 15.3 suggest major resistance at two points as the water moves through the SPAC: the movement of water into the plant from soil and its evaporation into the atmosphere from the surface of cells in the leaf. This is in accord with field observations that two primary factors determine whether plants are well supplied with water: (a) the rate at which water is supplied by the soil to the absorbing roots and (b) the rate at which water is evaporated from the plant leaves. Factors affecting the soil's ability to supply water were discussed in

FIGURE 15.3   Change in moisture tension as water moves from the soil through the root, stem, and leaf to the atmosphere. Note that water tension increases as the water moves through the system. [*Adapted from Hillel (1971).*]

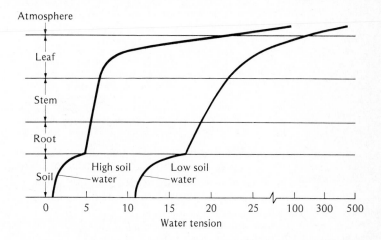

Chapter 3. Attention will now be given to the loss of water by evaporation and factors affecting this loss.

15.3   Evapotranspiration

Vapor losses from soils occur in two ways: (a) by the *evaporation* of water at the soil surface and (b) by *transpiration* from the leaf surfaces of water absorbed by the plants and translocated to the leaves. The combined loss resulting from these two processes, termed *evapotranspiration,* is responsible for most of the water removal from soils under normal field conditions. On irrigated soils located in arid regions, for example, it commonly accounts for the loss of 75–100 cm of water during the growing season of a crop such as alfalfa. Obviously, the phenomenon is of special significance to growing plants.

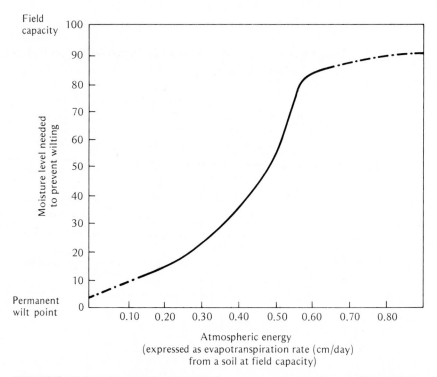

FIGURE 15.4   Moisture level needed to prevent turgor loss (wilting) of corn grown on a Colo silty clay loam under different conditions of atmospheric energy. The moisture level ranges from the permanent wilting point (0) to the field capacity (100). The atmospheric energy, which includes such factors as temperature and wind velocity, is expressed in terms of the evapotranspiration that it causes. The broken-line portions are extrapolations from actual measurements. [*Data adapted from Denmead and Shaw* (*1962*).]

The rate of water loss by either evaporation from the soil or transpiration is determined basically by differences in moisture potential identified as the *vapor pressure gradient*—the difference between the vapor pressure at the leaf or soil surface and that of the atmosphere. The vapor pressure gradient in turn is related to a number of other common climatic and soil factors, some of which will now be considered (Figure 15.4).

Radiant Energy. For every gram of water vaporized at 25°C, about 540 cal. of energy is required, whether the evaporation is from the soil directly or from leaf surfaces. The primary source of this energy is the sun (see Figure 4.12). So long as ample soil moisture is available, there is a close relationship between evaporation and the absorption of this radiant energy. In arid regions where cloud cover is sparse, solar radiation is high and evapotranspiration is encouraged. In contrast, in regions characterized by cloudy days, less solar radiation strikes the soil and plants, and the evaporative potential is not so great.

Atmospheric Vapor Pressure. The vapor pressure of the atmosphere along with the vapor pressure at the soil or plant surfaces determines the vapor pressure gradient and in turn the rate of evaporation from soils and plants. If the atmospheric vapor pressure is low compared to the vapor pressure at the plant and soil surfaces, evaporation takes place rapidly. If it is high, as on "humid" days, evaporation is slow. This accounts for relatively high vapor losses from irrigated soils in arid climates and the much lower losses in humid regions at comparable temperatures.

Temperature. The rate of evaporation of water is greatly influenced by temperature. Consequently, during warm or hot days, the vapor pressure at the leaf surfaces or the surface of a moist soil is quite high. Temperature does not have as correspondingly great an effect on the vapor pressure of the atmosphere. As a result, on hot days there is a sharp difference in vapor pressure between leaf or soil surfaces and the atmosphere (a higher vapor pressure gradient) and evaporation proceeds rapidly. Plants and especially soils may be warmer than the atmosphere on bright, clear days, further emphasizing the importance of temperature in controlling evapotranspiration.

Wind. A dry wind will continually sweep away moisture vapor from a wet surface. The moist air thus moved is replaced by air with a lower moisture content, thereby maintaining the vapor pressure gradient, and evaporation is greatly encouraged. The drying effect of even a gentle wind is noticeable even though the air in motion may not be at a particularly low humidity level. Hence, the capacity of a high wind operating under a steep vapor pressure gradient to intensify evaporation both from soils and plants is tremendous. Farmers of the Great Plains dread the *hot winds* characteristic of that region.

TABLE 15.2 Effect of Soil Moisture Level on
Evapotranspiration Losses[a]

*Where the surface moisture content was kept high, total
evapotranspiration losses were greater than when medium
level of moisture was maintained.*

Moisture condition of soil[b]	Evapotranspiration (cm)	
	Corn	Alfalfa
High	45	62
Medium	32	52

[a] Calculated from Kelly (1954).

[b] High moisture—irrigated when upper soil layers were 50%
depleted of available water. Medium moisture—irrigated when
upper layers were 85% depleted of available water.

Soil Moisture Supply. Evapotranspiration is higher where plants are grown
on soils at or near the field capacity than where soil moisture is low (Table
15.2). This is to be expected since at low soil moisture levels water uptake
by plants is restricted.

Soil physicists are agreed that the depth to which soils may be depleted
by capillarity and evaporation from the soil surface is far short of the 1–2 m
or even more sometimes postulated. Some investigators think that a 50–60
cm depth is probably a maximum range. In most cases, only the water of
the furrow slice sustains appreciable diminution by surface evaporation, the
lower layers being depleted primarily by root absorption (Figure 15.5).

15.4 Magnitude of Evaporation Losses

The combined losses by evaporation from the soil surface and by transpiration
account for the *consumptive* use of water, which is a measure of the total
water lost by evapotranspiration in producing crops. This is an important practical figure, especially in areas where irrigation must be employed to meet crop
needs.

Consumptive Use. As might be expected, there is a marked variation in the
consumptive use of water to produce different crops in different areas. All of
the factors previously considered as influencing evapotranspiration are operative. In addition, such plant characteristics as depth of rooting and length of
growing season become important.

Consumptive use while a crop is being produced may vary from as little

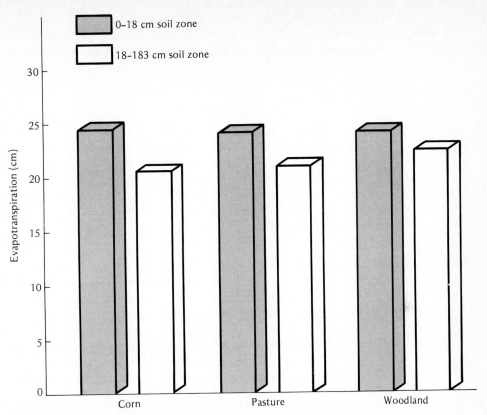

FIGURE 15.5 Evapotranspiration loss from the surface layer (0–18 cm) compared with that of the subsoil (18–183 cm). Note that more than half the water lost came from the upper 18 cm. Periods of measurement: corn, May 23–September 25; pasture, April 15–August 23; woodland, May 25–September 28. [*Calculated from Dreibelbis and Amerman (1965).*]

as 30 cm to as much as 220 cm or more. The low extreme might be encountered in cool mountain valleys where the growing seasons are short; the higher figure has been found in irrigated desert areas. The ranges commonly encountered are 35–75 cm in unirrigated, humid to semiarid areas and 50–125 cm in hot, dry regions where irrigation is used.

From a practical standpoint, daily consumption figures are in some cases more significant that those for the growing season. During the hot dry periods in the summer, daily consumptive use rates for corn, for example, may be as high as 1–1.25 cm. Even with deep soils having reasonably high capacities for storing available water, this rapid rate of moisture removal soon depletes the plant root zone of easily absorbed moisture. Sandy soils, of course, may lose most of their available moisture in a matter of a few days under these conditions. The importance of these vaporization losses is obvious.

Surface Evaporation Versus Transpiration. It is interesting to consider the comparative water losses from the soil surface and from transpiration. A number of factors determine these relative losses, include (a) plant cover in relation to soil surface; (b) efficiency of use of water by different plants; (c) proportion of time the crop is on the land, especially during the summer months; and (d) climatic conditions.

In humid regions, there is some evidence that vapor losses may be divided about equally between evaporation from the soil and transpiration. Naturally, this generalization would not hold in all cases since there are so many factors influencing vapor losses from both crops and soils.

Loss by evaporation from the soil is thought to be proportionately higher in drier regions than in humid areas. Such vapor loss is at least 60% of the total rainfall for the Great Plains area of the United States. Losses by transpiration account for about 35%, leaving about 5% for runoff.

15.5   Efficiency of Water Use

The crop production obtained from the use of a given amount of water is an important figure, especially in areas where moisture is scarce. This efficiency may be expressed in terms of (a) consumptive use in kilograms per kilogram of plant tissue produced or (b) transpiration in kilograms per kilogram of plant tissue produced. The second figure, called *transpiration ratio*, emphasizes the fact that large quantities of water are required to produce 1 kg of dry matter.

Transpiration Ratio. The transpiration ratio ranges from 200 to 500 for crops in humid regions, and almost twice as much for those of arid climates. The data in Table 15.3 give some idea of the water transpired by different crops.

TABLE 15.3   Transpiration Ratios of Plants as Determined by Different Investigators[a]

Crop	Harpenden, England	Munich, Germany	Dahme, Germany	Madison, Wisconsin	Pusa, India	Akron, Colorado
Barley	258	774	310	464	468	534
Beans	209	—	282	—	—	736
Buckwheat	—	646	363	—	—	578
Clover	269	—	310	576	—	797
Maize	—	233	—	271	337	368
Millett	—	447	—	—	—	310
Oats	—	665	376	503	469	597
Peas	259	416	273	477	563	788
Potatoes	—	—	—	385	—	636
Rape	—	912	—	—	—	441
Rye	—	—	353	—	—	685
Wheat	247	—	338	—	544	513

[a] Data compiled by Lyon et al. (1952).

FIGURE 15.6
Evapotranspiration ratios
(units of water to produce one
unit of dry matter) of different
crop plants grown at locations
differing in the saturation
deficit of the atmosphere. Corn
uses much less water per unit
of dry matter produced; peas
use the most. The difference is
greatest in drier climates
(areas with high saturation
deficits). [*Data from several
sources as plotted by
Bierhuizen and Slatyer
(1965).*]

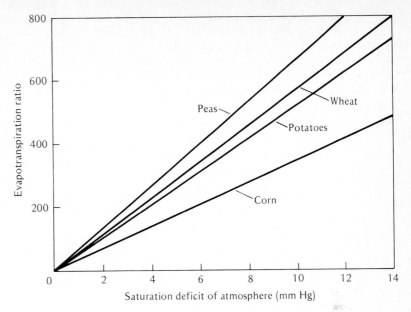

Much of the variation observed in the ratios arises from differences in climatic conditions. Thus, in areas of more intense sunshine, the temperature is higher, the humidity is lower, and the wind velocity is frequently greater. All this tends to raise the transpiration ratio. This is illustrated in Figure 15.6, which shows the transpiration ratio of four common crops as influenced by dryness of climate.

It is obvious from the transpiration ratios quoted in Table 15.3 that the amount of water necessary to mature the average crop is very large. For example, a representative crop of wheat containing 11,000 kg/ha of dry matter (about 5000 lb/A) and having a transpiration ratio of 500 will withdraw water from the soil during the growing season equivalent to about 25 cm of rain. The corresponding figure for corn, assuming the dry matter at 10,000 kg/ha and the transpiration ratio at 350, would be 35 cm. These amounts of water, in addition to that evaporated from the surface, must be supplied during the growing season. The possibility of moisture being the most critical factor in crop production is thus obvious.

Factors Influencing Efficiency.   The efficiency of water use in crop production is influenced by climatic, soil, and nutrient factors. The climatic factors have already received adequate attention (Section 15.3). Within a given climatic zone, the other effects can be seen.

In general, highest efficiency is found where optimum crop yields are being obtained. Conversely, where yields are limited by some factor or combination of factors, efficiency of water uptake is low and more water is required to produce a pound of plant tissue.

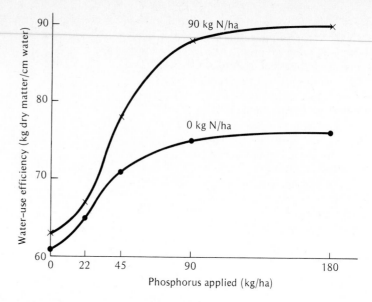

FIGURE 15.7 Water-use efficiency of wheat as affected by application of phosphorus fertilizer with and without nitrogen fertilizer (12 year average). Efficiency increased with increased fertilizer rates. [*Data from Black* (*1982*).]

Maintaining the moisture content of a soil too low or too high for optimum growth will likely increase the water required to produce 1 kg of dry matter. Similarly, the amount of available nutrients and their ratios also affect the economic utilization of water (Figure 15.7). The more productive the soil, the lower the transpiration ratio, provided the water supply is held at optimum. Therefore a farmer, in raising the productivity of his soil by drainage, liming, good tillage, and addition of farm manure and fertilizers, provides at the same time for a greater amount of plant production for every unit of water utilized. The total quantity of water taken from the soil, however, will probably be larger.

Legumes tend to have higher transpiration ratios than nonlegumes. This may account for the noted sensitivity of legumes to low soil moisture conditions.

15.6 Evaporation Control: Mulches and Cultivation

Any material used at the surface of a soil primarily to prevent the loss of water by evaporation or to keep down weeds may be designated as a mulch. Sawdust, manure, straw, leaves, crop residues, and other litter may be used successfully. Such mulches are highly effective in checking evaporation and are most practical for home garden use and for high-valued crops, including strawberries, blackberries, fruit trees, and such other crops as require infrequent, if any, cultivation. In recent years, however, tillage practices that leave a high percentage of crop residue on or near the surface are being used widely for field crops such as corn, wheat, and soybeans.

FIGURE 15.8 For crops with high cash value, plastic mulches are being used. The plastic is installed by machine (*above*) and at the same time the plants are transplanted (*below*). Plastic mulches help control weeds, conserve moisture, encourage rapid early growth, and eliminate need for cultivation. The high cost of plastic makes it practical only with the highest-value crops. [*Courtesy K. Q. Stephenson, Pennsylvania State University.*]

Paper and Plastic Mulches. Specially prepared paper and plastics have been used as mulches (Figure 15.8). This cover is spread and fastened down either between the rows or over the rows, the plants in the latter case growing through suitable openings. This mulch can be used only with crops planted in rows or in hills. So long as the ground is covered, evaporation and weeds are checked and, in some cases, remarkable crop increases have been reported. Unless the rainfall is torrential, the paper or plastic does not seriously interfere with the infiltration of the rain water.

The paper and plastic mulches have been employed with considerable success in the culture of pineapples in Hawaii, where the idea originated, and with other crops elsewhere. This type of mulch is also being used to some extent in truck farming and vegetable gardening in the United States. The cost of the cover and the difficulty of keeping it in place limit the use of these materials to high-value crops.

Crop Residue and Conservation Tillage.   There is growing recognition of the effectiveness of crop residues in reducing water losses from croplands, not only in the vapor form but also as runoff (Unger, 1980). To help assure better soil cover, advantage is taken of a number of new energy-saving soil tillage systems that minimize the number of land preparation and tillage operations. Collectively termed *conservation tillage* most of these systems leave part if not all of the residue from the previous crop on or near the soil surface (see Section 16.12). They replace or minimize the use of the traditional moldboard plow, which incorporates essentially all the residues into the soil.

Among the most widely used conservation tillage practices is *stubble mulch* tillage. In this method, used mostly in subhumid and semiarid regions, the residues from the previous crop are uniformly spread on the soil surface. The land is then tilled with an implement such as a sweep or underground blade, or a disk, which permits most of the crop residue to remain on the surface. Wheat stubble, straw, cornstalks, and similar crop residues are among the residues used.

The system that requires least soil manipulation and leaves essentially all the residues on the surface is termed *no tillage* (Figure 15.9). The new crop is planted directly in the sod or residues of the previous crop with no primary tillage. Herbicides are used to control weeds, making it possible for all crop residues to remain on the soil surface throughout the growing season.

A third widely used system is called *till planting*. It is a strip tillage system where residues are swept away from the immediate area of planting, but most of the soil and surface area remains untilled. Like the others mentioned, it leaves most of the soil covered with crop residues to help reduce vapor losses of water from soils.

Soil Mulch.   One of the early contentions about the control of evaporation was that the formation of a *natural* or *soil mulch* was a desirable moisture-conserving practice. Years of experimentation and practice have shown, however, that cultivating the surface soil to form a natural mulch does not necessarily conserve moisture, especially in humid regions. In fact, in some cases it may encourage moisture loss.

The effectiveness of a soil mulch is much affected by climatic conditions. In an area characterized by frequent rains during the growing season, a soil mulch probably will not conserve moisture and may in fact decrease it. Frequent surface stirring of the soil to provide a soil mulch hastens soil moisture loss by evaporation, and this loss may exceed any gains from the presence of the mulch. On the other hand, in a region with rather distinct wet and dry seasons, as in the tropics, a soil mulch can conserve moisture. The surface

(a)

(b)

FIGURE 15.9 (a) Planting soybeans in wheat stubble with no tillage. (b) In another field, corn has been planted in wheat stubble. In both cases the wheat residue will help reduce evaporation losses from the soil surface and will reduce erosion. [*Courtesy USDA Soil Conservation Service.*]

FIGURE 15.10 (a) Deep cultivation of intertilled crops such as corn may result in serious root pruning. Hence cultivation for the killing of weeds should be shallow (b), even early in the season. If weeds are under control and the structure of the soil is satisfactory, further intertillage usually is unnecessary. [*From Donald* (*1946*).]

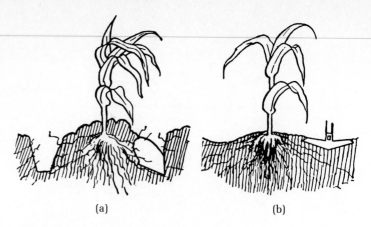

(a)   (b)

stirring to establish the mulch at the end of the wet season may need to be done only once or twice and the soil mulch thus established will remain during the entire dry period. Under such conditions considerable moisture conservation can occur.

Cultivation and Weed Control. Perhaps the most important reason for cultivation of the soil is to control weeds. Evapotranspiration of these unwanted plants can extract soil moisture far in excess of that used by the crop itself, especially a row crop (Figure 15.10). The widespread use of chemical herbicides, however, provides weed control without cultivation of the soil. Herbicides thus have become one of the most important tools in controlling evaporation. The necessity for tillage is thereby limited to situations where this practice is needed to maintain the proper physical conditions for plant growth.

15.7 Vaporization Control in Humid Regions—Minimum Tillage

Paper and plastic mulches are effective means of reducing vapor losses in gardens and nurseries in humid regions. But for the more extensively grown crops the use of conservation tillage systems that permit part or all of the crop residues to remain at or near the soil surface is the most practical means of controlling vapor losses in humid areas. While transpiration losses may not be reduced, the direct evaporation of water from the soil surface is surely minimized where there is soil cover, a condition which some conservation tillage systems permit. The extent to which conservation tillage systems have been adopted in the United States is shown in Table 15.4. This has favorable implications for reducing losses of both vapor and liquid forms of water (see Section 16.12).

The control of weeds is certainly an effective and practicable means in

TABLE 15.4 Estimate of Conservation Tillage in the United States[a]

Year	Conservation tillage (10^6 ha)
1973	11.9
1975	14.5
1977	19.2
1979	22.3

[a] U.S. Department of Agriculture data summarized by Crosson (1981).

humid regions of reducing the loss of water vapor from soils. Even though weed control may not decrease overall soil moisture loss, it does ensure a greater amount of water available for crop transpiration. This, of course, is the ultimate objective in the moisture management of field soils. The attention now being given to improved methods of weed control is well placed.

15.8   Vaporization Control in Semiarid and Subhumid Regions

In areas with average annual rainfall of 40–50 cm, so-called *dry-land farming* systems are used. Moisture conservation is a major requisite for the success of these systems.

Tillage Operations. The tillage operations practiced in dry-land agriculture should assure effective soil cover with crop residues. The methods employed include making rough furrows at right angles to the prevailing high winds, listing, strip cropping, and the use of stubble mulch when practicable. At times of rain, the land should be highly permeable and retentive and should be protected from water erosion if the rainfall is torrential. Vapor losses of water from the soil can be held in check with the maintenance of an effective stubble mulch.

Summer Fallow. Dry-land farming often includes a *summer fallow*, the idea being to catch and hold as much of one season's rain as possible and carry it over for use during the next. The soil profile is used as a large reservoir.

A common procedure for summer fallowing is to allow the previous year's stubble to stand on the land until the spring of the year. The soil is then

TABLE 15.5 Influence of Summer Fallow in Alternate Years on the Moisture at Wheat Seeding Time and the Yield of Wheat Fallowing[a]

	Available water at seeding time		Wheat yields	
	cm	in.	kg/ha	lb/A
Mandan, ND (av. 20 years)				
Wheat after fallow	17.9	7.08	4618[b]	4,120[b]
Wheat after wheat	6.3	2.48	2578	2,300
Hays, KS (av. 23 years)				
Wheat after fallow	22.2	7.96	1836	1638
Wheat after wheat	7.4	2.90	1170	1044
Garden City, KS (av. 13 years)				
Wheat after fallow	11.9	4.67	1042	930
Wheat after wheat	2.7	1.08	551	492

[a] Calculated from Thysell (1938) and Throckmorton and Meyers (1941).
[b] Total plant yields.

tilled before appreciable weed growth has taken place and the land is kept from weeds during the summer by occasional cultivation or by the use of herbicides. In this way, evaporation loss is limited largely to that taking place at the surface of the soil.

The effectiveness of summer fallow in carrying over moisture from one year to another is somewhat variable. A conservation of perhaps one fourth of the fallow season rainfall would be expected under most conditions. Even though this efficiency of moisture storage is lower than would be desirable, yields of succeeding crops have been augmented significantly by summer fallow (Table 15.5).

Here, then, is a situation where an appreciable amount of moisture may be conserved by the proper handling of the land. But it is a case specific for dry land areas and should not be confused with the situation in humid regions. In the latter, summer fallowing is detrimental and unnecessary.

15.9   Evaporation Control of Irrigated Lands

In the areas of low rainfall, or when precipitation is so distributed as to be insufficient for crops at critical times, *irrigation* water is commonly applied (Figure 15.11). In the United States, most of the irrigated land lies west of the hundredth meridian and in total embraces approximately 18 million ha. Moreover, irrigation in the more humid areas to the eastward is decidedly on the increase.

Control Principles.   The principles already discussed relating to the movement

FIGURE 15.11   Typical irrigation scene. The use of easily installed siphons reduces the labor of irrigation and makes it easier to control the rate of application of the water. Note the upward capillary movement of water along the sides of the rows. [*Courtesy USDA Soil Conservation Service.*]

and distribution of moisture through the soil, its losses, and its plant relationships, apply just as rigidly for irrigation water as for natural rainfall. Although the problems inherent in liquid water, such as runoff and erosion, infiltration, percolation, and drainage, are subject to better control under irrigation, they are met and solved in much the same way as for humid region soils.

In irrigated areas of dry-land regions, vapor losses are much more critical than even with contiguous soils under dry-land cropping. Conditions are ideal for evaporation and transpiration loss, the climate is dry, the sunshine intense, and wind velocities are often high. Furthermore, the soils are kept as moist as they are in well-drained humid region soils.

The only practical control of evaporation under these conditions is through irrigation practices. The surface of the soil should be kept only as moist as is needed for good crop production, yet the irrigation schedule should be such as to keep more than just the surface soil wet. Deep penetration of roots should be encouraged.

One unique method of reducing evaporation from irrigated lands is to concentrate the added water in the immediate root zone of the crop plant. This is done by trickling water out of pipes or tubes alongside each plant, without at the same time wetting the soil between the rows. This system,

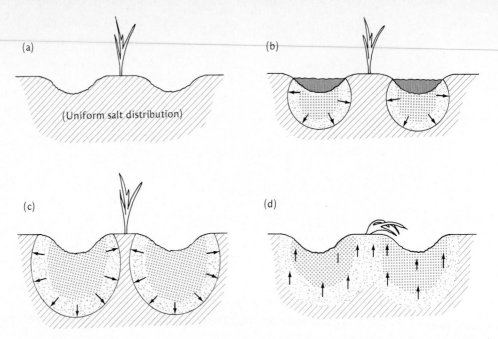

FIGURE 15.12 Effect of irrigation on salt movement in a saline soil. (a) Soil before irrigation. Notice the relatively uniform salt distribution. (b) Water running in the furrow. Some has moved downward in the soil, carrying soluble salts with it. (c) Soon after irrigation. Water continues to move downward and toward center of row. (d) A week or so after irrigation. Upward movement of water by capillarity carries salts toward the surface. Salts tend to concentrate in the center of the row and in some cases cause injury to plants. [*Concept obtained from Bernstein et al.* (*1955*).]

termed "trickle tube" or "ooze tube" irrigation, has the advantage of markedly reducing moisture loss by evaporation from the soil surface.

Salt Accumulation. A phenomenon closely correlated with evaporation and often encountered on arid-region lands under irrigation is the concentration of soluble salts at or near the soil surface. The upper horizons of *saline* and *alkali* soils may contain soluble salts in sufficient quantities to inhibit the growth of many cultivated plants (see Figure 6.14). Irrigation practices on these soils can improve or impair their usefulness as crop soils. By flood irrigation, some of the soluble salts can be temporarily washed down from the immediate surface. Subsequent upward movement of the water and evaporation at the soil surface may result in salt accumulation and concentration, which are harmful to young seedlings (Figure 15.12). Although little can be done to prevent evaporation and upward capillary movement of the salt-containing water, timing and method of irrigation may prevent the salts from being concentrated in localized areas.

15.10 Types of Liquid Losses of Soil Water

We now turn to the loss of water in the liquid form. Two types of liquid losses of water from soils are recognized: (a) the downward movement of free water (percolation), which frees the surface soil and upper subsoil of superfluous moisture, and (b) the runoff of excess water over the soil surface (Figure 15.1). Percolation results in the loss of soluble salts (leaching), thus depleting soils of certain nutrients. Runoff losses generally include not only water but also appreciable amounts of soil (erosion). Runoff and erosion will be considered in Chapter 16.

15.11 Percolation and Leaching—Methods of Study

Two general methods are available for the study of percolation and leaching losses—the use of an effective system of underground pipe or *tile drains* specially installed for the purpose, and the employment of *lysimeters* (from the Greek word *lysis*, meaning loosening, and *meter*, to measure). For the first method, an area should be chosen where the tile drain receives only the water from the land under study and where the drainage is efficient. The advantage of the tile method is that water and nutrient losses can be determined from relatively large areas of soil under normal field conditions.

The lysimeter method has been used more commonly to determine leaching losses. This method involves the measurement of percolation and nutrient losses under somewhat more controlled conditions. Soil is removed from the field and placed in concrete or metal tanks, or a small volume of field soil is

FIGURE 15.13 Cross section of monolith type of lysimeter at the Rothamsted Experiment Station, England.

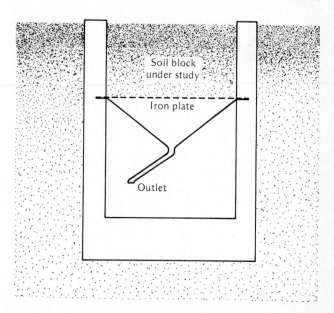

isolated from surrounding areas by concrete or metal dividers (Figure 15.13). In either case water percolating through the soil is collected and measured. The advantages of lysimeters over a tile drain system are that the variations in a large field are avoided, the work of conducting the study is not so great, and the experiment is more easily controlled.

15.12   Percolation Losses of Water

When the amount of rainfall entering a soil becomes greater than its water-holding capacity, losses by percolation will occur. Percolation losses are influenced by the amount of rainfall and its distribution, by runoff from the soil, by evaporation, by the character of the soil, and by the crop.

Percolation–Evaporation Balance.   The relationship among precipitation, runoff, soil storage, and percolation for representative humid and semiarid regions and for an irrigated arid-region area is illustrated in Figure 15.14. In the humid region, the rate of water infiltration into soils (precipitation less runoff) is commonly greater at least at some time of year than the rate of evapotranspiration. As soon as the soil field capacity is reached, percolation into the substrata occurs.

In the example shown in Figure 15.14, maximum percolation occurs during the winter and early spring, when evaporation is lowest. In contrast, during the summer little percolation occurs. Evapotranspiration that exceeds the infiltration during these months results in a depletion of soil water. Normal plant growth is possible only because of moisture stored in the soil the previous winter and early spring.

The general trends in the semiarid region are the same as for the humid region. Soil moisture is stored during the winter months and utilized to meet the moisture deficit in the summer. But because of the low rainfall, essentially no percolation out of the profile occurs. Water may move to the lower horizons but is absorbed by plant roots and ultimately lost by transpiration.

The irrigated soil in the arid region shows a unique pattern. Irrigation in the early spring, along with a little rainfall, provides more water than is being lost by evapotranspiration. The soil is charged with water and some percolation may occur. During the summer, fall, and winter months this stored water is depleted since the amount being added is less than the evapotranspiration.

Figure 15.14 depicts situations common in temperate zones. In the tropics one would expect evapotranspiration to be somewhat more uniform throughout the year, although it will vary from month to month depending on the distribution of precipitation. In very high rainfall areas of the tropics, the percolation and especially the runoff would be higher than shown in Figure 15.14. In irrigated areas of the arid tropics, the relationships would likely not be too different from those in temperate regions.

FIGURE 15.14 Generalized curves for precipitation and evapotranspiration for (a) a humid region, (b) a semiarid region, and (c) an irrigated arid region. Note the absence of percolation through the soil in the semiarid region. In each case water is stored in the soil. This moisture is released later when evapotranspiration demands exceed the precipitation. In the semiarid region evapotranspiration would likely be much higher if ample soil moisture were available. In the irrigated arid-region soil the very high evapotranspiration needs are supplied by irrigation. Soil moisture stored in the spring is utilized by later summer growth and lost through evaporation during the late fall and winter.

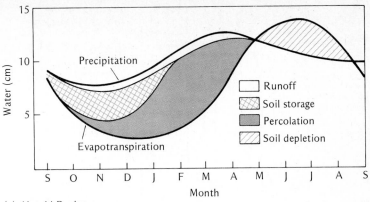

(a) Humid Region

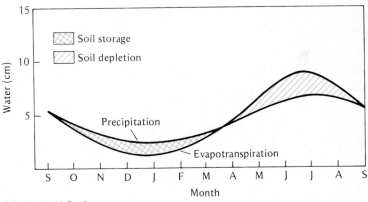

(b) Semiarid Region

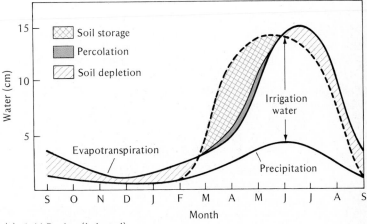

(c) Arid Region (irrigated)

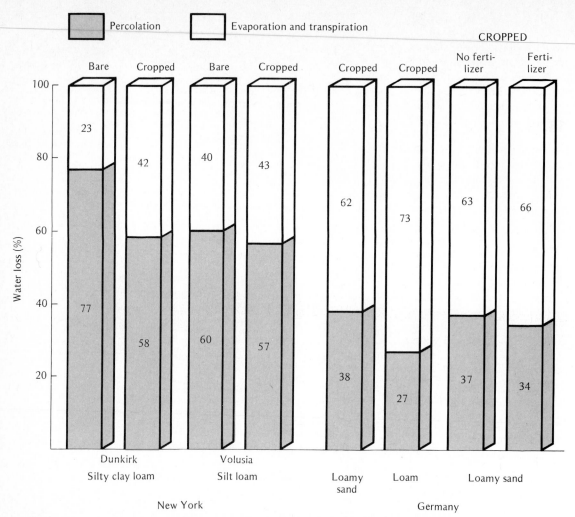

FIGURE 15.15 Losses of water from soils by percolation and by evaporation plus transpiration. Note that percolation was decreased when a crop was grown and when fertilizer was applied. High crop yields on the plots in Germany probably account for the high losses by evaporation and transpiration. Runoff was not allowed on these plots. [*New York data calculated from Lyon et al.* (*1930 and 1936*); *Germany data from Jürgen-Gschwind and Jung* (*1979*).]

Bare Versus Cropped Soil. The loss of water by percolation in cropped soil areas is generally less than that in bare areas unless surface runoff is excessively high from the bare soil. Data from lysimeter experiments in New York illustrate this point (Figure 15.15). The percentage of the 84 cm of precipitation lost by percolation was greater in the bare plots for both the Dunkirk and Volusia soils. In contrast, water lost by evaporation was greater in the cropped plots on which a rotation of corn, oats, wheat, and hay was grown. Percolation loss was higher on the moderately well-drained Dunkirk soil than on the poorly drained Volusia. In both soils the percolation loss was probably higher than would have been expected under field conditions, since in this experiment no surface runoff was permitted.

15.13 Leaching Losses of Nutrients

The loss of nutrients through leaching is determined by climatic factors and soil–nutrient interactions. In regions where water percolation is high, the potential for leaching is also high (Figure 15.16). Such conditions exist in the United States in the humid east and in the heavily irrigated sections of the west. In these areas percolation of excess water is the rule, providing opportunities for nutrient removal. In the semiarid Great Plains, little nutrient leaching occurs since there is little percolation. Some nutrient leaching takes place in the Corn Belt, although much less than in the states to the east. In all cases, the growing of crops reduces the loss of nutrients by leaching.

Nutrient–Soil Interaction. Soil properties have a definite effect on nutrient leaching losses. Sandy soils generally permit greater nutrient loss than do clays, because of the higher rate of percolation and lower nutrient-adsorbing power of the sandy soils. For example, soluble phosphorus is quickly bound chemically in soils with appreciable amounts of iron and aluminum oxides (see Section 10.7). Consequently, this element is lost very sparingly from medium- and fine-textured soils which often contain these minerals. Sulfates and nitrates react by anion exchange (Section 5.16) with iron and aluminum hydrous oxides. For this reason, sulfates and nitrates are less prone to leach from soils with red subsoils (high in iron oxides) than from soils in similar climates when red-colored iron oxides are not so prominent.

The leaching of cations added in fertilizers is affected by the soil's cation exchange capacity, soils with a high capacity tending to hold the added nutrients and prevent their leaching. At the same time, such soils are often naturally high in exchangeable cations thus providing a large reservoir of nutrients, a small portion of which are continually subject to leaching.

Some of the factors affecting the loss of nutrients by leaching are illustrated in Table 15.6, which includes data from lysimeter experiments at three locations. The loss of phosphorus is negligible in all cases. The leaching of nitrogen is dependent upon the crops grown and the amount of percolation occurring. The loss of cations is in rough proportion to their probable content in exchangeable form on the colloidal complex, calcium being lost in largest amounts and potassium least.

Nutrient Loss and Fertilization. The rates of fertilizer additions in the lysimeter experiments reported in Table 16.1 are modest compared to those common

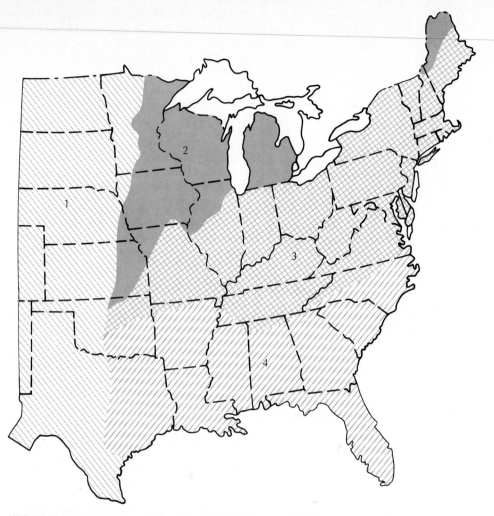

FIGURE 15.16 Eastern part of the United States, showing regions varying in their susceptibility to leaching during the winter months. In region 1 the losses are insignificant; in region 4 they are high. Regions 2 and 3 are intermediate. [*Modified from Nelson and Uhland* (*1955*).]

today. For this reason, the nutrient losses given in Table 15.6 are probably considerably lower, at least for some of the nutrients, than would be expected from a heavily fertilized area, especially if water percolation were high. For example, nitrogen applications at rates of 150 kg/ha and greater are common for cornfields. Likewise, crops such as vegetables and sugar cane are fertilized at rates that assure residual nutrients in the soil. These nutrients must react with the soil, be lost by microbial action (for example, volatilization of nitrogen), or be lost through leaching.

TABLE 15.6 Average Annual Loss of Nutrients by Percolation Through Soils from Four Different Areas[a]

The Monolith type of lysimeter was used, and runoff was allowed from Illinois and Wisconsin lysimeters only.

Condition	Loss per year (kg/ha)					
	N	P	K	Ca	Mg	S
Rotation on a Scottish soil (av. 6 yr)						
No treatment	8	Trace	10	56	17	—
Manure and fertilizers	7	Trace	9	63	18	—
Manure, fertilizers, and lime	9	Trace	9	89	21	—
Uncropped Illinois soils (av. 3½ yr)						
Muscatine (well drained)	86	—	1	101	52	12
Cowden (poorly drained)	7	—	1	12	4	2
Fayette silt loam, Wisconsin (av. 3 yr)						
Fallow	—	—	1	42	21	3
Cropped to corn	—	—	<1	16	7	1
New York soils (av. 10 yr, Dunkirk; 15 yr, Volusia)						
Dunkirk (bare)	77	Trace	81	446	71	59
Dunkirk (rotation)	9	Trace	64	258	49	48
Dunkirk (grass)	3	Trace	69	291	56	49
Volusia (bare)	48	Trace	72	362	46	39
Volusia (rotation)	7	Trace	64	280	30	37

[a] Data compiled by Buckman and Brady (1969).

The practice of fall application of nitrogen for crops to be planted the next spring offers considerable potential for leaching loss, especially in humid areas. Percolation losses are generally greatest in the spring prior to rapid growth. Generally nitrate losses would be expected to parallel the water lost by percolation. Similarly, some irrigated soils of arid areas that receive heavy nitrogen application may suffer considerable loss of nitrates through leaching during the intensive irrigation period (MacKenzie and Vietz, 1974).

The ion concentration in percolating water is generally highest when percolation first starts and decreases with time thereafter. For that reason, greater nutrient leaching loss will occur if a given amount of percolation takes place during several minor leaching events than if fewer much larger leaching events occur.

Because of the marked variability in climates, soils, cropping patterns, and rates and times of fertilizer application, it is not surprising to find a similar variability in the quantity of nutrients lost by leaching. In humid regions, however, the losses of at least the cations would be expected to be of the same order as the removal of these nutrients by crops (Table 15.7). Growing concern over excessive nutrient enrichment of streams and lakes forces reconsideration of practices that encourage the leaching of nutrients from soils.

TABLE 15.7 Comparison of the Average Annual Loss of Nutrients by Drainage from a Representative Humid Region Silt Loam, Cropped to a Standard Rotation, with the Nutrients Removed by an Average Well-Fertilized Rotation Crop (Corn, Wheat, Clover/Grass Hay).

	Loss per year (kg/ha)					
	N	P	K	Ca	Mg	S
Leached from a representative silt loam	30	Trace	35	115	25	12
Removed by average rotation crop	190	35	135	50	25	20

There are two prime reasons for concern over the loss of essential elements by leaching. First is the obvious concern for keeping these nutrients in the soil, thereby supplying essential elements to crop plants. A second and equally significant reason is to keep the nutrients out of streams, rivers, and lakes. Such bodies, overly enriched with nutrients, encourage the process of *eutrophication,* which involves excessive growth of algae and other aquatic species and the resulting deficiency of oxygen in the water. This can drastically adversely affect the population of fish and other aquatic animals.

Control of Nutrient Losses. Nutrient losses can be minimized by following a few simple rules. First, to the extent feasible, a crop should be kept on the land. Fall and winter cover crops can be grown following heavily fertilized cash crops such as corn, potatoes, and other vegetables. Second, fertilizer application rates should be no higher than can be clearly justified by scientific research trials. Third, in areas where water percolation from soils is common, the fertilizers should be added as close as feasible to the time of nutrient utilization by the crop plants. For example, it can be applied as a side dressing when the crop is growing vigorously rather than a broadcast application before planting. Even by following these suggestions, some leaching losses will occur. The aim should be to minimize these losses both for the sake of the farmer and for society generally.

15.14 Land Drainage[1]

If a poorly drained soil is to be utilized effectively, the excess soil moisture must be removed by the installation of a *land drainage system.* The objective is to lower the moisture content of the upper layers of the soil so that oxygen can be available to the crop roots and carbon dioxide can diffuse from these roots.

Land drainage is needed in select areas in almost every climatic region,

[1] For excellent reviews of various aspects of this subject see Van Schilfgaarde (1974).

but it is critical in areas such as the flat coastal plain sections of the eastern United States. Likewise, the fine-textured soils of river deltas, such as those of the Mississippi and the Nile, and lake-laid soils throughout the world commonly require some form of artificial drainage. Surprisingly enough, even irrigated lands of arid regions often require extensive drainage systems. This may be due to the need to remove excess salts or to prevent their buildup in irrigated areas.

Two general types of land drainage systems are used: (a) surface field drains and (b) subsurface drains. Each will be discussed briefly.

15.15   Surface Field Drains

The most extensive means of removing excess water from soils is through the use of surface drainage systems. Their purpose is to remove the water from the land before it infiltrates the soil. They may involve deep and rather narrow field ditches such as those used to remove large quantities of water from peat soil areas (Figure 15.17). More often, however, shallow ditches with gentle side slopes are used to help remove the water. They are commonly constructed with simple equipment and at not too great expense. If there is some slope on the land, the shallow ditches are usually constructed across the slope and across direction of planting and cultivating, thereby permitting the interception of runoff water as it moves down the slope (Figure 15.18).

Open drainage ditches have the advantage of large carrying capacities. For this reason they are an essential component of all drainage systems. In general, their cost per unit of water removed is relatively low. Disadvantages include cost of maintenance and interference with agricultural operations. Also, together with their banks, open ditches use valuable agricultural lands, which, if the area were underdrained, would be available for cropping.

Land Forming. Surface drain ditches are combined with *land forming* or smoothing to rapidly remove surface water from soils. Depressions or ridges

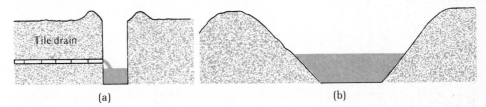

FIGURE 15.17   Two common types of drainage ditches. (a) A small ditch with vertical sides such as is used to drain organic-soil areas. Note the tile outlet from one side. (b) A larger sloping-sided ditch of the type commonly used to transport drainage water to a nearby stream. The source of drainage water may be either tile or open drain systems.

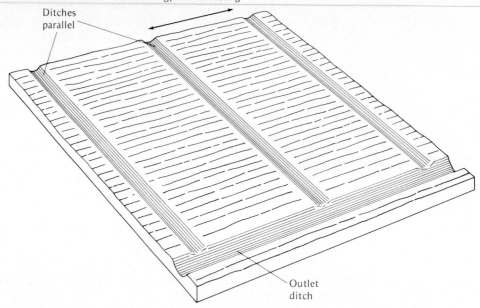

Ditches
parallel

Outlet
ditch

FIGURE 15.18 Example of an open ditch drainage system on a field with gentle slope. [*From Hughes* (*1980*); *used with permission of Deere & Company, Moline, IL.*]

that prevent water movement to the drainage outlet are filled in with precision using field leveling equipment (Figure 15.19). The resulting land configuration permits excess water to move slowly over the soil surface to the outlet ditch and then on to a natural drainage channel.

Land smoothing is a common practice in irrigated areas. Surface irrigation is made possible as is the removal of excess water by the outlet ditch. In humid areas the same methods are being put to use to remove excess surface water. In combination with selectively placed tile drains, this method of orderly water removal shows great promise.

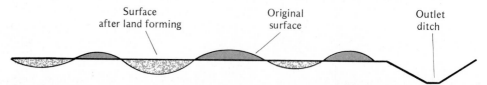

Surface
after land forming

Original
surface

Outlet
ditch

FIGURE 15.19 Land surface before and after land forming or smoothing. Note that soil from the ridges fills in the depressions. Land forming makes possible the controlled movement of surface water to the outlet ditch, which transports it to a nearby natural waterway.

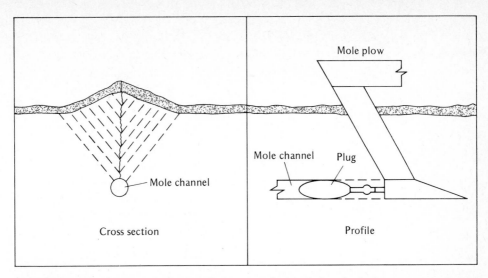

FIGURE 15.20 Diagrams showing how an underground mole drainage system is put in place. The plug is pulled through the soil, leaving a channel through which drainage water can move. [*From Hughes* (*1980*); *used with permission of Deere & Company, Moline, IL.*]

15.16 Subsurface or Underground Drains

Systems of underground channels to remove water from the zone of maximum water saturation provide the most effective means of draining land. The underground network can be created in several ways, including the following.

1. A *mole drainage* system can be created by pulling through the soil at the desired depth a pointed cylindrical plug about 7–10 cm in diameter. The compressed wall channel thus formed provides a mechanism for the removal of excess water (Figure 15.20).
2. A perforated plastic pipe can be laid underground using special equipment (Figure 15.21). Water moves into the pipe through the perforations and can be channeled to an outlet ditch.
3. A clay tile system made up of individual clay pipe units 30–40 cm long can be installed in an open ditch. The tiles are then covered with a thin layer of straw, manure, or gravel and the ditch is then refilled with soil.

The mole drain system is the least expensive to install but is also more easily clogged than the other two systems, which involve a more stable channel for water flow.

Operation of the Underground Systems. The same principles govern the function of all three types of underground drains. The tile, pipe, or mole channel

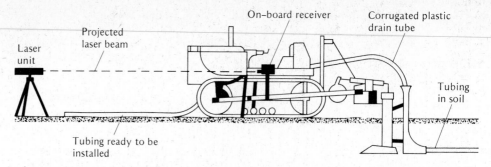

Laser unit

Projected laser beam

On–board receiver

Corrugated plastic drain tube

Tubing in soil

Tubing ready to be installed

FIGURE 15.21 Drawing showing how corrugated plastic tubing may be placed in the soil. The flexible tile is fed through a pipe as the tractor moves forward. Appropriate grade level is assured by a laser-beam-controlled mechanism. [*From Fouss* (1974); *used with permission of the American Society of Agronomy.*]

is placed in the zone of maximum water accumulation. Water moves from the surrounding soil to the channel, enters through perforations in a pipe, or joints between the tiles, and is then transported through the channel to an outlet ditch. An illustration of the water flow pattern to the channel from the surrounding soil is shown in Figure 15.22. Different field layouts for underground systems are illustrated in Figure 15.23.

Care needs be taken to place the tile or other underground channels in the zone of maximum water accumulation. The channels should be at least 75 cm below the soil surface because of the danger of damage from heavy machinery, and a depth of 1 m is commonly recommended. The distance between the channel lines will vary with the soil conditions (Table 15.8). On heavy clay soils, the lines may be as close as 9 or 10 m, but a distance of 15–20 m is more common.

The grade or fall in the underground drain system is normally 12–25 cm/50 m (about 3–6 in./100 ft), although a fall as great as 75 cm/50 m is used for rapid water removal.

FIGURE 15.22 Demonstration of the saturated flow patterns of water toward a drainage tile. The water, containing a colored dye, was added to the surface of the saturated soil and drainage was allowed through the simulated drainage tile shown on the extreme right. [*Courtesy G. S. Taylor, The Ohio State University.*]

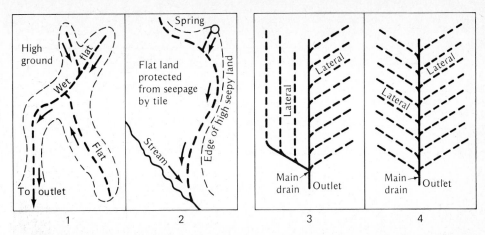

FIGURE 15.23 Tile drainage systems. The drawings show four typical systems for laying tile drains: (1) *natural,* which merely follows the natural drainage pattern; (2) *interception,* which cuts off water seeping into lower ground from higher lands above; (3) *gridiron* and (4) *fishbone* systems, which uniformly drain the entirety of an area. The aerial photo shows a freshly installed drainage system. Tile lines feed into main drains at the bottom of the photo, and the direction of flow is shown by the arrows. [*Photo courtesy USDA.*]

Special care must be given to protect the outlet of an underground drainage channel. If the outlet becomes clogged with sediment the whole system is endangered (Figure 15.24). It is well to embed the drainage outlet in a masonry concrete wall or block. The last 2.5–3 m of drain may even be replaced by a galvanized iron pipe or with sewer tile, thus ensuring against damage by frost.

TABLE 15.8 Suggested Spacing Between Tile Laterals for Different Soil and Permeability Conditions[a]

Soil	Permeability	Spacing	
		m	ft
Clay and clay loam	Very slow	9–18	30–70
Silt and silty clay loam	Slow to moderately slow	18–30	60–100
Sandy loam	Moderately slow to rapid	30–90	100–300
Muck and peat	Slow to rapid	15–61	50–200

[a] Modified from Beauchamp (1955).

The outlet may be covered by a gate or by wire in such a way as to allow the water to flow out freely, and to prevent the entrance of rodents in dry weather.

Drainage Around Building Foundations. Surplus water around foundations and underneath basement floors of houses and other buildings can cause serious economic damage. The removal of this excess water is commonly accomplished using underground drains similar in principle to those used for field drainage in agriculture. There must be free movement of the excess water to a tile drain placed alongside and slightly below the foundation or underneath the floor. The water must move rapidly through the tile drains to an outlet ditch or sewer. Care must be taken to assure that outlet openings are kept free from sediment or debris.

FIGURE 15.24 This tile half-filled with sediment has lost much of its effectiveness. Such clogging can result from poorly protected outlets or inadequate slope of the tile line. In some areas of the western United States iron and manganese compounds may accumulate in the tile even when the system is properly designed. [*Courtesy L. B. Grass, USDA Soil Conservation Service.*]

15.17   Benefits of Land Drainage

Granulation, Heaving, and Root Zone.   Draining the land promotes many conditions favorable to higher plants and soil organisms and provides greater stability of building foundations and roadbeds. "Heaving"—the ill effects of

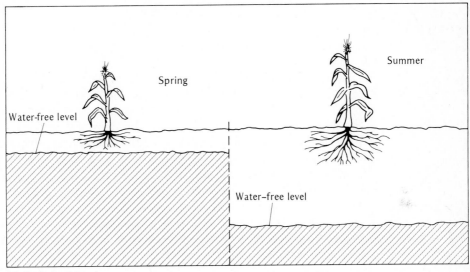

(a) Undrained Land

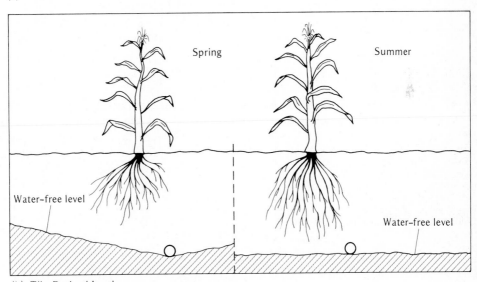

(b) Tile–Drained Land

FIGURE 15.25   Illustration of water levels of undrained and tile-drained land in the spring and summer. Benefits of the drainage are obvious. [*Redrawn from Hughes* (*1980*); *used with permission of Deere & Company, Moline, IL.*]

alternate expansion and contraction due to freezing and thawing of soil water—is alleviated (see Section 4.8). Such heaving can break foundations and seriously disrupt roadbeds. Also, it is the heaving of small-grain crops and the disruption of such taprooted plants as alfalfa and sweet clover that are especially feared (see Figure 4.11). By quickly lowering the water table at critical times, drainage maintains a sufficiently deep and effective root zone (Figure 15.25). By this means, the quantity of nutrients extractable by the plants is maintained at a higher level.

Soil Temperature. The removal of excess water also lowers the specific heat of soil, thus reducing the energy necessary to raise the temperature of the layers thus drained (see Section 4.10). At the same time, surface evaporation, which has a cooling effect, may be reduced. The two effects together tend to make the warming of the soil easier. The converse of the old saying that in the spring "a wet soil is a cold soil" applies here. Good drainage is necessary if the land is to be satisfactorily cultivated and is imperative in the spring if the soil is to warm up rapidly. At this season of the year, a wet soil may be 3–8°C cooler than a moist one.

Aeration Effects. Perhaps the greatest benefits of drainage, direct and indirect, relate to aeration. Good drainage promotes the ready diffusion of oxygen to, and carbon dioxide from, plant roots. The activity of the aerobic organism is dependent upon soil aeration, which in turn influences the availability of nutrients, such as nitrogen and sulfur. Likewise, the chance of toxicity from excess iron and manganese in acid soils is lessened if sufficient oxygen is present, since the oxidized states of these elements are highly insoluble. Removal of excess water from soils is at times just as important for plant growth as is the provision of water when soil moisture is low.

Timeliness of Field Operations. A distinct advantage of tile drain systems is the fact that drained land can be worked earlier in the spring. A well-functioning tile drain system permits the farmer to start spring tillage operations one to two weeks earlier than where no drainage is installed. Efficient time utilization in the spring of the year is critical in modern farming.

REFERENCES

Beauchamp, K. H. 1955. "Tile drainage—its installation and upkeep," *The Yearbook of Agriculture* (*Water*) (Washington, DC: U.S. Department of Agriculture), p. 513.

Bernstein, L., et al. 1955. "The interaction of salinity and planting practice on the germination of irrigated row crops," *Soil Sci. Soc. Amer. Proc.*, **19**:240–243.

Bierhuizen, J. F., and R. O. Slatyer. 1965. "Effect of atmospheric concentration of water vapor and CO_2 in determining transpiration–photosynthesis relationships of cotton leaves," *Agr. Meteor.*, 2:259.

Black, A. L. 1982. "Long-term N–P fertilizer and climate influences on morphology and yield components of spring wheat," *Agron. J.*, 74:651–657.

Buckman, H. O., and N. C. Brady. 1952 and 1969. *The Nature and Properties of Soils*, 5th ed., p. 221, and 7th ed. (New York: Macmillan).

Crosson, P. 1981. *Conservation and Conventional Tillage: A Comparative Assessment* (Ankeny, IA: Soil Cons. Soc. Amer.).

Denmead, O. T., and R. H. Shaw. 1962. "Availability of soil water to plants as affected by soil moisture content and meteorological conditions," *Agron. J.*, 45:385–390.

Donald, L. 1946. "Don't blame the weather," *Comm. Fert.*, Feb., 19–24.

Dreibelbis, F. R., and C. R. Amerman. 1965. "How much topsoil moisture is available to your crops?" *Crops and Soils*, 17:8–9.

Fouss, J. L. 1974. "Drain tube materials and installation," in J. Van Schilfgaarde (Ed.), *Drainage for Agriculture*, Agronomy Series No. 17 (Madison, WI: Amer. Soc. Agron.), pp. 147–177.

Haynes, J. L. 1954. "Ground rainfall under vegetative canopy of crops," *J. Amer. Soc. Agron.*, 6:67–94.

Hillel, D. 1971. *Soil and Water* (New York: Academic Press).

Hughes, H. S. 1980. *Conservation Farming* (Moline, IL: John Deere and Company).

Jürgen-Gschwind, S., and J. Jung. 1979. "Results of lysimeter trials at the Limburgerhof facility, 1927–77: The most important findings from 50 years experiments," *Soil Sci.*, 127:146–160.

Kelly, O. J. 1954. "Requirement and availability of soil water," *Adv. in Agron.*, 6:67–94.

Kozlowski, T. T. (Ed.). 1968–72. *Water Deficits and Plant Growth* (3 vols.) (New York: Academic Press).

Lyon, T. L., et al. 1930 and 1936. *Lysimeter Experiments*, III (Memoir 134) and IV (Memoir 194), Cornell Univ. Agr. Exp. Sta.

MacKenzie, A. J., and F. G. Vietz. 1974. "Nutrients and other chemicals in agricultural drainage waters," in J. Van Schilfgaarde (Ed.), *Drainage for Agriculture*, Agronomy Series No. 17 (Madison, WI: Amer. Soc. Agron.), pp. 489–511.

Nelson, L. B., and R. E. Uhland. 1955. "Factors that influence loss of fall-applied fertilizers and their probable importance in different sections of the United States," *Soil Sci. Soc. Amer. Proc.*, 19:492–496.

Philip, J. R. 1966. "Plant water relations: some physical aspects," *Ann. Rev. Plant Physiol.*, 17:245–268.

Science of Food and Agriculture. 1983. 1(1):28 (Ames, IA: Council on Agricultural Science and Technology).

Thysell, J. C. 1938. *Conservation and Use of Soil Moisture at Mandau, North Dakota*, Tech. Bull. 617 (Washington, DC: U.S. Department of Agriculture).

Throckmorton, R. I., and H. E. Meyers. 1941. *Summer Fallow in Kansas,* Bull. 293, Kansas Agr. Exp. Sta.

Unger, P. R., and T. M. McCalla. 1980. "Conservation Production Systems," *Adv. in Agron.,* **33**:2–58.

Van Doren, D. M. Jr., G. B. Triplett, and J. E. Henry. 1976. "Influence of long-term tillage, crop rotation, and soil type combinations on corn yield," *Soil Sci. Soc. Amer. J.,* **40**:100–105.

Van Schilfgaarde, J. (Ed.). 1974. *Drainage for Agriculture.* Agronomy Series No. 17 (Madison, WI: Amer. Soc. Agron.).

Soil Erosion and Its Control

No other soil phenomenon is more destructive worldwide than is soil erosion. It involves losing water and plant nutrients at rates far higher than those occurring through leaching. More tragically, however, it can result in the loss of the entire soil. Furthermore, the soil that is removed finds its way into streams, rivers, and lakes and becomes a pollution problem there. Erosion is serious in all climates, since wind as well as water can be the agent of removal. Water erosion carries away the larger amount of soil and is considered first.

16.1 Significance of Runoff and Soil Erosion[1]

Effects of Runoff. In some humid regions, loss by runoff may rise as high as 50–60% of the annual precipitation. In arid sections, it is usually lower unless the rainfall is of the torrential type, as it often is in the southwestern United States. Although the loss of the water itself is deplorable, the soil erosion that accompanies it is usually even more serious. The surface soil is gradually taken away. Many farmers, especially in the southern United States, are cultivating subsoils today, unaware that the surface soils have been stolen from underfoot. This means a loss not only of the natural fertility, but also of the nutrients that have been artificially added. Also, it is the finer portion of this soil that is always removed first, and this fraction, as already emphasized, is highest in fertility.

Nutrient Losses. The quantity of nutrients lost by erosion is high, as shown by the data in Table 16.1. Experiments have shown organic matter and nitrogen to be up to 5 times as high in eroded material as in the original soil. Comparable

TABLE 16.1 Nutrients Removed Annually in the Missouri Erosion Experiment[a] Compared to the Yearly Removal by an Average Field Crop

Condition	Nutrients removed per year (kg/ha)					
	N	P	K	Ca	Mg	S
Erosion removal						
Corn grown continuously	74	20	678	247	98	19
Rotation: corn, wheat, and clover	29	9	240	95	33	7
Crop removal						
Average for standard rotation	190	35	135	50	25	20

[a] The erosion data are averages of 2 yr only.

[1] For a recent interesting review of this subject see Batie (1983).

figures for phosphorus and potassium are 3 and 2, respectively. Moreover, a higher proportion of the nutrients in the erodate is usable by crop plants.

Erosion losses, even on a 4% slope, may easily exceed the removal of nutrients by crops occupying the land. This seems to be true especially for calcium, magnesium, and potassium. The data in Table 16.1 emphasize the continual loss of valuable nutrients by the process of erosion.

Extent of Erosion. Extensive soil erosion damage occurred in the United States long before its seriousness was widely recognized. Researchers had identified areas where erosion was rampant. But public recognition of this problem came only in the 1930s when soil erosion was emphasized by men such as H. H. Bennett, who later organized federal support for soil erosion control.

Although Bennett may have overemphasized the hazards of soil erosion,

0–11 Mg/ha per year

11–34 Mg/ha per year

34–56 Mg/ha per year

56 Mg/ha per year

FIGURE 16.1 Extent of water erosion (sheet and rill) in continental United States in 1977. Gully erosion is not included. For the average soil, erosion rates greater than 11 Mg/ha per year will result in soil productivity decline. [*From USDA (1975).*]

subsequent surveys have clearly shown that it is a serious problem. A 1967 national survey by the U.S. Department of Agriculture showed soil erosion to be a prominent problem on about 90 million ha of croplands, more than half the total of the cropland area. A second survey made ten years later identified areas of greatest soil loss by erosion (Figure 16.1) and showed an average annual erosion loss from croplands of 10.5 Mg/ha (4.7 tons/A). Losses ranged from less than 1 to more than 90 Mg/ha. It is likely that on at least one third of the U.S. cropland annual erosion losses exceed 11 Mg/ha (5 tons/A) per year, the maximum level that can be tolerated if soil productivity is to be maintained. About 50 million ha of cropland lose 11–22 Mg/year, while another 25 million ha lose more than 22 Mg/ha per year. Furthermore, erosion losses from other than agricultural sources are very high (Table 16.2). Some 60% of the sediment reaching streams in the United States is from stream roadbanks, patures and forests, and from intensive land use for other than agriculture. In addition to the sediment carried in streams, another 1.5 billion Mg of sediment is deposited each year in the nation's reservoirs.

TABLE 16.2 Sediment Sources and Their Contribution to Sediment in Streams in the United States[a]

Source	Total sediment (10⁶ Mg/yr)	Contribution to stream sediment (%)
Agricultural lands	680	40
Streambank erosion	450	26
Pasture and rangeland	210	12
Forest lands	130	7
Other federal lands	115	6
Urban	73	4
Roads	51	3
Mining	18	1
Other	14	1
Total	1741[b]	100

[a] From Brandt et al. (1972).

[b] Total sediment yield is generally reported as about 3.6 billion metric tons (3.6 × 10⁹ Mg) per year for the United States, 1.5 billion of which is deposited in reservoirs and 350 million in other low-lying areas near the source.

The effects of erosion are no less serious in countries other than the United States—China, India, Syria, and Palestine, to name but a few. Moreover, in ancient times also, erosion evidently was a menace in Greece, Italy, Syria, North Africa, and Persia. The fall of empires such as Rome was accelerated by the exhausting influence on agriculture of the washing away of fertile surface soils.

16.2 Accelerated Erosion—Mechanics

Water erosion is one of the most common geologic phenomena. It accounts in large part for the levelling of our mountains and the development of plains, plateaus, valleys, river flats, and deltas. The vast deposits that now appear as sedimentary rocks originated in this way. This normal erosion amounts to about 0.2–0.5 Mg/ha per year (0.1–0.2 tons/A per year). It operates slowly, yet inexorably. When erosion exceeds this normal rate and becomes unusually destructive, it is referred to as accelerated erosion. This is the water action of special concern to agriculture.

Two steps are recognized in accelerated erosion—the *detachment* or loosening influence, which is a preparatory action, and *transportation* by floating, rolling, dragging, and splashing. Freezing and thawing, flowing water, and rain impact are the major detaching agencies. Raindrop splash and especially running water, facilitate the carrying away of the loosened soil. In gullies, most of the loosening and cutting is due to water flow, but on comparatively smooth soil surfaces, the beating of raindrops causes most of the detachment.

Influence of Raindrops. Raindrop impact exerts three important influences: (a) it detaches soil; (b) its beating tends to destroy granulation; and (c) its splash, under certain conditions, effects an appreciable transportation of soil

(a)

(b)

FIGURE 16.2 (a) A raindrop and (b) splash that results when the drop strikes a wet bare soil. Such rainfall impact not only tends to destroy soil granulation and encourage sheet and rill erosion, but it also effects considerable transportation by splashing. A ground cover, such as sod, will largely prevent this type of erosion. [*Courtesy USDA Soil Conservation Service.*]

(Figure 16.2). So great is the force exerted by dashing rain that soil granules not only are loosened and detached, but may even be beaten to pieces. Under such hammering, the aggregation of a soil so exposed practically disappears. If the dispersed material is not removed by runoff, it may develop into a hard crust upon drying and prevent the emergence of seedlings of such crops as beans. Crust formation can be prevented if the soil is covered with organic residues or with live vegetation such as forests or grass sod.

Transportation of Soil—Splash Effects. In soil translocation, runoff water plays the major role. So familiar is the power of water to cut and carry that little more need be said regarding these capacities. In fact, so much publicity has been given to runoff that the public generally ascribes to it all of the damage done by torrential rainfall.

Under certain conditions, however, splash transportation is of considerable importance (Figure 16.2). On a soil subject to easy detachment, a very heavy torrential rain may splash as much as 225 Mg/ha of soil, some of the drops rising as high as 0.7 m and moving horizontally perhaps 1–2 m. On a slope or if the wind is blowing, splashing greatly aids and enhances runoff translocations of soil, the two together accounting for the total wash that finally occurs.

16.3 Types of Water Erosion

Three types of water erosion are generally recognized: *sheet, rill,* and *gully* (Figures 16.3 and 16.4). By *sheet erosion,* soil is removed more or less uniformly from every part of the slope. However, this type is often accompanied by tiny gullies irregularly dispersed, especially on bare land newly planted or in fallow. This is *rill erosion.* But where the volume of water is concentrated, the formation of large or small ravines by undermining and downward cutting occurs. This is called *gully erosion.* While all types may be serious, the losses due to sheet and rill erosion, although less noticeable, are undoubtedly the most important from the standpoint of field soil deterioration.

16.4 Factors Affecting Accelerated Erosion—Universal Soil-Loss Equation[2]

Decades of agricultural research coupled with centuries of farmers' experience have rather clearly identified the major factors affecting accelerated erosion. These factors are included in the *Universal Soil-Loss Equation* (USLE),

$$A = RKLSCP$$

[2] For an excellent discussion of this topic see Wischmeier and Smith (1978).

FIGURE 16.3 The perched stones and pebbles shown in the picture are mute evidence of sheet erosion—the higher the pedestals, the greater the erosion. The soil under each rock has been protected from the beating action of the rain. The pedestal in the center is about 3 inches in height. [*Courtesy USDA Soil Conservation Service*].

A, the predicted soil loss in metric tons per hectare per year, is the product of

R = rainfall and runoff
K = soil erodibility
L = slope length
S = slope gradient or steepness
C = cover and management
P = erosion-control practice

Working together these factors determine how much water enters the soil, how much runs off, and the manner and rate of its removal. A brief description of each factor and the extent of its influence on soil erosion will make it easy to ascertain how to control soil erosion.

16.5 Rainfall and Runoff Factor (*R*)

The rainfall and runoff factor, *R,* measures the erosive force of rainfall and runoff. Of the two phases of rainfall, amount of *total rainfall* and its *intensity,* the latter is usually the more important. A heavy annual precipitation received in a number of gentle rains may cause little erosion, while a lower yearly

FIGURE 16.4 Sheet and rill erosion on an unprotected slope. Serious gullying is imminent if protective measures are not instituted soon. Strip cropping or even terracing should be employed if cultivated crops are to be grown. Perhaps better, the field could be seeded and used as pasture or meadow. In any case a diversion ditch well up the slope would be advisable. [*Courtesy USDA Soil Conservation Service.*]

rainfall descending in a few torrential downpours may result in severe damage. This accounts for the marked erosion often recorded in semiarid regions.

The *seasonal distribution* of the rainfall is also critical in determining soil erosion losses. For example, in the northern part of the United States, precipitation that runs off the land in the early spring when the soils are still frozen may bring about little erosion. The same amount of runoff a few months later, however, often carries considerable quantities of soil with it. In any climate, heavy precipitation occurring at a time of year when the soil is bare is likely to cause soil loss. Examples of such conditions are seedbed preparation time and the time after the harvesting of such crops as beans, sugar beets, and early potatoes.

The *R* factor, sometimes called the rainfall erosion index, takes into account the erosive effects of storms. The total kinetic energy of each storm (related to intensity and total rainfall) plus the average rainfall during the

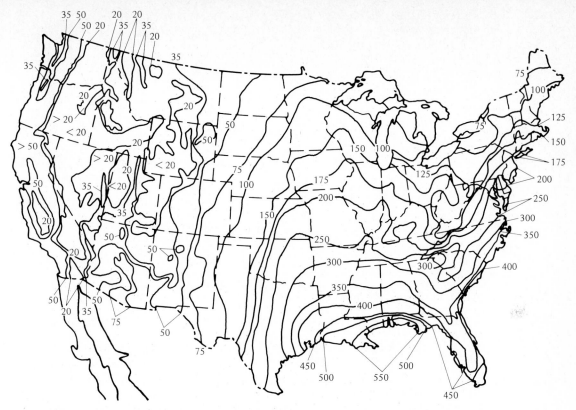

FIGURE 16.5 Average annual values of the rainfall erosion index in the United States. Note the high values in parts of the East and the generally low values in the West. [*From Wischmeier and Smith (1978).*]

30-min period of greatest intensity is considered. The sum of the indexes for all storms occurring during a year provides an annual index. An average of such indexes for several years is used in the universal soil-loss equation.

Rainfall indexes computed for the United States are shown in Figure 16.5. Note that they vary from less than 20 in the areas of the west to 550 along the coasts of the Gulf states.

16.6 Soil Erodibility Factor (*K*)

The soil erodibility factor, *K*, indicates the inherent erodibility of a soil. It gives an indication of the soil loss from a unit plot 22 m long with a 9% slope and continuous fallow culture. The two most significant soil characteristics influencing erosion are (a) infiltration capacity and (b) structural stability, which are closely related. The infiltration capacity is influenced greatly by

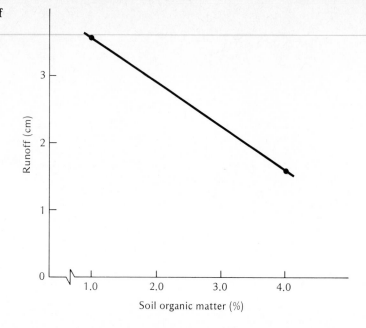

FIGURE 16.6 General relationship of the soil organic matter content to the runoff from 44 different Indiana soils, which received 2.5 in. of rain in 1 hr. [*From Wischmeier and Mannering* (*1965*).]

structural stability, especially in the upper soil horizons. In addition, soil texture, organic content (Figure 16.6), the kind and amount of swelling clays, soil depth, and the presence of impervious soil layers all influence the infiltration capacity.

The stability of soil aggregates affects the extent of erosion damage in another way. Resistance of surface granules to the beating action of rain saves soil even though runoff does occur. The marked granule stability of certain tropical clay soils high in hydrous oxides of iron and aluminum accounts for the resistance of these soils to the action of torrential rains. Downpours of a similar magnitude on temperate region clays would be disastrous.

The soil erodibility or *K* factor normally varies from near zero to about 0.6. It is low for soils into which water readily infiltrates, such as well-drained sandy soils or friable tropical clays high in hydrous oxides of iron and aluminum or kaolinite. Erodibility indexes of less than 0.2 are normal for these soils (Table 16.3). Soils with intermediate infiltration capacities and moderate soil structural stability generally have a *K* factor of 0.2–0.3. While the more easily eroded soils with low infiltration capacities will have a *K* factor of 0.3 or higher.

16.7 Topographic Factor (*LS*)

The topographic factor, *LS*, reflects the influence of length and steepness of slope. Table 16.4 gives *LS* figures for selected slope characteristics. It is the ratio of soil loss from the field in question to that of a unit plot with 9% slope,

TABLE 16.3 Computed *K* Values for Soils on Erosion Research Stations[a]

Soil	Source of data	Computed *K*
Dunkirk silt loam	Geneva, NY	0.69[b]
Keene silt loam	Zanesville, OH	0.48
Lodi loam	Blacksburg, VA	0.39
Cecil sandy clay loam	Watkinsville, GA	0.36
Marshall silt loam	Clarinda, IA	0.33
Hagerstown silty clay loam	State College, PA	0.31[b]
Austin silt	Temple, TX	0.29
Mexico silt loam	McCredie, MO	0.28
Cecil sandy loam	Clemson, SC	0.28[b]
Cecil sandy loam	Watkinsville, GA	0.23
Tifton loamy sand	Tifton, GA	0.10
Bath flaggy silt loam with surface stones >5 cm removed	Arnot, NY	0.05[b]

[a] From Wischmeier and Smith (1978).
[b] Evaluated from continuous fallow. All others were computed from rowcrop data.

22 m long, and continuously fallowed. The greater the steepness of *slope,* other conditions remaining constant, the greater the erosion due to increased velocity of water flow. Also, more water is likely to run off. Theoretically, a doubling of the velocity enables water to move particles 64 times larger, allows it to carry 32 times more material in suspension, and makes the erosive power in total 4 times greater.

TABLE 16.4 The Topographic Factor (*LS*) for Selected Combinations of Slope Length and Steepness[a]

Note that the factor increases with both percent slope and length of slope.

Slope (%)	Slope length (m)				
	15.35	30.5	45.75	61.0	91.5
2	0.163	0.201	0.227	0.248	0.280
4	0.303	0.400	0.471	0.528	0.621
6	0.476	0.673	0.824	0.952	1.17
8	0.701	0.992	1.21	1.41	1.72
10	0.968	1.37	1.68	1.94	2.37
12	1.280	1.80	2.21	2.55	3.13

[a] From Wischmeier and Smith (1978).

The *length* of the slope is of prime importance, since the greater the extension of the inclined area, the greater is the concentration of the flooding water. For example, research in southwestern Iowa showed that doubling the length of a 9% slope increased the loss of soil by 2.6 times and the runoff water by 1.8 times. This influence of slope is, of course, greatly modified by the size and general *topography* of the drainage area. Another modifying factor is the presence of *channels*, not only in the eroded area itself but in the watershed. The development of such channels controls the intensity of water concentration.

16.8 Cover and Management Factor (*C*)

The cover and management factor, *C*, indicates the influence of cropping systems and management variables on soil loss. Forests and grass are the best natural soil protective agencies known and are about equal in their effectiveness, but forage crops, both legumes and grasses, are next in effectiveness because of their relatively dense cover (Table 16.5). Small grains such as wheat and oats are intermediate and offer considerable obstruction to surface wash. Row crops such as corn, soybeans, and potatoes offer relatively little cover

TABLE 16.5 Area Under Different Land Use in the United States and the Percent With Water Erosion Greater Than 11 Mg/ha per Year[a]

Note the high rates with row crops and especially soybeans

Land use	Land area (1000 ha)	Land with water erosion greater than 11 Mg/ha per year (%)
Cropland		
All	167,288	23.5
Corn	37,832	33.6
Soybean	24,020	44.3
Cotton	6,713	33.6
Wheat	28,995	12.9
Pasture	53,846	10.0
Rangeland	165,182	11.0
Forestland		
Grazed	24,696	15.0
Ungrazed	124,696	2.0

[a] From USDA (1980a). 11 Mg/ha (5 tons/A) is the level above which the maintenance of reasonable soil productivity is very difficult.

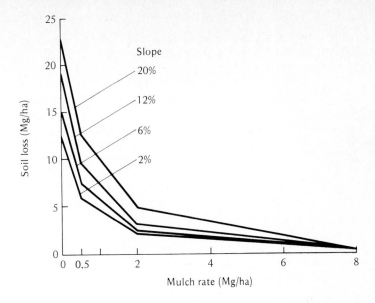

FIGURE 16.7 The effect of wheat straw mulch rate and soil slope on soil loss by erosion. [*From Lattanzi et al. (1979); used with permission of the Soil Science Society of America.*]

during the early growth stages and thereby encourage erosion. Most subject to erosion are fallowed areas where no crop is grown and all the residues have been incorporated into the soil. The effect on erosion of leaving wheat stubble as a surface mulch is illustrated in Figure 16.7.

The marked differences among crops in their ability to maintain soil cover emphasizes the need for appropriate crop rotation to reduce soil erosion. The inclusion of a close-growing forage crop in rotation with row crops will help control both erosion and runoff. Likewise, the use of so-called "conservation tillage" systems, which leave most of the crop residues on the surface greatly decreases erosion hazards.

The C value for a specific location is dependent upon a number of factors including the crop or crops being grown, crop stage, tillage, and other management factors. Technically, the C value is the ratio of soil loss under the conditions found in the field in question to that which would occur under clean tilled continuous fallow conditions. This ratio (C) will be high (approaching 1.0) with little soil cover such as bare soil in the spring before a crop canopy develops. It will be low (e.g., < 0.10) where large amounts of crop residues are on the land or in areas of dense forests. These C values are usually computed by experienced scientists with knowledge of the crop cover and management effects in a given area. Actual values are available through the state offices of the USDA Soil Conservation Service. Examples of selected C values are given in Table 16.6. Note the significant influence of tillage systems coupled with the soil cover that they provide.

TABLE 16.6 Crop Management or *C* Values for Different Crop Sequences in Northern Illinois[a]

Note the dominating effect of tillage systems and of the maintenance of soil cover. Values would differ slightly in other areas but the principles illustrated would pertain.

| Crop sequence[b] | Conventional tillage[c] | | Minimum tillage | | No tillage | |
| | Residue left | Residue removed | Residue level (kg) | | Residue level (kg) | |
			458–907	908–1816	458–907	908–1816
Continuous soybeans (Sb)	0.49	—	0.33	—	0.29	—
Continuous corn (C)	0.37	0.47	0.31	0.07	0.29	0.06
C–Sb	0.43	0.49	0.32	0.12	0.29	0.06
C–C–Sb	0.40	0.47	0.31	0.12	0.29	0.06
C–C–Sb–G–M	0.20	0.24	0.18	0.09	0.14	0.05
C–Sb–G–M	0.16	0.18	0.15	0.09	0.11	0.05
C–C–G–M	0.12	0.16	0.13	0.08	0.09	0.04

[a] Selected data from Walker (1980).
[b] Crop abbreviations: C = corn; Sb = soybeans; G = small grain (wheat or oats); M = meadow.
[c] Spring plowed; assumes high crop yields.

16.9 Support Practice Factor (*P*)

The support practice factor, *P*, reflects the benefits of contouring, strip cropping, and other supporting factors. It is the ratio of soil loss with a given support practice to the corresponding loss when crop culture is up and down the slope. On sloping fields the protection offered by surface coverage and crop management must be supported by other practices that help slow the runoff water and in turn reduce soil erosion. These support practices (factor *P*) include contour tillage, contour strip cropping, terrace systems, and grassed waterways (see Section 16.11).

Except on fields with only modest slopes, it is important that cultivation be across the slope rather than with it. This is called *contour tillage*. On long slopes subject to sheet and rill erosion, the fields may be laid out in narrow strips across the incline, alternating the tilled crops, such as corn and potatoes, with hay and grain. Water cannot achieve an undue velocity on the narrow strips of cultivated land, while the hay and grain definitely check the rate of runoff. Such a layout is called *strip cropping* and is the basis of much of the erosion control now advocated (Figure 16.8).

When the cross strips are laid out rather definitely on the contours, the system is called *contour strip cropping*. The width of the strips will depend primarily upon the degree of slope and the permeability of the land and its erodibility. Widths of 30–125 meters are common. Contour strip cropping often is guarded by diversion ditches and waterways between fields. When their grade is high, they should be sown with grass to prevent underwashing.

FIGURE 16.8 Aerial photograph of fields in Kentucky where strip cropping is being practiced. [*Courtesy USDA Soil Conservation Service.*]

When simpler methods of checking sheet erosion on cultivated lands are inadequate, it is advisable to resort to *terraces* constructed across the slope. These catch the water and conduct it away at a gentle grade. Opportunity is thus given for more water to soak in. The *Mangum terrace* and its modifications are now largely in use. Such a terrace is generally a broad bank of earth with gently sloping sides, contouring the field at a grade from 15 to 20 cm to 30 m. It is usually formed by moving soil to the terrace ridge from both the lower and upper sides. The interval between the successive embankments depends on the slope and erodibility of the land.

Since the terrace is low and broad, it may be cropped without difficulty and offers no obstacle to cultivating and harvesting machinery. It wastes little or no land and is quite effective if properly maintained. Where necessary, waterways are sodded. Such terracing is really a more or less elaborate type of contour-strip cropping.

The *P value* for a given support practice is the ratio of soil loss using that practice to the loss that would occur with up- and down-hill culture. Examples of P values for contour tillage and strip cropping at different slope gradients are shown in Table 16.7. Note that P values increase with slope and that they are low for stripcropping, illustrating the importance of this practice to erosion control.

TABLE 16.7 *P* Values for Contour-Farmed Terraced Fields in Relation to Slope Gradient[a]

Slope	Contour factor	Stripcrop factor
1–2	0.60	0.30
3–8	0.50	0.25
9–12	0.60	0.30
13–16	0.70	0.35
17–20	0.80	0.40
21–25	0.90	0.45

[a] From Wischmeier and Smith (1978).

16.10   Calculating Expected Soil Losses

Estimates of the expected soil loss in any location can be calculated using the Universal Soil-Loss Equation (*USLE*). Assume for example a location in central Iowa on a Marshall silt loam with an average slope of 4% and an average slope length of 30.5 m. Assume further that the land is clean tilled and fallowed.

From Figure 16.5 we note that the *R* factor is about 150 for this location. The *K* factor for a Marshall silt loam in central Iowa is 0.33 (Table 16.3), and the topographic factor (*LS*) from Table 16.4 is 0.400. The *C* factor is 1.0, since there is no cover or other management practice to discourage erosion. If we assume the tillage is up and down hill, the *P* value is also 1.0. Thus the anticipated soil loss can be calculated.

$$A = (150)(0.33)(0.400)(1.0)(1.0) = 19.8 \text{ tons/A or } 44.4 \text{ Mg/ha}$$

If the crop rotation involved wheat–hay–corn–corn and conservation tillage practices (e.g., minimum tillage) were used, a reasonable amount of residue would be left on the soil surface. Under these conditions the *C* factor may be reduced to about 0.1. Likewise, if the tillage and planting were done on the contour, the *P* value would drop to about 0.5 (Table 16.7). With these figures the soil loss becomes

$$A = (150)(0.33)(.400)(0.1)(0.5) = 0.99 \text{ tons/A or } 2.2 \text{ Mg/ha}$$

The benefits of good cover and management and support practices are obvious. While the figures cited were used to provide an example of the utility of the Universal Soil-Loss Equation, calculations can be made for specific locations. In the United States, pertinent factor values are generally available from state offices of the USDA Soil Conservation Service that can be used for erosion prediction in specific locations.

16.11   Sheet and Rill Erosion Control

From the preceding sections it is obvious that three objectives must be attained if sheet and rill erosion is to be held to a reasonable level. First, we must encourage as much of the precipitation as possible to run *into* the soil rather than *off* it. Second, the *sediment burden* of the runoff water should be kept low. And third, we must control the accumulation of the runoff water into channels, thereby minimizing its cutting action and the sediment burden it can carry.

The key to sheet and rill erosion control is found in factors included in the Universal Soil-Loss Equation. The control measures needed will be determined by the characteristic of the rainfall, by fundamental properties of the soil and by the slope gradient and length. Care must be exercised to keep vegetative cover on or near the soil surface. Close-growing crops such as grasses and small grains provide good cover, especially if adequate nutrients have been provided from fertilizer, lime, and manure. Furthermore, mechanical support practices such as *contour tillage* and *strip cropping* should continue to receive high priority on sloping fields. But perhaps the most significant and relatively inexpensive means of reducing sheet and rill erosion is the use of so-called "conservation tillage" practices. These will now be considered.

16.12   Conservation Tillage Practices[3]

Until fairly recently, conventional agricultural practice required rather extensive soil tillage. Most of the crop residues were incorporated into the soil using the moldboard plow. The soil surface was then tilled further using a harrow to provide a seedbed devoid of clods. Once row crops were planted, a cultivator was used, often several times, to keep the weeds down. Thus the soil was tilled repeatedly at great cost in terms of time and energy. More important, however, the soil was usually left bare immediately after plowing until later in the year when crop growth was sufficient to provide some ground coverage. This means that the soil was left unprotected during the spring of the year when runoff and erosion pressures were greatest.

Three developments of the past two decades have dictated drastic changes in tillage practices. First, herbicides capable of controlling most of the major weeds have become available at reasonable costs. This has reduced the need for cultivating and even for plowing in some cases. Second, dramatic increases in fuel costs forced tractor-dependent farmers to seek means of reducing their tillage operations. Third, public environmental concerns have forced a re-evaluation of soil erosion as a source of water pollution. These three developments have given farmers the opportunity to examine reduced-tillage methods, most

---

[3] For recent reviews of this topic see Unger and McCalla (1980), Crosson (1981), and SCSA (1983).

of which allow much less erosion than the conventional tillage systems. Because of this, these tillage practices are collectively known as *conservation tillage,* in contrast to the *conventional tillage* system described above.

Conservation Tillage Systems. In Table 16.8 some of the modern conservation tillage systems are listed along with the specific operations involved. In studying them, keep in mind that conventional tillage involves plowing, from one to three passes with a harrow, crop planting, and sometimes subsequent tillage with a cultivator. The conservation tillage systems all involve less tillage. Note that they vary from systems that merely reduce excess tillage to the "no tillage" system, which permits direct planting in the residue of the previous crop and utilizes only that localized tillage necessary to plant the seed. Figure 16.9a illustrates how one crop is planted under the residues of another and emphasizes how cover is kept on the land. Such a system minimizes opportunity for erosion.

One would expect better soil protection with any of the conservation tillage systems than with the conventional one. This expectation is confirmed by extensive field research that shows the conventional moldboard plow systems leave only 1–5% of the soil covered with crop residues. Reduced tillage systems (e.g., chiseling or disking) commonly leave 15–25% soil coverage, while with no-till systems we can expect 50–75% of the land to be so covered. These differences in land cover have marked effects on both soil erosion and runoff.

Runoff and Erosion Control. Since conservation tillage systems were initiated, hundreds of field trials have been run to ascertain the effects of these systems on runoff and erosion. As would be expected, tillage systems that maintain some soil cover with crop residues result in less soil erosion than that experienced with conventional tillage methods. Likewise surface runoff is generally decreased, although the differences are not as pronounced as with soil erosion. Typical of the results obtained are those presented in Figure 16.10. The minimum tillage (disk chisel) and no-till plots experienced far less erosion and signifi-

TABLE 16.8 Conservation Tillage Systems and the Field Operations Involved

Tillage systems	Operations
Stubble mulch	Sweeps, blades, or disks undercut residues, loosen soil, and kill weeds
Reduced tillage	
Fall plow or chisel, cultivate	Moldboard plow or chisel with minimum secondary tillage
Spring plow, plant in wheel track	Moldboard plow, no other tillage
Disk fall and/or spring and plant	Tandem disks, most residues left on surface
Till plant	No plowing; sweeps provide trash-free zone for planting
No tillage	No primary tillage; rotary tillers or coulters open soil for seed planting

FIGURE 16.9 Photographs illustrating no-tillage practice. (a) Corn is being planted in rye mulch on a farm in Georgia. Note where the coulter disks have cut through the rye and opened up slits in which the corn was planted. (b) In an Illinois field corn was planted in wheat stubble, and herbicide was used to control the weeds. In both cases the complete soil coverage provides good protection from erosion. [*Courtesy USDA Soil Conservation Service.*]

(a)

(b)

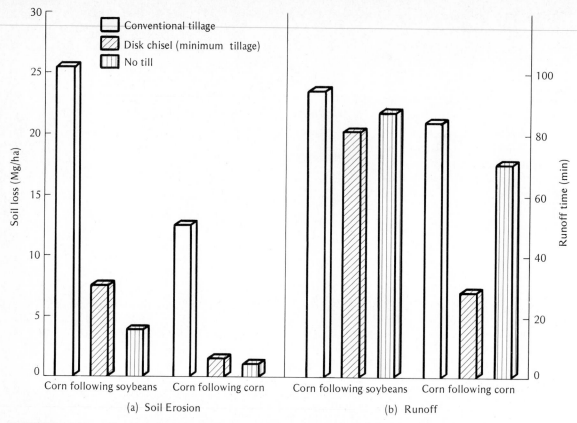

FIGURE 16.10 Effect of tillage systems on soil erosion and runoff from corn plots in Illinois following corn and following soybeans. Soil loss by erosion was drastically reduced by the conservation tillage practices. The period of runoff was reduced most by the disk chisel system where corn was grown after corn. The soil was a Catlin silt loam (Typic Argiudoll), 5% slope, up and down slope, tested in early spring. [*Data from Oschwald and Siemens (1976).*]

cantly less runoff than was found on the conventional fall-plowed plot. The practical significance of conservation tillage is obvious.

The need for conservation tillage is determined in part by the erosion potential of the soil on which crops are to be grown. Estimates shown in Table 16.9 illustrate the point. Note that on soils with high and very high erosion potentials estimated erosion losses where the conventional plow system is used are 2–4 times that considered tolerable if current soil productivity levels are to be maintained (11.2 Mg/ha). In contrast, soil losses, even from soils with high erosion potential, are estimated to be far below the tolerable limit where the no-till system is used. The high percentage of soil cover that the no-till system encourages (42–76%) is largely responsible for this difference.

TABLE 16.9 Estimates of Annual Soil and Nitrogen Losses as Affected by Three Tillage Systems for Soybeans Following Corn and for Corn Following Soybeans, Each Grown on Soils with Different Erosion Potentials[a]

Erosion potential[b]	Soil loss (Mg/ha)			Nitrogen loss (Kg/ha)		
	Plow[c]	Chisel	No-till	Plow	Chisel	No-till
			Soybeans following corn			
	(4% cover)[d]	(26% cover)	(76% cover)	(4% cover)	(26% cover)	(76% cover)
Low (½T)	5.6	1.9	0.2	11.4	5.6	2.8
Med. (1T)	11.2	3.7	0.3	18.9	8.7	3.2
High (2T)	22.4	7.5	0.6	31.9	14.1	3.9
Very high (4T)	44.8	14.9	1.2	54.5	23.4	5.2
			Corn following soybeans			
	(2% cover)	(14% cover)	42% cover)	(2% cover)	(14% cover)	(42% cover)
Low (½T)	6.2	3.4	0.8	4.4	2.9	1.5
Med. (1T)	12.4	6.8	1.7	7.5	4.9	2.1
High (2T)	24.8	13.6	3.4	13.0	8.2	3.2
Very high (4T)	49.5	27.2	6.7	22.6	14.1	5.1

[a] Estimated by Baker and Laflen (1983).

[b] T = soil loss tolerance of 11.2 Mg/ha (5 tons/A), thought to be the maximum allowable to maintain productivity of most soils.

[c] Plow = moldboard plow, disk, and plant; chisel = chisel, disk, and plant; no-till = plant in untilled land.

[d] Cover estimated assuming 90% residue cover for corn and 50% for soybeans, and that the plow system leaves 10% of the residue on the surface, the chisel 50%, and no-till 67%.

The minimum tillage system (chisel) provides intermediate soil coverage and soil and nutrient losses.

Table 16.9 also indicates that conservation tillage systems can significantly reduce losses of soil nitrogen. The finer fractions of soil, which are among the first to be carried away through erosion, contain most of this nitrogen. Similar results were noted for phosphorus.

Effect on Crop Yields. Conservation tillage systems generally provide yields equal to or higher than those from conventional tillage, provided the soil is not poorly drained and can be kept free of weeds through the use of chemicals. Typical of the results obtained from numerous field trials are those shown in Table 16.10. Long-term trials show yields from conservation tillage systems actually superior to those where conventional tillage is used.

On soils with restricted drainage, yields with conservation tillage are sometimes inferior to those from conventionally tilled areas (Figure 16.11). Reasons for these lower yields include lower soil temperatures in the spring on the

TABLE 16.10 Average Corn Grain Yields from Well-Drained Soils in Kentucky Under No Tillage and Conventional Tillage Systems[a]

Soil type	Years tested	Grain yields (kg/ha)	
		No tillage	Conventional tillage
Maury silt loam	8	9,136	8,932
Crider silt loam	5	9,886	8,318
Tilsit silt loam	5	7,705	7,705
Allegheny silt loam	3	10,977	10,909

[a] Phillips et al. (1980).

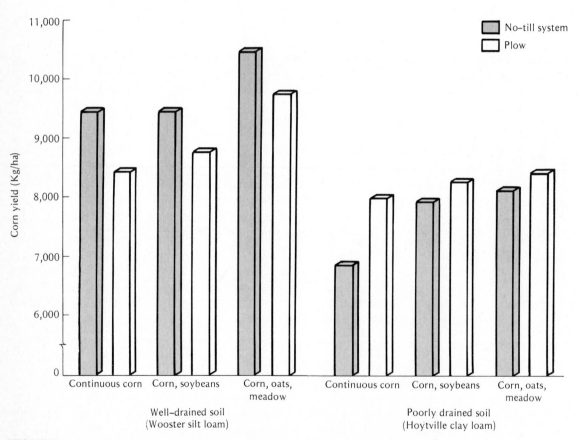

FIGURE 16.11 The effect of tillage systems on the yield of corn grown in different rotations on a well-drained and a poorly drained soil. Data averages of 5 years. [*Data from Van Doren et al.* (*1976*).]

conservation tillage plots, and the incidence of certain plant diseases, which may be higher under the somewhat higher moisture conditions that characterize these plots. Also certain weeds tend to be more of a problem on wet soils and rodent numbers may be higher where excess residues are left on the surface. These limitations may reduce the effectiveness of conservation tillage on poorly drained soils.

Effect on Soil Properties. Conservation tillage has variable effects on soil properties, depending on the particular system chosen. It is likely that those practices that reduce but do not entirely eliminate tillage will have little effect on soil properties. However, evidence suggests that no-till systems have some such effects. For example, no-tillage leaves the upper 10 cm of soil somewhat lower in total pore space than conventionally tilled soil (see Section 2.9). Also, the moisture level in this upper soil layer is generally somewhat higher under no-tillage because of the reduced evaporation caused by the residues left on the surface. This higher moisture level is likely associated with reduced oxygen levels in soil pores. Consequently, anaerobic conditions may prevail in local pockets in the soil. This may increase losses of gaseous nitrogen by denitrification, especially if the soil is not well drained (Figure 16.12). Coupled with greater leaching losses of nitrates sometimes observed on no-tilled areas, these gaseous losses can be of some concern.

Response to nitrogen fertilizer may also be affected by soil nitrogen reactions stimulated by the tillage system used. In some cases yield increases from low levels of nitrogen fertilizer applications are not as high under no-tillage systems as under conventional systems. But when higher levels of nitrogen are applied, the response under no tillage is fully as great if not greater than where the conventional "plow and cultivate" systems are used.

Since with the no-tillage system, fertilizers are usually placed on the surface, the acidifying effect of the applied nitrogen on the upper soil layers can be detrimental. For this and other reasons, it may be advisable occasionally to plow areas on which the no-tillage system is employed. This permits some mixing of the upper few centimeters of soil with the remainder of the plowed layer and reduces the hazard of too great acidity at the surface.

Localized anaerobic conditions and increased acidification due to nitrification may account for the increased organic matter content that has been observed in the surface horizon of some no-tilled plots (Figure 16.13). Apparently the rate of organic matter decomposition is not as high in no-tilled areas as where the conventional tillage has been used.

Labor and Energy Requirements. A primary reason for a farmer to adopt conservation tillage practices is their lower labor and energy requirements. While the time and energy savings would vary depending on the conservation tillage system chosen, a number of studies suggest labor requirements are cut in half by the farmers' switching to conservation tillage. This is not surpris-

FIGURE 16.12 Diagram showing differences in nitrogen reaction in soils under no-tillage compared with the conventional system of plowing. The no-till system tends to leave the soil somewhat less well aerated as compared with conventional tillage. [*From Doran (1982); used with permission of the American Society of Agronomy.*]

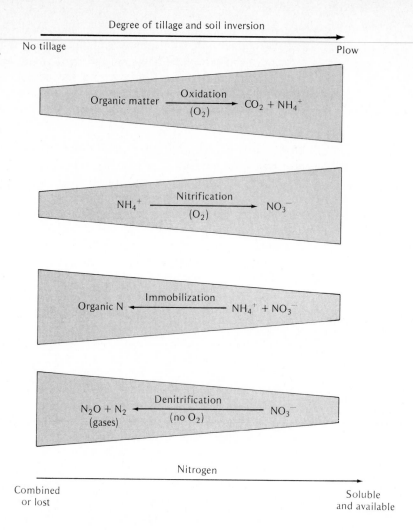

Degree of tillage and soil inversion

No tillage → Plow

Organic matter $\xrightarrow[\text{(O}_2)]{\text{Oxidation}}$ $CO_2 + NH_4^+$

NH_4^+ $\xrightarrow[\text{(O}_2)]{\text{Nitrification}}$ NO_3^-

Organic N $\xleftarrow[]{\text{Immobilization}}$ $NH_4^+ + NO_3^-$

$N_2O + N_2$ (gases) $\xleftarrow[\text{(no O}_2)]{\text{Denitrification}}$ NO_3^-

Nitrogen

Combined or lost → Soluble and available

ing since the number of passes over a field is greatly reduced by adopting these practices.

Fuel consumption for conservation tillage is likewise markedly lower than those for conventional tillage, comparative levels being only one-third to one-half those for the conventional systems. Even when the energy needed to manufacture the herbicides used to control weeds is taken into consideration, in most cases the conservation tillage systems require less total energy than do their conventional counterparts.

Probable Future Expansion. Conservation tillage practices have spread rapidly in the United States during the past 15 years. Whereas in 1965 on only 2–3% of the harvested cropland were these practices utilized, by 1979 this figure had risen to about 20%. Future projections suggest that this trend will

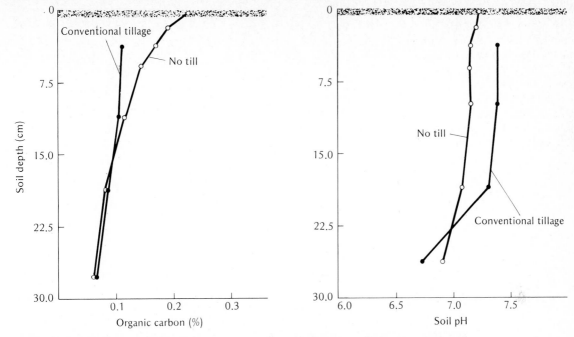

FIGURE 16.13 Comparative effects of 19 years of no-tillage and conventional tillage on the organic carbon content and pH of a Wooster silt loam. Keeping organic residues near the surface (no-tillage) increased the organic content of the upper 10 cm but reduced the soil pH in the upper 20 cm. The lower pH was likely due to increased nitrification where soil organic matter was higher. [*From Dick* (*1982*); *used with permission of the Soil Science Society of America.*]

continue and that by the year 2000 some three-fourths of the country's cropland will be under conservation tillage. One somewhat optimistic projection of future conservation tillage increases is shown in Figure 16.14. The expansion in area covered by conservation tillage practices is most heartening because of its probable constraint on soil erosion.

Up to this point consideration has been given to sheet and rill erosion. It is appropriate to have done so since most soil loss occurs from one of these processes. But in local areas gully erosion is of greatest import. This destructive process will now receive our attention.

16.13   Gully Erosion

Small gullies, though at first insignificant, soon enlarge into unsightly ditches or ravines. They quickly eat into the land above, exposing its subsoil and increasing the sheet and rill erosion, already undesirably active (Figure 16.4). If small enough, such gullies may be plowed in and seeded to grass.

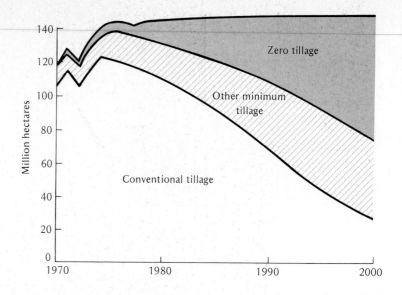

FIGURE 16.14 Projected changes in area under conventional, minimum, and zero tillage systems. While the projected increases in the two conservation tillage systems may be optimistic, there will probably continue to be marked expansion in areas so tilled. [*Redrawn from USDA (1975).*]

When the gully erosion is too active to be thus checked and the ditch is still small, dams of rotted manure or straw at intervals of 5–7 m are very effective. Such dams may be made more secure by strips of wire netting staked below them. After a time, the ditch may be plowed in and the site of the gully seeded and kept in sod permanently. It therefore becomes a *grassed waterway,* an important feature of most successful erosion control systems.

With moderate-sized gullies, larger dams of various kinds may be utilized. These are built at intervals along the channels. Piles of brush or woven wire tied to the soil by stakes driven into the ground make effective dams.

With very large gullies, dams of earth, concrete, or stone are often successfully used. Most of the sediment is deposited above the dam and the gully is slowly filled. The use of semipermanent check dams, flumes, and paved channels are also recommended on occasion. The only difficulty with engineering features if they are at all extensive is that the cost may exceed the benefits derived or even the value of the land to be served. They thus may prove uneconomical.

16.14   Wind Erosion—Its Importance and Control

Wind erosion, although most common in arid and semiarid regions, occurs to some extent in humid climates as well. It is essentially a dry-weather phenomenon and hence creates a moisture problem. All kinds of soils and soil materials are affected and at times their more finely divided portions are carried to great heights and for hundreds of miles.

In the great dust storm of May 1934, which originated in western Kansas,

Texas, Oklahoma, and contiguous portions of Colorado and New Mexico, clouds of powdery debris were carried eastward to the Atlantic seaboard and even hundreds of miles out over the ocean. Such activity is not a new phenomenon but has been common in all geologic ages. The wind energy that gave rise to the wind-blown deposits now so important agriculturally in the United States and other countries belongs in this category (see Figure 12.14, p. 411).

The destructive effects of wind erosion are often very serious. Not only is the land robbed of its richest soils, but crops are either blown away or left to die with roots exposed or are covered by the drifting debris. Even though the blowing may not be great, the cutting and abrasive effects upon tender crops is often disastrous. Figure 16.15 shows relative wind erosion forces at four locations in the United States.

Possibly 12% of the continental United States is somewhat affected by

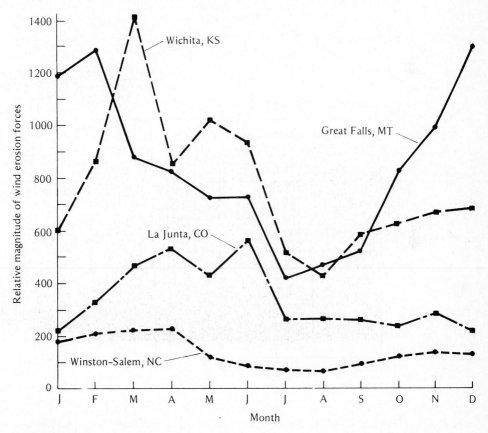

FIGURE 16.15 Monthly variation in the relative magnitude of wind erosion forces at four locations in the United States. One would expect little soil movement in Winston-Salem, NC, not only because of the low wind erosion forces but also because of the probable high soil moisture content. [*From Skidmore and Woodruff* (*1968*).]

wind erosion, 8% moderately so, and perhaps 2–3% greatly. The area subject to greatest wind action is the Great Plains (Figure 16.16). Here the mismanagement of plowed lands and the lowered holding power of the range grasses due to overgrazing have greatly exacerbated wind action. In dry years, as experience has shown, the results have been most deplorable.

Although most of the damage is confined to regions of low rainfall, some serious wind erosion occurs in humid sections. Sand dune movement is a good example. More important agriculturally, sandy soils and peat soils cultivated for many years are often affected by the wind. The drying out of the finely divided surface layers of these soils leaves them both extremely susceptible to wind erosion.

Wind erosion is not a concern limited to the United States. It is a worldwide problem. Large areas of the Soviet Union have been subject to severe damage by wind erosion, and declines in soil productivity and crop production have resulted. Likewise, in recent years large areas of Sub-Saharan Africa have been badly damaged by wind erosion. Overgrazing and other misuses of the fragile lands of arid and semiarid areas have brought starvation and human misery to millions of people in these areas. Together with water erosion it threatens humankind's capacity to sustain its food and fiber production.

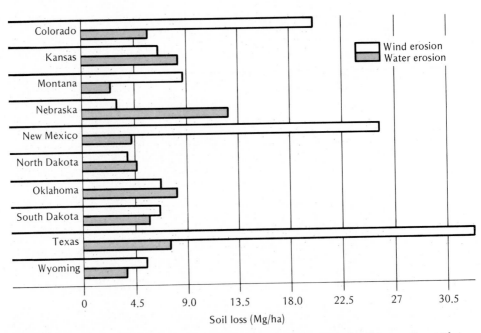

FIGURE 16.16 Average wind and water erosion rates on cropland in ten Great Plains States. In six states wind erosion rates exceed those of water erosion. In Texas, Colorado, and New Mexico wind erosion far exceeds the water-based losses. [*From USDA (1978)*.]

Mechanics of Wind Erosion. As was the case for water erosion, the loss of soil by wind movement involves two processes: (a) detachment and (b) transportation. The abrasive action of the wind results in some detachment of tiny soil grains from the granules or clods of which they are a part. When the wind is laden with soil particles, however, its abrasive action is greatly increased. The impact of these rapidly moving grains dislodges other particles from soil clods and aggregates. These particles are now ready for movement.

The transportation of the particles once they are dislodged takes place in several ways. The first and most important is that of *saltation*, or movement of soil by a short series of bounces along the surface of the ground (Figure 16.17). The particles remain fairly close to the ground as they bounce, seldom rising more than 30 cm or so. Depending on conditions, this process may account for 50–75% of the total movement.

Saltation also encourages *soil creep*, or the rolling and sliding along the surface of the larger particles. The bouncing particles carried by saltation strike the large aggregates and speed up their movement along the surface. Soil creep accounts for the movement of particles up to about 0.84 mm in diameter, which may amount to 5–25% of the total movement.

The most spectacular method of transporting soil particles is by movement in *suspension*. Here, dust particles of a fine sand size and smaller are moved parallel to the ground surface and upward. Although some of them are carried at a height no greater than a few yards, the turbulent action of the wind results in others being carried miles upward into the atmosphere and hundreds of miles horizontally. They return to the earth only when the wind subsides and when precipitation washes them down. Although it is the most obvious manner of transportation, suspension seldom accounts for more than 40% of the total and is generally no more than about 15%.

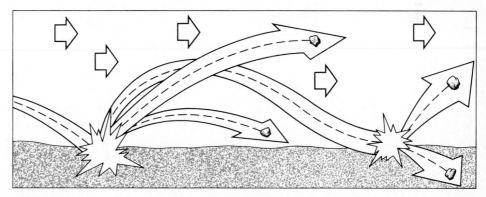

FIGURE 16.17 The process of saltation. Medium-sized particles (2.5–3.75 mm in diameter) bounce along the soil surface, striking and dislodging other particles as they move. They are too large to be carried long distances suspended in the air, but small enough to be transported by the wind. [*From Hughes* (*1980*); *used with permission of Deere & Company, Moline, IL.*]

Factors Affecting Wind Erosion. Susceptibility to wind erosion is related rather definitely to the moisture content of soils. Wet soils do not blow. The moisture content is generally lowered by hot dry winds to the wilting point and lower before wind erosion takes place (Figure 16.18).

Other factors that influence wind erosion are (a) wind velocity and turbulence, (b) soil surface condition, (c) soil characteristics, and (d) the nature and orientation of the vegetation. Obviously, the rate of wind movement, especially gusts having greater than average velocity, will influence erosion. Tests have shown that wind speeds of about 20 km/hr (12 mph) are required to initiate soil movement. At higher wind speeds, the soil movement is proportional to the cube of the wind velocity. Thus, the quantity of soil carried by wind goes up very rapidly as speeds above 30 km/hr are reached.

Wind turbulence will also influence the transporting capacity of the atmosphere. Although the wind itself has some direct influence in picking up fine soil, the impact of wind-carried particles on those not yet disengaged is probably more important.

Wind erosion is less severe where the soil surface is rough. This roughness can be obtained by proper tillage methods, which leave large clods or ridges on the soil surface. Leaving stubble mulch is perhaps an even more effective way of reducing wind-borne soil losses.

In addition to moisture content, several other soil characteristics influencing wind erosion are (a) mechanical stability of dry soil clods and aggregates, (b) stability of soil crust, and (c) bulk density and size of erodible soil fractions.

FIGURE 16.18 Effect of soil moisture and soil texture on the wind velocity required to move soils. Soil moisture and soil texture play a highly significant role in determining the susceptibility of soils to wind erosion. [*Redrawn from Bisal and Hsiek (1966).*]

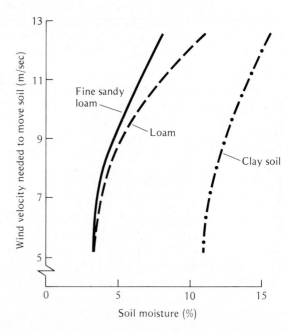

The clods must be resistant to the abrasive action of wind-carried particles. If a soil crust resulting from a previous rain is present, it too must be able to withstand impact without deteriorating. The importance of cementing agents here is quite apparent. For that reason sandy soils are sometimes easily eroded.

Soil particles of a given size and bulk density are more susceptible to erosion than are others. Apparently those particles about 0.1 mm in diameter are most erodible, those larger or smaller in size being less susceptible to movement. Thus fine sandy soils are quite susceptible to wind erosion. Particles about 0.1 mm in size apparently are responsible to a degree for the movement of larger or smaller particles that does occur. By saltation, most of the erodible particles bounce against larger particles, causing surface creep, and against smaller dust particles, resulting in movement in suspension.

Vegetation or a stubble mulch will reduce wind erosion hazards, especially if the rows run perpendicular to the prevailing wind direction. This effectively presents a barrier to wind movement. In addition, plant roots help bind the soil and make it less susceptible to wind damage.

Wind Erosion Equation.   As was the case for sheet and rill erosion, an equation relating to the quantity of wind erosion has been developed.

$$E = f(ICKLV)$$

The potential quantity of erosion per unit area E, is a function, f, of soil erodibility, I; a local wind erosion climate factor, C; the soil surface roughness, K; width of field, L; and quantity of vegetative cover, V. The influence of several of these factors is illustrated in Figure 16.19. Although wind and soil characteristics are generally beyond the farmer's control, the other factors are subject to control through the choice of cultural practices.

Control of Wind Erosion.   The factors just discussed give clues as to methods of reducing wind erosion. Obviously, if the soil can be kept moist there is little danger of wind erosion. A vegetative cover also discourages soil blowing, especially if the plant roots are well established. In dry-farming areas, however, sound moisture-conserving practices require summer fallow on some of the land, and hot, dry winds reduce the moisture in the soil surface. Consequently, other means must be employed on cultivated lands of these areas.

Conservation tillage practices described in Section 16.12 were used for wind erosion control long before they became popular as water erosion control practices. Keeping the soil surface rough and maintaining some vegetative cover is accomplished by using appropriate tillage practices. However, the vegetation should be well lodged into the soil lest it blow away. Stubble mulch has proved to be effective for this purpose (see Section 15.6). Tillage to provide for a cloddy surface condition should be at right angles to the prevailing winds. Likewise, strip cropping and alternate strips of cropped and fallowed land

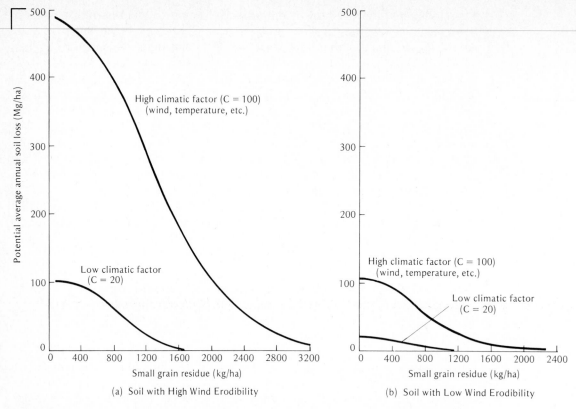

FIGURE 16.19 The effect of small grain residues on the potential wind erosion on soils with (a) high and (b) low erodibility. Strong dry winds and high temperatures encourage erosion, especially on the soil with the high wind erodibility characteristic. Surface residues can be used to help control this wind erosion. [*From Skidmore and Siddoway* (*1978*); *used with permission of the American Society of Agronomy.*]

should be perpendicular to the wind. Barriers such as three shelterbelts (Figure 16.20) are effective in reducing wind velocities for short distances and for trapping drifting soil.

In the case of the blowing of sands, sandy loams, and cultivated peat soils in humid regions, various control devices are used. Windbreaks and tenacious grasses and shrubs are especially effective (Figure 16.20). Picket fences and burlap screens, though less efficient as windbreaks than such trees as willows, are often preferred because they can be moved from place to place as crops and cropping practices are varied. Rye, planted in narrow strips across the field, is sometimes used on peat lands. All of these devices for wind erosion control, whether applied in arid or humid regions and whether vegetative or purely mechanical, are, after all, but phases of the broader problem of soil moisture control.

(a)

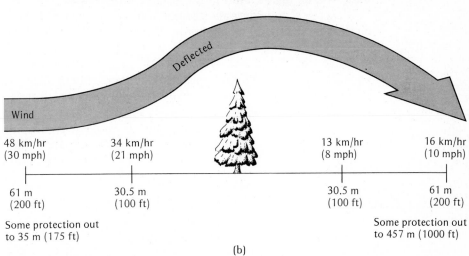

| 48 km/hr | 34 km/hr | | 13 km/hr | 16 km/hr |
| (30 mph) | (21 mph) | | (8 mph) | (10 mph) |

| 61 m | 30.5 m | | 30.5 m | 61 m |
| (200 ft) | (100 ft) | | (100 ft) | (200 ft) |

Some protection out
to 35 m (175 ft)

Some protection out
to 457 m (1000 ft)

(b)

FIGURE 16.20 (a) Shrubs and trees make good windbreaks and add beauty to a North Dakota farm homestead. (b) The effect of a windbreak on wind velocity. The wind is deflected upward by the trees and is slowed down even before reaching them. On the leeward side, further reduction occurs, the effect being felt as far as 457 m. [(a) *courtesy USDA Soil Conservation Service; (b) redrawn from Hughes (1980) and used with permission of Deere & Company, Moline, IL.*]

16.15   Soil Loss Tolerance[4]

The loss of any amount of soil by erosion generally is not considered beneficial. But years of field experience as well as scientific research indicate that some loss can be tolerated. Scientists of the U.S. Soil Conservation Service working in cooperation with field personnel throughout the country have developed tentative soil loss tolerance limits, known as *T-values,* for most of cultivated soils of the country. A tolerable soil loss (*T*-value) is considered as the maximum erosion that can take place on a given soil without degrading that soil's long-term productivity. As might be expected, there is insufficient research data to ascertain accurately the *T*-values for all soils. However, the Universal Soil-Loss Equation coupled with years of practical experience and the judgment of knowledgable soil scientists have made it possible to set these values for the more widely used soils.

The *T*-values for soils in the United States commonly range from 5 to 11 Mg/ha. They depend on a number of soil quality and management factors including soil depth, organic matter content, and the use of water control practices. At the present time 11 Mg/ha is the maximum *T*-value assigned for most soils in this country, and many have lower values.

The soil tolerance values set in the United States should be used with caution in other regions and particularly in the tropics. In some tropical soils the nutrient supplying power of the subsoil horizons are low. Since the subsoil becomes part of the plow layer in eroded soils, the *T*-value in the tropics may well be higher than those in the United States.

There is considerable controversy as to whether the current *T*-values should be increased or lowered. For soils with deep favorable rooting zones the current 11 Mg/ha *T*-value may be too low. Some scientists contend, for example, that for a soil with a rooting depth of 200 cm a *T*-value of 15 or even 20 Mg/ha would be appropriate. Other leaders are concerned, however, not only with soil productivity but with the sediment from eroded fields as it influences environmental quality including the quality of water for downstream users. These leaders contend that the *T*-values currently in use may be too high. In any case, the concept of *T*-values is useful if for no other reason than to force attention on practical means of reducing soil erosion.

16.16   Conservation Treatment in the United States

For years, farmers in the United States have received cost-sharing assistance in the establishment and use of conservation practices. That the adoption of these practices has helped reduce soil erosion loss can be seen from the data in Table 16.11. The adoption of each of the erosion control practices resulted

---

[4] For a recent discussion of this subject see Schertz (1983).

TABLE 16.11 Impact of United States Agricultural Conservation Program on Sheet and Rill Erosion by Type of Practice from 171 Sample Counties 1975–1978[a]

While reductions in erosion loss are noted in all cases, those involving vegetative cover were most effective.

Erosion control practice	No. of cases	Average erosion rates			
		Before assistance		After assistance	
		tons/A	Mg/ha	tons/A	Mg/ha
Diversions	429	21.8	48.9	11.2	25.1
Terraces	1754	14.2	31.8	4.9	11.0
Establish permanent cover	10315	11.9	26.7	3.6	8.1
Minimum tillage	119	9.7	21.7	3.8	8.5
Improve permanent cover	6978	7.9	17.7	3.4	7.6
Interim cover	2916	13.7	30.7	10.6	23.8
Strip cropping	172	8.0	17.9	4.9	11.0
Competitive shrub control	1011	3.0	6.7	2.0	4.5
Critical area treatment	217	31.3	70.2	30.3	67.9

[a] From USDA (1980b).

in significant decreases in erosion loss, in most cases to a level lower than 11 Mg/ha per year, the maximum soil loss tolerance (*T*-value) for most soils.

Data in Table 16.12 are disturbing since they show that in the past most of the cost-sharing assistance in the United States has been used on land suffering less than 11 Mg/ha soil loss per year. Unfortunately, areas that lose annually more than 30 Mg/ha of soil, and from which 84% of the total soil loss occurs, receive only 21% of the cost-sharing. These findings may well dictate changes in policies to ensure better erosion control in areas suffering very high erosion losses. They also reconfirm findings that on only a small

TABLE 16.12 The Percentage of Total Erosion Loss from Land with Differing Rates of Erosion per Hectare and the Proportion of Cost-Shared Erosion Practices on That Land[a]

Unfortunately, most of the cost sharing is on land with the least erosion hazards.

Erosion loss category (Mg/ha per yr)	Percent of total erosion loss	Percent of cost-shared erosion practices
0–11	2	52
11–31	14	27
>31	84	21

[a] From USDA (1980b).

portion of cropland in the United States are conservation treatments adequate. This applies to pastures, rangelands, and forest lands as well.

16.17   Summary of Soil Moisture Regulation

In the effective control of soil water, four more or less closely related phases must be considered. They are (a) weed transpiration, (b) surface evaporation, (c) percolation and leaching, and (d) runoff and erosion.

Cultivation traditionally has been the remedy recommended for weeds, but herbicides are being utilized more and more in connection with a suitable rotation. Surface evaporation is subject to some control through the use of crop residues and, in specific areas, of plastic film for cover. Keeping crops on the land as much as possible is the only practicable means of reducing leaching losses of nutrients.

Control measures for runoff and erosion involve organic matter maintenance to encourage infiltration, an effective vegetative cover, modern conservation tillage systems to ensure that coverage, and mechanical practices such as contour-strip cropping to allow an orderly removal of excess surface water.

REFERENCES

Baker, J. L., and J. M. Laflen. 1983. "Water quality consequences of conservation tillage," *J. Soil Water Cons.,* **38**:186–194.

Batie, S. S. 1982. *Soil Erosion—Crisis in America's Cropland* (Washington, DC: The Conservation Foundation).

Bisal, F., and J. Hsiek. 1966. "Influence of moisture on erodibility of soil by wind," *Soil Sci.,* **102**:143–146.

Brandt, Gerald H., et al. 1972. *Economic Analysis of Erosion and Sediment Control Methods for Watersheds Undergoing Urbanization,* A Report for the Office of Water Resources Research, U.S. Department of the Interior (Springfield, VA: National Technical Information Service, PB 209212).

Crosson, P. 1981. *Conservation and Conventional Tillage: A Comparative Assessment* (Ankeny, IA: Soil Cons. Soc. Amer.).

Dick, W. A. 1982. "Organic carbon, nitrogen and phosphorus concentrations and pH in soil profiles as affected by tillage intensity," *Soil Sci. Soc. Amer. J.,* **47**:102–107.

Doran, J. W. 1982. "Tilling changes soil," *Crops and Soils,* **34**:10–12.

Hughes, H. A. 1980. *Conservation Farming* (Moline, IL: Deere and Company).

Lattanzi, A. R., L. D. Meyer, and M. F. Baumgardner. 1979. "Influence of mulch rate and slope steepness in interrill erosion," *Soil Sci. Soc. Amer. Proc.* **38**:946–950.

Oschwald, W. R., and J. C. Siemens. 1976. "Conservation tillage: a perspective," SM-30, *Agron. Facts* (Urbana, IL: Univ. of Illinois).

Phillips, R. E., R. L. Blevins, G. W. Thomas, W. W. Frye, and S. H. Phillips. 1980. "No tillage agriculture," *Science,* **208**:1108–1113.

Schertz, D. L. 1983. "The basis for soil loss tolerance," *J. Soil Water Cons.,* **38**:10–14.

SCSA. 1983. Special issue on conservation tillage, *J. Soil Water Cons.,* **38**:134–319 (Ankeny, IA: Soil Cons. Soc. Amer.).

Skidmore, E. L., and F. H. Siddoway. 1978. "Crop residue requirements to control wind erosion," in W. R. Oschwalk (Ed.), *Crop Residue Management Systems,* ASA Special Publication No. 31 (Madison, WI: Amer. Soc. Agron., Crop Sci. Soc. Amer., and Soil Sci. Soc. Amer.).

Skidmore, E. L., and N. P. Woodruff. 1968. *Wind Erosion Forces in the United States and Their Use in Predicting Soil Loss,* Agr. Handbook 346 (Washington, DC: U.S. Department of Agriculture).

Unger, P. W., and T. M. McCalla. 1980. "Conservation tillage systems," *Adv. in Agron.,* **33**:1–58.

USDA. 1975. *Minimum Tillage: A Preliminary Technology Assessment,* Part II of a report for the Committee on Agriculture and Forestry, U.S. Senate (Washington, DC: Govt. Printing Office Publ. No. 57–398).

USDA, Soil Conservation Service. 1978. *1977 National Resource Inventories* (Washington, DC: U.S. Department of Agriculture).

USDA. 1980a. *Appraisal 1980 Soil and Water Resources,* Conservation Act Review Draft, Part I (Washington, DC: U.S. Department of Agriculture).

USDA. 1980b. *National Summary: Evaluation of the Agricultural Conservation Program* (Washington, DC: U.S. Department of Agriculture).

Van Doren, D. M., Jr., G. B. Triplett, Jr., and J. E. Henry. 1976. "Influence of long-term tillage, crop rotation and soil type combinations on corn yields," *Soil Sci. Soc. Amer. J.,* **40**:100–105.

Walker, R. D. 1980. "USLE, a quick way to estimate your erosion losses," *Crops and Soils,* **33**:10–13.

Wischmeier, W. J., and J. V. Mannering. 1965. "Effect of organic matter content of the soil on infiltration," *J. Soil Water Cons.,* **20**:150–152.

Wischmeier, W. J., and D. D. Smith. 1978. *Predicting Rainfall Erosion Loss— A Guide to Conservation Planning,* Agr. Handbook No. 537 (Washington, DC: U.S. Department of Agriculture).

Lime and Its Soil–Plant Relationships

Soil acidity and the nutritional conditions that accompany it result when there is a deficiency of adsorbed metallic cations (called bases) relative to hydrogen and aluminum. To decrease acidity the hydrogen and aluminum must be replaced by metallic cations. This is commonly done by adding oxides, hydroxides, or carbonates of calcium and magnesium.

These compounds are referred to as *agricultural limes*.[1] They are relatively inexpensive and leave no objectionable residues in the soil. They are vital to successful agriculture in most humid regions although their use is sometimes neglected.

17.1 Liming Materials

Oxide of Lime. Commercial oxide of lime is normally referred to as *burned lime, quicklime,* or often simply as the *oxide*. It is usually in a finely ground state and is marketed in paper bags. Oxide of lime is more caustic than limestones and may be somewhat disagreeable to handle.

Burned lime is produced by heating limestone in large commercial kilns. Carbon dioxide is driven off and the impure calcium and magnesium oxides are left behind. The essential reactions that occur when limestones containing calcite and dolomite are heated are as follows:

$$CaCO_3 + heat \rightarrow CaO + CO_2\uparrow$$
Calcite

$$CaMg(CO_3)_2 + heat \rightarrow CaO + MgO + 2\ CO_2\uparrow$$
Dolomite

The purity of burned lime sold for agricultural purposes varies from about 85 to 98%, 95% being a representative figure. The impurities of burned lime consist of the original impurities of the limestone, such as chert, clay, and iron compounds. In addition to the calcium oxide and magnesium oxides are small amounts of the hydroxides, since the oxides readily take up water from the air even when bagged. Also, contact with the carbon dioxide of the atmosphere will tend to produce carbonates.

Hydroxide of Lime. This form of lime is commonly, though improperly referred to as the *hydrate*. And since it is produced by adding water to burned lime, the hydroxides that result are often called *slaked lime*. The reaction is

$$CaO + MgO + 2\ H_2O \rightarrow Ca(OH)_2 + Mg(OH)_2$$

[1] For a recent discussion of lime and its use see Adams (in press).

Hydroxide of lime appears on the market as a white powder and is more caustic than burned lime. Like the oxide, it requires bagging. Representative samples generally show a purity of about 95%.

To maintain the concentration of this form of lime at a high point, the slaking often is not carried to completion. As a result, considerable amounts of the oxides are likely to remain. Moreover, hydroxide of lime carbonates rather readily. The carbonation of calcium and magnesium hydroxides occurs as follows.

$$Ca(OH)_2 + CO_2 \rightarrow CaCO_3 + H_2O$$

$$Mg(OH)_2 + CO_2 \rightarrow MgCO_3 + H_2O$$

Carbonation is likely to occur if the bag is left open and the air is moist. Besides the impurities, six important lime compounds are usually present: the oxides, the hydroxides, and the carbonates of calcium and magnesium. The hydroxides, of course, greatly predominate.

Carbonate of Lime. There are a number of sources of carbonate of lime. Of these, pulverized or ground limestone is the most common. There are also bog lime or marl, oyster shells, and precipitated carbonates. Also lime carbonates are by-products of certain industries. All of these forms of lime are variable in their content of calcium and magnesium.

The two important minerals carried by limestones are *calcite,* which is mostly calcium carbonate $(CaCO_3)_2$, and *dolomite,* which is primarily calcium magnesium carbonate $[CaMg(CO_3)_2]$. These occur in varying proportions. When little or no dolomite is present, the limestone is referred to as calcic or calcitic. As the magnesium increases, this grades into a *dolomitic limestone* and, finally, if very little calcium carbonate is present and the stone is almost entirely composed of calcium–magnesium carbonate and impurities, the term *dolomite*

FIGURE 17.1 Alfalfa was seeded in this field trial. In the foreground the soil was unlimed (pH = 5.2). In the background 49 Mg/ha (4 tons/A) of limestone was applied before seeding.

is used. Most of the crushed limestone on the market is calcic and dolomitic, although ground dolomite is available in certain localities.

Ground limestone is effective in increasing crop yields (Figure 17.1) and is used to a greater extent than all other forms of lime combined. It varies in purity from approximately 75 to 99%. The average purity of the representative crushed limestone is about 94%.

17.2 Chemical Guarantee of Liming Materials

Since the various forms of lime are sold on the basis of their chemical composition, the commercial guarantees in this respect become a matter of some importance. The oxide and hydroxide forms may bear composition guarantees stated in one or more of the following ways—the *conventional oxide content,* the *calcium oxide equivalent,* the *neutralizing power,* and *percentages of calcium and magnesium.* To facilitate the explanation and comparison of the various methods of expression, composition figures for commercial burned lime and hydroxide of lime are shown in Table 17.1.

Conventional Oxide and Calcium Oxide Equivalent. Since the *oxide* form of expression is so commonly used, this type of guarantee is designated here as the *conventional* method. The *calcium oxide equivalent,* as the term implies, is a statement of the strength of the lime in one figure, calcium oxide. The magnesium oxide is expressed in terms of calcium oxide equivalent and this figure is added to the percentage of calcium oxide present. This may be conveniently done by means of conversion factors. Thus, for the commercial oxide of Table 17.1, 18% magnesium oxide is equivalent to 25% calcium oxide (18 $\times$ 1.389 = 25), and 77 + 25 = 102 is the calcium oxide equivalent. This means that every 100 kg of the impure burned lime is equivalent in neutralizing capacity to 102 kg of *pure* calcium oxide.

To express one lime in chemically equivalent amounts of another, simply multiply by the appropriate ratio of molecular weights. Thus, the calcium oxide equivalent of magnesium oxide is obtained by multiplying by the CaO/MgO

TABLE 17.1 Composition of a Representative Commercial Oxide and Hydroxide of Lime Expressed in Different Ways

Forms of lime	Conventional oxide content (%)	Calcium oxide equivalent	Neutralizing power	Element (%)
Commercial oxide	CaO = 77	102.0	182.1	Ca = 55.0
	MgO = 18			Mg = 10.8
Commercial hydroxide	CaO = 60	76.7	136.9	Ca = 42.8
	MgO = 12			Mg = 7.2

molecular ratio, or 56/40.3 = 1.389. Other appropriate factors are calculated from molecular ratios, as follows.

$$\text{CaO eq. of CaCO}_3 = \frac{56}{100} = 0.56 \qquad \text{Mg eq. of MgCO}_3 = \frac{243}{843} = 0.288$$

$$\text{CaCO}_3 \text{ eq. of CaO} = \frac{100}{56} = 1.786 \qquad \text{Ca eq. of CaCO}_3 = \frac{40}{100} = 0.400$$

$$\text{CaO eq. of MgCO}_3 = \frac{56}{84.3} = 0.664 \qquad \text{Mg eq. of MgO} = \frac{24.3}{40.3} = 0.603$$

$$\text{CaCO}_3 \text{ eq. of MgCO}_3 = \frac{100}{84.3} = 1.186 \qquad \text{Ca eq. of CaO} = \frac{40}{56} = 0.714$$

Neutralizing Power and Elemental Expression. The neutralizing power, as the term is arbitrarily used in respect to lime, is nothing more than a statement of its strength in terms of calcium carbonate—its *calcium carbonate equivalent.* By multiplying by 1.786 in the case above, the calcium oxide equivalent of 102 becomes 182.1, the calcium carbonate equivalent. This means that every 100 kg of impure burned lime is equivalent in neutralizing capacity to 182.1 kg of pure calcium carbonate.

The *elemental* method of expression, though not so common as the others, is required by law in some states. It may be readily calculated from the conventional oxide guarantee. Or, if given alone, the other forms of statement may be derived from it.

Commercial limes practically always carry the conventional oxide guarantee and sometimes the elemental. Thus, the amount of magnesium as well as calcium present is indicated. This is an important consideration. In addition, one or even both of the other forms of guarantee may be given.

Limestone Guarantees. The guarantees on ground limestone differ in two respects from those of the oxide and hydroxide forms. Usually the separate percentages of calcium and magnesium carbonates are given. These are almost always accompanied by the percentage of *total carbonates* and sometimes

TABLE 17.2 Composition of a Representative Commercial Ground Limestone Expressed in Different Ways

Separate carbonates (%)	Total carbonates (%)	Calcium carbonate equivalent	Conventional oxide (%)	Calcium oxide equivalent	Element (%)
$CaCO_3 = 80$			$CaO = 44.80$		$Ca = 32.00$
	94	96.6		54.10	
$MgCO_3 = 14$			$MgO = 6.70$		$Mg = 4.03$

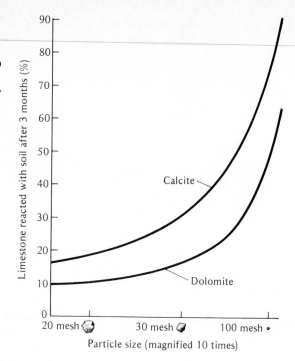

FIGURE 17.2 Relationship between particle size of calcite and dolomite and the rate of reaction of these minerals with the soil. Note that calcite particles of a given size react more rapidly than do corresponding dolomite particles. The coarse fractions of both minerals neutralize the soil acidity very slowly. [*Data calculated from Schollenberger and Salter* (*1943*).]

by the neutralizing power. In addition, one or more of the other three modes of guarantee may be used. By way of illustration, the six methods of expression are presented in Table 17.2 for a representative ground limestone.

Total Carbonates Versus Neutralizing Power. The total carbonate method of guarantee has the advantages of being simple and of requiring no chemical explanation. Furthermore, the total carbonates give perhaps a truer indication of the immediate comparative field value of limestones.

Those favoring the neutralizing-power method of expression stress its accuracy and the fact that it can be determined very easily and quickly. It gives a measure of the long-term neutralizing ability of limestones and provides recognition of the higher neutralizing power of the dolomite-containing stones. Since the rate of reaction of dolomite is significantly slower than calcite, however, there is some question as to whether this recognition is appropriate (Figure 17.2).

17.3 Fineness Guarantee of Limestone

The finer a limestone is, the more rapidly it will react with the soil. The oxide and hydroxide of lime usually appear on the market as almost impalpable powders so their fineness is always satisfactory. But different limestones may vary considerably in particle size as well as hardness. Consequently, a *fineness guarantee* is desirable. This is usually one of the requirements of laws controlling the sale of agricultural limestone. A mechanical analysis is made by the use of screens of different mesh, a 10-mesh sieve, for example, having ten openings to the linear inch (about 4 per cm). The proportion of the limestone that will pass through the various screens used constitutes the guarantee.

Interpretation of Guarantee.   Figure 17.3 shows that the finer grades of limestone are much more effective than the coarser grades in benefiting plant growth. Other data indicate that while the coarser lime is less rapid in its action, it remains in the soil longer and its influence should be effective for a greater period of years.

Fine Limestone.   Everything considered, a pulverized limestone, *all of which will pass a 10-mesh screen, and at least 50% of which will pass a 100-mesh sieve*, should give excellent results and yet be cheap enough to encourage its use. Such a lime is sufficiently pulverized to rate as a *fine* limestone. Some limestones are finer than this, 50 or 60% passing a 200-mesh screen, but the cost of grinding the stone to this very fine condition seldom can be justified.

A limestone that does not approximate the fineness designated above should be discounted to the extent to which it falls short. It may be necessary, for example, to consider 1.5 Mg of one limestone as equal to 1 Mg of another, even though their chemical analyses are the same.

FIGURE 17.3   Relationship between the particle size fraction of limestone and yield of crops in nine field experiments. [*Data from several sources; summarized by Barber (1967).*]

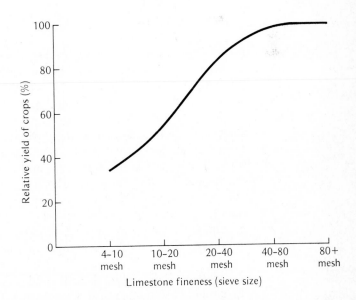

17.4   Changes of Lime Added to the Soil

When liming materials are added to a soil, two general changes occur: (a) the calcium and magnesium compounds applied dissolve in the presence of carbon dioxide, and (b) an acid colloidal complex will adsorb considerable amounts of calcium and magnesium ions.

Reaction with Carbon Dioxide.   When lime—whether the oxide, hydroxide, or the carbonate—is applied to an acid soil, the change as solution occurs is toward the bicarbonate form. This is because the carbon dioxide partial pressure in the soil, usually several hundred times greater than that in the atmospheric air, generally is high enough to cause a reaction with the hydroxide and even the carbonate. For the purely calcium limes, the reactions are

$$CaO + H_2O \rightarrow Ca(OH)_2$$
$$Ca(OH)_2 + 2\ H_2CO_3 \rightarrow Ca(HCO_3)_2 + 2\ H_2O$$
$$CaCO_3 + H_2CO_3 \rightarrow Ca(HCO_3)_2$$

Reaction with Soil Colloids.   All liming materials will react with acid soils, the calcium and magnesium replacing hydrogen on the colloidal complex. The adsorption in respect to calcium may be indicated as follows.

$$\begin{array}{l}H \\ H\end{array}\boxed{\text{Micelle}} + Ca(OH)_2 \rightleftharpoons Ca\boxed{\text{Micelle}} + 2\ H_2O$$

$$\begin{array}{l}H \\ H\end{array}\boxed{\text{Micelle}} + Ca(HCO_3)_2 \rightleftharpoons Ca\boxed{\text{Micelle}} + 2\ H_2O + 2\ CO_2\uparrow$$
$$\text{(in solution)}$$

$$\begin{array}{l}H \\ H\end{array}\boxed{\text{Micelle}} + CaCO_3 \rightleftharpoons Ca\boxed{\text{Micelle}} + H_2O + CO_2\uparrow$$
$$\text{(solid phase)}$$

As the above reactions of limestone proceed, carbon dioxide is freely evolved. In addition, the adsorption of the calcium and magnesium raises the percentage base saturation of the colloidal complex and the pH of the soil solution is pushed up correspondingly.

Compounds of Calcium and Magnesium in Limed Soil.   The calcium and magnesium supplied by a limestone will exist in the soil, at least for a time, in three forms: (a) as solid calcium and calcium–magnesium carbonates, (b) as exchangeable bases adsorbed by the colloidal matter, and (c) as dissociated cations in the soil solution mostly in association with bicarbonate ions. When the calcium and calcium–magnesium carbonates have all dissolved, the system becomes somewhat simpler, involving only the exchangeable cations, and those in soil solution, which are subject to loss by leaching.

Depletion of Calcium and Magnesium.   As the soluble calcium and magnesium compounds are removed, the percentage base saturation and pH are gradually

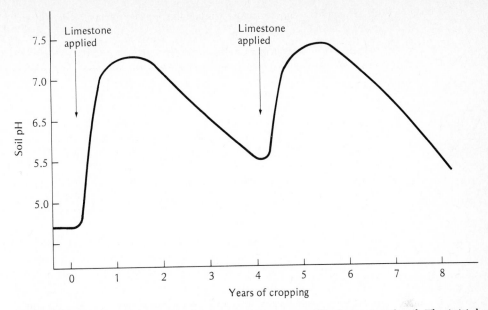

FIGURE 17.4 Influence of limestone applications on pH of a cropped soil. The initial rate of limestone application was assumed to be 8–10 Mg/ha (3.5–4.5 tons/A). Note that it takes about 1 year for most of the limestone to react. Leaching and crop removal deplete the calcium and magnesium supplied in the limestone, and in time the pH decreases until limestone is once again added.

reduced and eventually another application of lime is necessary. This then is the type of cycle through which much of the calcium and magnesium added to arable soils swings in humid regions (Figure 17.4).

17.5 Loss of Lime from Arable Soils

Calcium and magnesium are lost from soils in three ways: (a) by erosion, (b) by crop removal, and (c) by leaching. Since these three types of lime loss have already been discussed and evaluated in Chapter 15, it is a simple matter to draw such data together in tabular form (Table 17.3). Although the values shown do not represent losses from soils in general, they are quite significant when used comparatively.

The data for erosion removals are for a silt loam in Missouri cropped to a rotation of corn, wheat, and clover (Table 16.1, p. 534). The hectare annual removal by crops was calculated for a standard and representative rotation, assuming reasonable crop yields (Table 15.7, p. 522). The leaching losses are for a postulated representative silt loam under a rainfall of perhaps 90–100 cm (36–40 in.) and cropped to a standard farm rotation. All data are in kilograms

TABLE 17.3 Calcium and Magnesium Losses from Soil by Erosion, Crop Removal, and Leaching

Values are in kilograms per hectare per year.

Manner of removal	Calcium expressed as		Magnesium expressed as	
	CaO	CaCO₃	MgO	MgCO₃
By erosion, Missouri experiments, 4% slope	133	238	55	115
By the average crop of a standard rotation	70	125	42	88
By leaching from a representative silt loam	161	288	42	88
Total		651		291

per hectare per year expressed in the conventional *oxide* form and also in the more practical terms of *calcium* and *magnesium carbonates*.

The greater loss of calcium than magnesium no doubt is because the soil colloidal matter almost always carries a much larger amount of the calcium in an exchangeable condition. And since the average liming material supplies several times more calcium than magnesium, this loss ratio will in general be maintained and even accentuated in arable lands as liming proceeds.

This does not mean, however, that the magnesium in lime is of minor importance. Far from it. In fact, judging from the figures of Table 17.3, there should always be at least one third as much magnesium as calcium in the lime applied in order to meet the proportionate outgo of the two constituents. Other things being equal, it is generally wise to select a magnesium-containing limestone.

The figures quoted in Table 17.3 indicate that more than 600 kg/ha of limestone per year may be required to meet the loss from cropped soils in humid regions. This amounts to about 2.5–3.0 Mg/ha (1.1–1.3 tons/A) of carbonate of lime during a 4–5 year rotation, depending upon the kind of soil and other factors. Such a conclusion verifies the importance of lime in any scheme of fertility management in areas of medium to heavy rainfall.

17.6   Effects of Lime on the Soil

It has already been emphasized that the changes of lime in the soil are many and complicated. Therefore, the following presentation must necessarily be more or less general in nature. The better known effects of liming are (a) physical, (b) chemical, and (c) biological.

Physical Effects. Improved structure is somewhat encouraged when an acid soil is limed, although the influence is largely indirect. For example, the effects of lime upon soil microorganisms are significant, especially those concerned with the decomposition of the soil organic matter and the synthesis of humus. The genesis of the humus as well as its persistence greatly encourages granulation (see Section 2.10). The stimulating effect of lime on deep-rooted plants, especially legumes, is significant.

Chemical Effects. If a soil at pH 5.0 is limed to a more suitable pH value of about 6.5, a number of significant chemical changes occur. Many of these were described in Chapter 6 and will be outlined here to re-emphasize their importance.

1. The concentration of hydrogen ions will decrease (pH will increase).
2. The concentration of hydroxyl ions will increase.
3. The solubility of iron, aluminum, and manganese will decline.
4. The availability of phosphates and molybdates will increase.
5. The exchangeable calcium and magnesium will increase.
6. The percentage base saturation and pH will increase.
7. The availability of potassium may be increased or decreased depending on conditions.

Of the specific chemical effects of lime mentioned, the reduction in acidity is one commonly recognized. However, the indirect effects on nutrient availability and on the toxicity of certain elements are probably more important. Liming of acid soils enhances the availability and plant uptake of elements such as molybdenum, phosphorus, calcium, and magnesium. At the same time, it drastically reduces the concentration of iron, aluminum, and manganese, which under very acid conditions are likely to be present in toxic quantities.

Specific reference should be made to the effect of lime on reducing toxicities from heavy metals such as lead, zinc, cadmium, and nickel, commonly found in sewage sludge, an increasing amount of which is being added to soils. These elements are generally less soluble under alkaline than under acid conditions. This means they are less apt to be taken up by plants when the soil pH is near 7.0 or above. It also illustrates the hazard of applying sewage sludge high in these metals to highly acid soils.

Biological Effects. Lime stimulates the general-purpose heterotrophic soil organisms. This stimulation not only favors the formation of humus but also encourages the elimination of certain organic intermediate products that might be toxic to higher plants.

Most of the favorable soil organisms, as well as some of the unfavorable ones such as those that produce potato scab, are encouraged by liming. The formation of nitrates and sulfates in soil is markedly speeded up by liming an acid soil. The bacteria that fix nitrogen from the air, both nonsymbiotically and in the nodules of legumes, are especially stimulated by the application of lime. The successful growth of most soil microorganisms so definitely depends upon lime that satisfactory biological activities cannot be expected if calcium and magnesium levels are low.

17.7   Crop Response to Liming

Figure 6.12, page 210, specifies in a general way the levels at which various plants seem to grow most satisfactorily. Of the lime-loving plants with high pH requirements alfalfa, sweet clover, red clover, asparagus, cauliflower, and lettuce are representative. However, a surprisingly large proportion of crop plants is quite tolerant to the conditions presented by moderately acid soils. If such plants respond to lime, it is likely because of the stimulating influence of lime upon the legume that preceded them in the crop rotation.

Reasons for Response.   Plants respond to lime for several reasons: (a) direct nutritive or regulatory action of the calcium and magnesium, (b) removal or neutralization of toxic organic or inorganic compounds, (c) retardation of plant diseases, (d) increased chemical availability of plant nutrients, and (e) encouragement of microorganic activities. Since these factors function concurrently, crop response to liming is a complicated phenomenon.

The growth of a number of plants, including cranberries, blueberries, watermelons, laurel, and certain species of azaleas and rhododendrons, is definitely retarded by liming. Thus the plant to be grown as well as the soil on which it is grown determines liming needs.

17.8   Overliming

This leads to the question of *overliming*, the addition of lime until the soil pH is above that required for optimum plant growth. Under such conditions many crops that ordinarily respond to lime are detrimentally affected, especially during the first season following the application. With heavy soils and when only moderate amounts of lime are used, the danger is negligible. But on sandy soils low in organic matter and therefore lightly buffered, it is easy to injure certain crops, even with a relatively moderate application of lime.

The detrimental influences of excess lime already mentioned are

1. Deficiencies of available iron, manganese, copper, and/or zinc may be induced.
2. Phosphate availability may decrease because complex and insoluble calcium phosphates form.
3. The absorption of phosphorus by plants and especially its metabolic use may be restricted.
4. The uptake and utilization of boron may be hindered.
5. The drastic change in pH may in itself be detrimental.

With so many possibilities and with such complex interrelations to handle, it is easy to see why overliming damage in many cases has not been satisfactorily explained.

The use of lime in a practical way raises three questions: (a) Shall lime be applied? (b) Which form shall be used? (c) What shall be the rate of application? These will be considered in order.

17.9   Shall Lime Be Applied?

The concept that lime is a *cure-all*—almost certain to be beneficial—should now be discarded. The use of lime must be based on measured soil acidity and on crop requirements.

To determine the need for lime, the soil pH should be measured, using either a pH meter or by the less accurate indicator-dye method (Section 6.12). Representative subsoil as well as surface samples should be examined. Since the test is very easy and the results are quickly determined, the test is one of the most popular used for soil diagnosis.

Before a recommendation can be made, however, the general lime needs of the crop or crops to be grown should be considered. The final decision depends upon the proper coordination of these two types of information. A grouping of crops such as that shown in Figure 6.12 will greatly aid in deciding whether or not to lime.

17.10   Form of Lime to Apply

Five major factors should be considered in deciding on a specific brand of lime to apply.

1. Chemical guarantees of the limes under consideration.
2. Cost per ton applied to the land.
3. Rate of reaction with soil.
4. Fineness of the limestone.
5. Miscellaneous considerations (handling, storage, bag or bulk, and so on).

Guarantee and Cost Relations. By a purely arithmetical calculation based on factors 1 and 2, the cost of equivalent amounts of lime as applied to the land can be determined. These factors will show which lime will furnish the greatest amount of total neutralizing power for every dollar expended.

For instance, the *neutralizing power* ($CaCO_3$ equivalent) of two limes, a hydroxide and a ground limestone, are guaranteed at 135 and 95, respectively. The cost of applying 1 Mg of each to the land (all charges, including trucking and spreading) for purposes of calculation will be considered to be $40.00 for the hydrate and $16.00 for the carbonate. Obviously, it will require only

$^{95}/_{135}$, or 0.7 Mg of the hydroxide to equal 1 Mg of the carbonate. The cost of equivalent amounts of neutralizing power based on 1 Mg of limestone will therefore be $28.00 for the hydroxide of lime and $16.00 for the limestone. Unless a rapid rate of reaction is desired, the advantage in this case would definitely be with the limestone. Economic considerations account for the fact that about 95% of the agricultural liming material applied in the United States is ground limestone.

Rate of Reaction with Soil and Fineness. Burned and hydrated limes react with the soil much more rapidly than do the carbonate forms. For this reason, these caustic materials may be preferred where immediate reaction with the soil is required. In time, however, this initial advantage is nullified because of the inevitable carbonation of the caustic forms of lime.

In comparing carbonate-containing materials, it should be remembered that highly dolomitic limestones generally react more slowly with the soil than do those that are highly calcic. This difference is due to the comparatively slow rate of reaction of dolomite, which is supplied along with calcite in dolomitic limestones (Figure 17.2). When rapid rate of reaction is not a factor, however, dolomitic limestone is often preferred because significant quantities of magnesium are supplied by this material. Over a period of two or three rotations, soils treated with highly calcic limestone have been known to develop a magnesium deficiency even though the pH was maintained near 7.

FIGURE 17.5 Bulk application of limestone by specially equipped trucks is becoming more and more common. Because of the weight of such machinery, much limestone is applied to sod land and plowed under. In many cases this same method is used to spread commercial fertilizers. [*Courtesy Harold Sweet, Agway Inc., Syracuse, NY.*]

The fineness of the limestones under consideration is important, especially if the material is high in dolomite. If it is not sufficiently pulverized to rate as a *fine* lime, as defined in Section 17.3, allowance must be made for the lack of rapid acting material by increasing the rate of application.

Miscellaneous Factors. Several miscellaneous factors may at times be important. Handling caustic limes, even when bagged, is more disagreeable than working with limestone. The necessity for storage also comes in, since sometimes it is desirable to carry lime from one year to another. Limestone has the advantage here since it does not change in storage as the others do.

The question of purchasing the lime, especially the limestone, in bags or in bulk has become increasingly important in recent years. Spreading limestone in bulk by trucks has greatly reduced handling costs (Figure 17.5). This method tends, however, to limit the choice of when the material may be applied, since wet or plowed fields cannot be serviced by heavy machinery. The decision as to what method to use should not be finally approved until the nature of the soil and the probable response of the crops have again been reviewed.

17.11 Amounts of Lime to Apply

The amount of lime to apply is affected by a number of factors, including the following.

1. Soil:
    a. Surface; pH, texture and structure, amount of organic matter.
    b. Subsoil: pH, texture, and structure.
2. Crops to be grown.
3. Kind and fineness of lime used.
4. Economic returns in relation to cost of lime.

Soil Characteristics. The pH test is invaluable in making decisions because it gives some idea of the percentage base saturation of the soil and the need for lime. The texture and organic matter also are important since they are indicative of the adsorptive capacity of the soil and the strength of buffering (Figure 17.6). Naturally, the higher the buffer capacity of a soil, the greater must be the amount of lime applied to attain a satisfactory change in pH.

The subsoil also should be tested for pH and examined as to texture and structure. A subsoil pH markedly above or below that of the furrow slice may justify a reduction or an increase as the case may be in the acre rate of lime application.

Other Considerations. The other factors listed above have been discussed and their importance is self-evident. As to the kinds of lime, the three forms in respect to their effects on the soil are roughly in the ratio 1 Mg of representa-

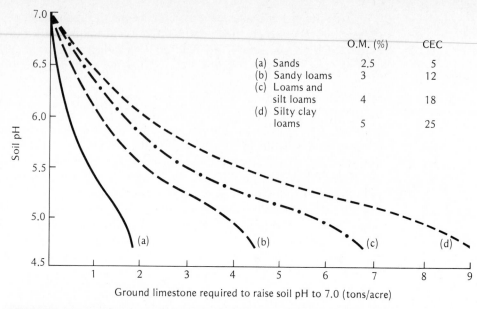

FIGURE 17.6 Relationship between soil texture and the amount of limestone required to raise the pH of New York soils to 7.0. Representative organic matter (O.M.) and cation exchange capacity (CEC) levels are shown. [*From Peech (1961)*.]

tive finely ground limestone to 0.7 Mg of commercial hydroxide to a little over 0.5 Mg of representative oxide. The fineness of limestone is as important as chemical composition. The experience factor emphasizes taking advantage of lime knowledge wherever found.

Suggested Amounts. In ordinary practice it is seldom economical to apply more than 7–9 Mg/ha (3–4 tons/A) of finely ground limestone to mineral soil at any one time unless it is very acid and the promise for increased crop yield exceptionally good. The data given in Table 17.4 serve as guides in practical liming operations for a four- or five-year rotation with average mineral soils. It is assumed that a crop with a medium lime requirement is the principal legume of the rotation. The recommendations are in terms of finely ground limestone. If the lime is coarser than the minimum quoted (see Section 17.3), or if an oxide or hydroxide is used, due allowance should be made.

17.12 Methods of Applying Lime

To obtain quick action, lime is best applied to plowed land and worked into the soil as the seedbed is prepared. It should be mixed thoroughly with the surface half of the furrow slice. However, it is often much more convenient and in the long run just as effective to apply the lime on the surface and

TABLE 17.4 Suggested Total Amounts of Finely Ground Limestone That Should Be Applied per Hectare–Furrow Slice of Mineral Soil for Alfalfa in Rotation[a]

	Limestone (kg/ha)	
Need for lime	Sandy loam	Silt loam
Moderate	2000–3000	3000–6000
High	3000–5000	6000–8000

[a] In calculating these rates, the assumption is made that the amounts suggested will be used as *initial* applications. After the pH of the soil has been raised to the desired level, smaller *maintenance* rates may be satisfactory.

plow it under. The spreading is usually done in the fall on sod land that is to be plowed later that autumn or the next spring. Applying lime on sod land reduces the possibility of extensive soil packing even if a heavy truck is used. The bulk spreading of lime, using large trucks is becoming increasingly popular (Figure 17.5).

The time of year when lime is applied is immaterial. The system of farming, the type of rotation, and related considerations are the deciding factors. Even winter application may be practiced.

Equipment Used. Although bulk spreading of limestone using large trucks is common, a smaller lime distributor may also be used, especially if only a small amount of lime is to be applied. The evenness of distribution is as important as the amount of lime used and should not be neglected.

The addition of small amounts of limestone, 350–550 kg/ha (310–490 lb/A), often gives remarkable results when drilled in with the crop being seeded. Even though the lime is not mixed thoroughly with the soil and there is little change in the pH of the furrow slice as a whole, the influence on the crop may be very favorable. Apparently the lime in this case is functioning more as a means of rectifying conditions within the crop and at its root–soil interfaces than as an amendment for the whole furrow slice.

Place in Rotation. Lime should be applied with or ahead of the crop that gives the most satisfactory response. Thus, in a rotation of corn, oats, fall wheat, and two years of alfalfa and timothy, the lime is often applied before planting the wheat in the fall. Its effect is thus especially favorable on the new legume seeding made in the wheat.

Application of lime to plowed land may result in some soil compaction if heavy machinery is used. For that reason, spreading on sod land is often

FIGURE 17.7 Important ways by which *available* calcium and magnesium are supplied to and removed from soils. The major losses are through leaching and erosion. These are largely replaced by lime and fertilizer applications. Fertilizer additions are much higher than is generally realized, owing to the large quantities of calcium contained in superphosphates.

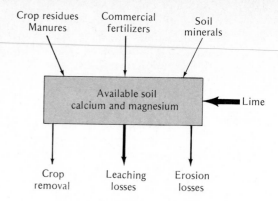

favored. In practice, the place of lime in the rotation is often determined by expediency since the vital consideration is, after all, the application of lime regularly and in conjunction with a suitable rotation of some kind.

17.13 Lime and Soil Fertility Management

The maintenance of satisfactory soil fertility levels in humid regions is dependent upon the judicious use of lime (Figure 17.7). The pH and associated nutritional condition of the soil is determined by the lime level. Furthermore, the heavy use of acid-forming nitrogen-containing fertilizers increases soil acidity, making even more important the maintenance of a favorable base status in soils.

REFERENCES

Adams, F. (Ed.). 1983. *Soil Acidity and Liming* (Madison, WI: Amer. Soc. Agron.).

Barber, S. A. 1967. "Liming materials and practices," in R. W. Pearson and F. Adams (Eds.), *Soil Acidity and Liming* (*Agronomy,* **12**:125–60) (Madison, WI: Amer. Soc. Agron.).

Peech, M. 1961. "Lime requirements vs. soil pH curves for soils of New York State," mimeographed (Ithaca, NY: Agronomy, Cornell University).

Schollenberger, C. J., and R. M. Salter. 1943. "A chart for evaluating agricultural limestone," *J. Amer. Soc. Agron.*, **35**:995–66.

Fertilizers and Fertilizer Management

Although the use of animal excrement on cultivated soils was common as far back as agricultural records can be traced, mineral salts have been systematically and extensively employed for the encouragement of crop growth hardly more than 100 years. They are now an economic necessity on many soils. Any inorganic salt, such as ammonium nitrate, or an organic substance, such as sewage sludge, purchased and applied to the soil to promote crop development by supplying plant nutrients is considered to be a commercial fertilizer.

18.1 The Fertilizer Elements

There are at least fourteen essential nutrient elements that plants obtain from the soil. Two of these, calcium and magnesium, are applied as lime in regions where they are deficient. Although not usually rated as a fertilizer, lime does exert a profound nutritive effect. Sulfur is present in several commercial fertilizers and its influence is considered important, especially in certain localities. This leaves three elements other than the micronutrients—*nitrogen, phosphorus,* and *potassium.* And since they are so commonly applied in commercial fertilizers, they are often referred to as the *fertilizer elements.*

18.2 Three Groups of Fertilizer Materials[1]

Fertilizer materials are classified into three major groups on the basis of the nutrient supplied: those supplying (a) nitrogen, (b) phosphorus, or (c) potassium. The classification is not so simple as this grouping would imply, however, since several fertilizer materials such as ammonium phosphate carry two of the elements of nutrition. The situation will be fully apparent when the fertilizer tables presented later are examined and the compounding of mixed goods considered. The amount of nitrogen, phosphate, and potash consumed in the United States in the past 20 years or so is shown in Figure 18.1. Note that the use of nitrogen has increased markedly in the past few years.

Each of the three groups will be discussed in a general way, listing the various fertilizers of each and mentioning the outstanding characteristics of the more important ones. Detailed descriptions of the various fertilizers and their properties are available in published form if additional information is desired.

[1] For more specific information on fertilizers see IFDC (1979).

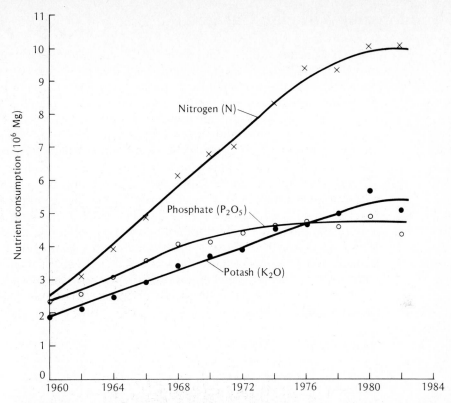

FIGURE 18.1 Primary nutrient content of fertilizers consumed in the United States 1960–82. Nitrogen use has increased very rapidly in the past 20 years. Note that by fertilizer convention the phosphorus and potassium are expressed in the oxide form, although they are actually present as P and K in fertilizer compounds. [*From USDA (1982b)*.]

18.3   Nitrogen Carriers—Two Groups

Nitrogen fertilizers may be divided for convenience into two groups: (a) *organic* and (b) *inorganic*. Organic carriers include materials such as cottonseed meal, guano, fish tankage, ammoniated peats, and cow manure. Because of their low content of nitrogen and high cost per unit of nitrogen supplied, these nitrogen carriers are not used extensively for field and vegetable crop production, supplying less than 2% of the total nitrogen added in commercial fertilizers. However, some of them are used as specialty fertilizers for lawns, flower gardens, and potted plants. Nitrogen is released slowly from these organic materials by microbiological action. This helps provide a continuing supply of the element during the warm summer months.

18.4   Inorganic Nitrogen Carriers

General Consideration.   Many inorganic carriers are used to supply nitrogen in mixed fertilizers. The most important of these, together with their compositions and sources, are listed in Table 18.1. Fortunately, there is a wide range in the nitrogen contents of the materials—from 3% in ammoniated superphosphate to 82% in anhydrous ammonia. Also, several chemical forms are represented, including ammonium and nitrate compounds as well as materials such as urea and cyanamid. Both of the latter upon hydrolysis in the soil yield NH_4^+ ions which can be taken up by plants or can be oxidized to nitrates.

The materials listed in Table 18.1 have one thing in common—they can all be produced synthetically starting with atmospheric nitrogen. This means that the quantity of nitrogen available to produce these compounds is limited only by the quantity present in the atmosphere. Beginning shortly after World War II (around 1945), these synthetic methods of supplying nitrogen played a significant role in reducing the cost of this element, historically so much more expensive than either potassium or phosphorus. In addition, synthetic processes have yielded a wide variety of materials not available when natural deposits alone were used.

TABLE 18.1   Nitrogen Carriers

Fertilizer	Chemical form	Source	Nitrogen (approx. %)
Sodium nitrate	$NaNO_3$	Chile saltpeter and synthetic	16
Ammonium sulfate	$(NH_4)_2SO_4$	By-product from coke and gas; also synthetic	21
Ammonium nitrate	NH_4NO_3	Synthetic	33
Cal-nitro and A.N.L.	NH_4NO_3 and dolomite	Synthetic	20
Urea	$CO(NH_2)_2$	Synthetic	42–46
Calcium cyanamid	$CaCN_2$	Synthetic	22
Anhydrous ammonia	Liquid NH_3	Synthetic	82
Ammonia liquor	Dilute NH_4OH	Synthetic	20–25
Nitrogen solutions	NH_4NO_3 in NH_4OH or urea in NH_4OH	Synthetic	27–32
Ammonium phosphate	$NH_4H_2PO_4$ and other ammonium salts	Synthetic	11 (48% P_2O_5)
Diammonium phosphate	$(NH_4)_2HPO_4$	Synthetic	21 (53% P_2O_5)
Ammonium polyphosphates	$(NH_4)_3HP_2O_7$ $NH_4H_2PO_4$	Synthetic	8–11 (10–37% P_2O_5)

A significant drawback to the synthesis of nitrogen fertilizers is the very high energy requirement for this process. More than 80% of the energy required to produce fertilizers in the United States is used for synthetic nitrogen production. Although the total energy required to produce the nitrogen fertilizers is only about 0.5% of the total energy consumed in the United States, rising costs of hydrocarbons, which are used to provide this energy, have significantly increased nitrogen fertilizer prices. The possibility of even higher costs in the future dictates efforts to efficiently use the synthetic nitrogen fertilizers.

Synthetic nitrogen carriers are assuming more and more importance. Most of the fertilizer nitrogen used in the United States today is carried by synthetics.

Ammonia and Its Solutions. Perhaps the most important of the synthetic processes is that in which ammonia gas is formed from the elements hydrogen and nitrogen.

$$N_2 + 3\,H_2 \rightarrow 2\,NH_3$$

Since this process requires high temperatures and pressures, it consumes large quantities of energy. Natural gas is the source of hydrogen for this reaction, and the atmosphere is the source of nitrogen. This reaction yields a compound that is the least expensive per unit of nitrogen of any listed in Table 18.1. Furthermore, the consumption of NH_3 in the United States far exceeds that of any other nitrogen carrier (Table 18.2).

Equally important, this is the first step in the formation of many other synthetic compounds. The ammonia gas is utilized in at least three ways. First, it may be liquefied under pressure, yielding anhydrous ammonia, large amounts of which are used as a separate material for direct application (Figure 18.2). Second, ammonia gas may be dissolved in water, yielding NH_4OH. This is often used alone (ammonia liquor) but is used more frequently as a solvent for other nitrogen carriers, such as NH_4NO_3 and urea, to give the "nitrogen solutions." In recent years, these solutions have become popular and are widely used. A third important use of ammonia gas is in the manufacture of other inorganic nitrogen fertilizers. A careful study of Figure 18.2 will help to establish the relationship between ammonia and the materials derived synthetically from it. These substances will be considered briefly.

Sulfate of Ammonia. Sulfate of ammonia is produced synthetically as shown in Figure 18.2 and is also a by-product of coke manufacture. This material has long been one of the most important nitrogen carriers, especially in the manufacture of mixed fertilizers. Its nitrogen is somewhat more expensive than that of the liquid forms and of urea. Its use is most satisfactory on soils well supplied with lime since it has a residual acidifying effect (see Section 18.11).

Sodium and Ammonium Nitrates. Nitric acid, the manufacture of which is possible from oxidized ammonia, is used in making both ammonium and sodium

TABLE 18.2 Supply of Selected Fertilizer Materials in the United States in 1981 and Quantities of Plant Nutrients Supplied[a]

Nitrogen carriers	Nutrient (10^3 Mg) N	
Ammonia[b]	5538	
Nitrogen solutions	2770	
Ammonium nitrate	1133	
Urea	1077	
Ammonium sulfate	338	
Phosphorus carriers	**P_2O_5**	**P**
Concentrated superphosphate	696	304
Ordinary superphosphate	158	69
Ammonium phosphates	2388	1043
Other	1238	540
Potassium carriers	**K_2O**	**K**
Muriate of potash	5343	4962
Sulfate of potash	219	203
Other	44	41

[a] Calculated from USDA (1982a).
[b] Includes anhydrous and aqueous ammonia.

FIGURE 18.2 How various fertilizer materials may be synthesized from ammonia. This gas is obtained as a by-product of coke manufacture and, in even larger quantities, by direct synthesis from elemental nitrogen and hydrogen. In recent years, ammonia and other synthetic materials shown have supplied most of our fertilizer nitrogen.

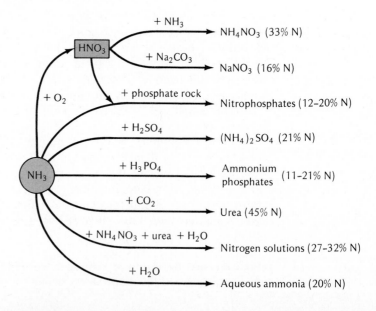

$+ NH_3 \rightarrow NH_4NO_3$ (33% N)
$+ Na_2CO_3 \rightarrow NaNO_3$ (16% N)
$+ O_2$ HNO_3
$+ $ phosphate rock $\rightarrow$ Nitrophosphates (12–20% N)
$+ H_2SO_4 \rightarrow (NH_4)_2SO_4$ (21% N)
NH_3 $+ H_3PO_4 \rightarrow$ Ammonium phosphates (11–21% N)
$+ CO_2 \rightarrow$ Urea (45% N)
$+ NH_4NO_3 +$ urea $+ H_2O \rightarrow$ Nitrogen solutions (27–32% N)
$+ H_2O \rightarrow$ Aqueous ammonia (20% N)

nitrates. Ammonium nitrate has the advantage of supplying both ammonium and nitrate ions. The percentage of nitrogen in ammonium nitrate carriers ranges from 20% for Cal-nitro and A.N.L. to about 33% for the higher grades of ammonium nitrate (Table 18.1).

Sodium nitrate is also obtained as a natural product, saltpeter, from salt beds in Chile. In the past, sodium nitrate was an important inorganic source of commercial nitrogen, but its high cost per unit of nitrogen is responsible for its minor role as a nitrogen source today.

Urea. Another promising synthetic is urea, a fertilizer containing almost three times as much nitrogen as nitrate of soda. The use of this material is increasing markedly not only in the United States but worldwide. It readily undergoes hydrolysis in the soil, producing ammonium carbonate.

$$CO(NH_2)_2 + 2 H_2O \rightarrow (NH_4)_2CO_3$$

The ammonium carbonate produced is ideal for rapid nitrification, especially if exchangeable bases are present in adequate amounts. Urea thus ultimately presents both ammonium and nitrate ions for plant absorption. Unfortunately, the ammonium carbonate is unstable at pH values above 7, releasing gaseous ammonia to the atmosphere. Consequently, it is best to incorporate urea into the soil rather than to leave it on the soil surface, especially if the soil is alkaline. Another serious objection, high deliquescence, has been largely overcome by coating its particles with dry powders.

Ammonium Phosphates. Of the fertilizers carrying both phosphorus and nitrogen, the ammonium phosphates are most important. Both mono- and diammonium phosphates are available. These compounds are made from phosphoric acid and ammonia (Figure 18.1). Since their phosphorus as well as their nitrogen is water soluble, these compounds are in demand where a high degree of water solubility is required.

Ammonium polyphosphates are used especially in liquid fertilizers. These materials are very high in phosphorus (58–61% P_2O_5; 25–27% P) and in addition contain 12–15% nitrogen.

Other Synthetic Nitrogen Carriers. Calcium cyanamid has declined in importance in the United States in recent years because of its relatively high unit cost of nitrogen. By acidulating rock phosphate with nitric rather than sulfuric or phosphoric acid, fertilizers called "nitrophosphates" are formed. They are apparently as effective as other materials in supplying nitrogen and are used extensively in the manufacture of complete fertilizers in Europe.

Slow-Release Nitrogen Carriers. For some purposes most nitrogen fertilizers have the disadvantage of too ready availability. For example, the homeowner wants a fertilizer for application to the lawn in the spring with the expectation that it will keep the grass green throughout the summer. Most nitrogen fertilizers do not live up to these expectations. They are quickly absorbed by the lawn grasses in the spring, and little is left for midsummer growth. Materials have been developed that at least partially meet such slow-release requirements.

Urea–formaldehyde complexes (ureaform) were among the first slow-release synthetic compounds produced. They contain about 38% nitrogen that is very slowly available. Other slow-release materials commercially available are crotonylidene diurea (CDU) and isobutylidene urea (IBDU), both of which contain about 32% nitrogen. Also, magnesium ammonium phosphate (8% N and about 40% P_2O_5) is a slow-release source of nitrogen. The rate of nitrogen release from these materials depends primarily on particle size, making possible a wide range of release rates. The major difficulty with these materials is their relatively high cost, which limits their use to specialty crops, lawns, and turfs.

Another way to reduce nitrogen release is to coat conventional fertilizer with substances that slow down their rate of solution and microbial attack. Waxes, paraffin, acrylic resins, and elemental sulfur are among the materials that have been used with some success. They slow down the rate of moisture penetration of the granule and the outward movement of the soluble nitrogen (Figure 18.3).

Nitrification Inhibitors. In recent years a number of compounds have been developed to prevent nitrification (see Section 9.11). Their purpose is to keep the nitrogen in the ammonium form so as to slow down the rate of leaching and possible loss of nitrogen by denitrification. They are mixed with the nitro-

FIGURE 18.3 Three methods of coating granules of nitrogen fertilizer with plastics, paraffins, or elemental sulfur to reduce the rate of release of the nitrogen. (a) Coating with a semipermeable membrane prevents movement of the nitrogen compound into the soil while permitting water movement inward. (b) An impermeable membrane with pin holes permits slow outward movement. (c) A membrane subject to microbial breakdown eventually releases the nitrogen compound. [*Modified from Parr (1972).*]

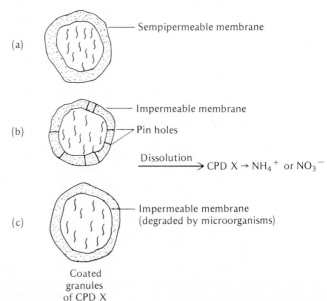

gen fertilizer or applied as a surface coating on individual pellets. Examples of these nitrification inhibitors are shown in Table 9.2, page 298. The compound most thoroughly researched in the United States is nitrapyrin (2-chloro-6-(trichloromethyl)pyridine), sold under the trade name N-serve. The effect of these compounds on crop yields varies with soil and climate, but is usually most beneficial in humid regions and when they are applied to clay soils.

18.5   Phosphatic Fertilizer Materials

The primary source of phosphorus fertilizers is rock phosphate, the essential component of which is the mineral apatite, $Ca_3(PO_4)_2 \cdot CaX$, where the X may be F, OH, Cl, etc. Since the phosphorus contained in apatite is at best slowly available, this mineral must be treated with phosphoric, sulfuric, or nitric acid to change the phosphorus into more readily available forms such as $CaHPO_4$ and $Ca(H_2PO_4)_2$. The overriding importance of phosphorus availability must be kept in mind.

Classification of Phosphate Fertilizers. For purpose of evaluation and sale, the various phosphorus compounds present in phosphatic fertilizers are classified in an arbitrary yet rather satisfactory way as follows.

1. Water-soluble—$Ca(H_2PO_4)_2$; $NH_4H_2PO_4$; K phosphates.⎫
2. Citrate-soluble (in 15% neutral ammonium ⎬ Available.
   citrate)—$CaHPO_4$. ⎭
3. Insoluble—phosphate rock $(Ca_3(PO_4)_2 \cdot CaX)$. ⎬ Unavailable.

It is well to note that the classification of the phosphates is in some degree artificial as well as arbitrary. For instance, *available* phosphates, because of

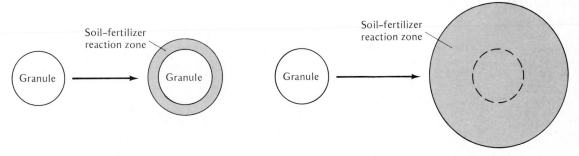

(a) Water–Insoluble Phosphate   (b) Water–Soluble Phosphate

FIGURE 18.4   Reaction of water-soluble and water-insoluble phosphates with soil. The water-insoluble granules react with only a small volume of soil in their immediate vicinity. Soluble phosphates move into the soil from water-soluble granules, reacting with iron, aluminum, and manganese in acid soils, with calcium in alkaline soils. [*From Engelstad* (*1965*).]

the reversion that occurs, are not strictly available once they contact the soil. The term "available" really refers to those phosphates that readily stimulate plant growth. Those phosphates that are less effective are rated as currently *unavailable*, yet in the soil they may eventually supply a certain amount of phosphorus to crops (Figure 18.4). The effectiveness of phosphorus materials differing in water solubility will be determined largely by the soil pH. Moderately high water solubility is desired for soils that are not too acid. For acid soils, materials with low water solubility seem to be quite satisfactory (Figure 18.5).

Phosphorus fertilizers are characterized in terms of their "available phosphate" content. Historically, the total and available phosphorus contents have been expressed in terms of oxide of phosphorus (P_2O_5) rather than the element (P). Most fertilizer control laws in the United States use the oxide method of

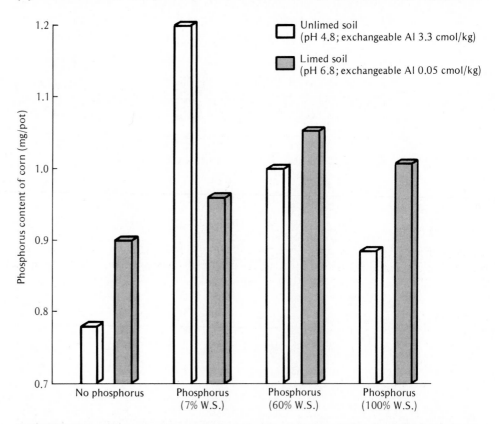

FIGURE 18.5 Effect of adding phosphate fertilizer (39 ppm P) having different degrees of water solubility (W.S.) on the phosphorus content of corn grown on a limed and an unlimed soil. Note that on the very acid unlimed soil, the highest phosphorus content was found with 7% water-soluble phosphate fertilizer, while on the limed soil the 60% water-soluble material gave the highest nutrient level. [*From McLean and Logan* (*1970*); *used with permission of the Soil Science Society of America.*]

TABLE 18.3 Phosphorus Carriers

Fertilizer	Chemical form	Available P_2O_5 (%)	Phosphorus (%)
Superphosphates	$Ca(H_2PO_4)_2$; $CaHPO_4$	16–50	7–22
Ammoniated superphosphate	$NH_4H_2PO_4$; $CaHPO_4$; $Ca_3(PO_4)_2$; $(NH_4)_2SO_4$	16–18 (3–4% N)	7–8
Ammonium phosphate	$NH_4H_2PO_4$ (mostly)	48 (11% N)	21
Ammonium polyphosphates	$(NH_4)_3HP_2O_7$; $NH_4H_2PO_4$	58–60 (12–15% N)	
Diammonium phosphate	$(NH_4)_2HPO_4$	46–53 (21% N)	20–23
Basic slag	$(CaO)_5 \cdot P_2O_5 \cdot SiO_2$	15–25	7–11
Steamed bone meal	$(Ca_3(PO_4)_2$	23–30	10–13
Rock phosphate	Fluor and chlor apatites	25–30	11–13
Calcium metaphosphate	$Ca(PO_3)_2$	62–63	27–28
Phosphoric acid	H_3PO_4	54	24
Superphosphoric acid	H_3PO_4; $H_4P_2O_7$	76	33

expression. In this chapter we will use both the oxide and elemental designations to permit the reader to relate to analyses of commercial fertilizer products. However, it should be stressed that it is the element phosphorus with which we are concerned, not the oxide. Furthermore, fertilizers do not contain phosphorus pentoxide as such, the oxide merely being used as a means of expressing the phosphorus level.

Superphosphate. The traditional phosphorus fertilizer material is superphosphate (Table 18.3). The ordinary grades containing 16–21% available P_2O_5 (7–9% P) are made by treating raw rock phosphate with suitable amounts of sulfuric acid (Figure 18.6). A large proportion of the phosphorus is thus changed to the primary phosphate form $(Ca(H_2PO_4)_2)$, although some is left in the secondary condition $(CaHPO_4)$.

By representing the complex raw rock by the simple formula $Ca_3(PO_4)_2$, the following conventional reactions may be used to show the changes that occur during the manufacture of ordinary 16–21% superphosphate.

$$\underset{\text{(insoluble)}}{Ca_3(PO_4)_2} + 2\,H_2SO_4 \rightarrow \underset{\text{(water soluble)}}{Ca(H_2PO_4)_2} + 2\,CaSO_4 + \text{impurities}$$

The acid is never added in amounts capable of completing this reaction. Consequently, some secondary phosphate—$CaHPO_4$, referred to as *citrate-soluble phosphoric acid,* is produced.

$$\underset{\text{(insoluble)}}{Ca_3(PO_4)_2} + H_2SO_4 \rightarrow \underset{\text{(citrate soluble)}}{2\,CaHPO_4} + CaSO_4 + \text{impurities}$$

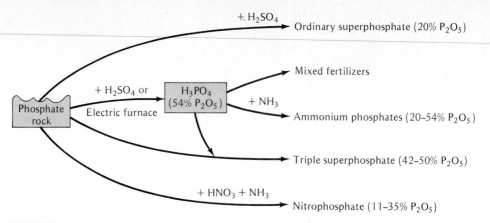

FIGURE 18.6 How several important phosphate fertilizers are manufactured. [*Concept from Travis Hignett, Tennessee Valley Authority.*]

Note that ordinary superphosphate consists of about 31% phosphates, 50% gypsum, and 19% impurities of various kinds. The total *phosphorus* is rather low, although there are significant quantities of sulfur and calcium present.

Triple superphosphate contains 40–47% available P_2O_5 (17–21% P). It differs from the ordinary type principally in that it contains more phosphorus and no gypsum. It is synthesized by treating a high-grade phosphate rock with phosphoric acid.

$$Ca_3(PO_4)_2 + 4 H_3PO_4 \rightarrow 3 Ca(H_2PO_4)_2 + \text{impurities}$$

This material is used in higher analysis fertilizers, although in recent years it has been replaced by diammonium phosphate for many mixtures.

Ammonium Phosphates. Ammonium phosphates are the most widely used phosphorus-containing fertilizers in the United States (Table 18.2). Diammonium phosphate, containing up to 21% nitrogen and 53% P_2O_5 (23% P), is used in largest quantities. It is popular as one of the materials used in bulk blending (see Section 18.10) and as a constituent of high-analysis fertilizers.

Diammonium phosphate is manufactured by producing phosphoric acid from rock phosphate and then reacting the acid with ammonia (Figure 18.6).

$$Ca_3(PO_4)_2 + H_2SO_4 \rightarrow 2 H_3PO_4 + 3 CaSO_4$$

$$H_3PO_4 + 2 NH_3 \rightarrow (NH_4)_2HPO_4$$

Other ammonium phosphate-containing materials include ammophos, mostly monoammonium phosphate (11% nitrogen and 48% phosphate), and ammoniated superphosphate, which contains 3–4% nitrogen and 16–18% phosphate.

Ammonium polyphosphates, made by ammoniating superphosphoric acid, are used extensively in the manufacture of liquid fertilizers (see Section 18.10). Liquid materials with analyses such as 11-34-0 are commonly used. They provide reasonably high analyses, prevent the precipitation of iron and other impurities, and keep most of the micronutrients in solution. In addition to orthophosphates (salts of H_3PO_4), the polyphosphates contain pyrophosphates, which are salts of $H_4P_2O_7$ (Table 18.3).

Basic Slag and Bone Meal.   Basic slag is a by-product of open-hearth steel production. It is commonly used in Europe, but is on the market only to a limited extent in the United States. Because of its alkalinity and the rather ready availability of its phosphorus, it is a desirable phosphatic fertilizer. It seems to be especially effective on acid soils, apparently because of its high content of calcium hydroxide.

Bone meal is an expensive form of phosphorus. Moreover, it is rather slowly available in the soil. Its use is limited.

Rock Phosphate.   Raw rock phosphate, because of its insolubility, must be finely ground if it is to react at all readily when applied to the soil. Its availability is markedly increased by the presence of decaying organic matter. Rock phosphate is the least soluble of the phosphatic fertilizers mentioned, the order being: ammonium phosphates and superphosphate, basic slag, bone meal, and raw rock. Although its chemical formula is given conventionally as $Ca_3(PO_4)_2$, rock phosphate is much more complicated than this formula would suggest. It apparently approaches fluor apatite $[3Ca_3(PO_4)_2 \cdot CaF_2]$ in its molecular makeup (Table 18.3). This, no doubt, accounts in part for its slow availability.

Finely ground rock phosphate is most effective when added to acid soils and to soils high in organic matter. Because of its low solubility, however, its use will continue to be as a source for the manufacture of other and more soluble forms.

High-Analysis Phosphates.   Mention should be made of two very high-analysis phosphate fertilizers: calcium metaphosphate $[Ca(PO_3)_2]$ with 62–63% available P_2O_5 (27–28% P) and superphosphoric acid with 76% P_2O_5 (33% P) (Table 18.3). Both hold great promise since they seem to be as effective, when used in equivalent amounts, as superphosphate. Their concentrations make them extremely attractive when transportation is a factor.

$Ca(PO_3)_2$, commonly called *meta-phos,* may be made by treating either phosphate rock or limestone with phosphorus pentoxide (Table 18.3). Unfortunately, high production costs of the phosphorus pentoxide have prevented the commercialization of this process.

Superphosphoric acid, a new synthetic product, is the highest phosphorus-containing material used in fertilizer manufacture today (Table 18.3). Its P_2O_5 content is 76% (33% P). It is made of a mixture of orthophosphoric, pyrophosphoric, and other polyphosphoric acids. This liquid can be used in the manufacture of liquid mixed fertilizers or to make a high-analysis superphosphate containing 54% P_2O_5 (24% P). In the formulation of liquid fertilizers, the polyphosphates help keep iron, aluminum, and micronutrients in solution.

Nitrophosphates. An important fertilizer-manufacturing process utilizes nitric acid instead of sulfuric or phosphoric acid to increase the solubility of phosphorus in rock phosphate (Figure 18.2). The products of this process are called nitrophosphates. Reactions such as the following occur (again simple $Ca_3(PO_4)_2$ is used to represent rock phosphate).

$$Ca_3(PO_4)_2 + 4\,HNO_3 \rightarrow Ca(H_2PO_4)_2 + 2\,Ca(NO_3)_2$$

$$Ca_3(PO_4)_2 + 6\,HNO_3 \rightarrow 2\,H_3PO_4 + 3\,Ca(NO_3)_2$$

The $Ca(NO_3)_2$ is later converted to NH_4NO_3 by interaction with ammonia and carbon dioxide.

$$Ca(NO_3)_2 + 2\,NH_3 + CO_2 + H_2O \rightarrow 2\,NH_4NO_3 + CaCO_3$$

Nitrophosphate processes are used extensively in Europe, where sulfuric acid sources are not abundant. Although the nitrophosphate products are somewhat lower in nutrient content than those coming from acidulation of rock phosphate with phosphoric acid, this is not a serious drawback in Europe. Since markets for fertilizers are near the manufacturing plants, shipping costs generally are not too high and high-analysis fertilizers are less essential than in the United States. Also, ammonium nitrate, one of the principal products of the nitrophosphate process, is popular in Europe, as are mixed fertilizers in contrast to single-element materials.

18.6   Potassium Fertilizer Materials

Potassium is obtained primarily by mining underground salt beds. Brine from salt lakes is also a source of some importance. Kainit and manure salts (Table 18.4) are the most common of the crude potash sources. The high-grade chloride and sulfate of potash are their refined equivalents.

All potash salts used as fertilizers are water soluble and are therefore rated as readily available. Unlike nitrogen salts, most potassium fertilizers, even if employed in large amounts, have little or no effect on the soil pH. Some discrimination is made, however, against potassium chloride (muriate) in respect to potatoes and especially tobacco, since large dosages are considered to lower the quality. Hence, when large amounts of potash are to be applied for tobacco, the sulfate form is preferred.

TABLE 18.4 Common Potassium Fertilizer Materials

The oxide means of expression (K₂O) is commonly used to identify the nutrient level, although the elemental expression (K) is technically more correct.

Fertilizer	Chemical form	K_2O (%)	K (%)
Potassium chloride[a]	KCl	48–60	40–50
Potassium sulfate	K_2SO_4	48–50	40–42
Potassium magnesium sulfate[b]	Double salt of K and Mg	25–30	19–25
Manure salts	KCl mostly	20–30	17–25
Kainit	KCl mostly	12–16	10–13
Potassium nitrate	KNO_3	44 (and 13% N)	37

[a] All of these fertilizers contain other potash salts than those listed.
[b] Contains 25% $MgSO_4$ and some chlorine.

Potassium magnesium sulfate, although rather low in potash, is used in parts of this country where magnesium is likely to be deficient. Because of magnesium's ready availability in this material, it is a more desirable source of magnesium than either dolomitic limestone or dolomite.

18.7   Sulfur in Fertilizers

There was little reason, until recently, to be concerned with sulfur deficiency in most soils of the United States. Additions of sulfur in rain and snow varied from a few kilograms per hectare annually to as much as 100 kg/ha near industrial centers. Just as important was the rather automatic additions of sulfur in the mixed fertilizers then in use. These almost invariably contained two sulfur-containing ingredients, ordinary superphosphate (which contains gypsum or calcium sulfate) and ammonium sulfate.

Two developments have made it necessary to reconsider the needs for fertilizer sulfur. Public concern over atmospheric pollution has resulted in some reductions in sulfur additions to agricultural lands near cities and industrial sites. Also, the trend in recent years toward high-analysis fertilizers has made it necessary to replace the superphosphate and ammonium sulfate in most mixed fertilizers with low-sulfur materials such as triple superphosphate, the ammonium phosphates, and anhydrous ammonia. In areas far removed from atmospheric sources of sulfur, such as eastern Washington and Oregon and parts of the southeast and the midwest, sulfur deficiencies have been noted. These are being met by adjusting the fertilizer constituents to include sulfur-containing materials. Thus, the automatic supply of sulfur through fertilizers can no longer be taken for granted. This element must be consciously added where deficiencies are apt to occur.

18.8   Micronutrients

The amount of micronutrients in fertilizers must be much more carefully controlled than the macronutrients. The difference between the deficiency and toxicity levels of a given micronutrient is extremely small. Consequently, micronutrients should be added only when their need is certain and when the amount required is known.

When a trace-element deficiency is to be corrected, especially if the case is urgent, a salt of the lacking nutrient often is added separately to the soil (Table 18.5). Copper, manganese, iron, or zinc generally is supplied as the sulfate, and boron is applied as borax. Molybdenum is added as sodium molybdate. Iron, manganese, and zinc are sometimes sprayed in small quantities on the leaves rather than being applied directly to the soil. Porous, glassy (fritted) silicates prepared by fusing compounds of boron, manganese, iron, and zinc with silica may also be used to supply these nutrients.

In recent years increased attention has been given to chelates as suppliers of iron, zinc, manganese, and copper (see Section 11.6). These materials are especially useful on soils of high pH where mineral sources would be quickly rendered unavailable. Because of their high cost, however, these materials are often used as foliar sprays, which permit much lighter application rates.

Micronutrients are commonly being applied in liquid fertilizers in areas where the quantities of these nutrients required has been carefully ascertained. The presence of polyphosphates in many of these fluid materials protects the micronutrients from precipitation as insoluble iron and other compounds. By including micronutrients in the liquid mixes application costs for these materials are kept low.

TABLE 18.5   Salts of Micronutrients Commonly Used in Fertilizers.[a]

Compound	Formula	Nutrient content (%)
Sodium borate (borax)	$Na_2B_4O_7 \cdot 10H_2O$	11
Cupric oxide	CuO	79
Copper sulfate	$CuSO_4 \cdot H_2O$	35
Ferric sulfate	$Fe_2(SO_4)_3 \cdot 9H_2O$	20
Ferrous sulfate	$FeSO_4 \cdot 7H_2O$	20
Manganous oxide	MnO	48
Manganous sulfate	$MnSO_4 \cdot H_2O$	24
Manganous-manganic oxide	Mn_3O_4	69
Ammonium molybdate	$(NH_4)_2MoO_4$	49
Sodium molybdate	Na_2MoO_4	46
Zinc oxide	ZnO	80
Zinc sulfate	$ZnSO_4 \cdot H_2O$	36

[a] Selected data from IFDC (1979).

18.9 Organic Sources of Nutrients

In addition to the inorganic sources of nutrients, important organic sources are also available. Historically, organic sources were the primary means of replenishing nutrients removed from soils. For centuries, animal and human wastes and crop residues have been returned to the soil to enhance crop production levels. Long before scientists identified the essential nutrients that these organic materials provided, farmers understood their value and took action to make use of them. Biblical references to the use of dung and other historical records of the use of leguminous crop residues attest to the long recognition of the value of these organic materials.

In recent years, some individuals and groups have expressed preferences for foodstuffs grown on soils to which only natural organic materials have been added. The demand for these foods is being met by farmers and gardeners who practice what some have termed "organic gardening." Although most of the natural organic materials are from manure or from crop residues grown on these farms, some commercially available organic materials are also used. Since Chapter 19 deals with farm manure and other organic wastes, brief reference will be made here only to commercially available organic fertilizers.

Commercial organic materials include dried poultry and cow manures, peats of various kinds, mixtures of peat and manure, composted organic residues, and dried domestic wastes. Ground bone meal, dried blood, oil seed meals, fish tankage, and other food processing wastes may be included. While the readily available nutrients in these materials are only a fraction of those in commercial inorganic fertilizers, and the cost of the nutrients they supply is high, these materials do provide slowly available essential elements and generally have beneficial effects on the physical condition of soils. If sufficient quantities of the organic materials are applied good crop yields can be obtained. Furthermore, since the prices for organically grown foods are generally higher than for those produced using inorganic chemicals, farmers and gardeners can justify the higher fertilizer costs.

18.10 Mixed Fertilizers

For years, farmers have used mixtures of materials that contain at least two of the "fertilizer elements," and usually all three. Materials such as those discussed in previous sections are mixed in proper proportion to furnish the desired amount of the nutrient elements. For example, an ammonia solution, triple superphosphate, muriate of potash, and a very small amount of an organic might be used if a complete fertilizer is desired. Such fertilizers supply about one-third of the total fertilizer nutrient consumption in the United States, the remainder coming from separate materials (Table 18.6). The nutrient use per hectare for various regions of the United States is shown in Figure 18.7.

TABLE 18.6 Total Nutrient Consumption in the United States for the Year Ending June 30, 1982, and the Percentage of the Nutrients Applied in Mixture and as Separate Materials[a]

Nutrient	Total consumption (Mg)	Applied in mixtures (%)	Applied separately (%)
Nitrogen (N)	10,051	21.1	78.9
Phosphorus (P)	1,910	88.1	11.9
(P_2O_5)	(4,371)		
Potassium (K)	4,228	45.9	54.1
(K_2O)	(5,093)		
Total N, P, K	16,187	35.5	64.5
(N, P_2O_5, K_2O)	(19,515)		

[a] Calculated from USDA (1982b).

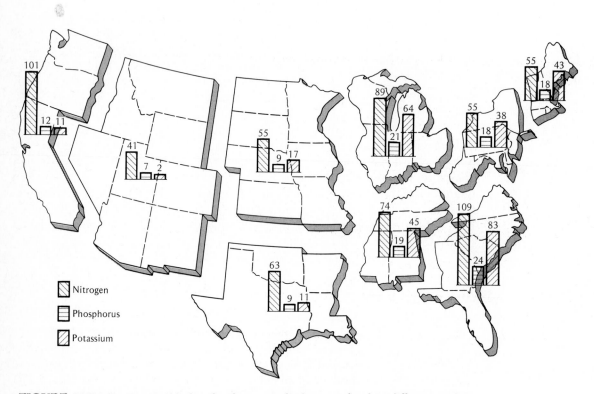

FIGURE 18.7 Nutrients added in fertilizers applied to cropland in different regions of the United States (represented as N, P, and K in kg/ha). While the South Atlantic uses more nutrients per hectare than the other regions, fertilizer usage has increased most drastically during the past few years in the central part of the country. [*Courtesy USDA.*]

Physical Condition. In addition to supplying nitrogen, phosphorus, and potassium in desired proportions, a solid commercial fertilizer should have a good *physical condition* and remain in this condition after storage. Mixtures of certain materials cannot be used because of their tendency to "set up" or harden. Of the fertilizer materials commonly found in mixed goods, ammonium nitrate, sulfate of ammonia, and potassium chloride are most likely to develop unsatisfactory physical conditions. The extreme deliquescence of some of the salts, especially urea and ammonium nitrate, tends to make mixed fertilizers sticky and thus reduce their drillability.

Methods of Ensuring Free Flowage. The free-flowing condition of a solid mixed fertilizer is maintained by (a) using moisture-resistant bags, (b) including certain moisture-absorbing materials in the fertilizer mix, and (c) granulating the mixed fertilizers. Separate materials such as ammonium nitrate have been pelleted for several years to prevent caking. In recent years the granulation of mixed fertilizers has become commonplace in this country.

In addition to having free-flowing properties, granulated fertilizers present certain other advantages. For example, they are less disagreeable to handle since they are comparatively free of small dust-like particles. The granulation also prevents the fertilizer from being carried by the wind, thus permitting more uniform spreading. Finally, granulation tends to reduce the rate of reaction of the fertilizer with the soil.

Bulk Blending. Thirty years ago, essentially all mixed fertilizers were bagged at the manufacturing plant and shipped to distribution points for sale to farmers. Today, bulk handling and blending of fertilizers have become more and more common, more than half of the fertilizer applied in the United States being so handled. Granular materials are shipped in bulk to a blending plant located within a 25- to 30-mile radius of the intended use of the fertilizer (Figure 18.8). The fertilizers are mixed to the customer's order and move directly to the farm, where they are spread on the field. Frequently the trucks used to do the spreading are owned and operated on a custom basis by the blending plant. More than 5000 such small plants are producing about 10 million Mg of fertilizer for spreading in different sections of the United States.

The most obvious advantage of bulk blending and handling is to reduce labor. Furthermore, costs of storing, production, transportation, and spreading are all lower, at least for the medium- to large-sized farms, than the conventional means of handling fertilizers. Also, the fertilizers are generally high in analysis, and chemical incompatibility of the fertilizer materials is usually not a problem. A variety of nutrient ratios can easily be formulated, including those containing micronutrients needed in specific field situations. In some cases, pesticides are mixed with the bulk fertilizers and applied to the soil.

A problem of some concern in bulk blending is the segregation of the materials, which could result in uneven distribution of the fertilizer nutrients in the field. The granules of all materials should be of about the same size

FIGURE 18.8 Bulk blending plant and bulk spreading trucks. (a) Different materials are blended in a small plant and loaded directly on a truck. (b) Truck in the field spreading the fertilizer. [*Courtesy Tennessee Valley Authority.*]

(a)

(b)

and density. Since this is not always possible, special mixing hoppers have been developed.

Fertilizer materials commonly used in the bulk blending process are urea, ammonium nitrate, ammonium sulfate, ammonium phosphates, triple superphosphate, and potassium chloride. These compounds are also commonly applied as separate materials rather than in mixtures.

Liquid Fertilizers. Another innovation in the formulating and handling of mixed fertilizers is the use of liquid fertilizers, sometimes termed *fluid* fertilizers. These materials are used extensively in California, where the practice started, and in the north central states. Nationally, about one third of the total fertilizers are applied in the liquid form. In 1980 about 3200 fluid mix plants were in operation. As with bulk blending, liquid fertilizers have the advantage of low labor costs since the materials are handled in tanks and pumped out for application. However, the cost per unit of nutrient element is usually higher, more sophisticated equipment is needed for storage and handling, and only relatively low analysis mixed fertilizers can be made. (For example, 5-10-10, 7-7-7, and 6-18-6 are common analyses.) Slightly higher analyses are possible by using superphosphoric acid or ammonium polyphosphates as a source of at least part of the phosphorus. Also, "suspension fertilizers," wherein a small amount of solids is suspended in the liquid fertilizer and 1–2% clay is added to keep the solids in suspension, permit higher analyses, such as 15-15-15. Suspension fertilizers comprise about 40% of the fluid market in the United States.

18.11   Effect of Mixed Fertilizers on Soil pH

Acid-Forming Fertilizers. Most complete fertilizers, unless specially treated, tend to develop an acid residue in soils. This is mainly due to the influence of certain of the nitrogen carriers that supply ammonia or produce ammonia when added to the soil. The major acidifying effect of ammonium ions is exerted when they are nitrified. Upon oxidation the ammonium compounds increase acidity, as shown by the following reaction.

$$NH_4^+ + 2\,O_2 \rightarrow 2\,H^+ + NO_3^- + H_2O$$

In addition to ammonium compounds, materials such as urea, which upon hydrolysis yield ammonium ions, are potential sources of acidity. The phosphorus and potash fertilizers commonly used have little effect upon soil pH unless they also contain nitrogen. The approximate acid-forming capacities of some fertilizer materials expressed in kilograms of calcium carbonate needed to neutralize the acidity produced by 20 kg of nitrogen supplied are as follows.

Ammonium sulfate	107	Ammonium nitrate	36
Ammo-phos	100	Cottonseed meal	29
Anhydrous ammonia	36	Castor pomace	18
Urea	36	High-grade tankage	15

The base-forming capacity expressed in the same terms, for certain fertilizers is

| Nitrate of soda | 36 | Tobacco stems | 86 |
| Calcium cyanamid | 57 | Cocoa meal | 12 |

It should be recognized that some nutrient-containing materials are added specifically to increase soil acidity. Thus, elemental sulfur and compounds such as iron or aluminum sulfates may be added to reduce the soil pH and to enhance the growth of plants such as azaleas and rhododendron (see Section 6.11).

Non-Acid-Forming Fertilizers. In some instances, the acid-forming tendency of nitrogen fertilizers is completely counteracted by adding dolomitic limestone to the mixture. Such fertilizers are termed *neutral* or *non-acid-forming* and exert little residual effect on soil pH. However, it is economically preferable to use acid-forming fertilizers and to neutralize the soil acidity with separate bulk applications of lime.

18.12   The Fertilizer Guarantee

Every fertilizer material, whether it is a single carrier or a complete ready-to-apply mixture, must carry a guarantee as to its content of nutrient elements. The exact form is generally determined by the state in which the fertilizer is offered for sale. The *total nitrogen* is usually expressed in its *elemental form* (N). The phosphorus is quoted in terms of *available phosphorus* (P) or *phosphoric acid* (P_2O_5); the potassium is stated as *water-soluble potassium* (K) or *potash* (K_2O). It is well to emphasize that there is no P_2O_5 or K_2O as such in fertilizers (see Section 18.5). The oxide means of expressing the nutrient contents of fertilizers is of historical origin and must be considered, since in most states laws require that the P and K contents be expressed in terms of the oxides. In any case, it is the phosphorus and potassium that the fertilizers supply that are of concern to us.

The guarantee of a simple fertilizer such as sulfate of ammonia is easy to interpret since the name and composition of the material are printed on the bag or tag. The interpretation of an analysis of a complete fertilizer is almost as easy. The simplest form of guarantee is a mere statement of the relative amounts of N, P, and K, or N, P_2O_5, and K_2O. Thus, an 8-16-16 fertilizer contains 8% *total nitrogen,* 16% *available* P_2O_5 (7% available P), and 16% *water-soluble* K_2O (13% water-soluble K). In some states, the percentages of *water-*

TABLE 18.7 Open Formula Guarantee of a 16-16-16 Fertilizer (Acid Forming)

| | Total N | Available | | Soluble | |
Materials (kg/Mg)		P_2O_5	P	K_2O	K
225 Diammonium phosphate	40	104	45		
267 Muriate of potash				160	133
122 Triple superphosphate		56	24		
386 Nitrogen solution	120				
1000 16-16-16 (total)	160	160	70	160	133

Nutrients (kg) spans the top of the Total, Available, and Soluble columns.

soluble nitrogen, water-insoluble nitrogen, and *available insoluble nitrogen* are required by law.

An *open formula* guarantee is used by some companies. Such a guarantee not only gives the usual chemical analysis but also a list of the various ingredients, their composition, and the pounds of each in a ton of the mixture. Such a guarantee is given in Table 18.7 for a 16-16-16 fertilizer with the nutrients expressed as N, P_2O_5, and K_2O.

Commercial fertilizers are sometimes grouped according to their nutrient *ratio*. For instance, a 5-10-10, an 8-16-16, a 10-20-20, and a 15-30-30 all have a 1:2:2 ration. These fertilizers should give essentially the same results when applied in equivalent amounts. Thus, 1000 kg of 10-20-20 furnishes the same amounts of N, P, and K as does 2000 kg of 5-10-10.

This grouping according to ratios is valuable when several analyses are offered and the comparative costs become the deciding factor as to which should be purchased.

18.13   Fertilizer Inspection and Control

To ensure product quality, laws controlling the sale of fertilizers are necessary. These laws apply not only to the ready-mixed goods but also to the separate carriers. Such regulations protect the public as well as the reliable fertilizer companies since goods of unknown value are kept off the market. The manufacturer is commonly required to print the following data on the bag or on an authorized tag.

1. Number of net pounds or kilograms of fertilizer to a package.
2. Name, brand, or trademark.
3. Chemical composition guaranteed.
4. Potential acidity in terms of pounds of calcium carbonate per ton.
5. Name and address of manufacturer.

FIGURE 18.9 Fertilizer bag on which is printed the information commonly required by law in most states. Note the acid-forming tendency of this particular fertilizer.

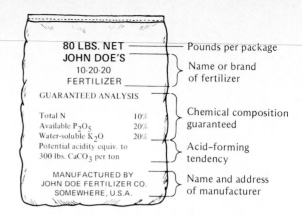

80 LBS. NET —————— Pounds per package

JOHN DOE'S
10-20-20
FERTILIZER —————— Name or brand of fertilizer

GUARANTEED ANALYSIS

Total N   10%
Available P_2O_5 20%   Chemical composition guaranteed
Water-soluble K_2O 20%
Potential acidity equiv. to
300 lbs. $CaCO_3$ per ton   Acid–forming tendency

MANUFACTURED BY
JOHN DOE FERTILIZER CO.   Name and address of manufacturer
SOMEWHERE, U.S.A.

An example of how this information often appears on the fertilizer bag is shown in Figure 18.9.

18.14 Fertilizer Economy

High Versus Low Analyses. Whether one is buying mixed fertilizers or the various separate carriers (ammonium nitrate, sulfate of ammonia, and the like), it is generally advantageous to obtain high-analysis goods. Price data indicate that the higher the grade of a mixed fertilizer, the lower is the nutrient cost. This is shown graphically in Figure 18.10. Obviously, from the standpoint of

FIGURE 18.10 Relative farm costs of nutrients in fertilizer of different concentrations. Note that the same amount of *nutrients* is supplied by the indicated quantities of each fertilizer. These nutrients are furnished more cheaply by the higher analysis goods, mainly because of the lower handling costs.

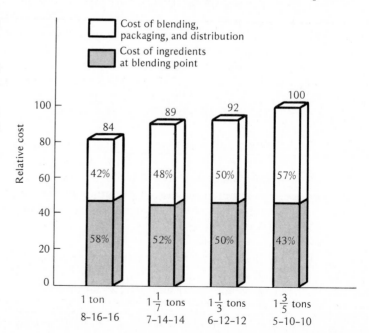

☐ Cost of blending, packaging, and distribution

▨ Cost of ingredients at blending point

Relative cost

100 —

84   89   92   100

80 —
        42%   48%   50%   57%

60 —

40 —
        58%   52%   50%   43%

20 —

0 —

1 ton $1\frac{1}{7}$ tons $1\frac{1}{3}$ tons $1\frac{3}{5}$ tons
8-16-16   7-14-14   6-12-12   5-10-10

economy, an 8-16-16 fertilizer furnishes more nutrients *per dollar* than a 4-8-8 or a 5-10-10. When improperly placed, the more concentrated goods may result in injury to young seedlings. Modern fertilizer-placement machinery, plus an increasing tendency to bulk-spread at least part of the fertilizer, has largely overcome these difficulties. Hence, the savings that result from the use of higher analysis materials seem to far outweigh any associated disadvantages that may now exist.

Relative Costs of Nitrogen, Phosphorus, and Potassium. Another price factor to remember in purchasing fertilizers is the relative costs of nitrogen, phosphorus, and potassium. In the past, nitrogen has generally been more expensive than the other two constituents. The comparative prices have changed, however, with the advent of synthetic nitrogen carriers. A unit of nitrogen supplied by anhydrous ammonia, for example, is less costly than a comparable amount of phosphorus (expressed as P_2O_5) supplied in superphosphate. Phosphorus per unit of P is generally more costly than a similar unit of nitrogen while potassium carriers are usually least expensive. However, there is considerable variation in the cost of different carriers for a given element. Thus, the manufactured cost of nitrogen in some solid carriers (e.g., sodium nitrate) is about double that in anhydrous ammonia. Likewise, sulfate of potash is considerably more expensive than an equivalent amount of potassium chloride.

Purchasing of Separate Materials. In many instances a saving can be effected by purchasing the separate materials and applying them to the soil. Anhydrous ammonia, nitrogen solutions, ammonium nitrate, and sulfate of ammonia are employed to advantage in orchards, as top dressings on meadows and pastures, and as a side dressing in vegetable production and the growing of such field crops as cotton.

The bulk handling of separate carriers as well as of mixtures has already been mentioned. In some sections of the country these and other developments have essentially revolutionized the production and distribution of fertilizers.

18.15   Movement of Fertilizer Salts in the Soil

For an understanding of the reasons for some of the specific methods of applying fertilizers (Section 18.16), the movement of fertilizer salts within the soil should be reviewed. As previously indicated, phosphorus compounds tend to move but little, except in the more sandy soils and some organic soils (Section 15.13). Consequently, for maximum effectiveness this element should be placed in the root zone. Surface applications, unless worked into the soil, do not supply the deeper roots with phosphorus. Also, because of phosphorus immobility, the total quantity of this element necessary for a given season can be applied at one time without fear of loss by leaching.

Potassium, on the other hand, and nitrogen, to an even greater extent, tend to move from their zones of placement. This movement is largely *vertical*, the salts moving up or down depending upon the direction of water movement. These translocations greatly influence the time and method of applying nitrogen and potassium. For example, it is often undesirable to supply nitrogen all in one annual application because of the leaching hazard. Yet this tendency of nitrates to move downward is an advantage when fertilizer is applied on the soil surface. Water movement in the latter case carries the dissolved nitrate salts down to the plant roots. Top dressings of nitrogen solutions and urea may present problems in some cases because of the danger of volatilization (see Section 9.10).

The movement of nitrogen and, to a lesser degree, potassium must be considered in the *placement* of the fertilizer with respect to the seed. If the fertilizer salts are located in a band directly under the seeds, the upward movement of nitrates and some of the potassium salts by capillary water often results in injury to the stand. Rain immediately after planting followed by a long dry spell encourages such damage to seedlings. Placing the fertilizer immediately above the seed or on the soil surface also may result in injury, especially to row crops. The possibility of such injuries should be kept in mind in reading the following section.

18.16 Methods of Applying Solid Fertilizers

The method of application is important and must not be overlooked. A fertilizer should be placed in the soil in such a position that it will serve the plant to the best advantage. This involves not only different zones of placement but also the time of year the fertilizer is to be applied. The methods of application will be discussed on the basis of the crops to be fertilized.

Row Crops. Cultivated crops such as corn, cotton, and potatoes are usually fertilized in the *hill* or the *row*, part or all of the fertilizer being applied at the time of planting. If placed in the hill, the fertilizer may be deposited slightly below and on one side or, better, on both sides of the seed. When applied to the row, the fertilizer usually is placed in a narrow *band* on one or both sides of the row, 5–8 cm away and a little below the seed level (Figure 18.11).

When the amount of fertilizer is large, as is often the case with vegetable crops, it is often wise to broadcast part of it and thoroughly work it into the soil before the planting is done. In some cases the crop is side-dressed with an additional amount of fertilizer applied on the surface alongside the plants later in the season.

Small Grains. With small grains such as wheat the drill is equipped with a fertilizer distributor, the fertilizer entering the soil more or less in contact with the seed. As long as the fertilizer is low in analysis and the amount applied does not exceed 300 or 400 kg/ha, germination injury is not serious.

FIGURE 18.11 Best fertilizer placement for row crops is to the side and slightly below the seed. This eliminates danger of fertilizer "burn" and concentrates the nutrients near the seed. [*Courtesy National Plant Food Institute, Washington, DC.*]

Higher rates, especially of high-analysis fertilizers, may result in serious injury if the seed and fertilizer are placed together. The more modern grain drills are equipped to place the fertilizer alongside the seed rather than in contact with it.

Pastures and Meadows. With meadows, pastures, and lawns, it is advisable to fertilize the soil well at the time of seeding. The fertilizer may be applied with the seed or, better, broadcasted, and worked thoroughly into the soil as the seedbed is prepared. The latter method is preferable, especially if the fertilization rate is heavy. During succeeding years it may be necessary to top-dress such crops with a suitable fertilizer mixture. This requires care. The amount of fertilizer applied and the time of treatment should be so regulated as to avoid injury to the foliage and to the root crowns of the plants.

Trees. Orchard trees usually are treated individually, the fertilizer being applied around each tree within the spread of the branches but beginning several feet from the trunk. The fertilizer is worked into the soil as much as possible. When the orchard cover crop needs fertilization, it is treated separately, the fertilizer being drilled in at the time of seeding or broadcast later.

Ornamental trees are often fertilized by what is called the *perforation* method. Numerous small holes are dug around each tree within the outer half of the branch-spread zone and extending well into the upper subsoil. Into these holes, which are afterward filled up, is placed a suitable amount of an appropriate fertilizer. This method of application places the nutrients within the root zone and avoids an undesirable stimulation of the grass that may be growing around ornamental trees.

Bulk Spreading. For economic reasons, much of the fertilizer used in the United States is spread by truck directly on the soil surface (Figure 18.8). Both mixed fertilizers and separate materials are so applied. This method eliminates the use of bags and greatly reduces the labor requirement. The fertilizer should be spread on the land as near planting time as possible since losses from the soil surface to the atmosphere can occur. Unfortunately, efficiency of use of the fertilizer for row crops is not as high as is the row placement. Even here, however, row placement at planting can be supplemented with bulk spreading.

18.17 Application of Liquid Fertilizers

The use of liquid materials as a means of fertilization is very important in certain areas of this country. Three primary methods of applying liquid fertilizers have been used: (a) direct application to the soil, (b) application in irrigation water, and (c) the spraying of plants with suitable fertilizer solutions.

Direct Application to Soil. The practice of making direct applications of anhydrous ammonia, nitrogen solutions, and mixed fertilizers to soils is rapidly increasing throughout the United States (see Section 18.10). Equipment has been designed specifically to handle the chemical in question. Corrosion-resistant tanks are used to transport and store the liquids. Pumps are used to transfer and apply aqueous ammonia, liquid mixes, and no-pressure solutions. Anhydrous ammonia and pressure solutions must be injected into the soil to prevent losses by volatilization. Depths of 15 and 5 cm, respectively, are considered adequate for these two materials.

Application in Irrigation Water. In the western part of the United States, particularly in California and Arizona, liquid fertilizers are sometimes applied in irrigation waters. Liquid ammonia, nitrogen solutions, phosphoric acid, and even complete fertilizers are allowed to dissolve in the irrigation stream or the overhead sprinkler system. The nutrients are thus carried into the soil in solution. Not only are application costs reduced but also relatively inexpensive nitrogen carriers can be used. Increased usage of these materials attests to the growing popularity of the irrigation method of application. Some care must be used, however, to prevent ammonia loss by evaporation.

Applied as Spray on Leaves. Micronutrients or urea can be sprayed directly onto plants. This type of fertilization is unique. It does not involve any extra procedures or machinery since the fertilizer is applied simultaneously with the insecticides. Soybeans and apple trees seem to respond especially well to urea, since much of the nitrogen is absorbed by the leaves. Moreover, what drips or is washed off is not lost, for it falls on the soil and may later be absorbed by plants.

18.18 Factors Influencing the Kind and Amount of Fertilizers to Apply

The agricultural value of a fertilizer is difficult to determine, since materials so easily subject to change are placed in contact with two wide variables, the *soil* and the *crop*. The soil and the added fertilizer react with each other chemically and biologically. The fixation of phosphorus is an example of the first; the microbial hydrolysis of urea illustrates the latter. The result may be an increase or, more often, a decrease in the effectiveness of the fertilizer. Allowance should be made for these reactions when deciding on the kind and amount of fertilizer to apply.

The *weather* has a tremendous effect on the soil, upon the crop, and upon the fertilizer applied. If there is either an excess or deficiency of moisture, full efficiency of the fertilizer cannot be expected. In fact, any factor that may tend to limit plant growth will necessarily reduce fertilizer efficiency and consequently the crop response to fertilization. It is only when other factors are not limiting that the appropriate amounts of fertilizers can be estimated with any degree of certainty.

In spite of the complexity of the situation, however, certain *guides* can be used in deciding the kind and amounts of fertilizers to be applied. The following are especially pertinent.

1. Kind of crop to be grown:
   a. Economic value.
   b. Nutrient removal.
   c. Absorbing ability.
2. Chemical condition of the soil in respect to
   a. Total nutrients.
   b. Available nutrients.
3. Physical state of the soil, especially as to
   a. Moisture content.
   b. Aeration.

The third factor in most cases has an indirect effect on fertilizer usage and has been discussed sufficiently elsewhere (Section 4.6). Consequently, only the first two will be considered here.

18.19 Kind and Economic Value of Fertilized Crop

Crops of high economic value such as vegetables justify larger fertilizer expenditures per kilogram of crop produced. Consequently, for these crops complete fertilizers are used in rather large amounts. As much as 2200 kg/ha of analyses such as 8-16-16 are often recommended.

TABLE 18.8 Most Profitable Nitrogen Rate for Corn Based on a Computer Model for Predicting Yield and Corn/Nitrogen Price Ratios[a]

Corn/nitrogen price ratio	Yield potential (bu/A)[b]			
	100 (6.3)	130 (8.2)	160 (10.0)	190 (12.0)
	Nitrogen (lb/A)[b]			
5:1	101 (90)	123 (110)	157 (140)	191 (170)
10:1	123 (110)	157 (140)	202 (180)	235 (210)
15:1	135 (120)	179 (160)	224 (200)	269 (240)
20:1	146 (130)	191 (170)	235 (210)	280 (250)
25:1	157 (140)	202 (180)	247 (220)	291 (260)

[a] Calculated from Michigan State University data quoted by Spies (1976).

[b] Metric units for yield potential (Mg/ha) and for nitrogen rate needed (kg/ha) are given in parentheses.

With crops having a low economic value, much lower rates of fertilizer application are generally advisable. The extra yields obtained by applying large amounts of fertilizer are not usually sufficient to pay for the additional plant nutrients used, especially with such crops as natural pastures and meadows.

With a given crop, the profitable rate of fertilizer to be applied will depend on the ratio of fertilizer costs to the value of the crop being produced. Table 18.8 illustrates the effect of corn/fertilizer price ratios on profitable rates of fertilizer.

It must always be remembered that the very highest yields obtainable under fertilizer stimulation are not always the ones that give the best return on the money invested (Figure 18.12). In other words, the law of diminishing returns is a factor in fertilizer practice regardless of the crop being grown. Therefore, the application of moderate amounts of fertilizer is to be urged for all soils until the maximum paying quantity that may be applied for any given crop is approximately ascertained.

If the nutrient removal by a given crop is high, fertilizer applications are usually increased to compensate for this loss. The extra fertilizer may be applied directly to the particular crop under consideration or to some preceding and more responsive crop in the rotation. It should not be implied, however, that in all cases an attempt should be made to return nutrients in amounts equivalent to those that have been removed. Sometimes advantage can be taken of the nutrient-supplying power of the soil. This source is usually insufficient, however, and extra fertilizer additions must be made.

Tremendous differences exist in the ability of plants to absorb nutrients from a given soil. For example, lespedeza and peanuts, both of which are

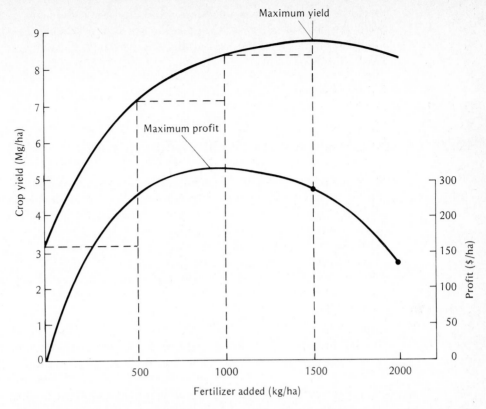

FIGURE 18.12 Relationship among rate of fertilizer addition, crop yield, and profit from adding the fertilizer. Note that the first 400 kg of fertilizer gave a much higher yield *increase* than the second 500 kg. Also note that profit is obtained at a lower fertilizer rate than the maximum yield.

legumes, can readily absorb adequate potassium under much lower soil potash levels than can alfalfa or soybeans. Consequently, responses to additions of potassium usually are much greater with the latter two crops. Obviously, crop characteristics deserve very careful study.

18.20   Chemical Condition of the Soil—Total Versus Partial Analyses

To determine the kind and quantity of fertilizer to add to soil, it is necessary to know what nutrient elements (or element) are deficient. Two general types of chemical analyses of soils are commonly used to obtain an estimate of nutrient deficiencies: *total* and *partial*. In a *total analysis* the entire amount of any particular constituent present in the soil is determined, regardless of its form of combination and its availability. Such data are of great value in studying soil formation and other phases of soil science. But they give little if any information on the availability of the essential elements to plants. Analyses suitable for measuring only the available portion of a given nutrient constituent must be employed. Such analyses are called *partial,* because only a portion of the total quantity of a soil constituent is determined.

18.21   Quick Tests for Available Soil Nutrients

Many procedures for the *partial analyses* of soils are available. In general, these procedures attempt to extract from the soil amounts of essential elements that are correlated with the nutrients taken up by plants. The large number of extracting solutions that have been employed in trying to measure nutrient availability is mute testimony of the difficulties involved. The extraction solutions employed have varied from strong acids such as H_2SO_4 to weak solutions of CO_2 in water. Buffered salt solutions such as sodium or ammonium acetate are the extracting agents now most commonly used.

General Consideration.   The group of tests most extensively used for nutrient availability are the *rapid* or *quick tests.* As the name implies, the individual determinations are quickly made, a properly equipped laboratory being able to handle thousands of samples a month. Since these tests are the only ones having practical possibilities for widespread use, their limitations as well as their merits will be critically reviewed.

A weak extracting solution such as buffered sodium acetate is generally employed in rapid tests and only a few minutes are allowed for the extraction. Most of the nutrients removed are those rather loosely held by the colloidal complex. The test for a given constituent may either be reported in general terms such as *low, medium,* or *high* or be specified in kilograms per hectare. Constituents most commonly tested for are those held largely in inorganic combination, phosphorus, potassium, calcium, and magnesium. Aluminum, iron, and manganese are sometimes included. While nitrogen is sometimes included, the expected release of this element through action of microorganisms is difficult to measure.

Limitations of Quick Tests.   Perhaps the first limitation is the difficulty of properly *sampling soils* in the *field.* Generally, only a very small sample is taken, perhaps a pint or so from one or more hectares of land. The chance of error, especially if only a few borings are made, is quite high. Therefore, a composite sample of at least 15–20 borings from the plow layer is suggested to increase the probability that a representative portion of the soil has been obtained.

A second limitation of these tests is the fact that it is essentially impossible to extract from a soil sample in a few minutes the amount of a nutrient, or even a constant proportion thereof, that a plant will absorb from that soil in the field during the entire growing season. The test results must be correlated with crop responses before reliable fertilizer recommendations can be made. The recommendations are made with regard to practical knowledge of the crop to be grown, the characteristics of the soil under study, and other environmental conditions. Two soils testing exactly the same in respect to a given constituent have been found to respond quite differently in terms of crop yields to identical fertilizer treatments in the field.

Need for Trained Personnel to Interpret Tests. Obviously, it is unwise to place any particular confidence in fertilizer recommendations if the tests are

RECORD OF AREA SAMPLED

1. Soil (By soil technologist) ☐ Profile and or ☐ Soil type

2. Position (Check ✓ one)
- ☐ 1st bottom
- ☐ bench (terrace)
- ☐ upland
- ☐ greenhouse
- ☐ muck or peat

3. Internal drainage (Check ✓ one)
- ☐ sluggish
- ☐ imperfect
- ☐ normal
- ☐ excessive

4. Slope (Check ✓ one)
- ☐ depression
- ☐ flat
- ☐ gently rolling
- ☐ rolling
- ☐ steep

5. Intended use (Check ✓ one)
- ☐ permanent pasture
- ☐ long-time meadow
- ☐ cropland rotation
- ☐ lawn
- ☐ tobacco
- ☐ sugar beet
- ☐ vegetable (home)
- ☐ vegetable (commercial)
- ☐ vegetable (greenhouse)
- ☐ other
- ☐ fruit (home)
- ☐ fruit (commercial)

Crop or Crops to Be Grown ☐

6. Lime
Has this area been limed in the last 4 years? (Check ✓ this unit)
☐ No ☐ Yes

When:
- ☐ this year
- ☐ last year
- ☐ 2 years ago
- ☐ 3 years ago

Rate (per acre)
- ☐ less than 1 ton
- ☐ 1 ton
- ☐ 2 tons
- ☐ more than 2 tons

7. Fertilizers and manure
If fertilizers or barnyard manure has been added to this area in the past 4 years, fill in below.

Year	Crop Grown	Fertilizers Added		Tons Manure per Acre
		Rate (per acre)	Analysis	

FIGURE 18.13 Satisfactory interpretation of a quick test analysis of a soil cannot be made without certain pertinent field data. This record sheet suggests the items that are most helpful to the expert in making a fertility recommendation. [*Modified from a form used by the Ohio Agricultural Experiment Station.*]

not interpreted correctly. The interpretation of quick test data is best accomplished by experienced and technically trained personnel, who fully understand the scientific principles underlying the common field procedures. In modern laboratories, the factors to be considered in making fertilizer recommendations are programmed into a computer and the interpretation is efficiently printed out for the farmer's or gardener's use.

As already suggested, considerable related information is needed to make the proper use of quick test data. Examples of desirable supplementary data are shown by the sample information sheet in Figure 18.13. This supplementary

REPORT ON SOIL TESTS
AUBURN UNIVERSITY
SOIL TESTING LABORATORY
Auburn University, AL 36849

NAME   ALABAMA RESIDENT   COUNTY   LEE
ADDRESS 118 MAIN STREET   DISTRICT 2
CITY   HOMETOWN, AL 36830   DATE   1/04/80

LAB. NO.	SENDER'S SAMPLE DESIGNATION AND	CROP TO BE GROWN	SOIL* GROUP	pH**	PHOSPHORUS P***	POTASSIUM K***	MAGNESIUM Mg***	LIME-STONE TONS/ACRE	N	P₂O₅	K₂O
									POUNDS PER ACRE		
23887	1	SOYBEANS	2	5.3	L 70	M 70	H 160	2.0	0	80	40

COMMENT 224–SOIL ACIDITY (LOW pH) CAN BE CORRECTED WITH EITHER DOLOMITIC OR CALCITIC LIME.

| 23888 | 2 | CORN | 1 | 5.6 | L 70 | M 70 | H 160 | 1.0 | 120 | 80 | 40 |

SEE COMMENT 224 ABOVE.

COMMENT 15–CORN ON SANDY SOILS MAY RESPOND TO NITROGEN RATES UP TO 150 LBS. PER ACRE. ON SANDY SOILS APPLY 3 LBS. ZINC (Zn) PER ACRE IN FERTILIZER AFTER LIMING OR WHERE pH IS ABOVE 6.0.

| 23889 | 3 | BAHIA | 1 | 6.0 | M 100 | H 140 | H 240 | 0.0 | 60 | 40 | 0 |

| 23890 | GARDEN | VEGETABLES | 3 | 5.2 | M 90 | M 70 | H 160 | 3.0 | 120 | 120 | 120 |

SEE COMMENT 224 ABOVE.

COMMENT 82–PER 100 FT. OF ROW APPLY 6 LBS. 8-8-8 (3 QUARTS) AT PLANTING AND SIDEDRESS WITH 4 LBS. 8-8-8 (2 QUARTS).

***ON SUMMERGRASS PASTURES APPLY P AND K AS RECOMMENDED AND 60 LBS. OF N BEFORE GROWTH STARTS. UP TO SEPTEMBER 1 REPEAT THE N APPLICATIONS WHEN MORE GROWTH IS DESIRED.

***1.0 TON LIMESTONE PER ACRE IS APPROXIMATELY EQUIVALENT TO 50 LBS. PER 1,000 SQ. FT.

***FOR CAULIFLOWER, BROCCOLI AND ROOT CROPS, APPLY 1.0 LB. OF BORON (B) PER ACRE. (FOR HOME GARDENS, 1 TABLESPOON BORAX PER 100 FT. OF ROW.)

THE NUMBER OF SAMPLES PROCESSED IN THIS REPORT IS 4.

*1. Sandy soils   **7.4 or higher Alkaline 6.5 or lower Acid
2. Loams & light clays   6.6-7.3 Neutral 5.5 or lower Very acid   APPROVED
3. Heavy clays (excluding Blackbelt)
4. Heavy clays of the Blackbelt   ***Rating & fertility (percent sufficiency)   SOIL TESTING FORM B

FIGURE 18.14 Example of a soil test report giving the soil test levels and the recommendations for lime and fertilizer application. [*From Cope et al.* (*1981*).]

information may be as helpful as the actual test results in making fertilizer recommendations. An example of a soil test report is shown in Figure 18.14.

Merits of Rapid Tests.   It must not be inferred from the preceding discussion that the limitations of rapid tests outweigh their advantages. When the precautions already described are observed, rapid tests are an invaluable tool in making fertilizer recommendations. Moreover, these tests will become of even greater importance insofar as they are correlated with the results of field fertilizer experiments. Such field trials undoubtedly will be expanded to keep pace with the increasing use of the rapid tests.

18.22   Broader Aspects of Fertilizer Practice

Fertilizer practice involves many intricate details regarding soils, crops, and fertilizers. In fact, the interrelations of these three are so complicated and far reaching that a practical grasp of the situation requires years of experience. The unavoidable lack of exactness in fertilizer decisions should be kept in mind.

Leaving the details and viewing the situation in a broad way, it seems to be well established that any fertilizer scheme should be built around the effective use of nitrogen. Applications of phosphorus and potassium should be made to balance and supplement the nitrogen supply, whether it be from the soil, crop residues, or added fertilizers.

A second aspect relates to economics. Farmers do not use fertilizers just to grow big crops or to increase the nutrient content of their soils. They do so to make a living. As a result, any fertilizer practice, no matter how sound it may be technically, that does not give a fair economic return will not long stand the test of time.

This point is emphasized by the data in Figure 18.15 from a field experiment in which different rates of fertilizer recommended by five different soil test laboratories were applied. Fertilizer applications based on soil tests calibrated with years of yield response data give by far the most economical return (Olsen et al., 1982). This "sufficiency level" concept used by laboratory E was more pertinent than two other concepts that governed the recommendations of the other four laboratories: (a) the "maintenance" concept, requiring essentially a replacement through fertilizers of all nutrients removed in the crop, and (b) the "cation saturation ratio" concept, requiring set Ca/K, Ca/Mg, and Mg/K ratios.

Not only must the soil, the crop, and the fertilizer receive careful study, but the crop rotation employed and its management should be considered. Obviously, the fertilizer applications must be correlated with the use of farm manure, crop residues, green manure, trace elements, and lime. Moreover, the residues of previous fertilizer additions must not be forgotten. In short, fertilizer

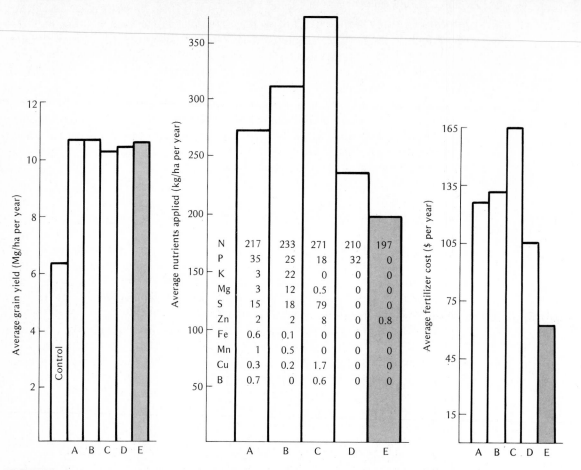

FIGURE 18.15 The yield of corn from plots near North Platte, ND, receiving five different rates of fertilizer as recommended by five soil-testing laboratories (A–E), the nutrient levels recommended by these laboratories, and the cost of the fertilizer applied each year. Note that the yields from all fertilized plots were about the same, even though markedly different rates of fertilizer were applied. The fertilizer rates recommended by laboratory E utilized a "sufficiency level" concept that was based on the calibration of soil tests with field yield responses. Those recommended by the other laboratories (A–D) were based either on a "maintenance" concept that required replacement of all nutrients removed by a crop or on the supposed need to maintain set cation ratios (Ca/Mg, Ca/K, and Mg/K). Obviously, the "sufficiency level" concept provided more economical results in this trial. Data averaged over the period 1974–80. The soil was a Cozad silt loam (Fluventic Haplustoll). [*From Olsen et al.* (*1982*); *used with permission of the American Society of Agronomy.*]

practice is only a phase, but a very important one, of the fertility management of soils.

REFERENCES

ASA. 1980. *Nitrification Inhibitors—Potentials and Limitations*, ASA Publication No. 38 (Madison, WI: Amer. Soc. Agron. and Soil Sci. Soc. Amer.).

Cope, J. T., C. E. Evans, and H. C. Williams. 1981. *Soil Test Fertilizer Recommendations for Alabama Crops*, Circular 251, Agr. Exp. Sta., Auburn Univ.

Englestad, O. P. 1965. "Phosphate fertilizers aren't all alike," *Crops and Soils*, **17**:14–15 (Aug.–Sept.)

IFDC. 1979. *Fertilizer Manual* (Muscle Shoals, AL: International Fertilizer Development Center.)

McLean, E. O., and T. J. Logan. 1970. "Sources of phosphorus for plants grown in soils with differing phosphorus fixation tendencies," *Soil Sci. Soc. Amer. Proc.*, **34**:907–911.

Olsen, R. A., K. D. Frank, P. H. Grabouski, and G. W. Rehm. 1982. "Economic and agronomic impacts of varied philosophies of soil testing," *Agron. J.*, **74**:492–499.

Parr, J. F. 1972. "Chemical and biochemical considerations for maximizing the efficiency of fertilizer nitrogen," in *Effects of Intensive Fertilizer Use on the Human Environment*, Soils Bull. 16 (Rome: FAO; published by Swedish International Development Authority).

Spies, Cliff D. 1976. "Toward a more scientific basis," *Proceedings of TVA Fertilizer Conference, July 27–28*, pp. 68–72.

USDA. 1982a. *The Fertilizer Supply, 1981–82*, ASCS-10, Emergency Preparedness Branch (Washington, DC: U.S. Department of Agriculture).

USDA. 1982b. *Commercial Fertilizers, Consumption for Year Ended June 30, 1982*, Sp. Cr. 7 (11–82), Statistical Reporting Service (Washington, DC: U.S. Department of Agriculture).

Recycling Nutrients Through Animal Manures and Other Organic Wastes

19

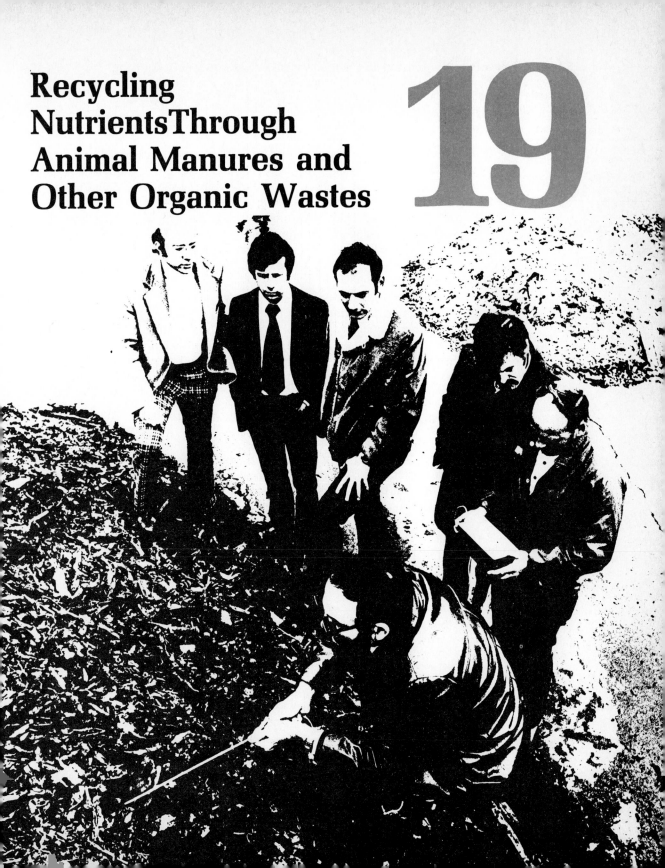

Only in recent years has agriculture depended upon inorganic fertilizers as primary sources of nutrients for crop production. Centuries before scientists discovered that specific chemical elements were essential for plant growth, farmers had learned the value of organic manures and other residues in enhancing crop production.[1] These organic materials are still a primary source of mineral elements, particularly among the resource-poor farmers of developing countries. Increased energy costs have have led to higher inorganic fertilizer prices, forcing even the more affluent nations to reconsider organic sources of plant nutrients.

Some of the organic residues and wastes available in the United States are listed in Table 19.1 along with the percentage of each that is used on the land. The largest quantity is crop residues, about which much has already been said. Returning these residues to the land, rather than burning them or otherwise preventing their being recycled, is a wise conservation practice. The next largest source of organic residues is animal manures.

19.1 Farm Manure Significance

For centuries the use of farm manure has been synonymous with a successful and stable agriculture. Not only does it supply organic matter and plant nutrients to the soil but it is associated with animal agriculture and with forage crops, which are generally soil protecting and conserving.

TABLE 19.1 Quantity of Major Organic Wastes in the United States and Percentage of These Materials That Are Used on the Land[a]

Organic waste	Annual production (thousand dry metric tons)	Used on land (%)[b]
Crop residues	431,087	68
Animal Manure	175,000	90
Municipal refuse	145,000	(1)
Logging and wood manufacture	35,714	(5)
Industrial organics	8,216	3
Sewage sludge and septage	4,369	23
Food processing	3,200	(13)

[a] From USDA (1980).
[b] Numbers in parentheses are estimates.

[1] For an excellent discussion of waste utilization, see Elliot and Stevenson (1977).

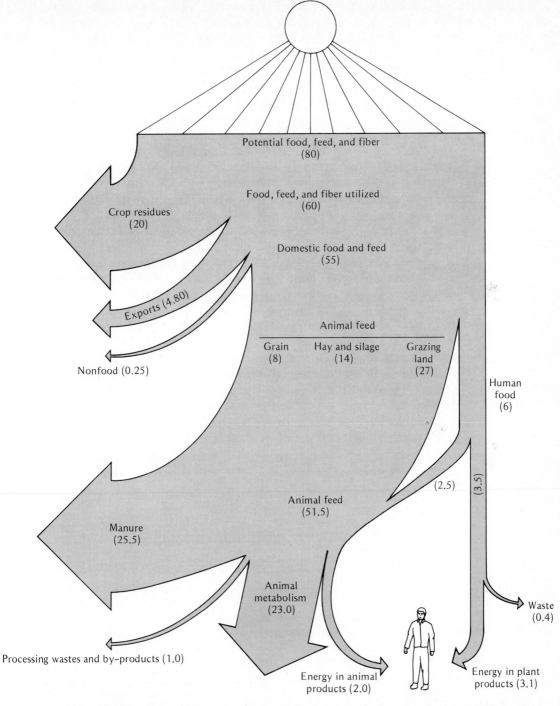

FIGURE 19.1 Diagram of the approximate energy flow for the United States food chain (in billions of joules) showing the high proportion of the energy ultimately found in animal manures. [*From Stickler et al.* (*1975*); *courtesy Deere & Company, Moline, IL.*]

In the United States, animal wastes offer both opportunities and challenges. The opportunities relate to the organic matter and associated nutrients available in animal manures. A high proportion of the solar energy captured by growing plants ultimately is embodied in farm manures (Figure 19.1). Crop and animal production and soil conservation are enhanced by its use.

The challenges relate to changes in animal production systems. In the past, animals were largely dispersed on the land, making possible the easy and economical application of manure to the nearby soils. Now, more than two-thirds of the beef production in the United States is under concentrated, confined conditions. In modern meat and poultry production factories, tens of thousands of beef and swine, and hundreds of thousands of poultry, are managed in centralized locations. Beef feedlots in the Missouri basin states and broiler- and egg-producing complexes in the Southeast and along the Atlantic Coast are examples of such large operations (Figure 19.2).

As these changes have provided for increased production efficiency, they have also provided huge mounds of animal manures, most of which are too distant from fields to be utilized economically. This has resulted in some environmental quality problems, which will receive attention in Chapter 20.

19.2   Quantity of Manure Produced

The quantity of manure available for land application depends on a number of factors, such as the species and size of the animals and the quantity and nature of feed consumed, and the nature and amount of bedding, as well as the type of housing or confinement of the animals. When comparisons are made on the basis of manure produced per unit weight of animal the production is remarkably similar from one animal to another (Table 19.2).

TABLE 19.2   Representative Annual Rates of Manure Production from Different Animals

Note the relative constancy of the annual rate of dry matter.

Animal	Annual production (Mg/1000 kg live weight)	
	Fresh excrement	Dry matter
Cattle	25.2	3.78
Poultry	11.2	4.28
Swine	26.4	3.96
Sheep	11.8	4.00
Horse	11.6	3.96

FIGURE 19.2 Aerial view of a 320-acre feedlot in Colorado wherein 100,000 cattle are on feed at one time. Problems of manure disposal and utilization are difficult to cope with under these conditions of concentrated animal feeding. Farm animals produce at least 10 times the biological waste produced by people in the United States. The challenge is to use this waste effectively for crop production and at the same time to minimize its potentially adverse effects on environmental quality. [*Courtesy Monfort Feed Lots, Inc., Greeley, CO.*]

Although the comparative production of manures is of some interest, the total quantity of manure voided by farm animals is of greater significance. In the United States this figure is more than 2 billion Mg annually. This is 10 times that produced by the human population in this country. More than 1 billion Mg of this manure are produced in feedlots or giant poultry or swine complexes where manure *disposal* as a problem tends to overshadow manure *utilization* as an opportunity (Figure 19.2). A 50,000-head beef feedlot operation, for example, produces about 90,000 Mg of manure annually after some decomposition and considerable moisture loss. At a conventional application rate of 10 tons per acre, 9000 acres of land would be required to utilize the manure. The enormity of the disposal problem is obvious even for operations one-half or one-third this size (see Section 20.9). Experienced farmers and gardeners view the disposal of animal manures as only a challenge, not a problem. They know of manure's value in enhancing crop production. They are concerned only with methods and costs of applying the wastes.

19.3 Chemical Composition

Manure as it is applied in the field is a combination of feces, urine, bedding (litter), and feed wastage. As might be expected, the chemical composition of this material varies widely from place to place depending upon factors such as (a) animal species, (b) age and condition of animals, (c) nature and amount of litter, and (d) the handling and storage of the manure before it is spread on the land.

Moisture and Organic Constituents. The moisture content of fresh manure is high, commonly varying from 60 to 85% (Table 19.3). This excess water is a nuisance if the fresh manure is spread directly on the land. If, on the other hand, the manure is handled and digested in a liquid form or slurry prior to application, the moisture content is not objectionable.

TABLE 19.3 Moisture and Nutrient Content of Manure from Farm Animals[a]

Animal	Feces/urine ratio	H_2O (%)	Nutrients (kg/Mg)				
			N	P_2O_5	P	K_2O	K
Dairy cattle	80:20	85	5.0	1.4	0.6	3.8	3.1
Feeder cattle	80:20	85	6.0	2.4	1.0	3.6	3.0
Poultry	100:0	62	15.0	7.2	3.1	3.5	2.9
Swine	60:40	85	6.5	3.6	1.6	5.5	4.5
Sheep	67:33	66	11.5	3.5	1.6	10.4	8.6
Horse	80:20	66	7.5	2.3	1.0	6.6	5.5

[a] Average values from a number of references.

Manures are essentially partially degraded plant materials. Animals utilize only about half of the organic matter they eat. It is not surprising, therefore, that the bulk of the solid matter in manures is composed of organic compounds very similar to those found in the feed the animals consumed. Thus, although most of the cellulose, starches, and sugars are decomposed, hemicellulose and lignin remain, as do lignoprotein complexes similar to those found in soil humus. These plant materials have been only partially degraded, however, as evidenced by the ready decomposition of at least the soluble components when they are added to soil or a digestion tank.

One other important organic component of animal manures is the live component—the soil organisms. Especially in ruminant animals (e.g., cattle and sheep) the manures are teeming with bacteria and other microorganisms. Between one fourth and one half of the fecal matter of ruminants may consist of microorganism tissue. Although some of the compounds may be the same as those in the original plant materials, others have been synthesized by the microorganisms. In any case, some of these organisms continue to break down constituents in the voided feces and participate in decomposition of the manure in storage.

Nitrogen and Mineral Element Contents. Portions of the nutrient elements consumed in animal feeds are found in the voided excrement. As a generalization, three-fourths of the nitrogen, four-fifths of the phosphorus, and nine-tenths of the potassium in ingested feeds are not retained by the animals. For this reason, animal manures are valuable sources of both macro- and micronutrients.

The quantities of nitrogen, phosphorus, and potassium that might be expected in different manures are given in Table 19.3. The range of other nutrients commonly found, in kilograms per metric tons, is as follows (Benne et al., 1961).

Calcium	2.4 –74.0	Boron	0.02 –0.12
Magnesium	1.6 – 5.8	Manganese	0.01 –0.18
Sulfur	1.0 – 6.2	Copper	0.01 –0.03
Iron	0.08– 0.93	Molybdenum	0.001–0.011
Zinc	0.03– 0.18		

As shown in Table 19.2, the ratio of feces to urine in farm manure varies from 2:1 to 4:1. On the average, a little more than *one-half of the nitrogen,* almost *all of the phosphorus,* and about *two-fifths of the potassium* are found in the solid manure (Figure 19.3). Nevertheless, this apparent superiority of the solid manure is offset by the ready availability of the constituents carried by the urine. Care must be taken in handling and storing the manure to minimize the loss of the liquid portion.

The data in Table 19.3 suggest a relatively low nutrient content in comparison with commercial fertilizer and an unbalanced nutrient ratio, being consider-

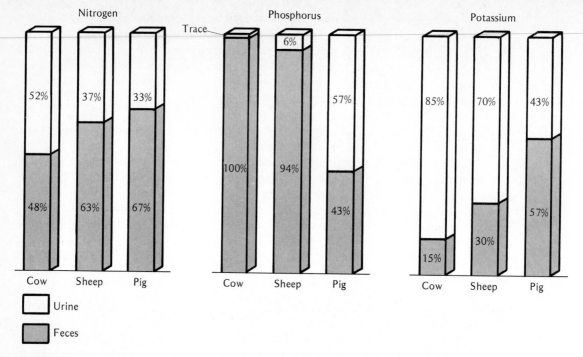

FIGURE 19.3 Distribution of nitrogen, phosphorus, and potassium between the solid and liquid portions of farm manure from cattle, sheep, and pigs. [*Calculated from data of Van Slyke* (*1932*).]

ably lower in phosphorus than in nitrogen and potassium. Modern commercial fertilizers commonly carry 20–40 times the nutrient content of manure and most have a phosphate content at least as great as the nitrogen and potassium. The nutrient imbalance requires correction through supplemental soil treatment.

Three Generalizations. From a practical point of view the characteristics shown in Table 19.3 suggest that 2 metric tons of an average farm manure supplies as much N, P_2O_5, and K_2O as 100 kg of 10-5-10 fertilizer. This is based on an average manure composition of 0.5% N, 0.25% P_2O_5, and 0.5% K_2O and points up the relatively low nutrient analysis of manures. However, since field rates of manure application are commonly 10–15 metric tons (Mg) per acre, the total nutrients supplied under practical conditions are substantial and sometimes more than needed for optimum growth.

A second generalization relates to nutrient availability in manures. In general, nearly one-half of the nitrogen, about one-fifth of the phosphorus, and about one-half of the potassium may be recovered by the first crop grown after manure is applied (Table 19.4). Thus, on the basis of readily available nutrients, 2 metric tons of average manure supplies 5 kg of N, 1 kg of P_2O_5, and 5 kg of K_2O.

TABLE 19.4 Effect of Handling Cow Manure on Yield and on Recovery of Nitrogen, Phosphorus, and Potassium by Corn Grown on Miami Silt Loan in a Greenhouse Experiment[a]

Type of manure	Yield (g/pot)	Recovery by crop (%)		
		N	P	K
No manure	11.0	—	—	—
Cow manure, 15 T/A				
Fresh	19.5	44.0	19.5	40.5
Fermented	19.5	42.0	22.5	49.5
Aerobic liquid	17.0	18.5	19.5	38.0
Anaerobic liquid	22.5	52.5	29.0	48.0

[a] From Hensler (1970).

The third practical generalization is the need to balance the N/P/K ratio by supplying phosphorus in addition to that contained in manures. This is done by adding phosphate fertilizer to the manure before it is spread or by adjusting the phosphorus content of commercial fertilizer applied in the field to supplement the manure applications.

Before turning to methods of storing manure, one more set of figures might be helpful—that is, the proportion of nutrients in a harvested crop that might be expected to survive animal removal and loss from the manure prior to its application to soil. The losses that feeds sustain during digestion and the waste of the manure in handling and storage permit the return to the soil of no more than about *25% of the organic matter, 40% of the nitrogen, 50% of the phosphorus,* and *45% of the potassium.* As will be seen in the next section, controlled manure digestion procedures will remove an even higher percentage of the organic matter and in some cases the nitrogen.

19.4   Storage, Treatment, and Management of Animal Manures

A generation ago, manure storage and application was a simple matter. Some farmers spread manure daily from their barns or allowed it to pile up until time and soil conditions permitted it to be spread. Animal feeding operations were small enough to permit the spreading in the spring of manure that had accumulated during the fall and winter. In any case, manure management was generally considered a private affair.

The coming of confined and concentrated animal management has drastically changed this situation. The marked concentration of animal manures has become the concern not only of the farmer but of rural nonfarm citizens as well as environmental ecologists generally. Together they have begun a revolution in the management of farm manures. Offensive odors, along with the possibilities of organic and nitrate pollution of streams and drinking water, are primary factors in stimulating the revolution.

Four general management systems are being used to handle farm manures: (a) collection and spreading of the fresh manure daily, (b) storage and packing in piles and allowing the manure to ferment before spreading, (c) aerobic liquid storage and treatment of the manure prior to application, and (d) anaerobic liquid storage and treatment prior to application. Since these methods of handling manure affect its biological value, each will be discussed briefly.

Applying Fresh Manure Daily.   This system is commonly used where stanchion dairy barns are employed. The manure is scraped or otherwise moved mechanically into spreaders, sometimes reinforced with superphosphate, and spread daily on the land. Daily spreading prevents nutrient loss by decomposition or volatilization, but subsequent loss of nutrients may occur if the manure is spread on frozen ground. Spring thaws carry off much potassium and nitrogen, thereby reducing crop response and increasing the probability of pollution of streams, ponds, and lakes. As much as 25% of the applied potassium can be lost in this manner.

Storage or Packing in Piles (Fermentation).   Manure from dairy barns and cattle feedlots may be allowed to accumulate under the animals over a period of time or may be removed to a pile nearby. In any case, if the manure is not allowed to dry out below 40% moisture (as it sometimes does in the Great Plains), fermentation will occur. Depending on the moisture content and degree of compaction, both anaerobic and aerobic breakdown takes place. The most abundant products of decay are carbon dioxide and water, along with considerable heat. Of perhaps greater practical significance, however, are reactions involving elements such as nitrogen, phosphorus, and sulfur. For example, hydrolysis of urea during the decomposition process yields ammonia.

$$CO(NH_2)_2 \ + \ 2\,H_2O \ \rightarrow \ (NH_4)_2CO_3$$

$$(NH_4)_2CO_3 \ \rightarrow \ 2\,NH_3\uparrow \ + \ CO_2\uparrow \ + \ H_2O$$

If conditions are favorable for nitrification, nitrates will appear in abundance. Either form of concentrated inorganic nitrogen can be lost and become a pollution hazard. The ammonia lost to the atmosphere may be captured by rain and snow and returned to the surface. The nitrates, being soluble and unadsorbed by the soil or the manure, are subject to leaching and to movement in runoff water.

Fermentation may provide a more satisfactory product for land application than fresh manure. For example, highly carbonaceous bedding or litter, which may be present in the fresh manure, is at least in part broken down during the fermentation process.

Aerobic Liquid Treatment. One of the more sophisticated animal-waste-treatment procedures is that of aerobic digestion of a liquid slurry containing the manure. The manure is stored in an aerated lagoon or in an oxidation ditch. Used primarily in swine operations, the manure may fall through slotted floors into the oxidation ditch or may be transported to a nearby lagoon. By vigorous stirring, oxygen is continuously incorporated into the system, bringing about continuous oxidation. Offensive odors are kept to a minimum, although some nitrogen is lost, probably as ammonia. The primary products of decomposition are carbon dioxide, water, and inorganic solids. Periodically the solids can be removed along with the resistant organic residues and applied to the land. A diagram showing an oxidation ditch and lagoon in use for swine is shown in Figure 19.4. This method of treatment minimizes pungent odors, but its costs of construction and operation are high and it permits some nutrient loss.

Anaerobic Liquid Treatment. This method is similar to its aerobic counterpart except that no gaseous oxygen is added to encourage aerobiosis of the liquid slurry. Consequently, the nature of the reactions is quite different, combined oxygen (e.g., SO_4^{2-} and NO_3^-) providing energy for much of the breakdown. The reactions are similar to those taking place in a septic tank. Sixty to 80% of the gaseous product is methane, the remainder being mostly carbon dioxide. Methane is thought to form through either of two types of biochemical reactions brought about by "methane bacteria."

$$CO_2 + 4\,H_2 \rightarrow CH_4\uparrow + 2\,H_2O$$

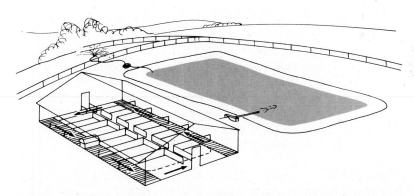

FIGURE 19.4 Use of an oxidation ditch and lagoon to handle wastes in a confinement swine building. A slotted floor around the outside of the building permits the wastes to drop into a continuous channel or ditch filled with water. A rotor on the left center of the building stirs oxygen from the air into the slurry and drives the mixture counterclockwise around the ditch. Aerobic organisms oxidize most of the organic wastes and some of the nitrogen is volatilized. The partially purified residue is pumped or allowed to flow into a lagoon, from which it can later be applied to the land by irrigation for final purification. [*Courtesy A. J. Muehling, University of Illinois.*]

and

$$CH_3COOH \rightarrow CH_4\uparrow + CO_2$$
Acetic acid

This method of treatment, along with the others, results in considerable loss of organic carbon. However, crop response to the treated product generally is as good as where fresh manure is used. Data in Table 19.4 illustrate this point. Only for the aerobic processed liquid was there evidence that nitrogen had been lost during the animal waste treatment.

Scientific developments during the past few years have markedly improved animal-waste-management technology. These technologies continue to provide products that are utilized in commercial agriculture.

19.5   Utilization of Animal Manures

Biologically, manure has many attributes. It supplies a wide variety of nutrients along with organic matter that improves the physical characteristics of soils. Its beneficial effects on plant growth are sometimes difficult to duplicate with other materials. At the same time, its bulkiness and low analysis reduce its competitive economic value. High labor and handling costs are responsible for this unfortunate situation. Even so, manure remains a most valuable soil organic resource. Economic considerations merely make it necessary to choose more carefully the soil and crop situation where manure is applied.

Responsive Crops.   Manure is an effective source of nutrients for most crops. However, those with a relatively high nitrogen requirement are most likely to respond to its application. Crops such as corn, sorghum, small grains, and the grasses respond well to manure as do vegetable and ornamental crops.

The rate of application of manure will depend upon the specific needs. However, rates of 20–35 Mg/ha (9–15 tons A) are commonly employed. In general, rates of application heavier than these give lower response per Mg of manure. Rates as high as 75–100 Mg/ha (33–45 tons/A) or even more can be justified but primarily in terms of manure disposal rather than utilization.

Special Uses.   There are a number of situations where manure is especially valuable. Denuded soil areas resulting from erosion or from land leveling for irrigation are good examples. Initial applications of 70–100 Mg/ha (31–34 tons/A) may be worked into the soil in the affected areas. These rates may be justified to supply organic matter as well as nutrients.

Special cases of micronutrient deficiency can be ameliorated with manure application. Such treatments are sometimes used when there is some uncertainty as to specific nutrient deficiency. Manure applications can be made with little concern for adding toxic quantities of the micronutrients.

The water-holding capacity of very sandy soils can be increased with heavy manure applications. Likewise, the structural stability and tilth of heavy plastic clays is sometimes dependent upon heavy organic applications. In both cases, the physical effects of manure justify its use.

Home gardeners commonly use manures at rates far in excess of those employed in commercial agriculture. Their aim is to provide a friable, easily tilled soil, and the cost is a secondary matter. Further, in planting trees and shrubs, one-time applications are common, therefore high initial rates of application are justified.

19.6 Long-Term Effects of Manures

The total benefits from manure utilization are sometimes not apparent from crop yields during the first or even second or third year following application. Although a portion of the nutrients and organic matter in manure is broken down and released during the first year or two, some is held in humus-like compounds subject to very slow decomposition. In this form, the elements are released only very slowly, rates of 2–4% per year being common. Thus, the components of manure that are converted to humus will have continuing effects on soils years after their application.

An example of the long-term effects of manure is shown in Figure 19.5. At the Rothamsted Experiment Station in England, starting in 1852, 34 metric tons of farm manure was applied each year for 19 years to a plot on which barley was grown continuously. Then in 1872 no more manure was applied, but the nitrogen content of the soil was monitored. For a period of more than 100 years the beneficial effect of the 19 years' manuring has been noted. Although such remarkable long-term effects may not be expected at all locations and on all soils, Figure 19.5 reminds us of the residual and highly beneficial effects of animal manures.

19.7 Urban and Industrial Wastes

Rising public concern about environmental pollution has forced great emphasis on the disposal of urban and industrial wastes. Land application is one promising means of such disposal. Although the primary reason for adding these wastes to land is to safely dispose of them, under some circumstances they can simultaneously enhance crop production. They supply a certain amount of plant nutrients and otherwise improve soil quality.

Types of Waste. Four major types of organic wastes are of significance in land application: (a) municipal garbage, (b) sewage effluents and sludges, (c) food processing wastes, and (d) wastes of the lumber industry. Because of

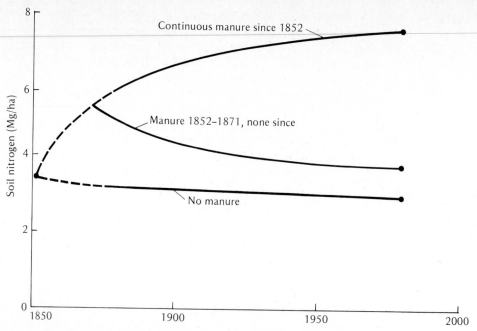

FIGURE 19.5 The effect of farm manure treatments on the nitrogen content of the top 23 cm of soil at the Rothamsted Experiment Station in England. Barley was grown continuously on all plots. Adding 35 Mg/ha of manure annually since 1852 greatly increased the soil nitrogen content. Note that the plots to which manure was applied from 1852 to 1871 and none since was reduced in nitrogen slowly over the next 100 years. The long lasting effect of manure is obvious. [*From Jenkinson and Johnston* (*1977*); *used with permission of the Rothamsted Experiment Station, Harpenden, England.*]

their uncertain chemical content, other industrial wastes may or may not be acceptable sources of organic matter and plant nutrients for land application.

Garbage. Municipal garbage is used to enhance crop production in China and other Asian countries, and in a few locations in Europe, where the municipal organic wastes are composted and later applied to the land. In the United States, most of the garbage is diposed of in other ways, little finding its way to the farmer's fields, for reasons that are mostly economic. The composted material contains only about 0.5% N, 0.4% P, and 0.2% K. Because of its low nutrient content and great bulk, the cost of transporting and handling municipal garbage compost generally makes it uneconomical in comparison with commercial inorganic fertilizers.

Food-Processing Wastes. Land application of food-processing wastes is being practiced in selected locations, but the practice is focused almost entirely on pollution abatement and not on crop production. Liquid wastes are commonly applied through sprinkle irrigation to permanent grass (Figure 19.6). Plant pro-

FIGURE 19.6 Agricultural processing wastes being sprayed on the land. A line of sprinklers from underground pipes applies the wastes coming from a nearby processing plant. [*Courtesy L. C. Gilde, Campbell Soup Co.*]

cessing schedules dictate the timing and rate of application, reducing the chances of harvest or grazing of the crop.

Sawdust. Sawdust and other organic wastes from the lumber industry have long been favorite sources of soil amendments, especially for the home gardener. These wastes are high in lignin and related materials, have very high C/N ratios, and therefore decompose very slowly. As a consequence, they are not ready suppliers of plant nutrients. In fact, if an alternate source of nitrogen is not applied with the sawdust, plants may show nitrogen deficiencies even when significant quantities of the sawdust have been added. The primary beneficial effect of the sawdust is on the physical properties of soil. Productive flower and vegetable home gardens are possible even on heavy textured soils if liberal and regular applications of sawdust are made.

19.8 Sewage Effluent and Sludge

Concerted efforts to prevent stream pollution by city sewage effluent have encouraged land application of either the sewage effluent or the solids emanating from these systems, sewage *sludge* (Figure 19.7). No other nonfarm source of organic waste for land application is more important for home use and farm crop production than is domestic sewage. A survey made in 1975 of 200 water-treatment plants in the United States showed that sewage is being spread on land by 25% of the plants, and that for 68% the spreading had started in only the past 10 years (Carroll et al., 1975).

In China and other Asian countries human wastes have long been applied directly on the land. The wastes, termed "night soil," are collected and applied to the soil with little or no processing, even for crops intended for human

FIGURE 19.7 Domestic sludge being removed from the Blue Plains wastewater treatment near Washington, DC, using vacuum filters. This material is available for direct application to the land. [*USDA photo by R. C. Bjork.*]

consumption. The Chinese have a remarkable record of recycling wastes, and agriculture provides them with excellent opportunities to make effective use of these wastes.

Liquid Sewage Effluents.   Sewage effluents have been applied to land for decades in Asia and Europe, and in selected sites in the United States. Today the city of Chicago is utilizing land application for disposal of treated sewage. The effluent is transported by barge to the disposal site, held temporarily in lagoons, and then applied to the soil by sprinkle irrigation. Using this disposal technique, nearly 7 million m³ of wastes each day are applied to the land. Corn, reed canary grass (*Phalaris arundinaceae*), alfalfa, and soybeans are among the crops grown on the land. Preliminary evidence suggests that crop production can benefit from the treatment.

Research at Pennsylvania State University has demonstrated the promise of land application of treated sewage for crop production. Effluent application rates of 2.5 and 5 cm per week were supplied to both cropland and forested areas (Kardos et al., 1974). Crop yields were essentially double those obtained without treatment. Corn and reed canary grass were able to absorb even larger quantities of nitrogen than was applied in the effluent, although more phospho-

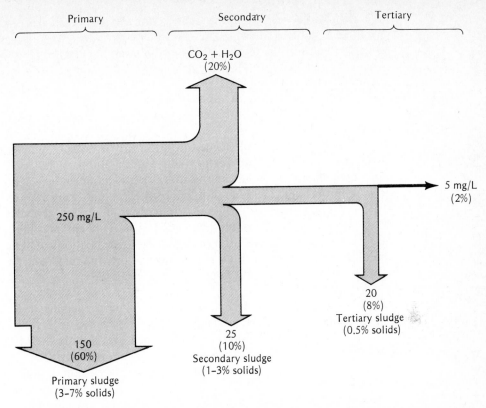

$CO_2 + H_2O$
(20%)

5 mg/L
(2%)

250 mg/L

20
(8%)
Tertiary sludge
(0.5% solids)

150
(60%)

25
(10%)
Secondary sludge
(1–3% solids)

Primary sludge
(3–7% solids)

FIGURE 19.8 Diagram showing the removal of suspended solids from wastes. Primary treatment permits the separation of most of the solid from raw sewage. Secondary treatment encourages oxidation of much of the organic matter and separation of more solids. Tertiary treatment usually involves the use of calcium, aluminum, or iron compounds to remove phosphorus from the sewage water. [*Redrawn from Loehr et al. (1979). Copyright © 1979 by Van Nostrand Reinhold Co., Inc.; used with permission.*]

rus was applied than was taken up. The sewage effluent came from domestic wastes with little industrial input. Consequently, there was no buildup of heavy metals such as cadmium or zinc, which effluent from industrial areas might have encouraged. About 52 ha of land such as that used in the Pennsylvania trials would be needed to accommodate the treated wastes from a city of 100,000 people. It is likely that sewage effluent will become more important in the future as a source of organic matter and nutrients for crop production.

Sewage Sludge. Sewage sludge is the solid by-product of domestic and/or industrial waste-water-treatment plants (Figure 19.8). It also has been spread on the land for decades, and its use will likely increase in the future. The product "Milorganite," a dried sludge sold by the Milwaukee Sewerage Commission, has been used widely in North America since 1927.

As might be expected, the composition of the sludge varies from one sewage treatment plant to another, depending on the nature of treatment the sewage receives and on the degree to which the organic material is allowed to digest. Levels of the heavy metals such as zinc, lead, copper, iron, manganese, and cadmium are determined largely by the degree to which industrial wastes have been mixed in with domestic wastes. If the content of these elements is too high, the sludge may be useless for agricultural purposes (see Section 20.6).

Data in Table 19.5 illustrate the variability in composition of sewage sludge from plants in seven midwestern and eastern states. In comparison with inorganic fertilizers, the sludges are generally low in nutrients, especially phosphorus and potassium. The median levels of N, P, and K are 2.5%, 1.8% and 0.24%, respectively. Obviously, rates of application of sludge far exceeding those for commercial fertilizers would be needed to provide comparable levels of essential elements, and especially for potassium. Applications of sludge would need to be supplemented with either manure or organic fertilizers to provide a nutrient balance needed for good crop production.

Practical Effects of Sludge Application. Typical responses of corn to sludge applications at different soil temperatures is shown in Figure 19.9. Note that the quantities of sludge applied were very high (up to 112 Mg/ha) compared with rates of animal manures commonly used. Although the sludge contained 3% nitrogen, most of this nitrogen was probably in the organic form and was not mineralized during the growing period. Yield increases were higher as the soil temperature was increased, suggesting that microbial action was responsible for nitrogen released.

TABLE 19.5 Median Level of Organic Carbon and Total N, P, and K in Sewage Sludge from Cities in Seven States[a]

State	No. of sewage plants	Median levels of			
		Organic C (%)	Total		
			N (%)	P (%)	K (%)
Wisconsin	38	35.8	5.4	2.7	1.13
Michigan	47	31.7	1.6	1.6	0.14
Indiana	14	22.7	3.1	1.9	0.33
Minnesota	19	29.9	5.3	3.5	0.30
New Jersey	13	—	2.6	1.7	0.16
New Hampshire	28	38.0	2.5	0.9	0.30
Ohio	15	—	3.8	—	0.15
All seven states	—	30.4	2.5	1.8	0.24

[a] Data from NC118 North Central Regional Project, published by McCalla et al. (1977).

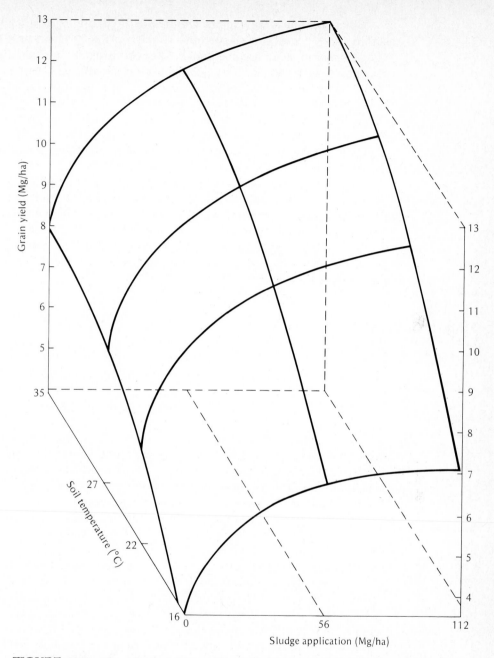

FIGURE 19.9 The effect of the application of sewage sludge on the yield of corn for grain at different soil temperatures. Response was obtained from only the first 56 tons at 16°C soil temperature but at higher temperatures further response was obtained to the 112 ton rate. [*Redrawn from Sheaffer et al. (1979); used with permission of the Journal of Environmental Quality.*]

Since the heavy sewage sludge applications are usually made on the soil surface, they exert some favorable effects on the physical properties of soil. They can serve as an organic mulch, thereby protecting the soil and conserving soil moisture. Unfortunately, however, considerable nitrogen can be lost by volatilization either as NH_3 gas or through denitrification (Figure 19.10). While these losses may be of no consequence to those who merely want to dispose of the wastes, they are of concern to the farmer, whose crops could effectively use the nitrogen if it were kept in the soil. For that reason some farmers are injecting the sludge into the soil or are plowing in at least a part of the applied sludge rather than leaving it on the soil surface.

Figure 19.9 also emphasizes the fact that heavy annual land applications of sewage sludge should result in increased organic matter and nitrogen contents of the soil. Unfortunately, these increases are reflected mostly in complex organic compounds that are only slowly decomposed. At the same time they do add to the large pool of organic matter and associated nitrogen, sulfur, and phosphoric compounds, which through the years can be recycled for plant and animal use.

19.9 Composts

Composting is the time-honored practice of partially rotting organic materials, of either plant or animal origin, by bacteria, fungi, protozoa, earthworms, and other soil organisms. The aerobic decomposition takes place in piles and in bins kept sufficiently moist to encourage the decay. Sewage sludge solids can be composted along with other municipal wastes such as leaves and prunings and, after drying, can be sold to gardeners. Home production of compost is more widespread and has increased in recent years due to restrictions on the burning of leaves and other domestic wastes in residential areas. This is a blessing in disguise since composting ensures the conservation of some nutrients which are lost when the residues are burned.

Almost any plant material can be composted. Leaves, weeds, lawn cuttings, small prunings, and garden wastes are the primary source of the organic materials. These can be supplemented with materials such as sawdust, vacuum cleaner dust, and fireplace or furnace ashes. To hasten decay, small quantities of fertilizer may be added along with a little soil to assure the presence of decaying organisms. The materials can be kept in a pile, in a wooden bin, or in an area surrounded by wire netting. The pile must be kept moist (50–70% water) but not too wet since the breakdown must be aerobic in nature. The materials are best packed down to help keep the pile from drying out. Figure 19.11 illustrates how a compost pile can be constructed.

The heat of combustion of organic material increases the temperature of the compost pile, temperatures of 50–72°C (122–161°F) being reached when decay is occurring rapidly. At these temperatures most weed seeds are killed along with most plant and animal disease organisms.

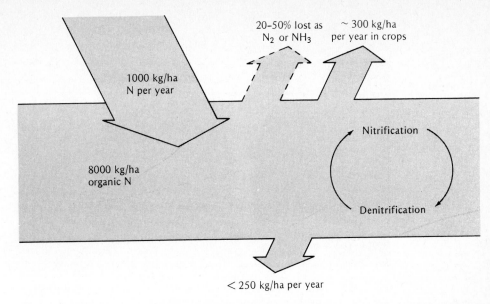

1000 kg/ha
N per year

20-50% lost as
N_2 or NH_3

~ 300 kg/ha
per year in crops

Nitrification

8000 kg/ha
organic N

Denitrification

< 250 kg/ha per year

FIGURE 19.10 Fate of nitrogen applied to land through sewage sludge. About 1000 kg of nitrogen is shown as being added each year. Between 20 and 50% is lost through volatilization as N_2 or NH_3, and approximately 300 kg/ha is removed in crops. Some residual nitrogen is left in the soil and adds to the approximately 8000 kg/ha of soil organic nitrogen. *[From Loehr et al. (1979). Copyright © 1979 by Van Nostrand Reinhold Co., Inc.; used with permission.]*

Since more organic materials may be added to the pile as residues and wastes become available, the materials at the bottom are usually most decomposed. Recently added wastes can be forked to the side so that more highly

FIGURE 19.11 Example of a backyard compost heap. Various wastes from the home and yard along with a little manure and soil are added in layers, kept moist, and partially decomposed by soil organisms.

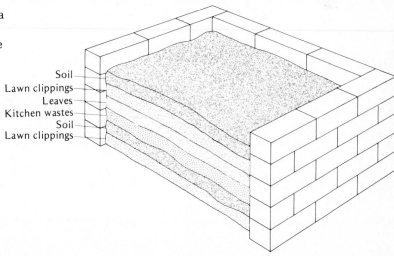

Soil
Lawn clippings
Leaves
Kitchen wastes
Soil
Lawn clippings

FIGURE 19.12 Compost piles in China. Crop wastes, manure, and soil are mixed together alongside fields and after partial decomposition of the organic matter added to the nearby fields.

decomposed materials can be removed for use. The material set aside can then be returned to the pile.

China has probably made more extensive use of composts than any other country. Crop residues, animal manures, and human wastes are mixed with soil and sludge from ponds and stream beds in elongated compost piles. The piles are then plastered on the outside with wet soil to prevent the rapid

TABLE 19.6 Effects of Composts on the Yield of Several Important Food and Feed Crops in China[a]

Crop	Yield (Mg/ha)	
	No compost	Compost added[b]
Corn	4,408	5,700 (30.4)
Potato	7,737	14,630 (38.0)
Sugarbeet	26,741	33,600 (15.2)
Wheat	2,336	3,230 (38.0)
Millet	2,257	3,341 (38.0)
Sorghum	1,664	3,078 (38.0)
Soybean	1,877	2,310 (30.4)

[a] From FAO (1977).
[b] Numbers in parentheses are rates of application in metric tons per hectare.

loss of moisture (Figure 19.12). Bamboo rods are used to make holes in the pile, thus assuring a supply of air throughout the decomposing mass. The large piles are used to provide compost not only for home gardens but for fields as well.

Crop Response to Composts. Properly prepared composts, if supplemented by small quantities of commercial fertilizers and lime when needed, are as effective as animal manures in enhancing crop production. Rates of application of 20–30 Mg/ha may be necessary since the nutrient element content is low. Furthermore, as with animal manures, the nutrients are not all readily available for plant uptake. The favorable response of different crops to compost in China is shown in Table 19.6.

Composted materials are sometimes used as mulches in vegetable or flower gardens (Figure 19.13). This practice provides not only plant nutrients but soil cover and moisture conservation as well. Indoor potted plants also thrive on mixtures high in composted materials.

19.10 Integrated Recycling of Wastes

For most of the industrialized countries, widespread recycling of organic wastes other than animal manures is a relatively recent phenomenon. In heavily popu-

FIGURE 19.13 Composts are widely used as mulches on flower beds and vegetable gardens. [*Courtesy USDA Soil Conservation Service.*]

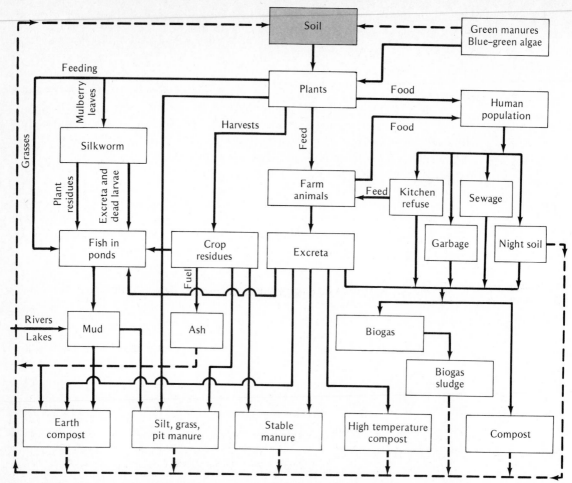

FIGURE 19.14 Recycling of organic wastes in the Peoples Republic of China. Note the degree to which the soil is used as a recipient of the wastes. [*Modified from FAO (1977).*]

lated areas of Asia, however, and particularly in China and Japan, such recycling has long been practiced. Figure 19.14 illustrates the many ways in which organic wastes are used in China. Agriculture is only one of the recipients of these wastes. Much is used for biogas production, as food for fish, and as a source of heat from compost piles to help warm homes, greenhouses, and household water. Most importantly, the plant nutrients and organic matter are recycled and returned to the soil for future plant utilization. Such conservation is likely to be practiced more widely in the future by other countries, including the United States.

REFERENCES

Benne, E. J., et al. 1961. *Ainman Manures,* Circ. 291, Michigan Agr. Exp. Sta.

Carroll, T. E., D. L. Maase, J. M. Geneo, and C. N. Ifeadi. 1975. *Review of Landspreading of Liquid Municipal Sewage Sludge,* EPA 670/2–75–049 (Cincinnati, OH: U.S. Environmental Protection Agency).

Dick, W. A. 1983. "Organic carbon, nitrogen, and phosphorus concentrations and pH in soil profiles as affected by soil tillage," *Soil Sci. Soc. Amer. J.,* **47**:102–107.

Elliott, L. F., and F. J. Stevenson (Eds.). 1977. *Soils for Management of Organic Wastes and Waste Waters* (Madison, WI: Soil Sci. Soc. Amer., Amer. Soc. Agron., and Crop Sci. Soc. Amer.).

FAO. 1977. *China: Recycling of Organic Wastes in Agriculture,* FAO Soils Bull. 40 (Rome: Food and Agriculture Organization of the United Nations).

Hensler, R. F. 1970. "Cattle manure: I. Effect of crops and soils; II. Retention properties for Cu, Mn, and Zn," Ph.D. Thesis, University of Wisconsin.

Jenkinson, D. S., and A. E. Johnston. 1977. "Soil organic matter in the Hoosfield barley experiment," Report of Rothamsted Experiment Station for 1976, Pt **2**:87–102.

Kardos, L. T., W. E. Sopper, E. A. Myers, R. R. Parizek, and J. B. Nesbitt. 1974. *Renovation of Secondary Effluent for Reuse as a Water Source,* Environ. Protect. Technol. Serv. EPA-660/2–74–016 (Cincinnati, OH: U.S. Environmental Protection Agency).

Loehr, R. C., W. J. Jewell, J. D. Novak, W. W. Clarkson, and G. S. Friedman. 1979. *Land Application of Wastes,* Vol. I (New York: Van Nostrand Reinhold Co.).

McCalla, T. M., J. R. Peterson, and C. Lue-Hing. 1977. "Properties of agricultural and municipal wastes," in L. F. Elliott and F. J. Stevenson (Eds.), *Soils for Management of Organic Wastes and Waste Waters* (Madison, WI: Soil Sci. Soc. Amer., Amer. Soc. Agron., Crop Sci. Soc. Amer.).

Sheaffer, C. C., A. M. Decker, R. L. Chaney, and L. W. Douglass. 1979. "Soil temperature and sewage sludge effects on corn yield and macronutrient content," *J. Environ. Qual.,* **8**:450–454.

Stickler, F. C., et al. 1975. *Energy from Sun to Plant to Man* (Moline, IL: Deere & Company).

USDA. 1980. *Appraisal, 1980 Soil and Water Resources Conservation Act, Review Draft,* Part I (Washington, DC: U.S. Department of Agriculture).

Van Slyke, L. L. 1932. *Fertilizers and Crop Production* (New York: Orange Judd), pp. 216–26.

Soils and Chemical Pollution

[*Preceding page*] There is growing concern that potentially toxic chemicals added to soils may be absorbed by plants in sufficient quantities to be harmful to humans and other animals. Here scientists are sampling forages growing in a former strip mine area to which large quantities of municipal sludge had been added. [*USDA photo by R. C. Bjork*.]

In recent years public attention has been focused increasingly on environmental pollution and its effects on humans and other creatures. Among the major pollutants are wastes from an urbanized and industrialized society along with the myriad of chemicals, new and old, needed to maintain an affluent society. The soil is a primary recipient, intended or otherwise, of many of these waste products and chemicals. Furthermore, once these materials enter the soil, they become part of a cycle that affects all forms of life. At least a general understanding of the pollutants themselves, their reactions in soils, and available means of managing, destroying, or inactivating them is essential.

Six general kinds of pollutants will receive attention. First are the thousands of *pesticide* preparations, most of which are used for agricultural purposes and all of which reach the soil. Second is a group of *inorganic pollutants*, such as mercury, cadmium, and lead, which have been discovered in toxic quantities as they move along the food chain. Third are the *organic wastes*, such as those from concentrated feedlots and food-processing plants, which will be considered along with municipal and industrial wastes, some of which may be dumped on soils. *Salts, radionuclides,* and *acid rain* are the remaining contaminants to be discussed.

20.1 Chemical Pesticides—Background

Human history is replete with stories of battles with pests. More than 10,000 species of insects, 600 weed species, 1500 plant diseases, and 1500 species of nematodes are known to be injurious, at least to some degree, to humans, plants, and animals. Various methods have been used to tip the scales of nature in humankind's favor. Crop varieties and breeds of animals have been developed that resist the pests. Tillage implements are used to control unwanted plants or weeds. And we have learned to rotate crops to prevent the buildup of pest organisms that thrive on a single crop species.

The use of chemicals to control pests has been practiced for centuries. For example, Greeks are said to have used sulfur to control certain plant diseases. When in the early nineteenth century Pasteur discovered that microbes were the cause of certain plant and animal diseases, chemical cures were sought and found. The use of copper-containing bordeaux mixture dates from soon after Pasteur's discoveries. Lime and sulfur mixtures and arsenic sprays have been used for half a century for the prevention of disease and insect damage in apple orchards. Naturally occurring insecticides such as rotenone and pyrethrins were also extensively used.

Synthetic Pesticides. It was the discovery of the insecticidal properties of DDT in 1939 and the herbicidal effects of 2,4-D in 1941 that truly began the chemical revolution in agriculture. These chemicals would kill pests and could be manufactured economically. Since the discovery of DDT and 2,4-D tens of thousands of such chemicals and formulations have been developed, tested, and put to use. In 1980, nearly 500 million metric tons of pesticides was applied in the United States, about half in agriculture. Some 900 chemicals in about 60,000 formulations are used to control pests.

Benefits from Pesticides. Perhaps the greatest benefit from pesticide use has been the millions of human lives saved from yellow fever, encephalitis, malaria, and other insect-borne diseases. Protection of crops and livestock has also brought economic benefits to society. Chemical weed control, for example, has in some cases virtually eliminated hand hoeing and even cultivation. More than any other factor, chemical weed control facilitates the adoption of minimum tillage practices so essential for soil conservation. In the United States, pesticides have been a prime factor in the agricultural revolution that has made it possible for less than 5% of the population to feed the other 95% and still export billions of dollars worth of farm products. Pesticides have also protected food from pest damage as it moves from the farm through processing and marketing channels to the dinner table.

Problems and Dangers from Pesticide Use. Three major problems threaten to limit the continued usefulness of pesticides. First, some pest organisms (particularly the insects) have developed resistance to the chemicals. This necessitates higher dosages or the development of new chemicals to replace those to which the pests are resistant. Second, some pesticides are not readily biodegradable and tend to persist for years in the environment. Although this characteristic may be advantageous in controlling some pests, it is a disadvantage as the chemical moves to other parts of the environment.

This leads to the third problem, the detrimental effects of the chemicals on organisms other than the target pests. As little as 1% of the pesticides applied may contact the target organism, much of the remainder moving into the soil. Soil flora and fauna may be adversely affected, as may be fish and other wildlife. This problem is compounded by the tendency of the chemicals to build up in organisms as movement up the food chain occurs. Birds and fish, being secondary and tertiary consumers, tend to concentrate these chemicals in their body tissue, in some cases to toxic levels. Damage to these creatures sounded the warning cry that we must know more about the ecological effects of pesticides if their use is to be continued. The use of those chemicals that are the greatest hazards has already been restricted or eliminated.

20.2 Kinds of Pesticides

Pesticides are commonly classified according to the target group of pest organisms: (a) insecticides, (b) fungicides, (c) herbicides (weedicides), (d) rodenticides, and (e) nematocides. Since the first three are used in largest quantities and are therefore more likely to contaminate soils, they will be given primary consideration.

Insecticides.   The quantity of insecticides in use in the United States has decreased somewhat over the past 10 years (Table 20.1), largely because of environmental concerns. Most of these chemicals are included in three general groups. The *chlorinated hydrocarbons,* such as DDT, were the most extensively used until the early 1970s. They had the advantages of low cost (especially DDT), general effectiveness, persistence, and relative low levels of toxicity to man. However, their low biodegradability and excessive persistence as well as toxicity to birds and fish have made them targets for environmentalists, resulting in restrictions on their use.

The *organophosphate* pesticides are generally biodegradable and thus less likely to build up in soils and water. At the same time, they are relatively much more toxic to humans than are the chlorinated hydrocarbons, so great care must be used in handling and applying them. The *carbamates* are also popular among most environmentalists because of their ready biodegradability and relatively low mammal toxicity.

TABLE 20.1   Classes of Pesticides Commonly Used in the United States and Examples of Each Group

Chemical group	Examples
Insecticides	
Chlorinated hydrocarbons	Aldrin, chlordane, heptachlor, toxaphene
Organophosphates	Diazinon, disulfoton, parathion, malathion
Carbamates	Carbaryl, carbofuran, methomyl
Pyrethrium	Permethrin
Fungicides	
Thiocarbamates	Ferbam, maneb
Mercurials	Ceresan
Others	Copper sulfate, chlorothalonil
Herbicides	
Phenoxyalkyl acids	2,4-D, 2,4,5-T, MCPA
Triazines	Atrazine, simazine, propazine
Phenylureas	Diuron, linuron, fluometuron
Aliphatic acids	Dalapon
Carbamates	Butylate, vernolate
Dinitroanilines	Trifluralin, benefin
Dipyridyl	Paraquat, diquat
Amides	Alachlor, propachlor, propanil, alanap
Benzoics	Amiben, dicambra

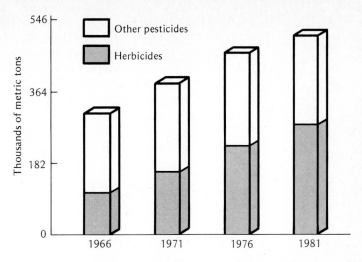

FIGURE 20.1 Increase in the use of pesticides in the United States since 1966. Note that essentially all the increase has been in herbicides. [*USDA as quoted by Sanders (1981).*]

Fungicides. The quantity of fungicides used in the United States is far less than that of either herbicides or insecticides. Even so, large areas receive these chemicals annually. Fungicides are used primarily to control the field diseases of fruits and vegetables. They are also used to counteract seed diseases of common crops and to protect harvested fruits and vegetables from decay and rot. Fungicides are used as wood preservatives and for the protection of clothing from mildew.

Herbicides. The quantity of herbicides used in the United States exceeds that of the other two types of pesticides (Figure 20.1). Starting with 2,4-D (chlorinated phenoxyacetic acid), dozens of chemicals in literally hundreds of formulations have been placed on the market. These include the *triazines* (rather specific for weed control in corn), *phenylureas, aliphatic acids, carbamates, dinitroanilines,* and *dipyridyls* (Table 20.1). As one might expect, this wide variation in chemical makeup provides an equally wide variation in properties. However, herbicides are generally biodegradable, and most of them are relatively low in mammal toxicity. Some are quite toxic to fish and perhaps to other wildlife, however, emphasizing once again the need to consider the indirect effects of these chemicals.

20.3 Behavior of Pesticides in Soils[1]

Pesticides are commonly applied to plant foliage or on the soil surface or are incorporated into the soil. In any case, a high proportion of the chemicals eventually moves into the soil, a fact that adds significance to studies of the fate of these chemicals in soil.

[1] For excellent recent reviews of this subject see Guenzi (1974) and Hill and Wright (1978).

There is great variability in the chemical structures of pesticides (Figure 20.2) and in the behavior of these chemicals in soil. A few characteristics of pesticides and how they affect the behavior and reactions of chemicals in soils will first be considered. Attention will be given to five possible fates of pesticides once they are added to soils: (a) the chemicals may vaporize into the atmosphere without chemical change, (b) they may be adsorbed by soils, (c) they may move downward through the soil in liquid or solution form and be lost from the soil by leaching, (d) they may undergo chemical reactions within or on the surface of the soil, and (e) they may be broken down by soil microorganisms.

Volatility. Pesticides vary greatly in their volatility and subsequent susceptibility to atmospheric loss. Some soil fumigants such as methyl bromide are selected because of their very high vapor pressure, which permits them to penetrate soil pores to contact the target organisms. This same characteristic encourages rapid loss to the atmosphere after treatment unless the soil is covered or sealed. A few herbicides (e.g., trifluralin) and fungicides (e.g., PCNB) are sufficiently volatile to make vaporization a primary means of their loss from soil. DDT and dieldrin are volatilized from soil in significant quantities even though their vapor pressures are quite low in comparison with the chemicals previously mentioned. This vaporization probably accounts at least in

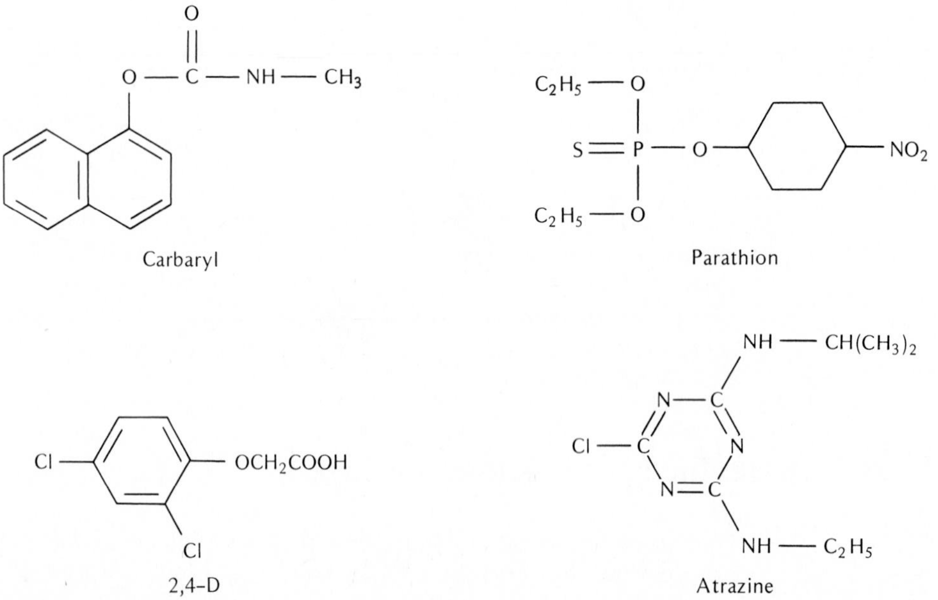

FIGURE 20.2 Structural formulas of some widely used pesticides. Carbaryl and parathion are insecticides; 2,4-D and atrazine are herbicides. This variety of structures dictates great variability in properties and reactivity in the soil.

part for the air transport of these chemicals to great distances from their point of application.

Earlier assumptions that disappearance of pesticides from soils was evidence of their breakdown are now known to be questionable. Loss of some chemicals to the atmosphere only to have them returned to the soil or surface waters in rain is now known to occur.

Adsorption. The tendency of pesticides to be adsorbed by soil is determined largely by the characteristics of the pesticides and of the soils to which they are added. The presence of certain functional groups, such as $-OH$, $-NH_2$, $-NHR$, $-CONH_2$, $-COOR$, and $-{^+NR_3}$, in the molecular structure of the chemical encourages adsorption, especially on the soil humus. Hydrogen bonding (see Section 3.1) and protonation (adding of H^+ to a group such as an amino group) probably promotes some of the adsorption. In general, the larger the size of the pesticide molecules, everything else being equal, the greater will be their adsorption.

The soil characteristic with which adsorption is most closely associated is the soil organic matter content. Apparently, the complexity of the humus fraction along with its nonpolar nature encourages adsorption of the pesticides. A few pesticides, such as the herbicides diquat and paraquat, which tend to form cations (strong bases), are also adsorbed by silicate clays. Adsorption by clays of pesticides that form bases tends to be pH dependent (Figure 20.3), maximum adsorption occurring at the pH level where protonation occurs. Some

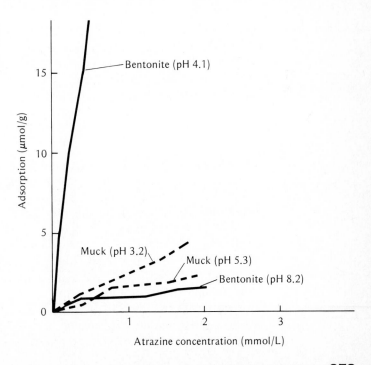

FIGURE 20.3 Effect of pH and type of colloid (bentonite versus muck) on the adsorption of atrazine, a common herbicide. [*From Harris and Warren* (*1964*).]

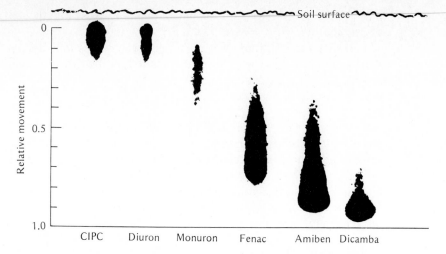

FIGURE 20.4 Comparative leaching of several herbicides in soils. The herbicides are radioactively tagged and move with the water. The movement is then measured by placing X-ray film in contact with a column of soil. [*USDA photo courtesy of C. S. Helling.*]

clay-adsorbed pesticides lose their biocidal properties until they are desorbed. This has some obvious practical implications.

Leaching. The tendency of pesticides to leach from soils is closely related to their potential for adsorption. Strongly adsorbed molecules are not likely to move down the profile (Figure 20.4). Likewise, conditions that encourage such adsorption will discourage leaching. Leaching is apt to be favored by water movement, taking place most readily in permeable sandy soils that are low in clay and organic matter. In general, herbicides seem to be more mobile than either fungicides or insecticides (Table 20.2).

Chemical Reactions. Upon contacting the soil some pesticides undergo chemical modification independent of soil organisms. For example, DDT, diaquat, and the triazines are subject to slow photodecomposition activated by solar radiation. Such degradation is relatively less important, however, than that which is catalyzed directly by the soil. Such catalysis takes place when certain pesticides are adsorbed by the silicate clay fraction, especially if the soils are acid. The adsorbed pesticides are subject to hydrolysis and subsequent degradation. The triazine herbicides (e.g., atrazine) and organophosphate insecticides (e.g., malathion) are examples of pesticides subject to adsorption and hydrolysis. While the complexities of molecular structure of the pesticides suggest different mechanisms of breakdown, it is important to realize that degradation independent of soil organisms does in fact occur.

TABLE 20.2 Comparative Ease with Which Pesticides Move in the Soil[a]

Mobility class 5 moves most rapidly, while class 1 is quite immobile. Note that herbicides generally are more mobile than the other (bold type) pesticides.

Mobility class

5	4	3	2	1
TCA	Picloram	Propachlor	Siduron	**Parathion**
Dalapon	Fenac	Fenuron	Prometryne	**Disulfoton**
2,3,6-TBA	MCPA	2,4,5-T	Propanil	Diquat
Tricamba	Amitrole	Propham	Diuron	Paraquat
Dicamba	Dinoseb	Fluometuron	Linuron	Trifuralin
Choramben		Monuron	Purazon	Benefin
		Atrazine	Vernolate	**Heptachlor**
		Simazine	Chlorpropham	**Aldrin**
		Propazine	**Azinphosmethyl**	**Chlordane**
			Diazinon	**Toxaphene**
				DDT

[a] Selected from Helling et al. (1971) who summarized data from others.

Microbial Metabolism. Biochemical degradation by soil organisms is perhaps the single most important method by which pesticides are removed from soils. Apparently the presence of certain polar groups on the pesticide molecules provides points of attack for the organisms. These groups include —OH, —COO⁻, —NH₂, and —NO₂.

DDT can be changed by certain bacteria and fungi to the related compounds DDD under anaerobic conditions (quite rapidly) and DDE in well-aerated soils (very slowly). Further decomposition, however, takes place very slowly. Other chlorinated hydrocarbons (such as aldrin, dieldrin, and heptachlor) are subject to only slow "partial decomposition" since organisms have not adapted themselves to the rapid destruction of such compounds. This helps account for the marked persistence of these compounds in soils and elsewhere.

The organophosphate insecticides are degraded quite rapidly in soils, apparently by a variety of organisms. Likewise, the most widely used herbicides, such as 2,4-D, the phenylureas, the aliphatic acids, and the carbamates, are readily attacked by a host of organisms. Exceptions are the triazines (for example, atrazine), which are degraded primarily by chemical action. Most organic fungicides are also subject to microbial decomposition, although the rate of breakdown of some is slow, causing troublesome residue problems.

Persistence in Soils. The persistence of pesticides in soils is a summation of all the reactions, movements, and degradations affecting these chemicals. Marked differences in persistence are the rule. For example, organophosphate insecticides may last only a few days in soils; the most widely used herbicide,

2,4-D, persists in soils for only two to four weeks; DDT and other chlorinated hydrocarbons may persist for from three to fifteen years or longer (Table 20.3). The persistence times of hundreds of other herbicides, fungicides, and insecticides generally are between the extremes cited. The majority of pesticides degrade rapidly enough to prevent buildup in soils. Those that do not do so have potential for environmental damage.

Continued use of the same pesticide on the same land can result in the rapid microbial breakdown of the chemical. For example, certain thiocarbamate herbicides that may be used year after year for continuous corn are degraded rapidly in some soils (Fox, 1983). While there is an advantage in relation to environmental quality, the breakdown is so rapid that the herbicide's effectiveness is greatly reduced.

Among the practices suggested to reduce pesticide levels in soils is the addition of easily decomposed organic matter. The growth of high-nitrogen cover crops or the additions of large quantities of animal manures should be helpful. Apparently degradation of even the most resistant pesticides is encouraged by conditions favoring overall microbial action. Other practices suggested to reduce pesticide levels are cropping to plants that accumulate the pesticide, and leaching the soil. Unfortunately, some of these procedures merely transfer the chemical from the soil to some other part of the environment, a process of dubious value.

This brief review of the behavior of pesticides in soils reemphasizes the complexity of the changes that take place when new and exotic chemicals

TABLE 20.3 Common Range of Persistence of a Number of Pesticides

Risks of environmental pollution are highest with those chemicals with greatest persistence.

Pesticide	Persistence
Arsenic	Indefinite
Chlorinated hydrocarbon insecticides (e.g., DDT, chlordane, dieldrin)	2–5 yr
Triazine herbicides (e.g., atrazine, simazine)	1–2 yr
Benzoic acid herbicides (e.g., amiben, dicamba)	2–12 mo
Urea herbicides (e.g., monuron, diuron)	2–10 mo
Phenoxy herbicides (2,4-D, 2,4,5-T)	1–5 mo
Organophosphate insecticides (e.g., malathion, diazinon)	1–12 wk
Carbamate insecticides	1–8 wk
Carbamate herbicides (e.g., barban, CIPC)	2–8 wk

are added to our environment. The wisdom is seen of evaluating as thoroughly as possible the ecological effects of new chemicals before their extensive use is permitted.

20.4   Effects of Pesticides on Soil Organisms

Since the purpose of pesticides is to kill organisms, it is not surprising that some of them are toxic to specific soil organisms. At the same time, the diversity of the soil organism population is so great that except for a few fumigants, most pesticides do not kill a broad spectrum of soil organisms. It is perhaps surprising that the extensive use of pesticides in the United States has not provided more extensive evidence of damage to soil organism numbers. Even so, there is evidence that some commonly used pesticides adversely affect specific groups of organisms, some of which carry out important processes in the soil.

Fumigants.   Fumigants are compounds used to free a soil of a given pest, such as nematodes. Those compounds have a more drastic effect on both the soil fauna and flora than do other pesticides. For example, 99% of the microarthropod population is usually killed by the fumigants DD and vampam, and it takes as long as 2 years for the population to recover. The recovery time for the microflora is generally much less, although the rate of recovery varies greatly among the affected organisms. Also, fumigation reduces the number of species of both flora and fauna, especially if the treatment is repeated, as is often the case where nematode control is attempted. At the same time, the total number of bacteria is frequently much greater following fumigation than before. This is probably due to the relative absence of competitors and predators following fumigation.

Effects on Soil Fauna.   The effects of pesticides on soil animals varies greatly from chemical to chemical and organism to organism. Nematodes are not generally affected except by specific fumigants. Mites are generally sensitive to most organophosphates and to the chlorinated hydrocarbons except for aldrin. Springtails vary in their sensitivity to both chlorinated hydrocarbons and organophosphates, some chemicals being quite toxic to these organisms.

Fortunately, many pesticides have only mildly depressing effects on earthworm numbers. Exceptions are most of the carbamates and some nematocides, which are quite toxic to these animals. The concentration of pesticides in the bodies of the earthworms are closely related to the levels found in the soil (Figure 20.5).

Pesticides have significant effects on the numbers of certain predators and in turn on the numbers of prey organisms. For example, an insecticide that reduces the numbers of predatory mites may stimulate numbers of spring-

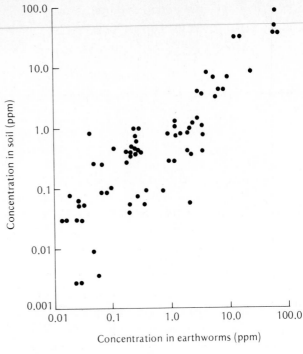

FIGURE 20.5 The effect of concentration of pesticides in soil on their concentration in earthworms. Birds eating the earthworms at any level of concentration would further concentrate the pesticides. [*Data from several sources gathered by Thompson and Edwards* (1974); *used with permission of the Soil Science Society of America.*]

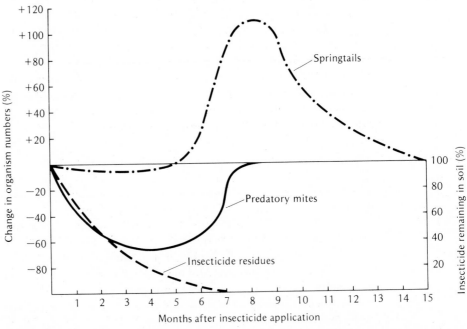

FIGURE 20.6 The direct effect of insecticide on predatory mites in a soil and the indirect effect of reducing mite numbers on the population of springtails (insect) that serve as prey for the mites. [*Replotted from Edwards* (1978). *Copyright Academic Press Inc. (London) Ltd.; used with permission.*]

tails that serve as prey for the mites (Figure 20.6). Such organism interaction is normal in most soils.

Effects on Soil Microorganisms.   The overall levels of bacteria in the soil are generally not too seriously affected by pesticides. However, organisms responsible for nitrification and nitrogen fixation are sometimes adversely affected. Insecticides and fungicides affect both processes more than do most herbicides, although some of the latter can reduce the numbers of organisms carrying out these two reactions.

Fungicides, especially those used as fumigants, have the most marked adverse effect on soil fungi and actinomycetes. Consequently they can slow down the degradation of soil organic matter in soil. Interestingly enough, however, the process of ammonification is often benefited by pesticide use.

The negative effects of most pesticides on soil microorganisms are temporary, and after a few days or a few weeks, organism numbers generally recover. But the exceptions are common enough to dictate caution in the use of the chemicals. Care should be taken to apply them only at recommended levels.

20.5   Contamination with Toxic Inorganic Compounds[2]

Public attention has been called in recent years to environmental contamination by a number of inorganic compounds, including those containing mercury, cadmium, lead, arsenic, nickel, copper, zinc, molybdenum, manganese, fluorine, and boron. To a greater or lesser degree, all of these elements are toxic to humans and other animals. Cadmium and arsenic are extremely poisonous; mercury, lead, nickel, and fluorine are moderately so, and boron, copper, manganese, and zinc are relatively lower in toxicity. Although the metallic elements (which exclude fluorine and boron) are not all normally included among the heavy metals, for simplicity this term is often used in referring to them. Table 20.4 provides background information on the uses, sources, and quantity mined annually for each of these elements.

Sources and Accumulation.   Several factors account for the emergence of inorganic chemical contamination as an important ecological problem. In the first place, modern technology requires the use of these elements in much larger quantities than in the past. The burning of fossil fuels, smelting, and other processing techniques release into the atmosphere tons of these elements, which can adversely affect surrounding vegetation. These "aerosol" dust particles may be carried for miles and later deposited on the vegetation and soil.

Lead is emitted to the atmosphere as one of the products of coal burning. It is also used as an additive to gasolines, as are nickel and boron. Upon combustion these elements are released to the atmosphere and carried to the

[2] For a review of this subject, see Lisk (1972).

TABLE 20.4 Estimated Quantities of Certain Inorganic Elements Used Annually, Their Major Uses, and Sources of Soil Contamination[a]

Chemical	Mined annually (10^3 Mg)	Major uses	Major sources of soil contamination
Arsenic	24	Medicines, pesticides, paints	Pesticides, industrial air pollution
Boron	91	Detergents, glass, fertilizer, gasoline additive	Gasoline combustion, irrigation water
Cadmium	9	Alloys, rust-proofing of steel, batteries, pigments	Smelting, sewage sludge, roasting and plating minerals, fertilizer impurities
Copper	3628	Coins, pipe, alloys, electrical wire	Industrial dusts, mine effluents, sewage treatment waters, fungicides
Fluorine	226	Refrigerants, spray-can propellants, fertilizers, pesticides	Fertilizers, pesticides, local air pollution
Lead	1814	Gasoline additives, batteries, solders, cable covering	Combustion of leaded gasoline, smelting, fertilizers, pesticides
Manganese	5443	Ferromanganese, batteries, chemicals, fertilizers	Mine seepage, fly ash, fertilizers
Mercury	3	Dental amalgams, drugs, fluorescent lights, electric switches, scientific instruments	Fungicides, atmospheric contamination from evaporation of metallic Hg
Nickel	272	Stainless steel and other alloys, gasoline additives	Fertilizers, gasoline combustion
Zinc	2772	Alloys, galvanized metals, brass, paints, cosmetics	Sewage effluents, industrial waste, fertilizers, pesticides

[a] From Bowen (1966).

soil through rain and snow. Borax is used as a detergent and in fertilizer, both of which commonly reach the soil. Superphosphate and limestone usually contain small quantities of cadmium, copper, manganese, nickel, and zinc. Cadmium is used in plating metals and in the manufacture of batteries. Arsenic was used for many years as an insecticide on cotton, tobacco, and fruit crops. It is still being used as a defoliant or vine killer and on lawns. The heavy metals are found as constituents in specific organic fungicides, herbicides, and insecticides, and tend to be concentrated in domestic and industrial sewage sludge. The process of recycling wastes further accentuates such concentration.

The quantities of most of the products in which these inorganic contaminants are used have increased notably in recent years, enhancing the opportu-

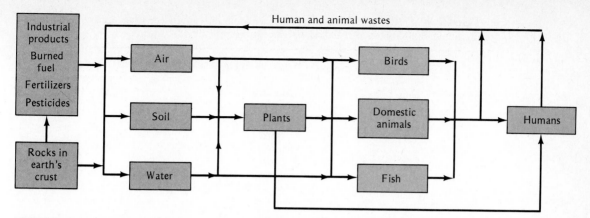

FIGURE 20.7   Sources of heavy metals and their cycling in the soil–water–air–organism ecosystem. It should be noted that the content of metals in tissue generally builds up from left to right, indicating the vulnerability of man to heavy metal toxicity.

nity for contamination. They are present in the environment in increasing amounts and are daily ingested by people either through the air or through food and water.

Detection of Heavy Metals.   Sophisticated analytical tools have made possible measurements of heavy metal contamination previously undetected. The discovery of the accumulation of one contaminant has prompted investigations of others, leading to a better understanding of the cycling of these elements in our environment. Also in the case of mercury, discovery of new means of solubilizing the element has elucidated the concept of its role as a contaminant.

Concentration in Organism Tissue.   Irrespective of their sources, toxic elements can and do reach the soil, where they become part of the life cycle of soil → plant → animal → human (Figure 20.7). Unfortunately, once the elements become part of this cycle they may accumulate in animal and human body tissue to toxic levels. This situation is especially critical for fish and other wildlife and for humans at the end of the food chain. It has already resulted in restrictions on the use for human food of certain fish and wildlife. Also, it has become necessary to curtail the release of these toxic elements in the form of industrial wastes.

20.6   Potential Hazards of Chemicals in Sewage Sludge

The domestic and industrial sewage sludges considered in Chapter 19 are major sources of potentially toxic chemicals. Nevertheless, land application of sewage sludge is increasing as other alternatives become more expensive or otherwise

TABLE 20.5 Range of Concentration in Soils and Plants of Inorganic Elements that Sometimes Occur as Environmental Contaminants[a]

| Element | Common range in concentration (ppm) | |
	Soils	Plants
Arsenic	0.1–40	0.1–5
Boron	2–100	30–75
Cadmium	0.1–7	0.2–0.8
Copper	2–100	4–15
Fluorine	30–300	2–20
Lead	2–200	0.1–10
Manganese	100–4000	15–100
Nickel	10–1000	1
Zinc	10–300	15–200

[a] From Allaway (1970).

unattractive. More than 1000 communities are using land application to dispose of at least part of their sewage sludge. At least 25% of the sewage sludge in this country is land applied. These organic wastes exert beneficial physical effects on soils and are significant sources of plant nutrients. But they commonly carry significant quantities of inorganic as well as organic chemicals that can have harmful environmental effects.

In Table 20.5 ranges in levels of several chemical elements are given for

TABLE 20.6 Comparative Concentrations of Several Potentially Toxic Elements in Municipal Sewage Sludge and Cow Manure[a]

| Element | In municipal sewage sludge (ppm) | | In cow manure (ppm) |
	Small village	Range from 15 larger cities[b]	
Antimony	3	4–44	0.5
Arsenic	3	4–30	4
Cadmium	7	9–444	1
Chromium	169	207–14,000	56
Copper	821	458–2890	62
Mercury	11	4–18	0.2
Manganese	128	32–527	286
Molybdenum	1	2–33	14
Nickel	36	51–562	29
Lead	136	329–7627	16
Zinc	560	601–6890	71

[a] Selected data from Furr et al. (1976).
[b] The 15 larger cities had significant industrial inputs.

sewage sludges from 15 municipal waste disposal plants in large cities in the United States. The plants all received effluent from industrial as well as domestic sources. Concentrations of most of these elements were much higher in sewage sludge from the large cities than in animal manure or in domestic sewage sludge from a small village. For example, the cadmium content of the sludge from one city was more than 400 times that of cow manure and 60 times that of domestic sludge from the village plant.

The likely major source of most of these chemicals is the industrial wastes being added to city sewage systems. Commonly a single treatment plant will serve not only domestic sources of sewage but 100 or more industrial sources. Though any one industrial unit may not account for a high proportion of the level of any one chemical, the combined additions from all sources result in the high levels shown in Table 20.6. When combined with the uncertainties as to the nature of many of the organic chemicals found in the sludge, these inorganic element levels dictate caution in the unmonitored application of sludge to croplands. The farmer must be assured that the levels of inorganic chemicals in sludge are not sufficiently high to be toxic to plants, or to humans and other animals who consume the plants (Figure 20.8).

Special attention will probably be needed to reduce soil contamination from annual land application of sewage sludges containing toxic quantities of heavy metals. The effect of such applications on heavy metal content of soils and of earthworms growing in the soil is illustrated in Table 20.7. The sludge-treated soil areas, as well as the earthworms growing on the soils, were higher in these elements than where sludge had not been applied. One would expect further concentration to take place in the tissue of birds and fish, many of which consume the earthworms. The potentially negative environmental impact of heavy metals in sewage sludge is obvious.

TABLE 20.7   The Effect of Sewage Sludge Treatment on the Content of Heavy Metals in Soil and in Earthworms Growing in the Soil[a]

Note the high concentration of cadmium and zinc in the earthworms.

| | Concentration of metal (ppm, dry wt.) | | | |
| | Soil | | Earthworms | |
Metal	Control	Sludge treated	Control	Sludge treated
Cd	0.1	2.7	4.8	57
Zn	56	132	228	452
Cu	12	39	13	31
Ni	14	19	14	14
Pb	22	31	17	20

[a] From Beyer et al. (1982).

FIGURE 20.8 Collecting soybean leaf samples from a field fertilized with sewage sludge. The samples are to be analyzed for their heavy metal content, thereby helping to establish the levels of these metals in sludge that can be tolerated. [*USDA photo by R. C. Bjork.*]

Soils are only a part of the biological cycle relative to heavy metals and other inorganic toxin contamination. At the same time, soils are the ultimate depositories of large quantities of these compounds. Furthermore, the variety of chemical reactions that these elements undergo in soils controls to a considerable extent their rate of cycling if not their removal from the cycle altogether. A brief summary of these reactions follows in the next section.

20.7   Behavior of Inorganic Contaminants in Soils

There is considerable variation in the level of these elements present in soils and plants. This is borne out by the data in Table 20.5, which give the ranges commonly found. These relative concentrations are of particular significance as the behavior of each of these elements is considered.

Four of the heavy metals, zinc, copper, manganese, and nickel, have similar chemical characteristics and undergo similar reations in soils, and so will be discussed as a group. Each of the other elements is sufficiently different in its properties to be given specific consideration.

Zinc, Copper, Manganese, and Nickel.   The reaction of these elements in soils is definitely affected by the pH, organic matter content, and the oxidation–reduction status of the soil. Ordinarily at pH values of 6.5 and above they tend to be only slowly available to plants, especially if they are present in their high-valent or oxidized forms. Consequently, most soils will tie up relatively large quantities of these elements if the soil pH is high and the drainage good.

The tendency of the cations of these elements to "chelate" in the presence of organic matter decidedly influences their behavior (see Section 11.6). The relative strength of chelation is generally copper > nickel > zinc > manganese. Since iron is more tightly adsorbed than any of them, its presence in a soluble form reduces the chelation tendency of all these elements. However, high pH and good drainage reduce the probability that soluble iron will be present in appreciable quantities.

Cadmium.   Cadmium is not essential for plant growth, but under certain conditions can accumulate in some plants to levels that are hazardous to animals and humans. Some sewage sludges contain enough cadmium to encourage such accumulation. The chemistry of cadmium reactions in soils is not well understood. But it is known that the uptake of this element is generally reduced by organic matter, silicate clay, hydrous oxides of iron and aluminum, and poor soil aeration. Also, cadmium uptake is high in acid soils and is reduced when the soil is limed. The element tends to accumulate in vegetative plant parts rather than in the seeds. The leaves of some vegetables such as lettuce may contain concentrations of cadmium undesirable for human or animal consumption even though no toxicity symptoms are obvious in the plants.

Mercury.   Research in Sweden and Japan as well as the United States has called attention to toxic levels of this element in certain species of fish. This situation stems from soil reactions whereby mercury is changed from insoluble inorganic forms not available to living organisms to organic forms that can be assimilated easily. Metallic mercury is first oxidized by the following chemical reaction in the sediment layer of lakes and streams:

$$Hg^0 \xrightarrow{[O]} Hg^{2+}$$

The divalent mercury is then converted by microorganisms to methylmercury, which is water soluble and can be absorbed through the food chain by fish. The methylmercury can be changed to dimethylmercury through biochemical reactions such as the following:

$$Hg^{2+} \rightarrow \underset{\text{Methylmercury}}{CH_3Hg^+}$$

$$CH_3Hg^+ \rightarrow \underset{\text{Dimethylmercury}}{CH_3HgCH_3}$$

Apparently the reactions will take place in either aerobic or anaerobic conditions. The methylmercury concentrates as it moves up the food chain, accumulating in some fish to levels that may be toxic to man.

Inorganic mercury compounds added to soils react quickly with the organic matter and clay minerals to form insoluble compounds. In this form the mercury is quite unavailable to growing plants. However, it can be reduced to metallic mercury, which is subject to volatilization and movement elsewhere in the environment. Mercury is not readily absorbed from soil by plants unless it is in the methylmercury form.

Lead. Interest in the soil as a source of lead for crop plants is heightened by the concern over airborne lead from automobile exhausts. The importance of this airborne source is verified by the concentrations of lead in plants and soils along heavily traversed highways (Figure 20.9). The airborne particles are moved far from the point of exhaust and are an important factor in determining the lead content of foods. Just how much lead is deposited directly on the leaf surface and how much is deposited on the soil and later taken up by the plants is not known. However, behavior of this element in soil would suggest that much of the lead in food crops comes from atmospheric contamination.

Soil lead is largely unavailable to plants, as evidenced by the small increases in lead content of plants following soil applications of the element. As with the other toxic metallic cations, lead is quite insoluble in soil, especially if the soil is not too acid. Most lead is found in the surface soil, indicating little if any downward movement. As might be expected, liming reduces the availability of the element and its uptake by plants.

Arsenic. Reasonably heavy applications of arsenical pesticides over a period of years, especially to orchard soils, have resulted in the accumulation of soil arsenic, in a few cases to toxic levels. Studies suggest that arsenic behaves

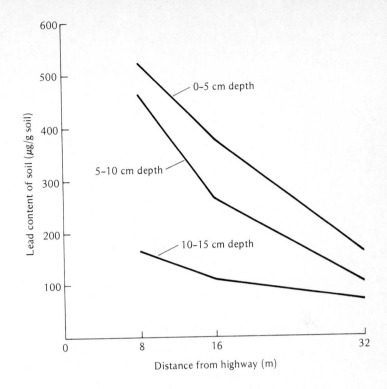

FIGURE 20.9 Lead content of soils at increasing distance from a heavily traveled highway near Beltsville, MD. Note that the lead is highest near the automobile traffic (source of lead) and tends to be concentrated in the upper depths. [*Adapted from Lagerwerff and Specht* (*1970*).]

in soils very much like phosphate. For this reason most of the applied arsenate is relatively unavailable for plant growth and uptake. Being present in an anionic form (e.g., AsO_4^{3-}), arsenic is adsorbed by hydrous iron and aluminum oxides. This adsorbed arsenate is replaceable from these oxides by phosphate through the process of anion exchange. The similarity in properties between phosphates and arsenates is important to remember.

In spite of the capacity of most soils to tie up arsenates, long-term additions of arsenical sprays have in a few instances resulted in decided toxicities to some sensitive plants (Figure 20.10). Even though the arsenic level in the plant tissue grown on such soils generally is not toxic to animals, normal plant growth is limited by excess arsenic in the soils. The arsenic toxicity can be reduced by applying to the soil sulfates of zinc, iron, or aluminum. These probably form insoluble arsenate compounds similar to those that form with phosphates.

Boron. Soil contamination by boron can occur from irrigation water high in this element or by excess fertilizer application. The boron can be adsorbed by organic matter and clays but is still available to plants except at high soil pH. Boron is relatively soluble in soils, toxic quantities being leachable especially from acid sandy soils. Boron toxicity is usually considered a localized problem and is probably much less important than a deficiency of the element.

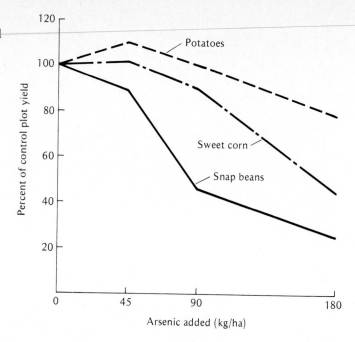

FIGURE 20.10 Effect of arsenic applications to a sandy soil on the yield of three vegetable crops. [*From Jacobs et al. (1970).*]

Fluorine. Fluorine toxicity is also generally localized; it is exhibited in drinking water for animals and in fluoride fumes from industrial processes. The fumes can be ingested directly by the animals or deposited on nearby plants. If the fluorides are adsorbed by the soil, their uptake by plants is restricted. The fluorides formed in soils are highly insoluble, the solubility being least if the soil is well supplied with lime.

20.8 Prevention and Elimination of Inorganic Chemical Contamination

Two primary methods of alleviating soil contamination by toxic inorganic compounds are (a) to eliminate or drastically reduce the soil application of the toxins, and (b) to so manage the soil and crop to prevent further cycling of the toxin.

Reducing Soil Application. The first method requires action to reduce unintentional aerial contamination from industrial operations and from automobile exhausts. Decision makers must recognize the soil as an important natural resource that can be seriously damaged if its contamination by accidental addition of inorganic toxins is not curtailed. Also, there must be judicial reductions in intended applications to soil of the toxins through pesticides, fertilizers, irrigation water, and solid wastes.

TABLE 20.8 Quantities of Several Metals Added in 15 Mg of Sewage Sludge and the Uptake of These Metals by Three Crops Grown on Plano Silt Loam[a]

Element	Amount applied (kg/ha)	Concentration in crop (%)		
		Rye	Sorghum–Sudan	Corn
Cu	22	0.16	0.17	0.05
Zn	45	0.35	1.14	3.39
Cd	1	0.13	0.37	0.08
Ni	11	0.05	0.20	0.06
Cr	28	0.01	0.04	0.01

[a] From Kelling et al. (1977).

Reducing Recycling.   Soil and crop management can help reduce the continued cycling of these inorganic chemicals. This is done primarily by keeping the chemicals in the soil rather than encouraging their uptake by plants. The soil becomes a "sink" for the toxins, thereby breaking the cycle of soil–plant–animal (humans) through which the toxin exerts its effect. The soil breaks the cycle by immobilizing the toxins. For example, most of these elements are rendered less mobile and less available if the pH is kept high. Liming of acid soils should expedite this immobility. Likewise, the draining of wet soils should be beneficial since the oxidized forms of the several toxic elements are generally less soluble and less available for plant uptake than are the reduced forms.

Heavy phosphate applications reduce the availability of these cations but may have the opposite effect on arsenic, which is found in the anionic form. Leaching may be effective in removing excess boron, although moving the toxin from the soil to water may not be of any real benefit. Also advantageous are differences in the abilities of crop species or varieties to extract the toxins. Data in Table 20.8 illustrate differences in uptake of several elements by different crop plants. Even though none of the crops tested take up a high proportion of the elements applied, some are more absorptive than others. "Accumulators" should be avoided if the harvests are to be fed to humans or domestic animals. Moreover, forage crops should be harvested at the maturity stage, when the concentration of the toxin is lowest. It is obvious that soil and crop management offers some potential for alleviating contamination by inorganic elements.

20.9   Organic Wastes

The pollution potential of organic wastes, urban and rural, has become a national and even international problem. In the United States, nearly 1 Mg of domestic organic wastes is generated per person each year. Food and fiber processing plants and other industrial operations produce millions of tons of

TABLE 20.9 Number of Cattle Feedlots with Less Than and More Than 1000 Head in Size and the Marketings from Them During the Period 1964–1970[a]

| | Under 1000 head | | Over 1000 head | |
Year	No. of lots (× 10³)	No. of marketings	No. of lots (× 10³)	No. of marketings
1964	223	11,094	1.7	7,050
1966	215	11,336	1.9	9,026
1968	197	11,775	2.0	10,461
1970	180	11,148	2.2	13,642

[a] From USDA (1980).

organic wastes, all of which must be disposed of. Increasingly land application is the favored procedure for disposing of these wastes. Farm animal wastes amount to nearly 2 billion Mg annually, about two-thirds of which is concentrated in large animal feedlots or other confined animal production units where thousands of animals are reared (Table 20.9). Runoff from these feedlots is high in biodegradable organic matter and nitrates (Figure 20.11). It has been

FIGURE 20.11 Nitrate nitrogen concentrations under a dairy corral as compared to those of a nearby cropped area and a control area to which no manure or irrigation water had been added. [*From Adriano et al. (1971).*]

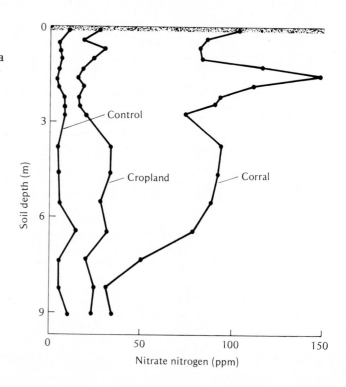

implicated in fish kills and is a source of nitrate pollution of streams and rivers. In dry climates it can become a wind-borne air pollutant. The odors of ammonia and organic compounds from some of these operations are also most offensive. The loss of ammonia is serious because nitrogen is one of the prized nutrients contained in the manure. The gaseous ammonia may be readsorbed by surrounding soils or by rainwater. Much of it eventually moves into nearby lakes, reservoirs, or streams.

Environmental considerations are leading to restrictions on the disposal of organic wastes, both urban and rural, into waterways or into the atmosphere by burning. The soil offers an alternative disposal sink in some instances which may be well worth considering. At the same time, care must be taken to prevent harmful contamination in the process.

Wastes are added to the land for two primary reasons. First, the wastes (e.g., domestic sewage and animal manures) can be added to currently cropped soils to increase their productivity (Section 19.8). Second, the soils can be used exclusively for waste disposal sites without reference to crop production. Both methods have advantages. The use of one or the other in a particular situation will depend on economic as well as biological factors.

20.10 Use of Organic Wastes for Crop Production

Organic wastes have long been used to enhance crop production, farm manure and prosperous agriculture having been closely associated for centuries. Now urban wastes such as sewage sludge are also being widely used in commercial agriculture. Materials such as shredded garbage (mostly paper) may have some potential for the future as a soil amendment.

There are obvious arguments in favor of using organic wastes for crop production. Crop yields can be increased and long-term soil productivity improved. In some situations the wastes can be surface-applied and thereby afford protection from erosion during critical seasons of the year. The spreading of manure from animal feedlots and poultry farms on nearby cropped lands is an example of effective organic-waste utilization.

Disadvantages. The primary disadvantages of using organic wastes to stimulate crop production are economic. These wastes are bulky and, compared to commercial inorganic fertilizers, low in nutrient content. Unless the land to be treated is nearby, the costs of handling and applying the wastes exceed their value. Furthermore, the nutrient composition of wastes, especially domestic sewage, is variable (Table 20.6). High carbonaceous wastes may require additional nitrogen for degradation, adding to the total cost of their use. Poultry manure may supply excess nitrogen and potassium, causing problems of both plant and animal nutrition. Magnesium-deficient forage resulting from heavy

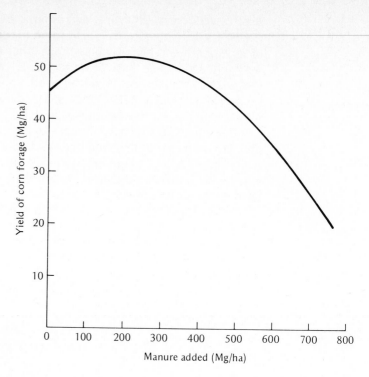

FIGURE 20.12 Effects of applying exceedingly large amounts of manure in 1969 on the forage yield of corn in 1970. Depressed yields at rates above 200 Mg/ha are probably due to high salt content in the manure. [*From Murphy et al. (1972).*]

poultry-manure applications in the southwestern United States is a case in point.

Since sewage sludge is often produced under anaerobic conditions, it may contain toxic reduced forms of heavy metals such as methylmercury. Care must be taken to be certain that application of such sewage does not occur.

The uninformed may make the assumption that crop production could effectively utilize all the organic wastes produced. Although this may be true biologically, economically it is not (Figure 20.12). Economics has prevented the widespread practice of composting of domestic wastes followed by land application. Even in Europe where this practice has been most successful in the past, only a minor part of the domestic wastes is composted.

Future Prospects. The lack of attractive alternatives for organic waste disposal has improved the economic feasibility of the use of soil for crop production. Applications of organic waste must pay for themselves through added crop yields. These materials are supplied to farmers at a low cost since adding them to the soil may well be the cheapest means of disposal. Also, if these wastes are free of contaminants, they can be used on specialty crops and home gardens where material cost is not so important. It is likely that land application of organic wastes for crop production will continue.

20.11 Soils as Organic Waste Disposal Sites

Soils have long been used as disposal "sinks" for municipal refuse. "Sanitary landfills" are widely employed to dispose of a variety of wastes from our towns and cities. These wastes include paper products, garbage, and nonbiodegradable materials such as glass and metals. These sites are often located in swampy lowland areas which eventually become built up by the dumping, creating upland areas for such uses as city parks and other facilities.

Unfortunately, sanitary landfills are sometimes not so sanitary. Leaching and runoff from these sites can contaminate both surface and ground waters. Contaminants include heavy metals as well as soluble and biodegradable organic materials. The environmental hazards associated with landfills place marked restrictions on their continued use.

Soils are being used successfully for disposal of domestic sewage sludge, food processing wastes, and selected industrial wastes. As pointed out in Section 19.8, the city of Chicago transports and applies liquid sewage sludge to a sandy soil area some 80 km from the city. Organic wastes from food processing plants are commonly applied, often by sprinkling, to nearby agricultural areas. Liquid poultry manure at rates of more than 50 Mg/ha has been applied on the surface and by incorporation into soil. Grassland vegetation is commonly permitted on the disposal sites and no attempt is made to harvest the areas.

Advantages. There are several advantages of soil "sinks" for organic waste disposal. In the first place, most soils have remarkable capacities to accommodate these wastes. The major biodegradable components are subject to breakdown by soil organisms. Inorganic components may react with soils and be effectively removed from the biological cycle. It is often more economical to use soils as waste disposal sites than to couple such disposal with crop production. Where there is no concern for the crop, rates of application are determined not by crop needs but by the capacity of the soil to utilize the wastes.

Disadvantages. The primary disadvantages of disposing of organic wastes in soils stem from exceeding the soil's capacity to accommodate the wastes. Examples include heavy metal contamination and excess nitrate leaching into ground waters. Toxicity from zinc, for instance, can prevent growth of grass or other plants used for ground cover. This in turn can lead to a reduced infiltration rate at the site, limiting the soils usefulness for further waste absorption.

The formation of nitrates from high nitrogen wastes and their removal from the soil by leaching can seriously affect ground water quality. Nitrate leaching from some feedlots or from nearby concentrated storage areas is an example (Figure 20.13). Unless these wastes are dispersed, the movement of nitrogen into drainage waters can be expected, the seriousness being greatest in humid and subhumid areas.

FIGURE 20.13 Mechanically loading manure from animals grown in a large feedlot. The manure to be utilized or disposed of from one of these large operations is equivalent to the human wastes from a small city. [*Courtesy Brookover Feed Yards, Inc., Garden City, KS.*]

Land disposal of sewage wastes may have some health implications. Such wastes may require treatment to prevent transmission of disease-causing agents to humans and other animals. This concern will continue to receive attention.

20.12   Soil Salinity

Contamination of soils with salts is one form of soil pollution primarily agricultural in origin. Furthermore, it is not a new problem. Ancient civilizations in both the New and Old Worlds crumbled because salts built up in their irrigated soils. The same principles govern the management of irrigated soils today and the same dangers exist of salt buildup and concomitant soil deterioration (see Section 6.19).

Salts accumulate in soils because more of them move into the plant rooting zone than move out. This may be due to application of salt-laden irrigation waters. And it may be caused by irrigating poorly drained soils. Salts move up from the lower horizons and concentrate in the surface soil layers.

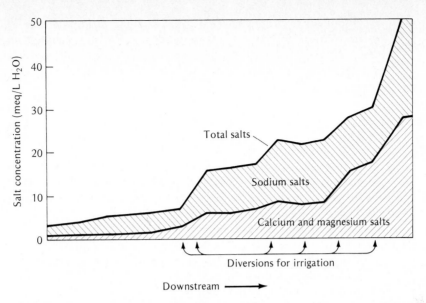

FIGURE 20.14 Changes in the amount and kind of salt in Sevier River, Utah, over a distance of 200 miles. Note the marked salt increase downstream from points of diversion of the water for irrigation purposes. [*From Thorne and Peterson* (*1967*); *copyright 1967 by the American Association for the Advancement of Science.*]

Example. The problem of salt concentration in some of the rivers of the western United States is illustrated in Figure 20.14. Removal of water from the river for irrigation purposes reduced total flow and probably increased the movement of salts back into the stream by drainage from irrigated lands. Consequently, in a distance of 330 km salt concentration in the river increased 20 times. Similar but perhaps less dramatic changes take place moving from the headwaters to the sea in other river systems. Noted examples are the Rio Grande and the Colorado systems. It is not difficult to understand why irrigation at the lower reaches of these rivers becomes hazardous.

Salts in Humid Regions. Some salt buildup occurs in rivers whose watersheds are in humid regions. The salts are found in heavily populated and industrialized areas where water is returned to streams following its domestic or industrial use. Sewage plant treatments commonly remove soluble inorganic salts only if they are known toxicants. Some sewage sludges have sufficiently high levels of salt to cause crop plant damage when the sludge is applied. When salts thus added are combined with salts leached from the watershed soils, the level may be at least as high as is found in rivers flowing through more arid areas.

The control of salinity depends almost entirely on water, its quality and management. In some areas, removal of excess water by the installation of good drainage systems is required. In local areas sulfur or gypsum applications can be used to help eliminate toxic sodium carbonate (see Section 6.19), but even in these areas, water management following chemical treatment is most vital. Water quality is determined by appropriate public policies and by individual farm practices such as those that assure good soil drainage and optimum irrigation practices.

20.13 Acid Rain[3]

Scientists in Europe and North America have called attention to marked increases in the acidity of precipitation over recent decades. Rainwater normally has a pH of about 5.6 owing to the presence of H_2CO_3 formed from the CO_2 in the atmosphere. But in or near areas with large-scale combustion of fossil fuel or with smelting of sulfide ores, the pH of precipitation may be as low as 4.0. In extreme cases of dense fog, the pH may drop to nearly 2.0, which is a potential hazard of much concern to environmentalists.

Acid precipitation, popularly called *acid rain,* is apparently due to the oxidation of nitrogen- and sulfur-containing gases that dissolve in the water vapor of the atmosphere to form nitric and sulfuric acids. Reactions such as the following are thought to occur.

$$2\,NO + O_2 \longrightarrow 2\,NO_2 \xrightarrow{H_2O} 2\,HNO_3 + HNO_2$$

Nitric   Nitrogen   Nitric   Nitrous
oxide   dioxide   acid   acid

$$2\,SO_2 + O_2 \longrightarrow 2\,SO_3 \xrightarrow{2\,H_2O} 2\,H_2SO_4$$

Sulfur   Sulfur   Sulfuric
dioxide   trioxide   acid

Figure 20.15 shows how these nitrogen and sulfur oxides can move into the atmosphere, be converted to inorganic acids, and return to the land in rain and snow. Such cycling may be responsible for lowering the pH of precipitation in the northeastern part of the United States and in eastern Canada. As illustrated in Figure 20.16, there are significant areas in the eastern United States and Canada where the pH of precipitation is 5 or less.

Effects on Soil pH. Acid rain is assumed to be responsible for increased acidity in some lakes of the Adirondacks and in other regions of the northeast. Since most fish will not tolerate pH levels below about 4.5, the resulting increased acidity of lake water is thought to have essentially eliminated most fish in some of the Adirondack lakes.

Effects of acid rain are more pronounced on the acidity of water than on soil acidity. Soils generally are sufficiently buffered to accommodate acid

[3] For a review of this subject, see NRC (1983).

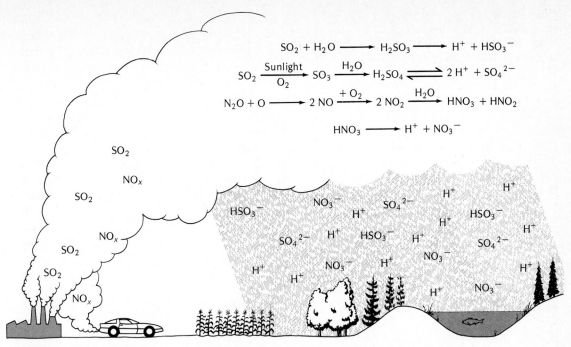

$$SO_2 + H_2O \longrightarrow H_2SO_3 \longrightarrow H^+ + HSO_3^-$$

$$SO_2 \xrightarrow[O_2]{Sunlight} SO_3 \xrightarrow{H_2O} H_2SO_4 \rightleftharpoons 2H^+ + SO_4^{2-}$$

$$N_2O + O \longrightarrow 2NO \xrightarrow{+O_2} 2NO_2 \xrightarrow{H_2O} HNO_3 + HNO_2$$

$$HNO_3 \longrightarrow H^+ + NO_3^-$$

FIGURE 20.15 Illustration of the formation of nitrogen and sulfur oxides from the combustion of fuel in sulfide ore processing and from motor vehicles. The further oxidation of these gases and their reaction with water to form sulfuric acid and nitric acid is shown. These help acidify rainwater, which falls on the soil as "acid rain." NO_x indicates a mixture of nitrogen oxides, primarily N_2O and NO. [*Modified from NRC (1983)*.]

rain with little or no increase in soil acidity on an annual basis. But continued inputs of acid rain at pHs of 4.0–4.5 would have significant effects on the pH of soils, especially those that are weakly buffered. This is also serious for soils that are already quite acid, since increased acidity could well make them even less fertile.

Alleviation of Adverse Effects. There are two obvious ways to alleviate the effects of acid rain on soils. First, the emission of sulfur and nitrogen oxides can be reduced drastically. This is a remedy already suggested, but economic and political decisions may delay its implementation. Second, the effects of the acid rain on soil pH can be overcome by adding lime. This solution may be less costly than the first, but just as difficult to achieve, since it will require commitment of resources by farmers, small and large alike. Also, many adverse effects of acid rain are being felt on forestlands and on steep mountainous areas where application of lime is difficult. Obviously, both potential solutions must receive attention.

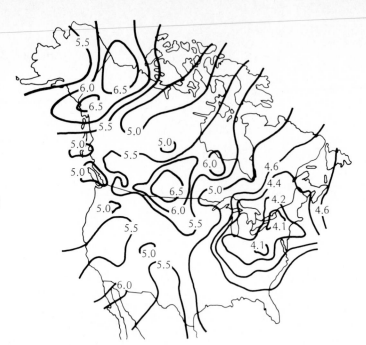

FIGURE 20.16 Map showing concentrations (pH) of acid rain in North America. Note the heaviest acid concentrations (lowest pH values) are found in the industrial northeastern areas of the United States. [*Redrawn from De Young (1982); data from Canadian and American government agencies. Reprinted with permission,* High Technology Illustrated Magazine, *Sept./Oct. 1982. Copyright © 1982 by High Technology/Technology Publishing Corporation, 38 Commercial Wharf, Boston, MA 02110.*]

20.14   Radionuclides in Soil

Nuclear fission in connection with atomic weapons testing provides another source of soil contamination. To the naturally occurring radionuclides in soil (e.g., ^{40}K, ^{87}Rb, and ^{14}C), a number of fission products have been added. However, only two of these are sufficiently long-lived to be of significance in soils: strontium-90 (half-life = 28 yr) and cesium-137 (half-life = 30 years). The average levels of these nuclides in soil in the United States are about 388 millicuries/km² (150 mC/mi²) for ^{90}Sr and 620 mC/km² (240 mC/mi²) for ^{137}Cs. A comparable figure for the naturally occurring ^{40}K is 51,800 mC/km² (20,000 mC/mi²). It is obvious that normal soil levels of the fission radionuclides are not high enough to be hazardous. Even during periods of weapons testing, the soil did not contribute significantly to the level of these nuclides in plants. Fallout from the atmosphere directly on the vegetation was the primary source. Consequently, only in the event of a catastrophic supply of fission products could toxic soil levels of ^{90}Sr and ^{137}Cs be expected. Fortunately, considerable research has been accomplished on the soil behavior and nutrient uptake characteristics of these two nuclides.

Strontium-90.   Strontium-90 behaves in soil much the same as does calcium, to which it is closely related chemically. It enters soil from the atmosphere in soluble forms and is quickly adsorbed by the colloidal fraction, both organic and inorganic. It undergoes cation exchange and is available to plants much as is calcium. The possibility that strontium is involved in the same plant reactions as calcium probably accounts for the fact that high soil calcium tends to decrease the uptake of ^{90}Sr.

Cesium-137. Although chemically related to potassium, cesium tends to be less readily available in many soils. This is apparently because ^{137}Cs is firmly fixed by vermiculite and related interstratified minerals. The fixed nuclide is nonexchangeable, much as is fixed potassium in some interlayers of clay. Plant uptake of ^{137}Cs from such soils is very limited. Where vermiculite and related clays are absent, as in some tropical soils, ^{137}Cs uptake is more rapid. In any case, the soil tends to be a damper on the movement of ^{137}Cs into the food chain of animals and man.

Radioactive Wastes.[4] Aside from radionuclides added to soils as a result of weapons testing, low-level radioactive waste materials are sometimes handled by burying them in soils. Even though the materials may be solids when placed in shallow-land burial pits, some dissolution and subsequent movement in the soil are possible. Plutonium, uranium, americium, neptunium, curium, and cesium are among the elements whose nuclides occur.

Nuclides in wastes vary greatly in water solubility, uranium compounds being quite soluble, compounds of plutonium and americium being relatively insoluble, and cesium compounds intermediate in solubility. Cesium, a positively charged ion, is adsorbed by soil colloids. Uranium is thought to occur as a negatively charged complex that is also adsorbed by soil. The charge on plutonium and americium appears to vary depending on the nature of the complexes these elements form in the soil.

There is considerable variability in the actual uptake by plants of these nuclides from soils, depending on properties such as pH and organic matter content. The uptake from soils by plants is generally lowest for plutonium, highest for neptunium, and intermediate for americium and curium. Crop fruits and seeds are generally much lower in these nuclides than are leaves, suggesting that human foods may be less contaminated by nuclides than forage crops.

Since soils are being used as burial sites for low-level radioactive wastes, care should be exerted to be certain the soil properties are such as to discourage leaching or significant plant uptake of the chemicals. Monitoring of selected sites will likely be needed to assure minimum transfer of the nuclides to other parts of the environment.

20.15 Three Conclusions

Three major conclusions may be drawn about soils in relation to environmental quality. First, since soils are valuable resources, they should be protected from environmental contamination, especially that which does permanent damage.

---

[4] This summary is based on papers on this subject in *Soil Science,* **132** (July 1981).

Second, because of their vastness and remarkable capacities to absorb, bind, and break down added materials, soils offer promising mechanisms for disposal and utilization of many wastes that otherwise may contaminate the environment. Third, products of soil reactions can be toxic to man and other animals if they move from the soil into the air and particularly into water.

To gain a better understanding of how soils might be used and yet protected in waste management efforts, soil scientists must devote a fair share of their research effort to environmental quality problems.

REFERENCES

Adriano, D. C., et al. 1971. "Nitrate and salt in soil and ground waters from land disposal of dairy manures," *Soil Sci. Soc. Amer. Proc.,* **35:**759–62.

Allaway, W. H. 1970. "Agronomic controls over the environmental cycling of trace elements," *Adv. in Agron.,* **20:**235–274.

Beyer, W. N., R. L. Chaney, and B. M. Mulhern. 1982. "Heavy metal concentration in earthworms from soil amended with sewage sludge," *J. Envir. Qual.,* **11:**381–385.

Bowen, H. J. M. 1966. *Trace Elements in Biochemistry* (New York: Academic Press).

De Young, H. G. 1982. "Acid rain regulators shift into low gear," *High Technology,* **2:**82–86.

Edwards, C. A. 1978. "Pesticides and the micro-fauna of soil and water," in I. R. Hill and S. J. Wright (Eds.), *Pesticide Microbiology* (London: Academic Press, pp. 603–622.

Fox, J. L. 1983. "Soil microbes pose problems for pesticides," *Science,* **221:**1029–1036.

Furr, A. K., A. W. Lawrence, S. S. C. Tong, M. C. Grandolfo, R. A. Hofstader, C. A. Bache, W. H. Gutenmann, and D. J. Lisk. 1976. "Multielement and chlorinated hydrocarbon analysis of municipal sewage sludges of American cities," *Envir. Sci. and Tech.,* **10:**683–687.

Guenzi, W. D. (Ed.). 1974. *Pesticides in Soil and Water* (Madison, WI: Soil Sci. Soc. Amer.).

Harris, C. I., and G. F. Warren. 1964. "Adsorption and desorption of herbicides by soil," *Weeds,* **12:**120–26.

Helling, C. S., P. C. Kearney, and M. Alexander. 1971. "Behavior of pesticides in soils," *Adv. in Agron.,* **23:**147–240.

Hill, I. R., and S. J. L. Wright (Eds.). 1978. *Pesticide Microbiology: Microbiological Aspects of Pesticide Behavior in the Environment* (New York: Academic Press).

Jacobs, L. W., et al. 1970. "Arsenic residue toxicity to vegetable crops grown on Plainfield sand," *Agron. J.,* **62:**588–591.

Kelling, K. A., D. R. Keeney, L. M. Walsh, and J. A. Ryan. 1977. "A field study of the agricultural use of sewage sludge: III. Effect on uptake and extractability of sludge-borne metals," *J. Envir. Qal.,* **6:**352–358.

Lagerwerff, J. V., and A. W. Specht. 1970. "Contamination of roadside soil and vegetation with cadmium, nickel, lead, and zinc," *Envir. Sci. and Tech.,* **4:**583–586.

Lisk, D. J. 1972. "Trace metals in soils, plants, and animals," *Adv. in Agron.,* **24:**267–325.

Murphy, L. S., et al. 1972. "Effects of solid beef feedlot wastes on soil conditions and plant growth," in *Waste Management Research Proceedings,* Cornell Agricultural Waste Management Conference, Ithaca, NY.

NRC. 1983. *Acid Deposition: Atmospheric Processes in Eastern North America,* a National Research Council Report (Washington, DC: National Academy of Sciences Press).

Sanders, H. J. 1981. "Herbicides," *Chem. and Eng. News,* Aug. 3, pp. 20–35.

Thorne, D. W., and H. B. Peterson. 1967. "Salinity in United States waters," in N. C. Brady (Ed.), *Agriculture and the Quality of Our Environment,* (Washington, DC: Amer. Assoc. for the Advancement of Science).

Thompson, A. R., and C. A. Edwards. 1974. "Effects of pesticides on nontarget invetebrates in freshwater and soil," in W. D. Guenzi (Ed.), *Pesticides in Soil and Water* (Madison, WI: Soil Sci. Soc. Amer., pp. 341–386.

USDA. 1980. *Agricultural Statistics 1980* (Washington, DC: U.S. Government Printing Office).

Soils and the World's Food Supply

Hunger is not new to the world. It has always been a threat to man's survival. At some place on earth through the centuries scarcity of food has brought misery, disease, and even death to man. But never in recorded history has the threat of starvation been greater than it is today. This threat is not due to the reduced capacity of the world to supply food. Indeed, this capacity is greater today than it has ever been and is continuing to grow at a reasonable rate. The problem lies in the even more rapid rate at which world population is increasing. World food production per person is at best holding its own. In selected areas it is declining.

21.1 Expansion of World Population[1]

Science is largely responsible for the marked expansion in world population growth, and especially that which has occurred in the developing nations. Until the near midpoint of the twentieth century, high birth rates in most of South America, Africa, and Asia were largely negated by equally high death rates. High infant mortality, poor health facilities, inadequate medical personnel, and disease-spreading insects took their toll. Population expansion was held in check.

During the past few decades advances in medical science and their application throughout the world have drastically changed this situation (Figure 21.1). Death rates have been drastically reduced, especially among the young. Pesticides have helped check mosquitoes and other disease-carrying pests. Medical services heretofore unheard of in remote areas of the developing nations have been initiated. The result is unprecedented population growth. The population is doubling every 18–27 years in the developing areas, where two thirds of the world's population now live. Experts predict the world's population will be between 6 and 7 billion by the year 2000—nearly double that of today. Furthermore, *more than 90% of this increase will occur in developing nations,* where food supplies are already critical and where the technology for increased food production is wholly inadequate. It is no wonder that the world food supply is considered by some to be mankind's most serious problem.

21.2 Factors Influencing World Food Supplies

The ability of a nation to produce food is determined by a multitude of variables. These include a complex of social, economic, and political factors, most of which affect the farmer's incentive to produce and the supply of production

[1] For world population projections see United Nations (1982).

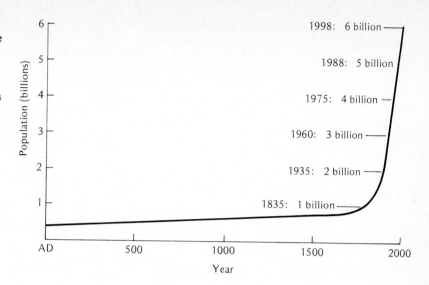

FIGURE 21.1 From the beginning of the human race until 1960, the world's population increased to 3 billion. Population estimates indicate that less than 40 years will be needed to provide the second 3 billion. Furthermore, most of the population growth will occur in countries already struggling to feed themselves. Projections for 1987 and 1997 from U.S. Bureau of the Census.

1998: 6 billion
1988: 5 billion
1975: 4 billion
1960: 3 billion
1935: 2 billion
1835: 1 billion

inputs. Also included are a number of physical and biological factors, such as the following.

1. The natural resources available, especially soil and water.
2. Available technology, including the knowledge of proper management of plants, animals, and soils.
3. Improved plant varieties and animal breeds which respond to proper management.
4. Supplies of production inputs such as fertilizers, insecticides, and irrigation water.

Each of these factors is affected by the area and quality of soils—their natural productivity and response to management. There is good reason to place satisfactory soil properties high on the list of requisites for an adequate world food supply.

21.3 The World's Land Resources

There is a total of more than 13 billion ha of land on the major continents (Table 21.1). However, most of it is not suited for cultivation. About half of it is completely nonarable. It is mountainous, too cold or too steep for tillage; it may be swampland; or it is desert country, too dry for any but the sparsest vegetation.

About one-fourth of the land area supports enough vegetation to provide grazing for animals, but for various reasons cannot be cultivated. This leaves only about 25% of the land with the physical potential for cultivation (Table

TABLE 21.1 World Land Area in Different Climatic Zones[a]

| Climatic zone | Area (10^9 ha) | | | |
	Potentially arable	Grazing	Nonarable	Total
Polar and subpolar	0	0	0.56	0.56
Cold temperate boreal	0.05	0.19	1.73	1.97
Cool temperate	0.91	1.00	1.00	2.91
Warm temperate subtropical	0.55	0.84	1.37	2.76
Tropical	1.67	1.63	1.65	4.95
Total	3.18	3.65	6.31	13.15

[a] From The President's Science Advisory Committee Panel on World Food Supply (1967), Vol. II, p. 23.

21.1). And only half of this potentially arable land is actually under cultivation. It is obvious that the kind of soils and their response to management may eventually hold the key to adequate food production, at least in some areas.

Continental Differences. Data in Table 21.2 suggest the role that soils may play in helping to meet the world's food requirements. While the *total* potentially arable land is more than double that being cultivated today, there is

TABLE 21.2 Population and Cultivated Land on Each Continent, Compared with Potentially Arable Land

| Continent | Population in 1979[a] (millions) | Area (10^6 ha) | | | Cultivated land per person (ha) | Potentially arable land in cultivation (%) |
		Total[b]	Potentially arable[b]	Cultivated[c]		
Africa	457	3,019	733	209	0.46	29
Asia	2,498	2,735	627	510	0.20	81
Australia and New Zealand	18	822	154	18	1.00	12
Europe	483	478	174	142	0.29	82
North America	363	2,108	465	272	0.75	58
South America	352	1,752	680	104	0.30	15
U.S.S.R.	264	2,234	356	232	0.89	65
Total	4,435	13,149	3,189	1,487	0.66	47

[a] From Population Reference Bureau (1979).
[b] From The President's Science Advisory Committee Panel on World Food supply (1967), Vol. II, p. 434.
[c] From FAO (1978).

great variation from continent to continent. In Asia and Europe where population pressures have been strong for years, most of the potentially arable land is under cultivation. In contrast, only 12% of the potentially arable land is cultivated in Australia and New Zealand. Comparable figures are 15% for South America and 29% for Africa. In these last four areas the physical potential for greater utilization of arable land is indeed great.

It is unfortunate that there is not better distribution of arable land in relation to population densities. The area of cultivated land *per person* is high in North America, the U.S.S.R., Australia, and New Zealand. It is low in Asia, Europe, and South America and not much higher in Africa. This does not present a serious problem in Europe or the more economically developed parts of Asia, They can readily purchase food from the countries with excess supply. Only the transportation, trade, and marketing problems must be overcome.

In developing countries of Asia, Africa, and Latin America the situation is much more critical. Their populations are increasing far more rapidly than their food production. Countries that formerly exported food crops now must import them. And their national economic growth rate is too slow to provide the resources to pay for the needed food. They must either be provided with food aid by their more fortunate neighbors or must increase dramatically their capacity to produce food.

Choices of Action.   There are two routes that nations may follow to utilize land to increase their food production: (a) they may clear and cultivate arable land that has heretofore not been tilled, or (b) they may intensify production on lands already under cultivation. Some nations, notably those in Europe and Asia, have only the latter choice. They have little opportunity to expand land under cultivation since most of their arable land is already in use. Only by increasing annual yields per hectare can they produce more food.

In areas outside Asia and Europe, the physical potential for increasing land under cultivation is great. For example, in Africa and South America where current land utilization per person is not high, there are more than 1 billion ha of arable land that are not now being cultivated. Unfortunately, much of this land is inaccessible to modern transportation. The cost of clearing the land, transporting the fertilizer and other needed inputs, and distributing the food produced from it is high. Also large areas on those continents have humid, tropical climates and tropical soils—the optimum management of which man has yet to learn.

For these reasons, in most areas of the world, intensification of land already under cultivation is the preferred immediate method of increasing food production. In time, however, as economic development, transportation, and knowledge of soil management progress, expansion of land under cultivation is almost certain to occur. Attention will now be directed to general kinds and acreage of soils found where food production is most critical.

21.4 Potential of Broad Soil Groups

In Table 21.3 are shown the areas dominated by different broad major soil groupings. Note that shallow soils and sand are most prevalent (3.5 billion ha) followed closely by a variety of Oxisols in tropical areas. There are vast areas of desert soils and associated dry areas (Aridisols). Spodosols and related soils make up the next most expansive grouping.

Although the total area of different soil groupings is interesting, the potential utilization is of much greater practical importance. For example, 80% of the 3.5 billion ha of shallow and sandy soils is considered nonarable, not fit even for grazing. In contrast, only about 9% of the Mollisols and related soils is classed as nonarable.

More than half of the dark-colored, base-rich soils such as the Mollisols are potentially arable. A similar proportion of Alluvial soils is so classed. The productivity of the dark-colored base-rich soils and especially the alluvial soils (mostly Entisols), has long been recognized. Alluvial soils are being utilized throughout the world, but are especially important in lowland rice culture. Their location with respect to river and ground water makes irrigation relatively easy, and their productivity is generally high. The Mollisols and related soils are the bulwark of agriculture in North America, the U.S.S.R., and parts of South America.

The arability of areas dominated by the light-colored, base-rich soils of dry areas (Aridisols) is determined to a considerable extent by the availability

TABLE 21.3   Areas of Broad Soil Groups in the Major Continents and the Percentages That Are Arable, Nonarable But Suitable for Grazing, and Nonarable[a]

Broad soil grouping	Total area (10⁶ ha)	Potentially arable (%)	Suitable for grazing (%)	Nonarable (%)
Light-colored, base-rich desert (Aridisols)	2.13	20	43	37
Dark-colored, base-rich (mostly Mollisols)	1.25	53	36	11
Moderately weathered, leached (Alfisols, some Inceptisols)	0.32	37	48	15
Highly weathered, leached				
Spodosols	1.96	16	26	58
Ultisols	0.39	33	51	16
Oxisols	2.50	42	29	29
Shallow soils and sands (mostly Inceptisols)	3.49	4	16	80
Alluvial (mostly Inceptisols)	0.59	54	29	17
Tundra (varies)	0.52	6	0	100
Total	13.15	24	28	48

[a] Modified from The President's Science Advisory Committee Panel on World Food Supply, Vol. II, p. 423.

of irrigation water. The soils are often quite fertile and will respond to management if supplemental water can be applied. The remarkable progress that has been made in expansion of irrigation in India and Pakistan illustrates the productivity potential of these soils. About 500 million ha are physically located where they can be irrigated if water can be stored and distributed economically.

The greatest potential for increasing land under cultivation is in tropical areas dominated by Oxisols and Ultisols. Scientists estimate that 40% of the more than 2.5 billion ha of these soils is potentially arable (Table 21.1), and another 0.7 billion ha is suitable for grazing. The potential here for increasing food production is enormous.

Potential by Continents. Table 21.4 gives an estimate of the potentially arable land by soil grouping and by continent. Again one is impressed with the total area of arable Oxisols and with the sizable arable areas of arid lands (Aridisols), Mollisols and related soils, and alluvial soils (mostly Entisols).

Europe and Asia are already utilizing most of their arable land. North America and Australia are excess-food-producing areas. For these reasons, attention will be focused on Africa and South America, where food shortages are occurring and where the soil resources are far underutilized.

Africa has both potential and limitations in its soil resources. Much of

TABLE 21.4   Estimates of Area of Potentially Arable Land by Soil Groups and Continents[a]

Broad soil grouping	Arable land[b] (10^6 ha)							
	Africa	Asia	Australia	Europe	North America	South America	U.S.S.R.	Total
Light-colored, base-rich desert (Aridisols)	137	101	45	12	20	36	85	437
Dark-colored base-rich (mostly Mollisols)	57	77	69	28	198	85	162	676
Moderately weathered, leached (Alfisols, some Inceptisols)	8	69	8	16	8	8	—	117
Highly weathered, leached								
Spodosols	—	20	1	113	101	—	89	325
Ultisols	8	36	4	—	77	4	—	129
Oxisols	417	101	12	—	16	514	—	1060
Shallow soils and sands (mostly Inceptisols)	57	32	8	8	16	16	8	145
Alluvials	53	194	0	4	32	32	12	320
Total	737	631	147	182	469	688	356	3210

[a] From The President's Science Advisory Committee Panel on World Food Supply (1967), Vol. II, p. 430. See Table 21.3 for approximate classification of soils according to *Soil Taxonomy*.

[b] Dash indicates soil group not present on the continent; 0 indicates essentially no potentially arable land in this group on the continent.

Region	Percent of total land area With limitation					
	Drought	Mineral stress[b]	Shallow depth	Water excess	Perma-frost	No serious limitation
North America	20	22	10	10	16	22
Central America	32	16	17	10	—	25
South America	17	47	11	10	—	15
Europe	8	33	12	8	3	36
Africa	44	18	13	9	—	16
South Asia	43	5	23	11	—	18
North and Central Asia	17	9	38	13	13	10
Southeast Asia	2	59	6	19	—	14
Australia	55	6	8	16	—	15
World	28	23	22	10	6	11

[a] Data compiled from FAO/UNESCO soil map of the world by Dent (1980).
[b] Nutritional deficiencies or toxicities related to chemical composition or mode of origin.

its land is in arid and semiarid areas where soil moisture is limiting and irrigation expansion is not promising. In the more humid regions of Africa large areas of Oxisols are present. Although these soils are potentially quite productive, special treatment is required. Similarly, large areas of Oxisols in South America have significant potential but definite limitations.

Table 21.5 illustrates the limitations of soils for agriculture in major geographic areas of the world. This table reconfirms the information suggested by the soil groupings in Table 21.4. A deficiency of soil moisture is probably the most significant worldwide constraint. This constraint is most notable in Australia, Africa, and South Asia (India, Pakistan).

Mineral deficiencies provide serious limitations in all areas but are most severe in Southeast Asia and South America. Shallow soils are problems in North, Central, and South Asia. Excess water is a more serious problem in Southeast Asia than in any other region. The heavy monsoon rains in this area account for this constraint.

Europe has the highest proportion of soils without serious limitation (36%). The comparable figure for Central and North America is more than 20%.

21.5   Problems and Opportunities in the Tropics

One cannot help but wonder why soils of the humid tropics have not been more widely exploited. They seemingly have many advantages over their temperate zone counterparts. The crop growing season is commonly year round.

TABLE 21.6 Distribution of General Soil Groups Found in the Humid Tropics Compared to Other Non-dryland Areas[a]

Note the high proportion of acid-infertile soils, especially in South America and Africa.

General soil grouping	Humid tropical soil areas (%)				Other dryland soil areas (%)
	Central and South America	Africa	Asia	World	
Acid, infertile soils (Oxisols and Ultisols)	82	56	38	63	10
Moderately fertile, well-drained soils (Alfisols, Vertisols, Mollisols, Andepts, Tropepts, Fluvents)	7	12	33	15	32
Poorly drained soils (Aquepts)	6	12	6	8	10
Very infertile sandy soils (Psamments, Spodosols)	2	16	6	7	13
Others	3	4	16	17	35[b]
Total	100	100	100	100	100

[a] From National Research Council (1982) and USDA data.
[b] Mostly mountainous areas.

In many areas ample moisture is available throughout the growing season. And some of the soils have physical characteristics far superior to the soils of the temperate zones. The hydrous oxide and kaolinitic types of clays that dominate in these areas permit cultivation under very high rainfall conditions.

A comparison of the distribution of some major soil groupings in the humid tropics and in other non-dryland areas of the world is shown in Table 21.6. The high proportion of acid infertile soils (Oxisols and Ultisols), especially in South America and Africa, helps explain why soils of these continents have not been more extensively utilized. These soils are low in nutrients and often are far from sources of limestone or fertilizers, making it difficult to increase their fertility.

Limiting Factors. The primary factor limiting more effective utilization of tropical soils is their infertility. In Table 21.7 are shown the major soil constraints on crop productivity in the Amazon region of South America. Note that 90% of the soils are deficient in phosphorus and more than two-thirds have toxic levels of aluminum due to their high acidity. The small reserves of essential elements in these soils are depleted quickly when cropping replaces natural vegetation.

Other reasons for failure to utilize some tropical areas more fully are economic and social in nature. Wet tropical areas are not always pleasant for humans or animals. The hot humid climate is uncomfortable, and it encourages diseases and pests that affect plants and animals, including humans.

TABLE 21.7 Main Soil Constraints in the Amazon Basin Under Native Vegetation[a]

Soil constraint	Millions of hectares	Percent of Amazon Basin
Phosphorus deficiency	436	90
Aluminum toxicity	315	73
Low potassium reserves	242	56
Poor drainage and flooding	116	24
High phosphorus fixation	77	16
Low cation exchange capacity	64	15
High erodibility	39	8
No major limitations	32	6
Steep slopes (>30%)	30	6
Laterite hazards	21	4

[a] After Cochrane and Sanchez (1981).

Most of those factors that favor domesticated crop and animal production also support competitors in the environment.

Another factor is the absence or inadequacy of quality transportation. There are often no highways or railroads connecting farm areas to the cities and towns where the customers are, and from which fertilizers, pesticides, and other inputs come. Agriculture will not likely succeed under these conditions.

Coupled with transportation are all other aspects of economic and social development. The package of inputs that has been so essential in the more developed areas of the world is no less essential for agriculture in the developing countries of the tropics. Capital is needed to clear the land, build the roads, construct the irrigation dams and canals, and to build the fertilizer and pesticide plants and distributing systems. Economic development in other segments of the economy is required if money is to be available to purchase the farmer's produce. Marketing systems must be developed to move perishable foods from farms to cities and to distribute them to hungry people. Last but not least, there must be research and education related specifically to the problems of the tropics.

Special Problems of Tropical Soils. Special problems are associated with soils of the tropics. In the first place, all too little is known about them and their management. Research on tropical soils in relation to their properties and potential for crop production is insignificant compared to that on temperate region soils. Knowledge of their characteristics makes possible identifying only the broadest categories of classification. More intensive study will undoubtedly show many different kinds of soils where now only a few can be identified.

The little we have learned about some tropical soils is encouraging. For example, some of the soils (Oxisols) of Hawaii and of the Philippines are excellent for pineapple and sugar cane production. They respond well to mod-

FIGURE 21.2 Exposed laterite area in central India. The pavement-like surface is barren and will not permit crop production. Sizable acreages of such Latosols (Oxisols) with laterite layers are found in India, Africa, South America, and Australia.

ern management and mechanization. In contrast, modern mechanized farming was a dramatic failure for the "Groundnut Scheme" carried out by the British in Tanganyika following World War II. In that case, apparently, exposing the cleared soil to tropical rains resulted in catastrophic erosion. Knowledge of the nature of the soil might well have prevented much of the project's $100 million loss.

There is a good likelihood that more intensive study will identify complexities among tropical soils similar to those known for temperate regions. We already know of much variability among soils of tropical areas. Some are deep and friable, easily manipulated and tilled. At the other extreme are the lateritic soils, which, when denuded of their upper horizon, expose layers that harden into a surface resembling a pavement (Figure 21.2). Such soils are essentially worthless from an agricultural point of vew.

The chemical characteristics of Oxisols differ drastically from those of soils of temperate regions. The high hydrous oxide content dictates enormous phosphate-fixing capacities. The low cation exchange capacities and heavy rainfall result in removal of not only macronutrients but micronutrients as well. Indeed, the level of technology needed to manage Oxisols is fully as high as that required for temperate zone soils.

Plantation Management Systems in the Tropics. The plantation system of agriculture has been successful in raising crops such as bananas, sugar cane, pineapples, rubber, coffee, and cacao. These crops are commonly grown for export and require considerable skill and financial inputs for their production. The plantation system generally imports the best available technology from developed countries and sometimes has associated with it sizable research staffs to gain new knowledge for improved technology. Although it has been generally successful in producing and marketing crops and animals, it provides little benefit for the indigenous producer. Consequently, social and political problems have plagued this system.

Shifting Cultivation. At the opposite extreme for the plantation approach are indigenous systems that require little from the outside and that have evolved mostly by trial and error of the native cultivators. One of the most widespread of these systems—that of *shifting cultivation*—will be described briefly to illustrate what the natives have learned from centuries of experience (Figure 21.3).

FIGURE 21.3 Nigerian farmer clearing land of natural vegetation in preparation for the planting of crops. The cleared underbrush and trees will be burned, and the ash derived therefrom will support crop production for a few years. The farmer will then move to another site, where the steps will be repeated. [*Courtesy International Institute of Tropical Agriculture, Ibadan, Nigeria.*]

While there are variations in the practice of shifting cultivation, in general it involves three major steps.

1. The cutting and burning of trees or other native plants, leaving their ashes on the soil. Sometimes only the vegetation in the immediate area is burned. In other cases this is supplemented with plants brought in from nearby areas.
2. Growing crops on the cleared area for a period of 1–5 yr, thereby utilizing nutrients left from the clearing and burning of native plants. Vegetation from outside the area may be brought in and burned between plantings.
3. Fallowing the area for a period of 5–12 yr, thereby permitting regrowth of the native trees and other plants that "rejuvenate" the soil. Nutrients are accumulated in the native plants and some, such as nitrogen, are released to the soil. The cutting and burning are repeated, and the cycle starts again.

Shifting cultivation is primarily a system of nutrient conservation, accumulation, and recycling. The native plants absorb available nutrients from the soil and from the atmosphere. Some are nitrogen fixers. Others are deep rooted and bring nutrients from lower horizons to the surface. All help protect the soil from the devastating effects of rain and sunshine. Also, the cropped area is generally small in size and is surrounded by native vegetation. This reduces the chances of gully formation or severe sheet erosion from runoff water.

There are other benefits of the system other than those relating to nutrients and erosion control. The burning is likely to destroy some weed seeds and even some unwanted insects and disease organisms. The short period of actual cropping (1–5 yr out of 10–20 yr) discourages the buildup of weeds, insects, and diseases harmful to the cultivated crops.

Although shifting cultivation seems primitive, it deserves careful study. In some tropical areas it is more successful than the seemingly more efficient temperate zone systems. Experimentation may permit improvement and alteration of the shifting cultivation system. Perhaps the nutrients accumulated in the native plants can be supplemented with compost or fertilizers. Seeding the fallowed area with plants selected for their beneficial effects rather than allowing natural invasion may also be a forward step.

Some soils in the tropics must have continuous vegetative cover to remain productive. If they dry out, and especially if erosion removes the surface layers, the doughy laterite layers beneath harden irreversibly, making plant growth impossible. This turn of events may be prevented by planting the crop desired among native or other crop plants without completely removing the latter. Since the plants involved are in most cases trees, this system has been termed the *mixed tree crop* system. The desired crop or crops are introduced by removing some of the existing plants and replacing them with crop plants. In time, a given area may be planted entirely to a number of crop plants. The essential feature of the system, however, is that at no time should the soil be free of vegetation. As primitive as this system may appear, up to now science has not been able to develop more successful alternatives for the management of laterite-containing soils (Figure 21.2).

Recent Soil Deterioration. Unparalleled increases in human populations have placed great strain on the traditional shifting cultivation systems in some areas of the tropics. The more than 250 million people who depend upon shifting cultivation for most of their food represent a dramatic increase in numbers during the past two decades. To provide food for their growing families, farmers have been forced to shorten the period of fallow, to recrop a given area after only 3–6 years fallow compared to 10 or more years in the past. As a conse-

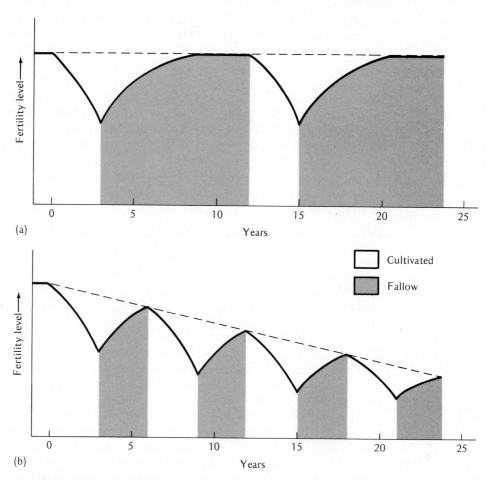

FIGURE 21.4 Changes in fertility levels under two shifting cultivation systems. (a) In the system with a long fallow period the natural vegetation is able to help "regenerate" the fertility level after cropping. (b) With a shorter fallow period so common in areas of high population pressure insufficient regeneration time is available and the fertility level declines rapidly.

quence insufficient time is allowed for soil rejuvenation between cropping periods. This leads to lower crop yields and greater exposure to erosion. The mechanism by which shortened fallow periods have reduced soil fertility is illustrated in Figure 21.4. Today's shorter fallow period is one of the reasons for a decline in per capita food production in Africa during the past decade.

Potential of Tropical Agriculture. In spite of our inadequate knowledge of tropical agriculture in general, one cannot help but be optimistic about the future of agriculture in the tropics. The basic requirements for maximum year-round production appear to be higher in the humid tropics than anywhere else. Total annual solar radiation and warm to hot climates provide unmatched photosynthetic potential. The unused soil resources are plentiful. There is an increasing tendency to grow several crops a year in a given land area of the tropics. This is of special importance to food production since the practice can be followed by small landholders who almost invariably produce some food crops.

Already researchers have developed new crop varieties especially adapted to the tropics. Adaptive research is identifying means of controlling pests, and fertilizer usage is becoming more common in tropical areas. The potential for food production there is enormous. Its realization depends only on people's ability to exploit this potential.

21.6   Requisites for the Future

The world's ability to feed itself depends upon many factors, not the least of which is improved agricultural technology in the developing nations of the world. This technology is in turn dependent largely on science and more specifically on research and education. Furthermore, the research and education must have direct relevance to the developing countries and not be a mere transplant of what is available in the more developed nations. Too often the mistake is made of assuming that the technology of western Europe or the United States can be transferred directly to the underdeveloped countries. Disastrous failures have shown the fallacy of this concept.

Package Approach. The most important feature is that improved technologies must provide a package of all the inputs, cultural techniques, incentives, etc., upon which a successful food production system depends. The law of limiting factors described briefly in Chapter 1 is applicable here. Each of the dozens of economic, social, political, and biological factors that affect a successful agricultural system must be considered.

Assuming for the moment that the social, economic, and political factors can be made reasonably favorable, those components of the biological package that have some bearing on soil utilization are now considered.

Crop Varieties. New crop varieties adapted to conditions in the developing areas along with methods of pest control are among the first elements of the package. For example, the dwarf wheats and rices, which are highly responsive to fertilizer applications, set a new potential for yields over wide geographic areas. The average yield of wheat in Mexico doubled as a result of introducing these new varieties, especially in irrigated areas. The production of wheat in India has tripled since the mid-1960s. Similarly, rice production in Indonesia has doubled in the past dozen years. Luckily, these rice and wheat varieties are adapted to these and other food-deficit countries. Disease and insect problems are being attacked through the development of resistant varieties.

Irrigation and Drainage. Water supply is a critical factor in crop production in most areas of the world. In some cases, only supplemental moisture is needed to meet temporary deficits or to lengthen the growing season. At other locations, irrigation must be looked to for the bulk of the growing season moisture. Such locations may be in a year-round dry climatic area that is traversed by rivers flowing from areas of higher precipitation. Or there may be a dry period during part of an otherwise wet year. There is a growing recognition that, coupled with irrigation, soil drainage systems are often essential.

Remarkable progress has been made in expanding irrigation. Worldwide, the area under irrigation in 1968 was about four times that in 1900, and it has grown at an annual rate of about 22% since 1960. In India, a country with the most serious of food problems, 39 million ha is under irrigation. This area is being expanded annually, not only with major dam and reservoir projects but also with "tube" wells that utilize groundwater and require only the simplest of distribution systems. Nevertheless, the potentially irrigable land in India is about double that currently irrigated. Similar potentials and plans for their realization exist for parts of South America, but in Africa south of the Sahara Desert wide-scale irrigation may not be economic.

Fertilizer. To move from yields common in subsistence agriculture to those dictated by today's food requirements, dramatic increase in supplies of fertilizer nutrients has occurred. From the early 1960s to 1979 fertilizer use rose eightfold in the developing countries. This has resulted in a demand for information on soil characteristics to determine the kinds of fertilizers that are needed. In highly leached areas evaluations are being made of micronutrient deficiencies as well as those of nitrogen, phosphorus, potassium, and lime. And attention is being given to fertilizer that resists rapid reaction with the soil or volatilization by microbial action.

While the use of manufactured fertilizers is being encouraged, economic considerations dictate the search for alternative sources, especially of nitrogen. Native and improved legumes can and should be used. And animal manures will help supplement the manufactured fertilizers.

Soil Management. In some areas there are serious soil management problems which need attention. For example, the "black cotton" soils of India and of

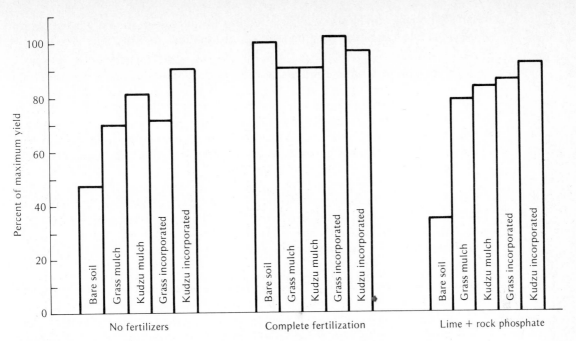

FIGURE 21.5 The effect of crop residues with and without fertilizers and lime on the yield of five consecutive crops grown in the Amazon Basin. Yields are expressed as percent of that obtained from complete fertilization on bare soil. [*From Wade and Sanchez (1983); used with permission of the American Society of Agronomy.*]

the Sudan (classified as Vertisols) are heavy textured and contain a high proportion of 2:1-type silicate clays. These soils are sticky when wet and hard when dry. The primitive implements and small draft animals used to plow and cultivate them do not permit timely soil manipulation. This means that much of the potential productive capacity of these soils is lost. The farmers are able to produce only a dry season crop each year since they cannot manage cultural operations during the monsoon (wet) season. Simple methods of land ridging without heavy equipment are being developed to permit crop production during the wet season, thereby doubling the number of crops grown each year. Methods of incorporating organic matter into the surface of these soils are also being sought. And tractors and other machines are being used to permit tillage of these soils.

Figure 21.5 demonstrates how organic residues can help increase crop yields. Even when used without fertilizer, the residues were effective either as a mulch or as incorporated organic matter.

Agricultural development activities have emphasized the critical need for the characterization of soils. For example, the salinity and alkalinity status can well determine the probable success of an irrigation project. Nutrient deficiencies can be identified, as can the potential for erosion and drainage problems. The time and effort being devoted to soil characterization are insignificant compared to the needs for this kind of information.

Soil Surveys. Reliable soil survey information is inadequate or unavailable in most of the developing areas. In part this is due to lack of information of the soil characteristics upon which a classification scheme can be based. More frequently it is due to ignorance of the significance and value of the soil survey. Planners sometimes look upon soils as soils, without regard to the vast differences that exist among them—differences that could affect markedly the plans that are made.

Soil surveys are of special significance in two ways. First, they make possible the extrapolation of research results from a given area to other areas where the same kinds of soil are found. Second, they provide one of the criteria to determine the economic feasibility of clearing and preparing for tillage lands that have as yet been unexploited.

Human Resources. A final requisite for increased food production is trained personnel. The range needed goes from basic scientists (from whose test tubes and field plots new technologies and perhaps new food products are to come) to the cultivators and their assistants. We must have researchers whose interest relates directly to the solution of the world food problem. We must have technicians, farm managers, field service personnel and individuals trained in processing and marketing trades. And we must have educators to teach not only the students but the farmers as well.

The fight to feed the world is not yet lost. But to win it will require technological and scientific inputs of a magnitude not yet realized. And among the most important of these inputs are those relating to soils and soil science.

REFERENCES

Cochrane, T. T., and P. A. Sanchez. 1981. "Land resources, soil properties and their management in the Amazon region: A state of knowledge report." in *Proceedings of Conference on Amazon Land Use Research.* (Cali, Colombia: Centro Internacional de Agricultura Tropical [CIAT]).

Dent, F. J. 1980. "Major production systems and soil-related constraints in Southeast Asia," in *Priorities for Alleviating Soil-related Constraints to Food Production in the Tropics* (Manila: International Rice Research Institute and Ithaca, NY: Cornell University), pp. 78–106.

FAO. 1978. *Production Yearbook*, Vol. 31 (Rome: United Nations).

National Research Council. 1982. *Ecological Aspects of Development in the Humid Tropics*, Report of the Committee on Selected Biological Problems in the Humid Tropics (Washington, DC: National Academy of Sciences Press).

Population Reference Bureau. 1979. *1979 World Population Data Sheet* (Washington, DC: Population Reference Bureau, Inc.).

The President's Science Advisory Committee Panel on World Food Supply. 1967. *The World Food Problem,* Vols. I–III (Washington, DC: The White House).

United Nations. 1982. *Demographic Indicators of Countries: Estimates and Projections as Assessed in 1980* (New York: United Nations).

Wade, M. K., and P. A. Sanchez. 1983. "Mulching and green manure applications for continuous crop production in the Amazon basin," *Agron. J.,* **75**:39–45.

GLOSSARY
of Soil
Science Terms[1]

A horizon. The surface horizon of a mineral soil having maximum organic matter accumulation, maximum biological activity and/or eluviation of materials such as iron and aluminum oxides and silicate clays.

ABC soil. A soil with a distinctly developed profile, including A, B, and C horizons.

absorption, active. Movement of ions and water into the plant root as a result of metabolic processes by the root, frequently against an activity gradient.

absorption, passive. Movement of ions and water into the plant root as a result of diffusion along a gradient.

AC soil. A soil having a profile containing only A and C horizons with no clearly developed B horizon.

accelerated erosion. *See* erosion.

acid rain. Atmospheric precipitation with pH values less than about 5.6, the acidity being due to inorganic acids such as nitric and sulfuric that are formed when oxides of nitrogen and sulfur are emitted into the atmosphere.

acid soil. A soil with a pH value <7.0. Usually applied to surface layer or root zone, but may be used to characterize any horizon. *See also* reaction, soil.

acidity, active. The activity of hydrogen ion in the aqueous phase of a soil. It is measured and expressed as a pH value.

acidity, exchange (reserve). Exchangeable hydrogen, aluminum, and other cations that, when removed from the exchange complex, are sources of soil solution hydrogen ions. This acidity must be neutralized to bring an acid soil to neutrality or to some predetermined higher pH. Usually expressed in centimoles per kilogram (cmol/kg) of soil.

actinomycetes. A group of organisms intermediate between the bacteria and the true fungi that usually produce a characteristic branched mycelium. Includes many, but not all, organisms belonging to the order of Actinomycetales.

activated sludge. Sludge that has been aerated and subjected to bacterial action.

adhesion. Molecular attraction that holds the surfaces of two substances (e.g., water and sand particles) in contact.

adsorption. The attraction of ions or compounds to the surface of a solid. Soil colloids adsorb large amounts of ions and water.

adsorption complex. The group of substances in soil capable of adsorbing other materials. Colloidal particles account for most of this adsorption.

aerate. To impregnate with gas, usually air.

[1] This glossary was compiled and modified from several sources including the following: *Glossary of Soil Science Terms,* Soil Sci. Soc. Amer., Madison, WI, 1975; *Resource Conservation Glossary,* Soil Cons. Soc. Amer., Anheny, IA, 1976; *Soil Taxonomy,* U.S. Department of Agriculture, Washington, DC, 1976; and *Soil, the 1957 Yearbook of Agriculture,* U.S. Department of Agriculture, Washington, DC, 1957.

aeration, soil. The process by which air in the soil is replaced by air from the atmosphere. In a well-aerated soil, the soil air is very similar in composition to the atmosphere above the soil. Poorly aerated soils usually contain more carbon dioxide and correspondingly less oxygen than the atmosphere above the soil.

aerobic. (1) Having molecular oxygen as a part of the environment. (2) Growing only in the presence of molecular oxygen, as aerobic organisms. (3) Occurring only in the presence of molecular oxygen (said of certain chemical or biochemical processes, such as aerobic decomposition).

aggregate (soil). Many soil particles held in a single mass or cluster such as a clod, crumb, block, or prism.

agric horizon. *See* diagnostic subsurface horizons.

agronomy. A specialization of agriculture concerned with the theory and practice of field-crop production and soil management. The scientific management of land.

air-dry. (1) The state of dryness (of a soil) at equilibrium with the moisture content in the surrounding atmosphere. The actual moisture content will depend upon the relative humidity and the temperature of the surrounding atmosphere. (2) To allow to reach equilibrium in moisture content with the surrounding atmosphere.

air porosity. The proportion of the bulk volume of soil that is filled with air at any given time or under a given condition, such as a specified moisture tension. Usually the large pores; that is, those drained by a tension of less than approximately 100 cm of water. *See also* moisture tension.

albic horizon. *See* diagnostic subsurface horizons.

Alfisols. *See* soil classification.

alkali soil. (1) A soil with a high degree of alkalinity (pH ≥ 8.5) or with a high exchangeable sodium content (15% or more of the exchange capacity), or both. (2) A soil that contains sufficient alkali (sodium) to interfere with the growth of most crop plants. *See also* saline-sodic soil; sodic soil.

alkaline soil. Any soil that has pH >7. Usually applied to surface layer or root zone but may be used to characterize any horizon or a sample thereof. *See also* reaction, soil.

alluvial soil. (Obsolete) A soil developing from recently deposited alluvium and exhibiting essentially no horizon development or modification of the recently deposited materials.

alluvium. A general term for all detrital material deposited or in transit by streams, including gravel, sand, silt, clay, and all variations and mixtures of these. Unless otherwise noted, alluvium is unconsolidated.

Alpine Meadow soils. A great soil group of the intrazonal order, comprised of dark soils of grassy meadows at altitudes above the timberline [1949 classification system.]

alumino-silicates. Compounds containing aluminum, silicon, and oxygen as main constituents. An example is microcline, $KAlSi_3O_8$.

amendment, soil. Any substance such as lime, sulfur, gypsum, and sawdust used to alter the properties of a soil, generally to make it more productive. Strictly speaking, fertilizers are soil amendments, but the term is used most commonly for materials other than fertilizers.

amino acids. Nitrogen-containing organic acids that couple together to form proteins. Each acid molecule contains one or more amino groups ($-NH_2$) and at least one carboxyl group ($-COOH$). In addition, some amino acids contain sulfur.

ammonification. The biochemical process whereby ammoniacal nitrogen is released from nitrogen-containing organic compounds.

ammonium fixation. The adsorption of ammonium ions by the mineral or organic fractions of the soil in forms that are relatively insoluble in water and relatively nonexchangeable.

anaerobic. (1) Without molecular oxygen. (2) Living or functioning in the absence of air or free oxygen.

Ando soils. (Obsolete) A zonal group of dark-colored soils high in organic matter developed in volcanic ash deposits.

anion. Negatively charged ion; ion that during electrolysis is attracted to the anode.

anion exchange capacity. The sum total of exchangeable anions that a soil can adsorb. Expressed as centimoles per kilogram (cmol/kg) of soil (or of other adsorbing material such as clay).

anthropic epipedon. *See* diagnostic surface horizons.

antibiotic. A substance produced by one species of organism that, in low concentrations, will kill or inhibit growth of certain other organisms.

Ap. The surface layer of a soil disturbed by cultivation or pasturing.

apatite. A naturally occurring complex calcium phosphate that is the original source of most of the phosphate fertilizers. Formulas such as $3Ca_3(PO_4)_2 \cdot CaF_2$ illustrate the complex compounds that make up apatite.

argillic horizon. *See* diagnostic subsurface horizons.

Aridisols. *See* soil classification.

association, soil. *See* soil association.

Atterberg limits. Atterberg limits are measured for soil materials passing the No. 40 sieve.

  shrinkage limit (SL). The maximum water content at which a reduction in water content will not cause a decrease in the volume of the soil mass. This defines the arbitrary limit between the solid and semisolid states.

  plastic limit (PL). The water content corresponding to an arbitrary limit between the plastic and semisolid states of consistency of a soil.

  liquid limit (LL). The water content corresponding to the arbitrary limit between the liquid and plastic states of consistency of a soil.

autotroph. An organism capable of utilizing carbon dioxide or carbonates as the sole source of carbon and obtaining energy for life processes from the oxidation of inorganic elements or compounds such as iron, sulfur, hydrogen, ammonium, and nitrites, or from radiant energy. *Contrast with* heterotroph.

available nutrient. That portion of any element or compound in the soil that can be readily absorbed and assimilated by growing plants. ("Available" should not be confused with "exchangeable.")

available water. The portion of water in a soil that can be readily absorbed by plant roots. Considered by most workers to be that water held in the soil against a pressure of up to approximately 15 bars. *See also* field capacity; moisture tension.

azonal soils. (Obsolete) Soils without distinct genetic horizons. A soil order under the 1949 classification system.

B horizon. A soil horizon usually beneath the A that is characterized by one or more of the following: (a) a concentration of silicate clays, iron and aluminum oxides, and humus, alone or in combination; (b) a blocky or prismatic structure, and (c) coatings of iron and aluminum oxides that give darker, stronger or redder color.

bar. A unit of pressure equal to one million dynes per square centimeter (10^6 dynes/cm^2).

base saturation percentage. The extent to which the adsorption complex of a soil is saturated with exchangeable cations other than hydrogen and aluminum. It is expressed as a percentage of the total cation exchange capacity.

BC soil. A soil profile with B and C horizons but with little or no A horizon. Most BC soils have lost their A horizons by erosion.

bedding. (Engineering) Arranging the surface of fields by plowing and grading into a series of elevated beds separated by shallow depressions or ditches for drainage.

bedrock. The solid rock underlying soils and the regolith in depths ranging from zero (where exposed by erosion) to several hundred feet.

bench terrace. An embankment constructed across sloping fields with a steep drop on the downslope side.

biodegradable. A material subject to degradation by biochemical processes.

biomass. The amount of living matter in a given area.

bleicherde. The light-colored, leached A2 horizon of Spodosols.

blocky soil structure. *See* soil structure types.

blown-out land. Areas from which all or almost all of the soil and soil material has been removed by wind erosion. Usually unfit for crop production. A miscellaneous land type.

Bog soil. A great soil group of the intrazonal order and hydromorphic suborder. Includes muck and peat. [1949 classification system.]

border-strip irrigation. *See* irrigation methods.

bottomland. *See* floodplain.

breccia. A rock composed of coarse angular fragments cemented together.

broad-base terrace. A low embankment with such gentle slopes that it can be farmed, constructed across sloping fields to reduce erosion and runoff.

Brown Earths. Soils with a mull horizon but having no horizon of accumulation of clay or sesquioxides. (Generally used as a synonym for "Brown Forest soils" but sometimes for similar soils acid in reaction.)

Brown Forest soils. A great soil group of the intrazonal order and calcimorphic suborder, formed on calcium-rich parent materials under deciduous forest, and possessing a high base status but lacking a pronounced illuvial horizon. [1949 classification system.]

Brown Podzolic soils. A zonal great soil group similar to Podzols but lacking the distinct A2 horizon characteristic of the Podzol group. [1949 classification system.]

Brown soils. A great soil group of the temperate to cool arid regions, composed of soils with a brown surface and a light-colored transitional subsurface horizon over calcium carbonate accumulation. They develop under short grasses. [1949 classification system.]

Brunizem. Synonymous with *Prairie soils.* [1949 classification system.]

buffer compounds, soil. The clay, organic matter, and compounds such as carbonates and phosphates that enable the soil to resist appreciable change in pH.

bulk density, soil. The mass of dry soil per unit of bulk volume, including the air space. The bulk volume is determined before drying to constant weight at 105°C.

buried soil. Soil covered by an alluvial, loessal, or other deposit, usually to a depth greater than the thickness of the solum.

C horizon. A mineral horizon generally beneath the solum that is relatively unaffected by biological activity and pedogenesis and is lacking properties diagnostic of an A or B horizon. It may or may not be like the material from which the A and B have formed.

calcareous soil. Soil containing sufficient calcium carbonate (often with magnesium carbonate) to effervesce visibly when treated with cold 0.1 N hydrochloric acid.

calcic horizon. *See* diagnostic subsurface horizons.

caliche. A layer near the surface, more or less cemented by secondary carbonates of calcium or magnesium precipitated from the soil solution. It may occur as a soft thin soil horizon, as a hard thick bed just beneath the solum, or as a surface layer exposed by erosion.

cambic horizon. *See* diagnostic subsurface horizons.

capillary conductivity. (Obsolete) *See* hydraulic conductivity.

capillary porosity. The small pores, or the bulk volume of small pores, which hold water in soils against a tension usually >60 cm of water. *See also* moisture tension.

capillary water. (Obsolete) The water held in the "capillary" or *small* pores of a soil, usually with a tension >60 cm of water. *See also* moisture tension.

carbon cycle. The sequence of transformations whereby carbon dioxide is fixed in living organisms by photosynthesis or by chemosynthesis, liberated by respiration and by the death and decomposition of the fixing organism, used by heterotrophic species, and ultimately returned to its original state.

carbon/nitrogen ratio. The ratio of the weight of organic carbon (C) to the weight of total nitrogen (N) in a soil or in organic material.

"Cat" clays. Wet clay soils high in reduced forms of sulfur that upon being drained become extremely acid due to the oxidation of the sulfur compounds and the formation of sulfuric acid.

catena. A sequence of soils of about the same age, derived from similar parent material, and occurring under similar climatic conditions, but having different characteristics due to variation in *relief* and in *drainage.*

cation. A positively charged ion.

cation exchange. The interchange between a cation in solution and another cation on the surface of any surface-active material such as clay or organic matter.

cation exchange capacity. The sum total of exchangeable cations that a soil can adsorb. Sometimes called "total-exchange capacity," "base-exchange capacity," or "cation-adsorption capacity." Expressed in centimoles per kilogram (cmol/kg) of soil (or of other adsorbing material such as clay).

cemented. Indurated; having a hard, brittle consistency because the particles are held together by cementing substances such as humus, calcium carbonate, or the oxides of silicon, iron, and aluminum.

channery. Thin, flat fragments of limestone, sandstone, or schist up to 6 in. (15 cm) in major diameter.

chelate. (Greek, claw) A type of chemical compound in which a metallic ion is firmly combined with a molecule by means of multiple chemical bonds.

Chernozem. A zonal great soil group consisting of soils with a thick, nearly black or black, organic matter-rich A horizon high in exchangeable calcium, underlain by a lighter colored transitional horizon above a zone of calcium carbonate accumulation; occurs in a cool subhumid climate under a vegetation of tall and midgrass prairie. [1949 classification system.]

chert. A structureless form of silica, closely related to flint, that breaks into angular fragments.

Chestnut soil. A zonal great soil group consisting of soils with a moderately thick, dark-brown A horizon over a lighter colored horizon that is above a zone of calcium carbonate accumulation. They develop under mixed tall and short grasses in a temperate to cool and subhumid to semiarid climate. [1949 classification system.]

chisel, subsoil. A tillage implement with one or more cultivator-type feet to which are attached strong knifelike units used to shatter or loosen hard, compact layers, usually in the subsoil, to depths below normal plow depth. *See also* subsoiling

chlorosis. A condition in plants relating to the failure of chlorophyll (the green coloring matter) to develop. Chlorotic leaves range from light green through yellow to almost white.

chroma. The relative purity, strength, or saturation of a color; directly related to the dominance of the determining wavelength of the light and inversely related to grayness; one of the three variables of color. *See also* hue; Munsell color system; value, color.

class, soil. A group of soils having a definite range in a particular property such as acidity, degree of slope, texture, structure, land-use capability, degree of erosion, or drainage. *See also* soil structure; soil texture.

classification, soil. *See* soil classification.

clastic. Composed of broken fragments of rocks and minerals.

clay. (1) A soil separate consisting of particles <0.002 mm in equivalent diameter. (2) A soil textural class containing >40% clay, <45% sand, and <40% silt.

clay mineral. Naturally occurring inorganic material (usually crystalline) found in soils and other earthy deposits, the particles being of clay size, that is, <0.002 mm in diameter.

claypan. A dense, compact, slowly permeable layer in the subsoil having a much higher clay content than the overlying material, from which it is separated by a sharply defined boundary. Claypans are usually hard when dry, and plastic and sticky when wet.

clod. A compact, coherent mass of soil produced artificially, usually by the activity of man by plowing, digging, and so on, especially when these operations are performed on soils that are either too wet or too dry for normal tillage operations.

coarse texture. The texture exhibited by sands, loamy sands, and sandy loams except very fine sandy loam.

cobblestone. Rounded or partially rounded rock or mineral fragments 3–10 in. in diameter.

cohesion. Holding together: force holding a solid or liquid together, owing to attraction between like molecules. Decreases with rise in temperature.

colloid, soil. (Greek, glue-like) Organic and inorganic matter with very small particle size and a correspondingly large surface area per unit of mass.

colluvium. A deposit of rock fragments and soil material accumulated at the base of steep slopes as a result of gravitational action.

color. *See* Munsell color system.

columnar soil structure. *See* soil structure types.

compost. Organic residues, or a mixture of organic residues and soil, that have been piled, moistened, and allowed to undergo biological decomposition. Mineral fertilizers are sometimes added. Often called "artificial manure" or "synthetic manure" if produced primarily from plant residues.

concretion. A local concentration of a chemical compound, such as calcium carbonate or iron oxide, in the form of a grain or nodule of varying size, shape, hardness, and color.

conduction. The transfer of heat by physical contact between two or more objects.

conductivity, hydraulic. *See* hydraulic conductivity.

conifer. A tree belonging to the order Coniferae, usually evergreen, with cones and needle-shaped or scale-like leaves and producing wood known commercially as "softwood."

conservation tillage. *See* tillage, conservation.

consistence. The combination of properties of soil material that determine its resistance to crushing and its ability to be molded or changed in shape. Such terms as loose, friable, firm, soft, plastic, and sticky describe soil consistence.

consumptive use. The water used by plants in transpiration and growth, plus water vapor loss from adjacent soil or snow, or from intercepted precipitation in any specified time. Usually expressed as equivalent depth of free water per unit of time.

contour. An imaginary line connecting points of equal elevation on the surface of the soil. A contour terrace is laid out on a sloping soil at right angles to the direction of the slope and nearly level throughout its course.

contour strip cropping. Layout of crops in comparatively narrow strips in which the farming operations are performed approximately on the contour. Usually strips of grass, close-growing crops, or fallow are alternated with those in cultivated crops.

convection. The transfer of heat through a gas or solution because of molecular movement.

corrugated irrigation. *See* irrigation.

cover crop. A close-growing crop grown primarily for the purpose of protecting and improving soil between periods of regular crop production or between trees and vines in orchards and vineyards.

creep. Slow mass movement of soil and soil material down relatively steep slopes primarily under the influence of gravity, but facilitated by saturation with water and by alternate freezing and thawing.

crop rotation. A planned sequence of crops growing in a regularly recurring succession on the same area of land, as contrasted to continuous culture of one crop or growing different crops in haphazard order.

crotovina. A former animal burrow in one soil horizon that has been filled with organic matter or material from another horizon (also spelled "krotovina").

crumb. A soft, porous, more or less rounded natural unit of structure from 1 to 5 mm in diameter. *See also* soil structure types.

crushing strength. The force required to crush a mass of dry soil or, conversely, the resistance of the dry soil mass to crushing. Expressed in units of force per unit area (pressure).

crust. A surface layer on soils, ranging in thickness from a few millimeters to perhaps as much as 3 cm, that is much more compact, hard, and brittle, when dry, than the material immediately beneath it.

crystal. A homogeneous inorganic substance of definite chemical composition bounded by plane surfaces that form definite angles with each other, thus giving the substance a regular geometrical form.

crystal structure. *See* lattice structure.

crystalline rock. A rock consisting of various minerals that have crystallized in place from magma. *See also* igneous rock; sedimentary rock.

cultivation. A tillage operation used in preparing land for seeding or transplanting or later for weed control and for loosening the soil.

deciduous plant. A plant that sheds all its leaves every year at a certain season.

deflocculate. (1) To separate the individual components of compound particles by chemical and/or physical means. (2) To cause the particles of the *disperse phase* of a colloidal system to become suspended in the *dispersion medium*.

denitrification. The biochemical reduction of nitrate or nitrite to gaseous nitrogen, either as molecular nitrogen or as an oxide of nitrogen.

desalinization. Removal of salts from saline soil, usually by leaching.

desert crust. A hard layer, containing calcium carbonate, gypsum, or other binding material, exposed at the surface in desert regions.

Desert soil. A zonal great soil group consisting of soils with a very thin, light-colored surface horizon, which may be vesicular and is ordinarily underlain by calcareous material; formed in arid regions under sparse shrub vegetation. [1949 classification system.]

desorption. The removal of sorbed material from surfaces.

diagnostic horizons. (As used in *Soil Taxonomy*): Combinations of specific soil characteristics that are indicative of certain classes of soils. Those that occur at the soil surface are called epipedons; those below the surface, diagnostic subsurface horizons.

diagnostic subsurface horizons. The following diagnostic subsurface horizons are used in *Soil Taxonomy*.

agric horizon. A mineral soil horizon in which clay, silt, and humus derived from an overlying cultivated and fertilized layer have accumulated. The wormholes and illuvial clay, silt, and humus, occupy at least 5% of the horizon by volume.

albic horizon. A mineral soil horizon from which clay and free iron oxides have been removed or in which the oxides have been segregated to the extent that the color of the horizon is determined primarily by the color of the primary sand and silt particles rather than by coatings on these particles.

argillic horizon. A mineral soil horizon characterized by the illuvial accumulation of layer-lattice silicate clays.

calcic horizon. A mineral soil horizon of secondary carbonate enrichment that is more than 15 cm thick, has a calcium carbonate equivalent of more than 15%, and has at least 5% more calcium carbonate equivalent than the underlying C horizon.

cambic horizon. A mineral soil horizon that has a texture of loamy very fine sand or finer, contains some weatherable minerals, and is characterized by the alteration or removal of mineral material. The cambic horizon lacks cementation or induration and has too few evidences of illuviation to meet the requirements of the argillic or spodic horizon.

duripan. A mineral soil horizon that is cemented by silica, to the point that air-dry fragments will not slake in water or HCl. A duripan may also have accessory cement such as iron oxide or calcium carbonate. *See also* hardpan.

gypsic horizon. A mineral soil horizon of secondary calcium sulfate enrichment that is more than 15 cm thick.

natric horizon. A mineral soil horizon that satisfies the requirements of an argillic horizon, but that also has prismatic, columnar, or blocky structure and a subhorizon having more than 15% saturation with exchangeable sodium.

oxic horizon. A mineral soil horizon that is at least 30 cm thick and characterized by the virtual *absence* of weatherable primary minerals or 2:1 lattice clays and the *presence* of 1:1 lattice clays and highly insoluble minerals such as quartz sand, hydrated oxides of iron and aluminum, low cation exchange capacity, and small amounts of exchangeable bases.

petrocalcic horizon. A continuous, indurated calcic horizon that is cemented by calcium carbonate and, in some places, with magnesium carbonate. It cannot be penetrated with a spade or auger when dry, dry fragments do not slake in water, and it is impenetrable to roots.

petrogypsic horizon. A continuous, strongly cemented, massive gypsic horizon that is cemented by calcium sulfate. It can be chipped with a spade when dry. Dry fragments do not slake in water and it is impenetrable to roots.

placic horizon. A black to dark reddish mineral soil horizon that is usually thin but that may range from 1 mm to 25 mm in thickness. The placic horizon is commonly cemented with iron and is slowly permeable or impenetrable to water and roots.

salic horizon. A mineral soil horizon of enrichment with secondary salts more soluble in cold water than gypsum. A salic horizon is 15 cm or more in thickness.

spodic horizon. A mineral soil horizon characterized by the illuvial accumulation of amorphous materials composed of aluminum and organic carbon with or without iron.

diagnostic surface horizons. The following diagnostic surface horizons are used in *Soil Taxonomy* and are called *epipedons.*

anthropic epipedon. A surface layer of mineral soil that has the same requirements as the mollic epipedon but that has more than 250 ppm of P_2O_5 soluble in 1% citric acid, or is dry more than 10 months (cumulative) during the period when not irrigated. The anthropic epipedon forms under long-continued cultivation and fertilization.

histic epipedon. A thin organic soil horizon that is saturated with water at some period of the year unless artificially drained and that is at or near the surface of a mineral soil.

mollic epipedon. A surface horizon of mineral soil that is dark colored and relatively thick, contains at least 0.58% organic carbon, is not massive and hard or very hard when dry, has a base saturation of more than 50% when measured at pH 7, has less than 250 ppm of P_2O_5 soluble in 1% citric acid, and is dominantly saturated with bivalent cations.

ochric epipedon. A surface horizon of mineral soil that is too light in color, too high in chroma, too low in organic carbon, or too thin to be a plaggen, mollic, umbric, anthropic, or histic epipedon, or that is both hard and massive when dry.

plaggen epipedon. A man-made surface horizon more than 50 cm thick that is formed by long-continued manuring and mixing.

umbric epipedon. A surface layer of mineral soil that has the same requirements as the mollic epipedon with respect to color, thickness, organic carbon content, consistence, structure, and P_2O_5 content, but that has a base saturation of less than 50% when measured at pH 7.

diatoms. Algae having siliceous cell walls that persist as a skeleton after death; any of the microscopic unicellular or colonial algae constituting the class Bacillariaceae. They occur abundantly in fresh and salt waters and their remains are widely distributed in soils.

diatomaceous earth. A geologic deposit of fine, grayish, siliceous material composed chiefly or wholly of the remains of diatoms. It may occur as a powder or as a porous, rigid material.

diffusion. The transport of matter as a result of the movement of the constituent particles. The intermingling of two gases or liquids in contact with each other takes place by diffusion.

disintegration. The breakdown of rock and mineral particles into smaller particles by physical forces such as frost action.

disperse. (1) To break up compound particles, such as aggregates, into the individual component particles. (2) To distribute or suspend fine particles, such as clay, in or throughout a dispersion medium, such as water.

diversion dam. A structure or barrier built to divert part or all of the water of a stream to a different course.

diversion terrace. *See* terrace.

drain, to. (1) To provide channels, such as open ditches or drain tile, so that excess water can be removed by surface or by internal flow. (2) To lose water (from the soil) by percolation.

drainage, soil. The frequency and duration of periods when the soil is free of saturation with water.

drift. Material of any sort deposited by geological processes in one place after having been removed from another. Glacial drift includes material moved by the glaciers and by the streams and lakes associated with them.

drumlin. Long, smooth cigar-shaped low hills of glacial till, with their long axes parallel to the direction of ice movement.

dryland farming. The practice of crop production in low-rainfall areas without irrigation.

duff. The matted, partly decomposed organic surface layer of forest soils.

duripan. *See* diagnostic subsurface horizons; hardpan.

dust mulch. A loose, finely granular, or powdery condition on the surface of the soil, usually produced by shallow cultivation.

ectotrophic mycorrhiza (ectomycorrhiza). A mycorrhizal association in which the fungal hyphae form a compact mantle on the surface of the roots and extend into the surrounding soil. Mycelial strands extend inward between cortical cells, but not into these cells. Associated primarily with certain trees. *See also* endotrophic mycorrhiza.

edaphology. The science that deals with the influence of soils on living things, particularly plants, including man's use of land for plant growth.

electrokinetic potential. In a colloidal system, the difference in potential between the immovable layer attached to the surface of the dispersed phase and the dispersion medium.

eluviation. The removal of soil material in suspension (or in solution) from a layer or layers of a soil. (Usually, the loss of material in *solution* is described by the term "leaching.") *See also* leaching.

endotrophic. Nourished or receiving nourishment from within, as fungi or their hyphae receiving nourishment from plant roots in a mycorrhizal association.

endotrophic mycorrhiza (endomycorrhiza). A mycorrhizal association in which the fungal hyphae penetrate directly into root hairs, other epidermal cells, and occasionally into cortical cells. Individual threads extend from the root surface outward into the surrounding soil. (*See also* vesicular-arbuscular mycorrhiza.

Entisols. *See* soil classification.

eolian soil material. Soil material accumulated through wind action. The most extensive areas in the United States are silty deposits (loess), but large areas of sandy deposits also occur.

epipedon. A diagnostic surface horizon that includes the upper part of the soil that is darkened by organic matter, or the upper eluvial horizons or both. [*Soil Taxonomy.*]

erosion. (1) The wearing away of the land surface by running water, wind, ice, or other geological agents, including such processes as gravitational creep. (2) Detachment and movement of soil or rock by water, wind, ice, or gravity. The following terms are used to describe different types of water erosion:

accelerated erosion. Erosion much more rapid than normal, natural, geological erosion; primarily as a result of the influence of the activities of humans or, in some cases, of animals.

gully erosion. The erosion process whereby water accumulates in narrow channels and, over short periods, removes the soil from this narrow area to considerable depths, ranging from 1–2 feet to as much as 75–100 feet.

natural erosion. Wearing away of the earth's surface by water, ice or other natural agents under natural environmental conditions of climate, vegetation, and so on, undisturbed by man. Synonymous with *geological erosion.*

rill erosion. An erosion process in which numerous small channels of only several centimeters in depth are formed; occurs mainly on recently cultivated soils. *See also* rill.

sheet erosion. The removal of a fairly uniform layer of soil from the land surface by runoff water.

splash erosion. The spattering of small soil particles caused by the impact of raindrops on very wet soils. The loosened and separated particles may or may not be subsequently removed by surface runoff.

essential element. A chemical element required for the normal growth of plants.

eutrophic. Having concentrations of nutrients optimal (or nearly so) for plant or animal growth. (Said of nutrient solutions or of soil solutions.)

eutrophication. A means of aging of lakes whereby aquatic plants are abundant and waters are deficient in oxygen. The process is usually accelerated by enrichment of waters with surface runoff containing nitrogen and phosphorus.

evapotranspiration. The combined loss of water from a given area, and during a specified period of time, by evaporation from the soil surface and by transpiration from plants.

exchange acidity. The titratable hydrogen and aluminum that can be replaced from the adsorption complex by a neutral salt solution. Usually expressed as centimoles per kilogram (cmol/kg) of soil.

exchange capacity. The total ionic charge of the adsorption complex active in the adsorption of ions. *See also* anion exchange capacity; cation exchange capacity.

exchangeable-sodium percentage. The extent to which the adsorption complex of a soil is occupied by sodium. It is expressed as follows.

$$ESP = \frac{\text{exchangeable sodium}}{\text{cation exchange capacity}} \times 100$$
$$\quad\quad\quad\;\; \text{(cmol/kg soil)}$$

fallow. Cropland left idle in order to restore productivity, mainly through accumulation of water, nutrients, or both. Summer fallow is a common stage before cereal grain in regions of limited rainfall. The soil is kept free of weeds and other vegetation, thereby conserving nutrients and water for the next year's crop.

family, soil. In soil classification, one of the categories intermediate between the great group and the soil series. Families are defined largely on the basis of physical and mineralogical properties of importance to plant growth. [*Soil Taxonomy.*]

fertility, soil. The status of a soil with respect to the amount and availability to plants of elements necessary for plant growth.

fertilizer. Any organic or inorganic material of natural or synthetic origin added to a soil to supply certain elements essential to the growth of plants. The major types of fertilizers include:

bulk blended fertilizers. Solid fertilizer materials blended together in small blending plants delivered to the farm in bulk, usually spread directly on the fields by truck or other special applicator.

granulated fertilizers. Fertilizers that are present in the form of rather stable granules of uniform size, which facilitate ease of handling the materials and reduce undesirable dusts.

liquid fertilizers. Fluid fertilizers that contain essential elements in liquid forms either as soluble nutrients or as liquid suspensions or both.

mixed fertilizers. Two or more fertilizer materials mixed together. May be as dry powders, granules, pellets, bulk blends, or liquids.

fertilizer grade. The guaranteed minimum analysis, in percent, of the major plant nutrient elements contained in a fertilizer material or in a mixed fertilizer. (Usually refers to the percentage of $N–P_2O_5–K_2O$, but a simpler expression recommended by scientific societies is the percentage of N–P–K.)

fertilizer requirement. The quantity of certain plant nutrient elements needed, in addition to the amount supplied by the soil, to increase plant growth to a designated optimum.

fibric materials. *See* organic soil materials.

field capacity (field moisture capacity). The percentage of water remaining in a soil two or three days after having been saturated and after free drainage has practically ceased.

fine-grained mica. A silicate clay having a 2:1-type lattice structure with much of the silicon in the tetrahedral sheet having been replaced by aluminum and with considerable interlayer potassium which binds the layers together and prevents interlayer expansion swelling and limits interlayer cation exchange capacity.

fine texture. Consisting of or containing large quantities of the fine fractions, particularly of silt and clay. (Includes clay loam, sandy clay loam, silty clay loam, sandy clay, silty clay, and clay textural classes.)

first bottom. The normal floodplain of a stream.

fixation. (1) For other than elemental nitrogen: the process or processes in a soil by which certain chemical elements essential for plant growth are converted from a soluble or exchangeable form to a much less soluble or to a nonexchangeable form; for example, potassium, ammonium and phosphate "fixation." (2) For elemental nitrogen: process by which gaseous elemental nitrogen is chemically combined with hydrogen to form ammonia.

biological nitrogen fixation. Occurs at ordinary temperatures and pressures. It is commonly carried out by certain bacteria, algae and actinomycetes which may or may not be associated with higher plants.

chemical nitrogen fixation. Takes place at high temperatures and pressures in manufacturing plants; produces ammonia, which is used to manufacture most fertilizers.

flagstone. A relatively thin rock or mineral fragment 15–38 cm in length commonly composed of shale, slate, limestone, or sandstone.

flocculate. To aggregate or clump together individual, tiny soil particles, especially fine clay, into small clumps or floccules. Opposite of *deflocculate* or *disperse*.

floodplain. The land bordering a stream, built up of sediments from overflow of the stream and subject to inundation when the stream is at flood stage. Sometimes called bottomland.

flora. The sum total of the kinds of plants in a area at one time.

fluor apatite. A member of the apatite group of minerals containing fluorine. Most common mineral in rock phosphate.

fluvioglacial. *See* glaciofluvial deposits.

foliar diagnosis. An estimation of mineral nutrient deficiencies (excesses) of plants based on examination of the chemical composition of selected plant parts, and the color and growth characteristics of the foliage of the plants.

fragipan. Dense and brittle pan or subsurface layer in soils that owe their hardness mainly to extreme density or compactness rather than high clay content or cementation. Removed fragments are friable, but the material in place is so dense that roots penetrate and water moves through it very slowly.

friable. A soil consistency term pertaining to the ease of crumbling of soils.

frigid. *See* soil temperature classes.

fritted micronutrients. Sintered silicates having total guaranteed analyses of micronutrients with controlled (relatively slow) release characteristics.

fulvic acid. A term of varied usage but usually referring to the mixture of organic substances remaining in solution upon acidification of a dilute alkali extract from the soil.

furrow irrigation. *See* irrigation methods.

fungi. Simple plants that lack a photosynthetic pigment. The individual cells have a nucleus surrounded by a membrane, and they may be linked together in long filaments called *hyphae,* which may grow together to form a visible body.

genesis, soil. The mode of origin of the soil, with special reference to the processes responsible for the development of the solum, or true soil, from the unconsolidated parent material.

geological erosion. *See* erosion, natural.

gilgai. The microrelief of soils produced by expansion and contraction with changes in moisture. Found in soils that contain large amounts of clay that swells and shrinks considerably with wetting and drying. Usually a succession of microbasins and microknolls in nearly level

areas or of microvalleys and microridges parallel to the direction of the slope.

glacial drift. Rock debris that has been transported by glaciers and deposited, either directly from the ice or from the melt-water. The debris may or may not be heterogeneous.

glacial till. *See* till.

glaciofluvial deposits. Material moved by glaciers and subsequently sorted and deposited by streams flowing from the melting ice. The deposits are stratified and may occur in the form of outwash plains, deltas, kames, eskers, and kame terraces.

gley soil. (Obsolete) Soil developed under conditions of poor drainage resulting in reduction of iron and other elements and in gray colors and mottles.

granular structure. Soil structure in which the individual grains are grouped into spherical aggregates with indistinct sides. Highly porous granules are commonly called crumbs. A well-granulated soil has the best structure for most ordinary crop plants. *See also* soil structure types.

gravitational potential. *See* soil water potential.

gravitational water. Water that moves into, through, or out of the soil under the influence of gravity.

Gray-Brown Podzolic soil. A zonal great soil group consisting of soils with a thin, moderately dark A1 horizon and with a grayish brown A2 horizon underlain by a B horizon containing a high percentage of bases and an appreciable quantity of illuviated silicate clay; formed on relatively young land surfaces, mostly glacial deposits, from material relatively rich in calcium, under deciduous forests in humid temperate regions. [1949 classification system.]

Gray Wooded soils. A zonal great soil group consisting of soils with a thin A1 horizon over a light-colored, bleached (A2) horizon underlain by a B horizon containing a high percentage of bases and an appreciable quantity of illuviated silicate clay. These soils occur in subhumid to semiarid, cool climatic regions under coniferous, deciduous, or mixed forest cover. [1949 classification system.]

great group. *See* soil classification.

great soil group. Any one of several broad

groups of soils with fundamental characteristics in common. Examples are Chernozems, Gray-Brown Podzolic, and Podzol. [1949 classification system.]

green manure. Plant material incorporated with the soil while green, or soon after maturity, for improving the soil.

groundwater. Subsurface water in the zone of saturation that is free to move under the influence of gravity.

Ground-Water Laterite soil. A great soil group of the intrazonal order and hydromorphic suborder, consisting of soils characterized by hardpans or concretional horizons rich in iron and aluminum (and sometimes manganese) that have formed immediately above the water table. [1949 classification system.]

Ground-Water Podzol soil. A great soil group of the intrazonal order and hydromorphic suborder, consisting of soils with an organic mat on the surface over a very thin layer of acid humus material underlain by a whitish-gray leached layer, which may be as much as 60–90 cm in thickness, and is underlain by a brown, or very dark-brown, cemented hardpan layer; formed under various types of forest vegetation in cool to tropical humid climates under conditions of poor drainage. [1949 classification system.]

gully erosion. *See* erosion.

gypsic horizon. *See* diagnostic subsurface horizon.

Half-Bog soil. A great soil group, of the intrazonal order and hydromorphic suborder, consisting of soil with dark-brown or black peaty material over grayish and rust mottled mineral soil; formed under conditions of poor drainage under forest, sedge, or grass vegetation in cool to tropical humid climates. [1949 classification system.]

halomorphic soil. A suborder of the intrazonal soil order, consisting of saline and alkali soils formed under imperfect drainage in arid regions and including the great soil group Solonchak or Saline soils, Solonetz soils, and Soloth soils. [1949 classification system.]

hardpan. A hardened soil layer, in the lower A or in the B horizon, caused by cementation of soil particles with organic matter or with materials such as silica, sesquioxides, or calcium

carbonate. The hardness does not change appreciably with changes in moisture content and pieces of the hard layer do not slake in water. *See also* caliche; claypan; duripan.

harrowing. A secondary broadcast tillage operation that pulverizes, smooths, and firms the soil in seedbed preparation, controls weeds, or incorporates material spread on the surface.

heaving. The partial lifting of plants, buildings, roadways, fenceposts, etc., out of the ground, as a result of freezing and thawing of the surface soil during the winter.

heavy metals. Metals with densities >5.0.

heavy soil. (Obsolete in scientific use) A soil with a high content of the fine separates, particularly clay, or one with a high drawbar pull, hence difficult to cultivate.

hemic materials. *See* organic soil materials.

heterotroph. An organism capable of deriving energy for life processes only from the decomposition of organic compounds and incapable of using inorganic compounds as sole sources of energy or for organic synthesis. *Contrast with* autotrophic.

histic epipedon. *See* diagnostic surface horizons.

Histosols. *See* soil classification.

horizon, soil. A layer of soil, approximately parallel to the soil surface, differing in properities and characteristics from adjacent layers below or above it. *See also* diagnostic subsurface horizons; diagnostic surface horizons.

hue. One of the three variables of color. It is caused by light of certain wavelengths and changes with the wavelength. *See also* chroma; Munsell color system; value, color.

humic acid. A mixture of variable or indefinite composition of dark organic substances, precipitated upon acidification of a dilute alkali extract from soil.

Humic Gley soil. Soil of the intrazonal order and hydromorphis suborder that includes Wiesenboden and related soils, such as Half-Bog soils, which have a thin muck or peat O2 horizon and an A1 horizon. Developed in wet meadow and in forested swamps. [1949 classification system.]

humification. The processes involved in the decomposition of organic matter and leading to the formation of humus.

humin. The fraction of the soil organic matter that is not dissolved upon extraction of the soil with dilute alkali.

humus. That more or less stable fraction of the soil organic matter remaining after the major portion of added plant and animal residues have decomposed. Usually it is dark in color.

hydraulic conductivity. An expression of the readiness with which a liquid such as water flows through a solid such as soil in response to a given potential gradient.

hydrologic cycle. The circuit of water movement from the atmosphere to the earth and return to the atmosphere through various stages or processes, as precipitation, interception, runoff, infiltration, percolation, storage, evaporation, and transpiration.

hydromorphic soils. A suborder of intrazonal soils, all formed under conditions of poor drainage in marshes, swamps, seepage areas, or flats. [1949 classification system.]

hydrous mica. *See* fine-grained mica.

hydroxyapatite. A member of the apatite group of minerals rich in hydroxyl groups. A nearly insoluble calcium phosphate.

hygroscopic coefficient. The amount of moisture in a dry soil when it is in equilibrium with some standard relative humidity near a saturated atmosphere (about 98%), expressed in terms of percentage on the basis of oven-dry soil.

hyperthermic. *See* soil temperature.

igneous rock. Rock formed from the cooling and solidification of magma, and which has not been changed appreciably since its formation.

illite. A fine-grained mica. *See* fine-grained mica.

illuvial horizon. A soil layer or horizon in which material carried from an overlying layer has been precipitated from solution or deposited from suspension. The layer of accumulation.

immature soil. A soil with indistinct or only slightly developed horizons because of the relatively short time it has been subjected to the various soil-forming processes. A soil that has not reached equilibrium with its environment.

immobilization. The conversion of an element from the inorganic to the organic form in microbial tissues or in plant tissues, thus rendering the element not readily available to other organisms or to plants.

impervious. Resistant to penetration by fluids or by roots.

Inceptisols. *See* soil classification.

indurated (soil). Soil material cemented into a hard mass that will not soften on wetting. *See also* consistence; hardpan.

infiltration. The downward entry of water into the soil.

infiltration rate. A soil characteristic determining or describing the *maximum* rate at which water *can* enter the soil under specified conditions, including the presence of an excess of water.

inoculation. The process of introducing pure or mixed cultures of microorganisms into natural or artificial culture media.

intergrade. A soil that possesses moderately well-developed distinguishing characteristics of two or more genetically related great soil groups.

interlayer. (mineralogy) Materials between layers within a given crystal, including cations, hydrated cations, organic molecules, and hydroxide groups or sheets.

intrazonal soils. Soils with more or less well-developed soil characteristics that reflect the dominating influence of some local factor of relief, parent material, or age over the normal effect of climate and vegetation. [1949 classification system.]

ions. Atoms, groups of atoms, or compounds that are electrically charged as a result of the loss of electrons (cations) or the gain of electrons (anions).

iron-pan. An indurated soil horizon in which iron oxide is the principal cementing agent.

irrigation efficiency. The ratio of the water actually consumed by crops on an irrigated area to the amount of water diverted from the source onto the area.

irrigation methods. Methods by which water is artificially applied to an area. The methods and the manner of applying the water are as follows.

border-strip. The water is applied at the upper end of a strip with earth borders to confine the water to the strip.

center pivot. Automated sprinkler irrigation achieved by automatically rotating the sprinkler pipe or boom, supplying water to the sprinkler heads or nozzles, as a radius from the center of the field to be irrigated.

check-basin. The water is applied rapidly to relatively level plots surrounded by levees. The basin is a small check.

corrugation. The water is applied to small, closely-spaced furrows, frequently in grain and forage crops, to confine the flow of irrigation water to one direction.

drip. A planned irrigation system where all necessary facilities have been installed for the efficient application of water directly to the root zone of plants by means of applicators (orifices, emitters, porous tubing, perforated pipe, etc.) operated under low pressure. The applicators may be placed on or below the surface of the ground.

flooding. The water is released from field ditches and allowed to flood over the land.

furrow. The water is applied to row crops in ditches made by tillage implements.

sprinkler. The water is sprayed over the soil surface through nozzles from a pressure system.

subirrigation. The water is applied in open ditches or tile lines until the water table is raised sufficiently to wet the soil.

wild-flooding. The water is released at high points in the field and distribution is uncontrolled.

isomorphous substitution. The replacement of one atom by another of similar size in a crystal lattice without disrupting or changing the crystal structure of the mineral.

kame. A conical hill or ridge of sand or gravel deposited in contact with glacial ice.

kaolinite. An aluminosilicate mineral of the 1:1 crystal lattice group; that is, consisting of one silicon tetrahedral sheet alternating with one aluminum octahedral sheet.

labile. Descriptive of a substance in soil that readily undergoes transformation or is readily available to plants.

lacustrine deposit. Material deposited in lake water and later exposed either by lowering of the water level or by the elevation of the land.

land. A broad term embodying the total natural environment of the areas of the earth not covered by water. In addition to soil, its attributes include other physical conditions such as mineral deposits and water supply; location in relation to centers of commerce, populations, and other land; the size of the individual tracts or holdings; and existing plant cover, works of improvement, and the like.

land capability classification. A grouping of kinds of soil into special units, subclasses, and classes according to their capability for intensive use and the treatments required for sustained use. One such system has been prepared by the USDA Soil Conservation Service.

land classification. The arrangement of land units into various categories based upon the properties of the land or its suitability for some particular purpose.

land-use planning. The development of plans for the uses of land that, over long periods, will best serve the general welfare, together with the formulation of ways and means for achieving such uses.

laterite. An iron-rich subsoil layer found in some highly weathered humid tropical soils that, when exposed and allowed to dry, becomes very hard and will not soften when rewetted. When erosion removes the overlying layers the laterite is exposed and a virtual pavement results.

Latosol. A suborder of zonal soils including soils formed under forested, tropical, humid conditions and characterized by low silica sesquioxide ratios of the clay fractions, low base exchange capacity, low activity of the clay, low content of most primary minerals, low content of soluble constituents, a high degree of aggregate stability, and usually a red color. [1949 classification system.]

lattice structure. The orderly arrangement of atoms in a crystalline material.

layer. (clay mineralogy) A combination in silicate clays of (tetrahedral and octahedral) sheets in a 1:1, 2:1, or 2:2 combination.

leaching. The removal of materials in solution from the soil by percolating waters. *See also* eluviation.

legume. A pod-bearing member of the Leguminosae family, one of the most important and widely distributed plant families. Includes many valuable food and forage species, such as peas, beans, peanuts, clovers, alfalfas, sweet clovers, lespedezas, vetches, and kudzu. Nearly all legumes are nitrogen-fixing plants.

Liebig's law. The growth and reproduction of an organism are determined by the nutrient substance (oxygen, carbon dioxide, calcium, etc.) that is available in minimum quantity, the *limiting factor*.

light soil. (Obsolete in scientific use) A coarse-textured soil; a soil with a low drawbar pull and hence easy to cultivate. *See also* coarse texture; soil texture.

lime (agricultural). In strict chemical terms, calcium oxide. In practical terms, it is a material containing the carbonates, oxides and/or hydroxides or calcium and/or magnesium used to neutralize soil acidity.

lime requirement. The mass of agricultural limestone, or the equivalent of other specified liming material required to raise the pH of the soil to a desired value under field conditions.

limestone. A sedimentary rock composed primarily of calcite ($CaCO_3$). If dolomite ($CaCO_3 \cdot MgCO_3$) is present in appreciable quantities, it is called a dolomitic limestone.

limiting factor. *See* Liebig's law.

liquid limit (LL). *See* Atterberg limits.

Lithosols. A great soil group of azonal soils characterized by an incomplete solum or no clearly expressed soil morphology and consisting of freshly and imperfectly weathered rock or rock fragments. [1949 classification system.]

loam. The textural class name for soil having a moderate amount of sand, silt, and clay. Loam soils contain 7–27% clay, 28–50% silt, and <52% sand.

loamy. Intermediate in texture and properties between fine-textured and coarse-textured soils. Incudes all textural classes with the words "loam" or "loamy" as a part of the class name, such as clay loam or loamy sand. *See also* loam; soil texture.

loess. Material transported and deposited by wind and consisting of predominantly silt-sized particles.

Low Humic Gley soils. An intrazonal group of somewhat poorly to poorly drained soils with very thin surface horizons moderately high in organic matter over gray and brown mineral horizons, which are developed under wet conditions. [1949 classification system.]

luxury consumption. The intake by a plant of an essential nutrient in amounts exceeding what it needs. Thus if potassium is abundant in the soil, alfalfa may take in more than it requires.

lysimeter. A device for measuring percolation and leaching losses from a column of soil under controlled conditions.

macronutrient. A chemical element necessary in large amounts (usually 50 ppm in the plant) for the growth of plants. Includes C, H, O, N, P, K, Ca, Mg, and S. ("Macro" refers to quantity and not to the essentiality of the element.) *See also* micronutrient.

marl. Soft and unconsolidated calcium carbonate, usually mixed with varying amounts of clay or other impurities.

marsh. Periodically wet or continually flooded areas with the surface not deeply submerged. Covered dominantly with sedges, cattails, rushes, or other hydrophytic plants. Subclasses include freshwater and salt-water marshes.

matric potential. *See* soil water potential.

mature soil. A soil with well-developed soil horizons produced by the natural processes of soil formation and essentially in equilibrium with its present environment.

maximum water-holding capacity. The average moisture content of a disturbed sample of soil, 1 cm high, which is at equilibrium with a water table at its lower surface.

mechanical analysis. (Obsolete) *See* particle-size analysis; particle-size distribution.

medium texture. Intermediate between fine-textured and coarse-textured (soils). It includes the following textural classes: very fine sandy loam, loam, silt loam, and silt.

mellow soil. A very soft, very friable, porous soil without any tendency toward hardness or harshness. *See also* consistence.

mesic. *See* soil temperature classes.

metamorphic rock. A rock that has been greatly altered from its previous condition through the combined action of heat and pressure. For example, marble is a metamorphic rock produced from limestone, gneiss is produced from granite, and slate from shale.

micas. Primary aluminosilicate minerals in which two silica tetrahedral sheets alternate with one alumina octahedral sheet. They separate readily into thin sheets or flakes.

microfauna. That part of the animal population which consists of individuals too small to be clearly distinguished without the use of a microscope. Includes protozoans and nematodes.

microflora. That part of the plant population which consists of individuals too small to be clearly distinguished without the use of a microscope. Includes actinomycetes, algae, bacteria, and fungi.

micronutrient. A chemical element necessary in only extremely small amounts (<50 ppm in the plant) for the growth of plants. Examples are B, Cl, Cu, Fe, Mn, and Zn. ("Micro" refers to the amount used rather than to its essentiality.) *See also* macronutrient.

microrelief. Small-scale local differences in topography, including mounds, swales, or pits that are only a meter or so in diameter and with elevation differences of up to 2 m. *See also* gilgai.

mineralization. The conversion of an element from an organic form to an inorganic state as a result of microbial decomposition.

mineral soil. A soil consisting predominantly of, and having its properties determined predominantly by, mineral matter. Usually contains <20% organic matter, but may contain an organic surface layer up to 30 cm thick.

minimum tillage. *See* tillage, conservation.

minor element. (Obsolete) *See* micronutrient.

moderately coarse texture. Consisting predominantly of coarse particles. In soil textural classification, it includes all the sandy loams except the very fine sandy loam. *See also* coarse texture.

moderately fine texture. Consisting predominantly of intermediate-sized (soil) particles or with relatively small amounts of fine or coarse particles. In soil textural classification, it includes clay loam, sandy loam, sandy clay loam, and silty clay loam. *See also* fine texture.

moisture equivalent. The weight percentage of water retained by a previously saturated sample of soil 1 cm in thickness after it has been subjected to a centrifugal force of 1000 times gravity for 30 min.

moisture tension (pressure). The equivalent negative pressure in the soil water. It is equal to the equivalent pressure that must be applied to the soil water to bring it to hydraulic equilibrium, through a porous permeable wall or membrane, with a pool of water of the same composition.

mole drain. Unlined drain formed by pulling a bullet-shaped cylinder through the soil.

mollic epipedon. *See* diagnostic surface horizons.

Mollisols. *See* soil classification.

montmorillonite. An aluminosilicate clay mineral in the smectite group with a 2:1 expanding crystal lattice, with two silicon tetrahedral sheets enclosing an aluminum octahedral sheet. Considerable expansion may be caused by water moving between silica sheets of contiguous layers.

mor. Raw humus; a type of forest humus layer of unincorporated organic material, usually matted or compacted or both; distinct from the mineral soil, unless the latter has been blackened by washing in organic matter.

moraine. An accumulation of drift, with an initial topographic expression of its own, built within a glaciated region chiefly by the direct action of glacial ice. Examples are ground, lateral, recessional, and terminal moraines.

morphology, soil. The constitution of the soil including the texture, structure, consistence, color, and other physical, chemical, and biological properties of the various soil horizons that make up the soil profile.

mottling. Spots or blotches of different color or shades of color interspersed with the dominant color.

muck. Highly decomposed organic material in which the original plant parts are not recognizable. Contains more mineral matter and is usually darker in color than peat. *See also* muck soil; peat.

muck soil. (1) a soil containing 20–50% organic matter; (2) An organic soil in which the organic matter is well decomposed.

mulch. Any material such as straw, sawdust, leaves, plastic film, and loose soil that is spread upon the surface of the soil to protect the soil and plant roots from the effects of raindrops, soil crusting, freezing, evaporation, etc.

mulch tillage. *See* tillage, conservation.

mull. A humus-rich layer of forested soils consisting of mixed organic and mineral matter. A mull blends into the upper mineral-layers without an abrupt change in soil characteristics.

Munsell color system. A color designation system that specifies the relative degree of the three simple variables of color: hue, value, and chroma. For example: 10YR 6/4 is a color (of soil) with a hue = 10YR, value = 6, and chroma = 4. These notations can be translated into several different systems of color names as desired. *See also* chroma; hue; value, color.

mycorrhiza. The association, usually symbiotic, of fungi with the roots of seed plants. *See also* ectotrophic mycorrhiza; endotrophic mycorrhiza; vesicular-arbuscular mycorrhiza.

natric horizon. *See* diagnostic subsurface horizon.

necrosis. Death associated with discoloration and dehydration of all or parts of plant organs, such as leaves.

nematodes. Very small worms abundant in many soils and important because some of them attack and destroy plant roots.

neutral soil. A soil in which the surface layer, at least to normal plow depth, is neither acid nor alkaline in reaction. *See also* acid soil; alkaline soil; pH; reaction, soil.

nitrification. The biochemical oxidation of ammonium to nitrate, predominately by autotrophic bacteria.

nitrogen assimilation. The incorporation of nitrogen into organic cell substances by living organisms.

nitrogen cycle. The sequence of chemical and biological changes undergone by nitrogen as it moves from the atmosphere into water, soil, and living organisms, and upon death of these organisms (plants and animals) is recycled through part or all of the entire process.

nitrogen fixation. The biological conversion of elemental nitrogen (N_2) to organic combinations or to forms readily utilizable in biological processes.

nodule bacteria. *See* rhizobia.

noncalcic brown soils. The zonal group of soils with slightly acid, light pinkish or light reddish brown A horizons over light reddish brown or dull red B horizons developed under mixed grass and forest vegetation in a subhumid, wet-dry climate. See horizon, soil; zonal soil. [1949 classification system.]

no-tillage. *See* tillage, conservation.

nucleic acids. Complex compounds found in plant and animal cells; may be combined with proteins as nucleoproteins.

O horizon. Organic horizon of mineral soils.

ochric epipedon. *See* diagnostic surface horizons.

order, soil. *See* soil classification

organic soil. A soil that contains at least 20% organic matter (by weight) if the clay content is low and at least 30% if the clay content is as high as 60%.

organic fertilizer. By-product from the processing of animal or vegetable substances that contain sufficient plant nutrients to be of value as fertilizers.

organic soil materials. (As used in *Soil Taxonomy* in the United States): (1) Saturated with water for prolonged periods unless artificially drained and having 18% or more organic carbon (by weight) if the mineral fraction is more than 60% clay, more than 12% organic carbon if the mineral fraction has no clay, or between 12 and 18% carbon if the clay content of the mineral fraction is between 0 and 60%. (2) Never saturated with water for more than a few days and having more than 20% organic carbon. Histosols develop on these organic soil materials. There are three kinds of organic materials.

fibric materials. The least decomposed of all the organic soil materials, containing very high amounts of fiber that are well preserved and readily identifiable as to botanical origin.

hemic materials. Intermediate in degree of decomposition of organic materials between the less decomposed fibric and the more decomposed sapric materials.

sapric materials. The most highly decomposed of the organic materials, having the highest bulk density, least amount of plant fiber, and lowest water content at saturation.

ortstein. An indurated layer in the B horizon of

Podzols in which the cementing material consists of illuviated sesquioxides (mostly iron) and organic matter.

osmotic. A type of pressure exerted in living bodies as a result of unequal concentration of salts on both sides of a cell wall or membrane. Water will move from the area having the least salt concentration through the membrane into the area having the highest salt concentration and, therefore, exerts additional pressure on this side of the membrane.

osmotic potential. *See* soil water potential.

oven-dry soil. Soil that has been dried at 105°C until it reaches constant weight.

oxic horizon. *See* diagnostic subsurface horizon.

oxidation ditch. An artificial open channel for partial digestion of liquid organic wastes in which the wastes are circulated and aerated by a mechanical device.

Oxisols. *See* soil classification.

pans. Horizons or layers, in soils, that are strongly compacted, indurated, or very high in clay content. *See also* caliche; claypan; duripan; fragipan; hardpan.

parent material. The unconsolidated and more or less chemically weathered mineral or organic matter from which the solum of soils is developed by pedogenic processes.

particle density. The mass per unit volume of the soil particles. In technical work, usually expressed as grams per cubic centimeter (g/cm³) or metric tons per cubic meter (Mg/m³).

particle size. The effective diameter of a particle measured by sedimentation, sieving, or micrometric methods.

particle-size analysis. Determination of the various amounts of the different separates in a soil sample, usually by sedimentation, sieving, micrometry, or combinations of these methods.

particle-size distribution. The amounts of the various soil separates in a soil sample, usually expressed as weight percentages.

peat. Unconsolidated soil material consisting largely of undecomposed, or only slightly decomposed, organic matter accumulated under conditions of excessive moisture. *See also* organic soil materials; peat soil.

peat soil. An organic soil containing more than 50% organic matter. Used in the United States to refer to the stage of decomposition of the organic matter, "peat" referring to the slightly decomposed or undecomposed deposits and "muck" to the highly decomposed materials. *See also* muck; muck soil; peat.

ped. A unit of soil structure such as an aggregate, crumb, prism, block, or granule, formed by natural processes (in contrast with a clod, which is formed artificially).

pedon. The smallest volume that can be called "a soil." It has three dimensions. It extends downward to the depth of plant roots or to the lower limit of the genetic soil horizons. Its lateral cross section is roughly hexagonal and ranges from 1 to 10 m² in size depending on the variability in the horizons.

peneplain. A once high, rugged area that has been reduced by erosion to a low, gently rolling surface resembling a plain.

penetrability. The ease with which a probe can be pushed into the soil. May be expressed in units of distance, speed, force, or work depending on the type of penetrometer used.

percolation, soil water. The downward movement of water through soil. Especially, the downward flow of water in saturated or nearly saturated soil at hydraulic gradients of the order of 1.0 or less.

permafrost. (1) Permanently frozen material underlying the solum. (2) A perennially frozen soil horizon.

permanent charge. The net negative (or positive) charge of clay particles inherent in the crystal lattice of the particle; not affected by changes in pH or by ion-exchange reactions.

permanent wilting point. *See* wilting point.

permeability, soil. The ease with which gases, liquids, or plant roots penetrate or pass through a bulk mass of soil or a layer of soil.

petrocalcic horizon. *See* diagnostic subsurface horizon.

petrogypsic horizon. *See* diagnostic subsurface horizon.

pF. (Obsolete) The logarithm of the soil moisture tension expressed in centimeters height of a column of water.

pH, soil. The negative logarithm of the hydrogen-ion activity of a soil. The degree of acidity (or alkalinity) of a soil as determined by means of a glass, quinhydrone, or other suitable electrode or indicator at a specified moisture content or soil–water ratio, and expressed in terms of the pH scale.

pH-dependent charge. That portion of the total charge of the soil particles that is affected by, and varies with, changes in pH.

phase, soil. A subdivision of a soil series or other unit of classification having characteristics that affect the use and management of the soil but that do not vary sufficiently to differentiate it as a separate series. A variation in a property or characteristic such as degree of slope, degree of erosion, and content of stones.

photomap. A mosaic map made from aerial photographs with physical and cultural features shown as on a planimetric map.

phyllosphere. The leaf surface.

physical properties (of soils). Those characteristics, processes, or reactions of a soil that are caused by physical forces and that can be described by, or expressed in, physical terms or equations. Examples of physical properties are bulk density, water-holding capacity, hydraulic conductivity, porosity, pore-size distribution, and so on.

placic horizon. *See* diagnostic subsurface horizons.

plaggen epipedon. *See* diagnostic surface horizons.

Planosol. A great soil group of the intrazonal order and hydromorphic suborder consisting of soils with eluviated surface horizons underlain by B horizons more strongly eluviated, cemented, or compacted than associated normal soil. [1949 classification system.]

plant nutrients. *See* essential elements.

plastic limit (PL). *See* Atterberg limits.

plastic soil. A soil capable of being molded or deformed continuously and permanently, by relatively moderate pressure, into various shapes. *See also* consistence.

platy. Consisting of soil aggregates that are developed predominantly along the horizontal axes; laminated; flaky.

plinthite (brick). A highly weathered mixture of sesquioxides or iron and aluminum with quartz and other diluents that occurs as red mottles and that changes irreversibly to hardpan upon alternate wetting and drying.

plow layer. The soil ordinarily moved when land is plowed; equivalent to *surface soil.*

plow pan. A subsurface soil layer having a higher bulk density and lower total porosity than layers above or below it, as a result of pressure applied by normal plowing and other tillage operations.

plow-plant. *See* tillage, conservation.

plowing. A primary broad-base tillage operation that is performed to shatter soil uniformly with partial to complete inversion.

Podzol. A great soil group of the zonal order consisting of soils formed in cool-temperate to temperate, humid climates, under coniferous or mixed coniferous and deciduous forest, and characterized particularly by a highly leached, whitish gray A2 horizon. [1949 classification system.]

podzolization. A process of soil formation resulting in the genesis of Podzols and podzolic soils.

polypedon. (As used in *Soil Taxonomy* in the United States): Two or more contiguous pedons, all of which are within the defined limits of a single soil series. In early stages of the development of the classification scheme this was called a soil individual.

pore size distribution. The volume of the various sizes of pores in a soil. Expressed as percentages of the bulk volume (soil plus pore space).

porosity, soil. The volume percentage of the total soil bulk not occupied by solid particles.

Prairie soils. A zonal great soil group consisting of soils formed under temperate to cool-temperate, humid regions under tall grass vegetation. [1949 classification system.]

primary mineral. A mineral that has not been altered chemically since deposition and crystallization from molten lava.

primary tillage. *See* tillage, primary.

prismatic soil structure. A soil structure type with prism-like aggregates that have a vertical axis much longer than the horizontal axes.

productivity, soil. The capacity of a soil for producing a specified plant or sequence of plants under a specified system of management. Productivity emphasizes the capacity of soil to produce crops and should be expressed in terms of yields.

profile, soil. A vertical section of the soil through all its horizons and extending into the parent material.

protein. Any of a group of nitrogen-containing organic compounds formed by the polymerization of a large number of amino acid molecules and that, upon hydrolysis, yield these amino acids. They are essential parts of living matter and are one of the essential food substances of animals.

puddled soil. Dense, massive soil artificially compacted when wet and having no regular structure. The condition commonly results from the tillage of a clayey soil when it is wet.

rain, acid. *See* acid rain.

reaction, soil. The degree of acidity or alkalinity of a soil, usually expressed as a pH value.

Extremely acid	<4.5
Very strongly acid	4.5–5.0
Strongly acid	5.1–5.5
Medium acid	5.6–6.0
Slightly acid	6.1–6.5
Neutral	6.6–7.3
Mildly alkaline	7.4–7.8
Moderately alkaline	7.9–8.4
Strongly alkaline	8.5–9.0
Very strongly alkaline	9.1 and higher

Red Desert soil. A zonal great soil group consisting of soils formed under warm-temperate to hot, dry regions under desert-type vegetation, mostly shrubs. [1949 classification system.]

Reddish Brown soils. A zonal group of soils with a light brown surface horizon of a slightly reddish cast which grades into dull reddish brown or red material heavier than the surface soil, then into a horizon of whitish or pinkish lime accumulation; developed under shrub and short-grass vegetation of warm-temperate to tropical regions of semiarid climate. [1949 classification system.]

Reddish Brown Lateritic soils (of U.S.). A zonal group of soils with dark reddish brown granular surface soils, red friable clay B horizons, and red or reticulately mottled lateritic parent material, developed under humid tropical climate with wet-dry seasons and tropical forest vegetation. [1949 classification system.]

Red-Yellow Podzolic soils. A zonal great soil group consisting of soils formed under warm-temperate to tropical, humid climates, under deciduous or coniferous forest vegetation and usually under conditions of good drainage. [1949 classification system.]

regolith. The unconsolidated mantle of weathered rock and soil material on the earth's surface; loose earth materials above solid rock. (Approximately equivalent to the term "soil" as used by many engineers.)

Regosol. Any soil of the azonal order without definite genetic horizons and developing from or on deep, unconsolidated, soft mineral deposits such as sands, loess, or glacial drift. [1949 classification system.]

Rendzina. A great soil group of the intrazonal order and calcimorphic suborder consisting of soils with brown or black friable surface horizons underlain by light gray to pale yellow calcareous material; developed from soft, highly calcareous parent material under grass vegetation or mixed grasses and forest in humid and semiarid climates. [1949 classification system.]

residual material. Unconsolidated and partly weathered mineral materials accumulated by disintegration of consolidated rock in place.

rhizobia. Bacteria capable of living symbiotically with higher plants, usually in nodules on the roots of legumes, from which they receive their energy, and capable of converting atmospheric nitrogen to combined organic forms; hence, the term *symbiotic nitrogen-fixing bacteria.* (Derived from the generic name *Rhizobium.*)

rhizosphere. That portion of the soil in the immediate vicinity of plant roots in which the abundance and composition of the microbial population are influenced by the presence of roots.

rill. A small, intermittent water course with steep sides; usually only a few inches deep and hence no obstacle to tillage operations.

rill erosion. *See* erosion.

riprap. Broken rock, cobbles, or boulders placed on earth surfaces, such as the face of a dam or the bank of a stream, for protection against the action of water (waves); also applied to brush or pole mattresses, or brush and stone, or other similar materials used for soil erosion control.

rock. The material that forms the essential part of the earth's solid crust, including loose incoherent masses such as sand and gravel, as well as solid masses of granite and limestone.

rotary tillage. *See* tillage, rotary.

runoff. The portion of the precipitation on an area that is discharged from the area through stream channels. That which is lost without entering the soil is called *surface runoff* and that which enters the soil before reaching the stream is called *groundwater runoff* or *seepage flow* from groundwater. (In soil science "runoff" usually refers to the water lost by surface flow; in geology and hydraulics "runoff" usually includes both surface and subsurface flow.)

salic horizon. *See* diagnostic subsurface horizons.

saline-sodic soil. A soil containing sufficient exchangeable sodium to interfere with the growth of most crop plants and containing appreciable quantities of soluble salts. The exchangeable-sodium percentage is >15, the conductivity of the saturation extract >2 ds/m (at 25°C), and the pH is usually 8.5 or less in the saturated soil.

saline soil. A nonsodic soil containing sufficient soluble salts to impair its productivity.

salinization. The process of accumulation of salts in soil.

saltation. Particle movement in water or wind where particles skip or bounce along the stream bed or soil surface.

sand. A soil particle between 0.05 and 2.0 mm in diameter; a soil textural class.

sapric materials. *See* organic soil materials.

savanna (savannah). A grassland with scattered trees, either as individuals or clumps. Often a transitional type between true grassland and forest.

second bottom. The first terrace above the normal flood plain of a stream.

secondary mineral. A mineral resulting from the decomposition of a primary mineral or from the reprecipitation of the products of decomposition of a primary mineral. *See also* primary mineral.

sedimentary rock. A rock formed from materials deposited from suspension or precipitated from solution and usually being more or less consolidated. The principal sedimentary rocks are sandstones, shales, limestones, and conglomerates.

self-mulching soil. A soil in which the surface layer becomes so well aggregated that it does not crust and seal under the impact of rain but instead serves as a surface mulch upon drying.

separate, soil. One of the individual-sized groups of mineral soil particles—sand, silt, or clay.

series, soil. *See* soil classification.

sewage sludge. Settled sewage solids combined with varying amounts of water and dissolved materials, removed from sewage by screening, sedimentation, chemical precipitation, or bacterial digestion.

shear. Force, as of a tillage implement, acting at right angles to the direction of movement.

sheet. (minerology) A flat array of more than one atomic thickness and composed of one or more levels of linked coordination polyhedra. A sheet is thicker than a plane and thinner than a layer. Example: tetrahedral sheet, octahedral sheet.

sheet erosion. *See* erosion.

shelterbelt. A wind barrier of living trees and shrubs established and maintained for protection of farm fields. Syn. windbreak.

shrinkage limit (SL). *See* Atterberg limits.

Sierozem. A zonal great soil group consisting of soils with pale-grayish A horizons grading into calcareous material at a depth of 30 cm or less, and formed in temperate to cool arid climates under a vegetation of desert plants, short grass, and scattered brush. [1949 classification system.]

silica/alumina ratio. The molecules of silicon dioxide (SiO_2) per molecule of aluminum oxide (Al_2O_3) in clay minerals or in soils.

silica/sesquioxide ratio. The molecules of silicon dioxide (SiO_2) per molecule of aluminum oxide (Al_2O_3) plus ferric oxide (Fe_2O_3) in clay minerals or in soils.

silt. (1) A soil separate consisting of particles between 0.05 and 0.002 mm in equivalent diameter. (2) A soil textural class.

silting. The deposition of water-borne sediments in stream channels, lakes, reservoirs, or on flood plains, usually resulting from a decrease in the velocity of the water.

site index. A quantitative evaluation of the productivity of a soil for forest growth under the existing or specified environment.

slag. The product of smelting, containing mostly silicates; the substances not sought to be produced as matte or metal and having a lower specific gravity.

slick spots. Small areas in a field that are slick when wet, due to a high content of alkali or exchangeable sodium.

slope. The degree of deviation of a surface from horizontal, measured in a numerical ratio, percent, or degrees.

smectite. A group of silicate clays having a 2:1-type lattice structure with sufficient isomophorous substitution in either or both the tetrahedral and octahedral sheets to give a high interlayer negative charge and high cation exchange capacity and to permit significant interlayer expansion and consequent shrinking and swelling of the clay. Montmorillonite, beidellite, and saponite are in the smectite group.

sodic soil. A soil that contains sufficient sodium to interfere with the growth of most crop plants, and in which the exchangeable sodium percentage is 15 or greater.

sodium adsorption ratio (SAR).

$$SAR = \frac{[Na^+]}{\sqrt{[Ca^{2+}] + [Mg^{2+}]}}$$

where the cation concentrations are in millimoles per liter (mmole/L).

soil. (1) a dynamic natural body composed of mineral and organic materials and living forms in which plants grow. (2) The collection of natural bodies occupying parts of the earth's surface that support plants and that have properties due to the integrated effect of climate and living matter acting upon parent material, as conditioned by relief, over periods of time.

soil air. The soil atmosphere; the gaseous phase of the soil, being that volume not occupied by solid or liquid.

soil alkalinity. The degree or intensity of alkalinity of a soil, expressed by a value >7.0 on the pH scale.

soil amendment. Any material, such as lime, gypsum, sawdust, or synthetic conditioner, that is worked into the soil to make it more amenable to plant growth.

soil association. A group of defined and named taxonomic soil units occurring together in an individual and characteristic pattern over a geographic region, comparable to plant associations in many ways.

soil classification. (*Soil Taxonomy*) The systematic arrangement of soils into groups or categories on the basis of their characteristics. The categories of the system used in the U.S. since 1966 are described briefly.

order. The category at the highest level of generalization in the soil classification system. The properties selected to distinguish the orders are reflections of the degree of horizon development and the kinds of horizons present. The ten orders are as follows.

Alfisols. Soils with gray to brown surface horizons, medium to high supply of bases, and B horizons of illuvial clay accumulation. These soils form mostly under forest or savanna vegetation in climates with slight to pronounced seasonal moisture deficit.

Aridisols. Soils of dry areas. They have pedogenic horizons, low in organic matter, that are never moist as long as 3 consecutive months. They have an ochric epipedon and one or more of the following diagnostic horizons: argillic, natric, cambic, calcic, petrocalcic, gypsic, salic, or a duripan.

Entisols. Soils that have no diagnostic pedogenic horizons. They may be found in virtually any climate on very recent geomorphic surfaces.

Histosols. Soils formed from materials high in organic matter. Histosols with essentially no clay must have at least 20% organic matter by weight (about 78% by volume). This minimum organic matter content rises with increasing clay content to 30% (85% by volume) in soils with at least 50% clay.

Inceptisols. Soils that are usually moist with pedogenic horizons of alteration of parent materials but not of illuviation. Generally, the direction of soil development is not yet evident from the marks left by various soil-forming processes or the marks are too weak to classify in another order.

Mollisols. Soils with nearly black, organic-rich surface horizons and high supply of bases. They have mollic epipedons and base saturation greater than 50% in any cambic or argillic horizon. They lack the characteristics of Vertisols and must not have oxic or spodic horizons.

Oxisols. Soils with residual accumulations of inactive clays, free oxides, kaolin, and quartz. They are mostly in tropical climates.

Spodosols. Soils with subsurface illuvial accumulations of organic matter and compounds of aluminum and usually iron. These soils are formed in acid, mainly coarse-textured materials in humid and mostly cool or temperate climates.

Ultisols. Soils that are low in bases and have subsurface horizons of illuvial clay accumulations. They are usually moist, but during the warm season of the year some are dry part of the time.

Vertisols. Clayey soils with high shrink-swell potential that have wide, deep cracks when dry. Most of these soils have distinct wet and dry periods throughout the year.

suborder. This category narrows the ranges in soil moisture and temperature regimes, kinds of horizons, and composition, according to which of these is most important.

great group. The classes in this category contain soils that have the same kind of horizons in the same sequence and have similar moisture and temperature regimes.

subgroup. The great groups are subdivided into central concept subgroups that show the central properties of the great group, intergrade subgroups that show properties of more than one great group, and other subgroups for soils with atypical properties that are not characteristic of any great group.

family. Families are defined largely on the basis of physical and mineralogic properties of importance to plant growth.

series. The soil series is a subdivision of a family and consists of soils that are similar in all major profile characteristics.

soil complex. A mapping unit used in detailed soil surveys where two or more defined taxonomic units are so intimately intermixed geographically that it is undesirable or impractical, because of the scale being used, to separate them. A more intimate mixing of smaller areas of individual taxonomic units than that described under *soil association*.

soil conditioner. Any material added to a soil for the purpose of improving its physical condition.

soil conservation. A combination of all management and land-use methods that safeguard the soil against depletion or deterioration caused by nature and/or humans.

soil correlation. The process of defining, mapping, naming, and classifying the kinds of soils in a specific soil survey area, the purpose being to ensure that soils are adequately defined, accurately mapped, and uniformly named.

soil erosion. *See* erosion.

soil fertility. *See* fertility, soil.

soil genesis. The mode of origin of the soil, with special reference to the processes or soil-forming factors responsible for the development of the solum, or true soil, from the unconsolidated parent material.

soil geography. A subspecialization of physical geography concerned with the areal distributions of soil types.

soil horizon. *See* horizon, soil.

soil management. The sum total of all tillage operations, cropping practices, fertilizer, lime, and other treatments conducted on or applied to a soil for the production of plants.

soil map. A map showing the distribution of soil types or other soil mapping units in relation to the prominent physical and cultural features of the earth's surface.

soil mechanics and engineering. A subspecialization of soil science concerned with the effect of forces on the soil and the application of engineering principles to problems involving the soil.

soil microbiology. A subspecialization of soil science concerned with soil-inhabiting microorganisms and with their relation to agriculture, including both plant and animal growth.

soil moisture potential. *See* soil water potential.

soil monolith. A vertical section of a soil profile removed from the soil and mounted for display or study.

soil morphology. The physical constitution, particularly the structural properties, of a soil profile as exhibited by the kinds, thickness, and arrangement of the horizons in the profile, and by the texture, structure, consistency, and porosity of each horizon.

soil profile. A vertical section of the soil from the surface through all its horizons, including C horizons. *See also* horizon, soil.

soil organic matter. The organic fraction of the soil that includes plant and animal residues at various stages of decomposition, cells and tissues of soil organisms, and substances synthesized by the soil population. Commonly determined as the amount of organic material contained in a soil sample passed through a 2-mm sieve.

soil porosity. *See* porosity, soil.

soil productivity. *See* productivity, soil.

soil reaction. *See* reaction, soil; pH, soil.

soil salinity. The amount of soluble salts in a soil, expressed in terms of percentage, parts per million (ppm), or other convenient ratios.

soil separates. *See* separate, soil.

soil series. *See* soil classification.

soil solution. The aqueous liquid phase of the soil and its solutes consisting of ions dissociated from the surfaces of the soil particles and of other soluble materials.

soil structure. The combination or arrangement of primary soil particles into secondary particles, units, or peds. These secondary units may be, but usually are not, arranged in the profile in such a manner as to give a distinctive characteristic pattern. The secondary units are characterized and classified on the basis of size, shape, and degree of distinctness into classes, types, and grades, respectively.

soil structure classes. A grouping of soil structural units or peds on the basis of size from the very fine to very coarse.

soil structure grades. A grouping or classification of soil structure on the basis of inter- and intra-aggregate adhesion, cohesion, or stability within the profile. Four grades of structure, designated from 0 to 3, are recognized.

0: *structureless*—no observable aggregation.

1: *weakly* durable peds.

2: *moderately* durable peds.

3: *strong,* durable peds.

soil structure types. A classification of soil structure based on the shape of the aggregates or peds and their arrangement in the profile, including platy, prismatic, columnar, blocky, subangular blocky, granulated, and crumb.

soil survey. The systematic examination, description, classification, and mapping of soils in an area. Soil surveys are classified according to the kind and intensity of field examination.

soil temperature classes. (*Soil Taxonomy*) Classes are based on mean annual soil temperature and on differences between summer and winter temperatures at a depth of 50 cm.

1. Soils with 5°C and greater difference between summer and winter temperatures are classed on the basis of mean annual temperatures.

   frigid: <8°C mean annual temperature.

   mesic: 8–15°C mean annual temperature.

   thermic: 15–22°C mean annual temperature.

   hyperthermic: >22°C mean annual temperature.

2. Soils with <5°C difference between summer and winter temperatures are classed on the basis of mean annual temperatures.

   isofrigid: <8°C mean annual temperature.

   isomesic: 8–15°C mean annual temperature.

   isothermic: 15–22°C mean annual temperature.

   isohyperthermic: >22°C mean annual temperature.

soil textural class. A grouping of soil textural units based on the relative proportions of the various soil separates (sand, silt, and clay). These textural classes, listed from the coarsest to the finest in texture, are sand, loamy sand, sandy loam, loam, silt loam, silt, sandy clay loam, clay loam, silty clay loam, sandy clay, silty clay, and clay. There are several subclasses of the sand, loamy sand, and sandy loam classes based on the dominant particle size of the sand fraction (e.g., loamy fine sand, coarse sandy loam).

soil texture. The relative proportions of the various soil separates in a soil.

soil type. The lowest unit in the natural system of soil classification; a subdivision of a soil series and consisting of or describing soils that are alike in all characteristics including the texture of the A horizon. [1949 classification system.]

soil water potential. (total) A measure of the difference in the free energy state of soil water and that of pure water. Technically it is defined as "that amount of work that must be done per unit quantity of pure water in order to transport reversibly and isothermically an infinitesimal quantity of water from a pool of pure water, at a specified elevation and at atmospheric pressure to the soil water (at the point under consideration)." This *total* potential consists of the following potentials.

    matric potential. That portion of the total soil water potential due to the attractive forces between water and soil solids as represented through adsorption and capillarity. It will always be negative.

    osmotic potential. That portion of the total soil water potential due to the presence of solutes in soil water. It will generally be negative.

    gravitational potential. That portion of the total soil water potential due to differences in elevation of the reference pool of pure water and that of the soil water. Since the soil water elevation is usually chosen to be higher than that of the reference pool, the gravitational potential is usually positive.

Sol Brun Acide. A zonal group of soils developed under forest vegetation with thin A1 horizon, a paler A2 horizon that is poorly differentiated from the B2 horizon, a B2 horizon with uniform color from top to bottom, weak subangular blocky structure, and lacking evidence of silicate clay accumulation. The sola are strongly to very strongly acid.

solodized soil. A soil that has been subjected to the processes responsible for the development of a Soloth and having at least some of the characteristics of a Soloth. [1949 classification system.]

Solonchak soils. An intrazonal group of soils with high concentrations of soluble salts in relation to those in other soils, usually light-colored, without characteristic structural form, developed under salt-loving plants, and occurring mostly in a subhumid or semiarid climate. [1949 classification system.]

Solonetz soils. An intrazonal group of soils having surface horizons of varying degrees of friability underlain by dark hard soil, ordinarily with columnar structure (prismatic structure with rounded tops). This hard layer is usually highly alkaline. Such soils are developed under grass or shrub vegetation, mostly in subhumid or semiarid climates. [1949 classification system.]

soluble-sodium percentage (SSP). The proportion of sodium ions in solution in relation to the total cation concentration, defined as

$$SSP = \frac{\text{Soluble sodium conc. (cmol/L)}}{\text{Total cation conc. (cmol/L)}} \times 100$$

solum (plural: **sola**). The upper and most weathered part of the soil profile; the A and B horizons.

splash erosion. *See* erosion.

spodic horizon. *See* diagnostic subsurface horizons.

Spodosols. *See* soil classification.

sprinkler irrigation. *See* irrigation methods.

stratified. Arranged in or composed of strata or layers.

strip cropping. The practice of growing crops that require different types of tillage, such as row and sod, in alternate strips along contours or across the prevailing direction of wind.

structure, soil. *See* soil structure.

stubble mulch. The stubble of crops or crop residues left essentially in place on the land as a surface cover before and during the preparation of the seedbed and at least partly during the growing of a succeeding crop.

subirrigation. *See* irrigation methods.

subsoil. That part of the soil below the plow layer.

subsoiling. Breaking of compact subsoils, without inverting them, with a special knife-like instrument (chisel), which is pulled through the soil at depths usually of 30–60 cm and at spacings usually of 1–2 m.

summer fallow. The tillage of uncropped land during the summer in order to control weeds and store moisture in the soil for the growth of a later crop.

surface runoff. *See* runoff.

surface soil. The uppermost part of the soil, ordinarily moved in tillage, or its equivalent in uncultivated soils. Ranges in depth from 7 to 25 cm. Frequently designated as the "plow layer," the "Ap layer," or the "Ap horizon."

symbiosis. The living together in intimate association of two dissimilar organisms, the cohabitation being mutually beneficial.

talus. Fragments of rock and other soil material accumulated by gravity at the foot of cliffs or steep slopes.

taxonomy, soil. The science of classification of soils; laws and principles governing the classifying the soil. *See also* soil classification.

tensiometer. A device for measuring the negative pressure (or tension) of water in soil *in situ;* a porous, permeable ceramic cup connected through a tube to a manometer or vacuum gauge.

tension, soil-moisture. *See* moisture tension.

terrace. (1) A level, usually narrow, plain bordering a river, lake, or the sea. Rivers sometimes are bordered by terraces at different levels. (2) A raised, more or less level or horizontal strip of earth usually constructed on or nearly on a contour and designed to make the land suitable for tillage and to prevent accelerated erosion by diverting water from undesirable channels of concentration.

tertiary waste treatment. Waste-water treatment beyond the secondary or biological stage that includes removal of nutrients such as phosphorus and nitrogen and of a high percentage of suspended solids.

texture. *See* soil texture.

thermal analysis (differential thermal analysis). A method of analyzing a soil sample for constituents, based on a differential rate of heating of the unknown and standard samples when a uniform source of heat is applied.

thermic. *See* soil temperature.

thermophilic organisms. Organisms that grow readily at temperatures above 45°C.

tile, drain. Pipe made of burned clay, concrete, or ceramic material, in short lengths, usually laid with open joints to collect and carry excess water from the soil.

till. (1) Unstratified glacial drift deposited directly by the ice and consisting of clay, sand, gravel, and boulders intermingled in any proportion. (2) To plow and prepare for seeding; to seed or cultivate the soil.

tillage. The mechanical manipulation of soil for any purpose; but in agriculture it is usually restricted to the modifying of soil conditions for crop production.

tillage, conservation. Any tillage sequence that reduces loss of soil or water relative to conventional tillage, including the following systems.

minimum tillage. The minimum soil manipulation necessary for crop production or meeting tillage requirements under the existing soil and climatic conditions.

mulch tillage. Tillage or preparation of the soil in such a way that plant residues or other materials are left to cover the surface; also called *mulch farming, trash farming, stubble mulch tillage, plowless farming.*

no-tillage system. A procedure whereby a crop is planted directly into a seedbed not tilled since harvest of the previous crop; also zero tillage.

plow-planting. The plowing and planting of land in a single trip over the field by drawing both plowing and planting tools with the same power sources.

sod planting. A method of planting in sod with little or no tillage.

subsurface tillage. Tillage with a special sweep-like plow or blade that is drawn beneath the surface, cutting plant roots and loosening the soil without inverting it or without incorporating residues of the surface cover.

wheel track planting. A practice of planting in which the seed is planted in tracks formed by wheels rolling immediately ahead of the planter.

tillage, conventional. The combined primary and

secondary tillage operations normally performed in preparing a seedbed for a given crop grown in a given geographic area.

tillage, primary. Tillage that contributes to the major soil manipulation.

tillage, rotary. An operation using a power-driven rotary tillage tool to loosen and mix soil.

tilth. The physical condition of soil as related to its ease of tillage, fitness as a seedbed, and its impedance to seedling emergence and root penetration.

top dressing. An application of fertilizer to a soil after the crop stand has been established.

toposequence. A sequence of related soils that differ, one from the other, primarily because of *topography* as a soil-formation factor.

topsoil. (1) The layer of soil moved in cultivation. *See also* surface soil. (2) Presumably fertile soil material used to top-dress roadbanks, gardens, and lawns.

trace element. (Obsolete) *See* micronutrient.

truncated. Having lost all or part of the upper soil horizon or horizons.

tuff. Volcanic ash usually more or less stratified and in various states of consolidation.

tundra. A level or undulating treeless plain characteristic of arctic regions.

tundra soils. (1) Soils characteristic of tundra regions. (2) A zonal great soil group, consisting of soils with dark-brown peaty layers over grayish horizons mottled with rust and having continually frozen substrata; formed under frigid, humid climates, with poor drainage, and native vegetation of lichens, moss, flowering plants, and shrubs. [1949 classification system.]

type, soil. *See* soil type.

Ultisol. *See* soil classification.

umbric epipedon. *See* diagnostic surface horizons.

unsaturated flow. The movement of water in a soil that is not filled to capacity with water.

value, color. The relative lightness or intensity of color and approximately a function of the square root of the total amount of light. One of the three variables of color. *See also* chroma; hue; Munsell color system.

varnish, desert. A glossy sheen or coating on stones and gravel in arid regions.

Vertisols. *See* soil classification.

vesicular-arbuscular mycorrhiza. A common endomycorrhizal association produced by phycomycetous fungi of the genus *Endogone*. Host range includes many agricultural and horticultural crops. *See also* endomycorrhiza.

virgin soil. A soil that has not been significantly disturbed from its natural environment.

volume weight. (Obsolete) *See* bulk density, soil.

waterlogged. Saturated with water.

water potential, soil. *See* soil water potential.

water-stable aggregate. A soil aggregate stable to the action of water such as falling drops, or agitation as in wet-sieving analysis.

water table. The upper surface of ground water or that level below which the soil is saturated with water.

water table, perched. The surface of a local zone of saturation held above the main body of groundwater by an impermeable layer of stratum, usually clay, and separated from the main body of groundwater by an unsaturated zone.

water use efficiency. Dry matter or harvested portion of crop produced per unit of water consumed.

weathering. All physical and chemical changes produced in rocks, at or near the earth's surface, by atmospheric agents.

wilting point (permanent wilting point). The moisture content of soil, on an oven-dry basis, at which plants wilt and fail to recover their turgidity when placed in a dark humid atmosphere.

xerophytes. Plants that grow in or on extremely dry soils or soil materials.

zero tillage. *See* tillage, conservation.

zeta potential. *See* electrokinetic potential.

zonal soil. A soil characteristic of a large area or zone. [1949 classification system.]

INDEX

Calcium
  adsorbed, prominence of, 149
  amount in soils, 22, 28, 489
  amounts exchangeable in
    mineral soils, 26, 181
    peats, 488
  compounds used as lime, 572
  effect on
    boron absorption, 206, 582
    microbial activity, 209, 245, 582
    nitrification, 292, 581
    phosphate availability, 207–208,
      337, 581–82
    potassium availability, 357, 413,
      479, 581
  an essential element, 21
  flocculating effects, 185
  forms in soils, 28
  gains and losses of, 588
  ionic form in soil, 28
  loss by erosion, 534
  loss in drainage, 521, 522
  pH and exchangeable calcium,
    149, 181, 206
  removal by crops, 522
Calcium carbonate
  in liming materials, 573
  solubility influenced by CO_2, 392
Calcium cyanamid, as a fertilizer,
    592, 595
Calcium metaphosphate, as a
    fertilizer, 599, 601
Calcium oxide equivalent of lime,
    574–575
Cal-nitro as a fertilizer, 592, 595
Capability classes (see Land
    capability classification)
Capillary movement of soil water
  explained, 76–79
  supplying plants with water, 104
Capillary water in soil
  amounts in soils, 99
  tension limits, 101
Carbon
  an essential element, 21
  ratio to nitrogen
    in plants, 269
    in soil, 269, 488
Carbon cycle, 262
Carbon dioxide (see also Carbonic
    acid)
  diffusion of, 112
  generation in farm manure, 637
  generation in soil, 261

place in the carbon cycle, 262
  in soil air, 17, 113–14
  source of carbon for plants, 261
  used by autotrophic bacteria, 244
Carbon/nitrogen ratio
  conservation of nitrogen, 269–72
  effect of climate on, 269
  effect on nitrates, 271, 292, 488
  explanation of, 269
  maintenance of soil humus, 271
  of mineral soils, 269
  of peat soils, 488
  of plants, 269
Carbonate of lime (see Lime)
Carbonic acid
  cation exchange reaction, 174
  developed in soil solution, 261
  effect on soil pH, 198
  reaction with lime, 578
  release of calcium, 174, 392
"Cat" clays, (see Acid sulfate soils)
Catenas, soil, 461–462
Cation exchange, 173–75
  availability of nutrients, 181–82
Cation exchange capacity
  calculation of, 179
  factors influencing, 178
  means of expressing, 175
  of hydrous oxide clays, 177
  of peats, 486, 488
  of representative mineral soils,
    179
  of silicate clays, 156, 157, 158, 177
  of soil humus, 179
  of various mineral soils, 177
Cations
  adsorbed by colloids, 24, 148
  flocculating capacity, 186
  hydration, 145, 185
  present in soil solution, 24, 174
Cellulose in plant residues, 256, 257
Centimoles of charge, 175
Cesium, 685
Chelates, 374–76
Chemical analysis
  complete analysis of soil, 22
  for fertilizer needs, 619–20
  for organic carbon, 272
  of farm manure, 632
  of fertilizers, 592, 599, 603, 604, 612,
    614
  of liming materials, 574–76
  of peat soils, 488–90
  of plant tissue, 255–56

of silicate clays, 162–64
of soil air, 113
of soil separates, 41
rapid tests for soil, 620
representative arid region soil, 22
representative humid region soil,
  22
Chemical weathering, 391
Chernozems (see also Mollisol), 430,
  463
Chestnut soils, 463
Chlorine
  added in rain, 369, 376
  amounts in soil, 369
  an essential element, 21, 369, 376
Chlorite, a soil mineral, 158, 166
Classification of soil (see Soil
  classification)
Clays (see Hydrous oxide clays;
  Silicate clays)
Climate
  cause of variation in loess soils,
    410
  effect on
    soil nitrogen, 274
    soil organic matter, 274
    transpiration ratio, 504
    vegetation, 417
    weathering, 394, 416
  factor in soil formation, 416
Clostridium in soils, 309
Cohesion in soil, 146
  of clay, 39, 184
  of peat, 485
  of water, 76
Colloids, soil (see also Humus;
    Hydrous oxide clays; Silicate
    clays)
  acid-salt-like nature, 148
  availability of nutrients from, 31,
    181
  cation exchange capacity of, 175–
    79
  cation exchange of, 173
  charges on, 169
  control of soil pH by, 196
  dispersion of, 184, 217
  flocculation of, 58, 186
  general organization of, 25, 146
  humus-clay complex of, 17
  kinds of clay, 153–60
  organic colloids, 147, 486
  reaction with lime, 578
  seat of soil activity, 18

Hectare–furrow slice, 51
  energy of, 260
  nutrients in, 22
  weight of
    mineral soils, 51
    peat soils, 485
Hematite, 388
Hemicellulose, in plant residues, 256–57
Herbicides, 655, 657, 660
Heterotrophic bacteria, 244, 270
Histosol soil order, 434, 451, 452, 453, 476
Horizons, diagnostic, 432
Horizons of soil (see Soil profile)
Humic Gley soils, 463
Humid vs. arid region soils, 22, 181
Humus, soil (see also Soil organic matter)
  adsorption of cations, 148, 267
  cation exchange capacity, 16, 148, 177, 267
  characteristics of, 1, 5, 147–48, 266–68
  C/N ratio, 269
  definition of, 258, 266
  determination of, 272
  direct influence on plants, 268
  effect of farm manures on, 278, 639
  energy of, 260
  genesis of, 258, 264
  importance of, 18–19, 266, 268
  lignin (modified), 265
  physical properties of, 266–67
  polyuronides in, 265
  ratio to nitrogen, 274
  stability of, 268
Hydrated lime
  changes in soil, 578
  guarantee, 574
  source and nature of, 572–573
Hydration
  of cations, 74, 146
  of clays, 74
  in weathering, 391
Hydraulic conductivity, 91
Hydrogen, an essential element, 21
Hydrogen bonding
  in dry minerals, 158
  in soil water, 75
Hydrogen ion
  as active acidity, 201–202
  adsorbed by colloidal complex, 145, 148, 196

cation exchange by, 173
effect on
  microorganisms, 209
  plants, 209
as exchange acidity, 201–202
flocculating effect by, 186
pH expression of, 29–30
source of, 190
strength of adsorption, 149
Hydrolysis
  alkalinity developed by, 217
  of proteins, 288
  in weathering, 391
Hydrous micas (see Fine-grained micas)
Hydrous oxide clays
  cation exchange capacity, 146, 177
  general organization, 146
  an important type of clay, 142, 145–46
  origin of, 167
  phosphorus fixation by, 337
  physical properties, 146
  positive charge on particles, 172
Hydroxide of lime, 572
Hygroscopic water
  amount in soils, 99
  nature of, 100

Igneous rocks, 386
Illite (see Fine-grained micas)
Illuvial horizon in soil, 425
Inceptisol soil order, 434, 443, 445, 452, 453, 454, 463
Indicators, determination of pH, 211, 212
Infiltration (see Percolation)
Insecticides, 655, 656, 658, 661
Insects in soils, 225, 226, 227
Interglacial periods, 405–406
Ionic exchange (see Anion exchange; Cation exchange)
Iron
  amount in soils, 22
  availability of, 125, 372, 581, 582
  cause of mottling, 125
  chelates of, 374
  an essential element, 21
  fertilizers of, 380–81, 604
  influence on phosphate availability, 208, 337
  ionic forms in soil, 370
  and minerals in weathering, 392

oxidation and color, 125
pH and active iron, 372
role of, 366, 367
Iron oxides (see Hydrous oxides)
Irrigation
  adding fertilizer by, 616
  alkali control under, 220
  control of salts by quality of water, 220
  evaporation control under, 512

Kainit as a fertilizer, 602
Kaolinite in soils
  cation exchange capacity, 156, 177
  genesis, 167
  geographical distribution in U.S., 168–69
  mineralogical organization, 153–54
  physical properties, 154, 183

Lacustrine deposits (see Glacial-lacustrine deposits)
Land-capability classification, 467, 472
Land resources, world, 691
Laterite, 448
Latosols, 448
Layered silicate clay (see Silicate clays)
Leaching
  and nutrient losses, 519–22
  in reclaiming saline sodic soils, 220
  in soil formation, 392, 423
Lead, as pollutant, 672
Legumes
  bacteria in root nodules, 303
  effect on soil nitrogen, 306
  fertility maintenance by, 306, 312
  fixation of soil nitrogen by, 303–305
  importance of, 303–304
  influence of soil reaction on, 209, 210, 582
Lignin
  as plant constituent, 256
  protection of soil nitrogen, 256
  rate of decomposition, 257
  role in humus formation, 265
  stability of, 257
Lime
  amounts to apply, 585–86
  burned lime, 572

calcium carbonate equivalent of, 575

calcium oxide equivalent of, 574, 575

carbonate of, 573

chemical effects of, 581

chemical guarantees, 570

conversion factors for, 575

crop response to, 582

diagram of gain and loss, 588

effect on
  microorganisms, 245, 581
  nitrification, 292, 581
  nitrogen fixation, 304, 309, 581
  phosphate availability, 340, 341, 581, 582
  potassium, 357, 581
  soil structure, 580

fineness guarantee of limestone, 576

forms of, 572–73

forms to apply, 583

hydrated lime, 572

losses from soil, 579

methods of application, 586

neutralizing power, 574, 575, 576

overliming, 206, 582

oxide of lime, 576

reactions with the soil, 578

removal of, by crops, 579

slaked lime, 572

total carbonates of, 575

when to apply, 587

Limestone
  calcic, 573, 576, 584
  chemical composition of, 393, 575
  chemical guarantee, 574–76
  dolomitic, 573, 575, 576, 584
  fineness of, 576, 577
  rate of reaction with soil, 578, 584
  as source of minerals, 387, 388

Limiting factors, principle of, 20

Loam, a soil class, 42

Loess, 410–12

Loss of nutrients from soils
  by cropping, 522, 580
  by drainage, 519–22, 580
  by erosion, 534–35, 580
  by volatilization, 294, 295

Low-humic Gley soils, 463

Luxury consumption of potassium, 352

Lysimeters
  description of types, 515
  loss of nutrients through, 519–22

Macronutrients (see also Calcium, Magnesium; Nitrogen; Phosphorus; Potassium; Sulfur), 24–28, 284–362, 590–603

Macroorganisms, 229–33

Macropore spaces
  affected by
    cultivation, 56, 58, 126
    rotation, 61
  importance of, in aeration and percolation, 54–56, 126

Magnesium
  content of mineral soils, 22, 26
  content of peat soils, 489
  diagram, gains and losses of, 588
  in dolomite, 573, 578, 584
  exchangeable in
    mineral soils, 25, 26, 28
    peat soils, 487
  forms in soil, 26, 28
  loss by
    erosion, 534, 580
    leaching, 521, 522, 580
  as nutrient, 21, 584
  pH and availability of, 207, 581

Manganese
  active in acid soil, 125
  amount in soils, 22, 369
  chelates of, 374
  an essential element, 21, 364
  in farm manure, 633
  in fertilizers, 604
  influence on phosphorus availability, 336
  ionic form in soil, 125, 370, 372
  as pollutant, 665, 666, 671
  reduction–oxidation of, 125, 372
  relationship to pH, 206, 581, 582
  role of, 366, 367
  suppression by overliming, 582

Manure (see Animal manure; Green manure)

Manure salts, 603

Marine soil materials, 403

Marl, bog lime, 490

Matric potential of soil water, 80

Maximum retentive capacity of soil for water, 97

Mechanical analysis (see Particle-size analysis)

Mercury, as pollutant, 671–72

Metamorphic rock, 387

Meta-phos, a fertilizer, 512

Mica minerals
  as potassium sources, 305
  structure of, 157
  weathering of, 166, 396

Micelle
  of humus, 147
  of hydrous oxides, 146
  meaning of term, 151
  of peat, 486
  of silicate clays, 151

Microcline
  liberation of potash from, 391
  reaction of acid humus on, 267
  a soil mineral, 391
  as source of clay, 166, 391

Micronutrients, 364–83
  affected by
    overliming, 415, 484
    pH, 388, 389, 490–93, 497
  amount in soils, 23, 489
  anion availability, 376
  cation availability, 371
  chelates of, 376
  deficiencies of, 22, 376, 540
  essential elements, 21, 364
  in fertilizers, 381, 604
  forms in soil, 369, 370
    ionic, 370
  management needs, 379
  oxidation–reduction of, 264, 491
  in peat soils, 494
  role of, 486

Milliequivalents, 175

Minerals (see Soil minerals)

Minor elements (see Micronutrients)

Mixed layer silicate clays, 159

Microorganisms (see Soil organisms)

Moisture (see Soil water)

Moisture equivalent, 99

Mole drain, 525

Mollisol soil order, 434, 443, 444, 445, 447, 452, 453, 459, 460, 463

Molybdenum
  amount in soils, 369
  essential element, 21, 364
  in farm manure, 633
  in fertilizers, 381, 604

744   Index

structure, fundamentals of, 150
surface area, 145, 156
swelling of, 184
types in soil, 153–60
Silicon
amounts in soils, 22
in silicate clays, 151, 153–60
Silt (*see* Soil particles)
Slaked lime, 572
Slope, effect on
erosion, 542–44
soil temperature, 130
Smectite clays
cation exchange capacity, 156
genesis, 166, 167
geographic distribution, 168
mineralogical organization, 154–56
physical properties of, 40–41, 183–87
Sod
check on erosion, 544
effect on
drainage losses, 521
organic matter accumulation, 280
Sodic soils, 217
Sodium
adsorbed by colloids, 217
detrimental effects of salts, 217
as nutrient, 21
in saline-sodic and sodic soils, 215, 217
Sodium nitrate
effect on soil reaction, 610
as fertilizer, 593, 595
Soil(s) (*see also* following entries
and specific soil names)
a soil vs. *the* soil, 9, 428
active portion of, 18
chemical composition of, 21, 488
edaphological definition, 7
four components of, 13
general biological nature, 17
individual soils concept, 9, 427–29
vs. land, 467
light vs. heavy, 37
minerals of, 14, 23, 387, 388
mineral vs. organic, 12, 476
parent materials of, 386–413, 419
pedological definition, 7
profile of, 9, 423–27
specific heat of, 132
subsoil vs. surface soil, 10

volume composition of, 13
weight of, 51, 485
Soil acidity (*see also* Lime; pH of
soil)
compounds suitable for reduction
of, 572
effect of fertilizer on, 198, 609
effect on
nitrification, 209, 292, 581
phosphate fixation, 336, 341
exchangeable vs. active, 201
important correlations with, 206, 209
methods of increasing, 214
problems of, 213
relation of carbon dioxide to, 198
relation to higher plants, 30, 209, 219
Soil aggregates (*see* Aggregation of
soil particles)
Soil air (*see* Aeration of soil; Air of
soil)
Soil alkali (*see* Alkaline and saline
soils)
Soil analysis (*see* Chemical
analysis; Particle-size
analysis)
Soil association, 460–61
Soil bacteria (*see* Bacteria in soil;
Soil organisms)
Soil buffering
capacity of, 203
curves of, 204, 205
explained, 202
importance of, 205
of peat soils, 487
Soil catenas, 461, 462
Soil class
determination of, 43
explained, 42–43
names used, 43
Soil classification, 416–64
Alfisols, 434, 446, 447, 453, 454
Aridisols, 434, 450, 453, 454, 604
azonal groups, 463
bases of, 431
categories used, 433–34
Chernozems, 430, 463
diagnostic horizons, 432
Entisols, 434, 436, 443, 694, 695
great groups, 433, 455–56
Histisols, 434, 451, 452, 453, 476
Inceptisols, 434, 443, 445, 452, 453

Mollisols, 434, 443, 444, 445, 447,
452, 453, 459, 460, 463
new comprehensive system, 430–60
nomenclature of, 434–436
orders of, 436
Oxisols, 434, 448, 452, 453
peat soils, 451, 481
phases of, 433, 460
soil series, 433, 435, 458, 459, 465
soil survey maps, nature and use,
465, 467
Spodosols, 434, 445, 450, 452, 453
suborders, 433, 435, 454–55
Ultisols, 434, 445, 447–48, 452, 453
Vertisols, 434, 445, 449, 452, 453
"wet and dry," 455
Soil colloidal matter (*see* Colloids,
soil)
Soil color (*see* Color of soils)
Soil compaction, 58
Soil consistence, 63
Soil development (*see* Soil
formation)
Soil erosion
accelerated type, 257
conservation tillage practices,
549–57
contour strip cropping, 546
cover and management factor, 544
effects of crops on, 544
factors governing rate, 538–39
gully control, 557
loss of
nutrients by, 534
soil by, 534
magnitude of, 534, 535–36
mechanics of, 537
normal erosion, 537
raindrop impact, 537–38
rainfall and runoff factor, 539
sheet and rill, 538
soil erodibility factor, 541
strip cropping control, 549
support practice factor, 546
terraces, 547
topographic factor, 542
Universal Soil Loss Equation, 538
in U.S., map of, 535
by water, 534–58
by wind, 558–65
windbreaks, 564, 565
wind erosion equation, 563
Soil family, 433, 435, 457, 458, 459